ROADS, RAILWAYS, BRIDGES and TUNNELS ENGINEERING

[*For Diploma, Degree, A.M.I.E., U.P.S.C. and other Professional Examinations*]

By

Professor T.D. Ahuja

(Director)

Institute of Engineering and Rural Technology, Allahabad

G. S. Birdi

Formerly Principal, Jawahar Lal Nehru Polytechnic, Mahmudabad (Sitapur)

STANDARD BOOK HOUSE

unit of: **RAJSONS PUBLICATIONS PVT. LTD.**

1705-A, Nai Sarak, PB.No. 1074, Delhi-110006 Ph.: +91-(011)-23265506
Show Room: 4262/3, First Lane, G-Floor, Gali Punjabian, Ansari Road, Darya Ganj, New Delhi-110002 Ph.: +91-(011) 43751128 Tel Fax : +91-(011)43551185, Fax: +91-(011)-23250212

E-mail: sbh10@hotmail.com www.standardbookhouse.in

Published by:
RAJINDER KUMAR JAIN
Standard Book House
Unit of: Rajsons Publications Pvt. Ltd.
1705-A, Nai Sarak, Delhi - 110006
Post Box: 1074
Ph.: +91-(011)-23265506 Fax: +91-(011)-23250212

Showroom:
4262/3, First Lane, G-Floor, Gali Punjabian
Ansari Road, Darya Ganj
New Delhi-110002
Ph.: +91-(011)-43751128, +91-(011)-43551185
E-mail: sbhl0@ hotmail.com
Web: www.standardbookhouse.in

First Published : 1967
Second Edition : 1968
Third Edition : 1972
Fourth Edition : 1975
Fifth Edition : 1978
Sixth Edition : 1980
Seventh Edition : 1982
Eighth Edition : 1985
Ninth Edition : 1987
Tenth Edition : 1989
Eleventh Edition : 1993
Twelveth Edition : 1996
Thirteenth Edition : 2000
Fourteenth Edition : 2005
Fifteenth Edition : 2009
Sixteenth Edition : 2012
Seventeenth Edition : 2015
Eighteenth Edition : 2017
Nineteenth Edition : 2022

Price: **Rs. 380.00**

ISBN: 978-81-89401-33-7

Typeset by:
N D Enterprises, Delhi.

Printed by:
R.K. Print Media Company, New Delhi

Preface to the Sixteenth Edition

We are thankful to the fellow teachers, eminent engineers and students for their warm and cordial reception accorded to the fifteenth edition, which got exhausted in the short span of time, and encouraged us with an opportunity to bring out this sixteenth edition.

The book has been thoroughly revised, keeping in view the latest trends of the various examinations and suggestions received from various corners. At the end of each chapter Review Questions have been given, mostly asked in the examinations of various Technical Board and Universities.

Two new chapters in the Road Engineering part of the book have been added: Street Lighting and Airport Engineering.

Most of the old figures have been redrawn to make them more clear and understandable.

We are thankful to Shri R.K. Jain, Proprietor of M/s Standard Book House, for taking keen interest and timely publication of the book.

Suggestions and comments for the improvement of forthcoming edition of the book and the comprehension of the readers will be gratefully acknowledged.

We are confident that our readers will welcome this new edition with renewed interest and appreciation.

T.D. AHUJA
G.S. BIRDI

Preface to the First Edition

The present work on Roads, Railways and Bridges is mainly intended for final year students of Civil Engineering, preparing for diploma examination of various Boards of Technical Education of India. There are many standard works separately on Roads, Railways and Bridges by Indian and foreign authors. All the same, it is a well-known fact that many good books miss their mark when they are not restricted to a particular class of readers. A book particularly written for a particular class of readers, can be proved more useful and beneficial to the students for whom it is intended, as then the readers will not bother about selecting portions for their studies. The present book is in pursuance of the urgent need of diploma students.

It has been attempted to present the subject-matter in a lucid and easily understandable manner, with the help of numerous sketches which are self-explanatory and can be easily reproduced by the students. At the end of each chapter important questions, mostly asked in the examinations of various Technical Boards and Universities, along with specific hints and answers, where felt necessary, have been given. This will help readers to grasp the subject-matter easily.

In preparing this work, a number of standard works and Journals on the subject have been freely consulted, for which we are grateful to the authors and publishers. We are thankful to our colleagues and friends for their encouragement and help. We are specially indebted to Shri G.S. Sharma for helping us in preparing standard drawings of railway tracks, etc. of Indian Railways.

The publishers, M/s Standard Book House, have taken keen interest in early publication and get-up of the book, for which we are grateful.

In spite of all our efforts, some errors and mistakes might have crept in. Any error or misprint, if pointed out and any suggestion for the improvement of the book, will be thankfully acknowledged.

T.D. Ahuja

Allahabad Polytechnic **G.S. Birdi**

March, 1967

Contents

PART 1 ROADS

PART 2 RAILWAYS

PART 1

Railway Engineering

Introduction

GENERAL

Roads are ladder of all sort of development and civilisation. Road Engineering is one of the important branches of engineering. It deals with the construction, design and maintenance of roads of different kinds. The first chapter has been particularly devoted to the definition of common terms used in road engineering, types of roads and importance of roads for a country. The importance of roads in a country are comparable to the veins in the human body. Just as the veins supply the blood to the different parts of a body, so are the roads in a country, they convey men, material and information to the different parts of a country. By going through this chapter the reader will have a fair idea about the various technical terms used in road engineering and the various types of roads existing in India.

1.1 DEFINITIONS

1. Road. A way for vehicles and for other types of traffic over which they may lawfully travel. It includes the entire area comprising the roadway and all structures pertaining to the road within the limits of the defined boundary or 'right-of-way'.

2. Highway. It is main and larger in width and length by all means of road system.

3. Roadway. It is the portion of the road (included within the construction limits) usually used for traffic. It includes carriage way and the shoulders.

4. Carriageway. It is the portion of the roadway designed and constructed for vehicular traffic.

5. Right-of-way. (i) The land secured and reserved for development of a road and all structures pertaining to the road, (ii) the privilege of use of a way, acquired by law for the traffic, custom or usage.

6. National Highway. These are the most important roads connecting capital cities of different states or provinces. It is notified by Ministry of Road and Transport of government of India.

7. State Highway. These are the main roads within a state connecting important towns of the state.

8. Major District Roads. These are less important roads within the district headquarters, boundaries.

9. Bye-pass Road. The road constructed around the city or town in such a way, to avoid congested areas or other obstructions in the movement of thorough traffic.

10. Loop Road. It is a route formed by a road or a series of roads to avoid an obstruction or provide an alternative way for traffic.

11. Ring Road. It is the circumferential road constructed around an urban area to enable free flow of traffic.

12. Radial Road. It is a road providing direct link between the centre and outskirts of an urban area.

13. Drive way. It is the way to secure access from a road to private property.

14. Service-Road. It is a subsidiary road constructed between a road and buildings or properties facing thereon and connected only at the selected points with the main road. Usually water, sewer and gas pipes are laid through this road and connections given to the buildings.

15. Fair-weather Road. It is a road which can be used for the traffic during dry weather only. In monsoons this road is closed for traffic.

16. Island. It is a central or subsidiary area in a carriage way, generally at road junctions, shaped and placed so as to constrain and conrtol the movement of the traffic.

17. Transition Length. It is the length of the transition curve connecting a straight-length of a road with another main curve which may be circular or transitional.

18. Sub-way. It is the underground passage or tunnel to permit the movement of traffic, or to accommodate service pipes, sewers, cables etc.

19. Fly-over. It is the road junction so designed so that traffic streams are divided to enable them to pass over or under each other.

20. Traffic Density. It is the number of vehicles using the road per hour during peak periods and is the average of several peak days. The daily traffic is approximately ten times the maximum hourly traffic.

21. Formation. It is the final ground surface after completion of earthwork.

22. Formation Width. It is the finished top width of earth work in fill or cut to lay the road structure. It is also known as 'roadway'.

23. Base Course. It is the first layer of road structure laid over the soil formation.

24. Base-coat. It is the intermediate course between the base course and the wearing coat.

25. Sub-crust. It is an intermediate layer which acts as a cushion between the foundation and the pavement.

26. Crust. This is the total thickness of all the layers laid on earth formation.

27. Wearing Coat. It is the topmost coat of the road, over which the traffic moves.

28. Carpet. It is the wearing coat of bitumen or tar concrete having thickness more than 25 mm.

29. Pavement. It is the hard crust placed on the soil formation.

30. Creteways. It is a carriageway in which a cement concrete wearing surface is provided for the wheel tracks only. Usually sugar mills construct the creteways for collection of sugarcane from the interior of rural areas.

31. Edging. It is the bricks or blocks of stones embedded along the edges of a carriageway pavement for its protection from damage by traffic.

32. Berm, Haunch or Shoulder. This is the portion of the roadway just beyond the edges of a carriageway, which is used by the traffic occasionally for passing, during overtaking etc. The strip of the roadway between side drain and the lower edge of the bank is also included in it.

33. Camber. It is also known as transverse slope of road way. It is the convexity given to the curved cross-section of the carriageway for draining water. It may be defined as crossfall of the road.

34. Crown. The cross-sectional highest point of the road is termed as crown.

35. Super-elevation, Banking or Cant. It is the inward tilt or transverse inclination given to the cross-section of the road at the curve positions, to reduce the effect of the centrifugal force on the moving vehicles.

36. Footpaths. Footpaths are particularly provided in the case of urban roads, and are 15 cm to 20 cm higher than the road surface. Footpaths are generally made of bricks or cement concrete. Shoulders are generally in level with the road surface, having a slope towards the drain side. Shoulders and footpaths provide lateral strength to the road and prevent the edges of the road from wear and tear.

In some urban roads, running through congested areas, some portion of the road-way is reserved for cycle traffic and is called *cycle track,* while some portion, generally in the centre, for high speed vehicles such as motor cars, trucks etc. and is called *motor way* or *express way.* Footpaths are reserved for pedestrians only.

1.2 MEANS OF COMMUNICATIONS

Communication is defined as the means of moving men and material or postal information etc. from one place to another. There are four means of communication :

1. Roadways
2. Railways
3. Waterways
4. Airways

1. Roadways. These are the means of communication on earth surface. Roadways help in carrying large number of passengers and goods from one place to another. They play a very vital role in the development of a country. Road transport is a means of detailed distribution between homes, shops, factories etc. Roads thus form an integral part of the country to facilitate transport from one place to another. This is the easiest and most common means of communication to all the persons.

2. Railways. Railways are a very good means of communication for men, material and information. Railways are steel tracks laid on the ground, over which the train moves. In the case of railways the tractive resistance between the steel rail and steel tyre is reduced to $\frac{1}{6}$th or $\frac{1}{5}$th of the pneumatic tyre on a modern highway. The steel track can take heavy axle loads nearly three times as heavier as the road and the trains can run at a higher speed.

3. Waterways. Waterways are known to men for a very long time. The existence of large sheets and depths of water are taken advantage of, for transport of men and material by means of boats and ships. The cost of transport is very moderate as compared with the cost of mechanical transport by road or rail. But these means of transport are limited or restricted to only those places, where there is enough water. This method of transport is very slow as compared to other methods.

4. Airways. This method of communication is the quickest but it is very costly. Hence the method is confined to the rich and the aristocracy. Elaborate arrangements are being made for this sort of communications. Roads and railways work as feeders without the help of which this means of communication is not possible.

Hence we see from the above explanations that roads are of utmost importance for all other types of communication. Roadways are inter-communicating links. The entire development of a country depends on the efficient and widespread network of roads throughout the country.

1.3 HISTORY OF HIGHWAY DEVELOPMENT

Roads have been put to use from a very early time. The first hard surface was constructed in 3500 B.C. in Mesopotamia. A stone paved road was located as early as 1500 B.C. on the Island of Crete in Mediterranean Sea. A hard surface road was constructed between Babylon and Egypt in 539 B.C. Actual development of the roads started in the 18th century in France and England. McAdam in England used crushed stone as road surfacing material (1576–1836) which method is still used, though in a modified form. The road development was very slow in USA till nineteenth century. Road development got a momentum in the first two decades of the 20th century with the improvement of motor vehicles which proved to be easiest method of transporting men and material from one place to another.

1.4 DEVELOPMENT OF ROADS IN INDIA

History tells us that Indians had knowledge of the science of Road Construction long, long ago. Excavation of Mohanjodaro and Harappa has revealed that even in 3500 B.C., Indians knew this science. In about 600 B.C., a metallic road 6 m to 7.5 m wide existed in Rajgir near Patna. The road was made of stone. In about 300 B.C. Kautilya got constructed a National Highway connecting North West Frontier Province and Patna. Chandra Gupta Maurya had opened a special department for looking after the construction and maintenance of roads. In about 269 B.C. during the regime of Ashoka, there was a good network of roads in India. In the days of Ashoka, trees were planted along the sides of the roads and rest houses were constructed at a distance of 6 to 10 miles along the road side.

Muslim ruler Mohamed Tughlaq constructed a road connecting Delhi to Daultabad. Shershah became famous for the construction of several roads in India. The longest road constructed in his time was from Lahore to

Sonargaon (West Bengal) at present this road is known Grand Trunck Road (GT).

During the Mughal period about 24 long roads connecting different parts of the country were constructed.

In the British period Lord William Bentinck was the first who revived the idea of road construction. It was only during the time of Lord Dalhousie that a Central P.W.D. was formed to look after the work of road construction. Lord Mayo and Lord Ripon contributed a lot in the construction of roads.

With the development of railways, the road construction received a serious setback. Road construction was given a secondary importance. But the circumstances changed after World War I. Motor transport came to the forefront. By the time the existing roads were badly deteriorated and they could not cater with the increased tyre traffic. The Central Government became conscious of this. In 1930, a Central Road Organisation was set up and in 1935 a Transport Advisory Committee was formed. In 1931, the Road Congress met to discuss the development of roads in India. In 1931, a semi-official body called the Indian Road Congress was set up to give recommendations for the development of roads.

After the Second World War I, on the recommendations of I.R.C. the conference of all Chief Engineers was held in December, 1943 in Nagpur. It drew a 10-year plan for the construction and development of nearly 5,29,600 km of all kinds of roads and bridges at an estimated cost of Rs. 448 crores. This plan is known as Nagpur Plan.

In September 1950, the Central Road Research Institute was started in Delhi. This institute is financed and controlled by Central Transport Ministry. It gives technical advice to the State and Central Governments on various aspects of road construction and development. The Government of India through its five-year plans is taking keen interest in the development of all kinds of roads.

1.5 CLASSIFICATION OF ROADS

According to the Nagpur Plan, the roads in India are classified as under :

1. National Highways
2. State or Provincial Highways
3. District Roads
   (a) Major District Roads
   (b) Minor or other District Roads
4. Village Roads

1. National Highways. These are the important roads of the country connecting important towns and capital cities of different states and important cities to ports etc. They are running throughout the length and breadth of the country. Roads connecting neighbouring countries are also called National Highways.

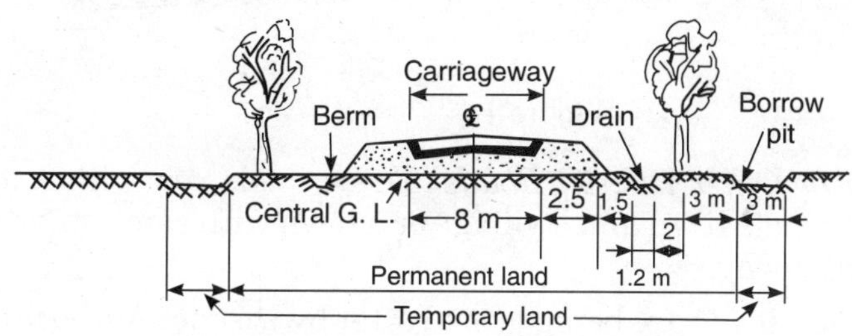

Figure 1.1 *National highway*

The National Highway should have two-lane traffic at least 8 m wide with at least 2 m wide shoulders on each side. However, this has been modified taking in view the present need and increased vehicles on the roads.

The construction and maintenance responsibility of the roads is of the Central Government Department, Ministry of Road Transport Highway, National Highway Authorities of India such as Military Engineering Service (MES), Border Roads Organisation (BRO), Central Public Work Departments (CPWD) etc.

2. State Highways. These are the important roads of a particular state, which connect important cities and towns within a particular state. They also connect important cities with National Highways. The provincial or state highways should also have 8 m wide carriageway with 2 m wide shoulders on each side. The State Government Department of Public Works look after the construction and development of these roads. The central government give grants for the development of the state highways.

3. District Roads. These are roads which transverse each district, serving areas of production and market and connect them with each other or with other highways and railways. District roads are of two types :

(a) *Major District Roads.* These roads connect the areas of production and market-places, either with a state highway or railways. They also connect important places and towns with a district headquarters, within a particular district.

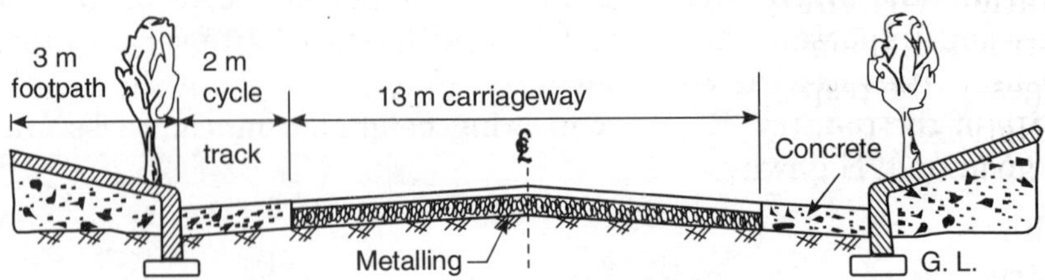

Figure 1.2 *District road*

(b) *Minor or Other District Roads.* These roads run within a particular town, connecting a town and a village or a town with some other roads, such as a State Highway.

The District Roads are being looked after by District Authorities with the help of state government departments. The roads in the built-up areas are being constructed and maintained by Municipal Boards and Corporations.

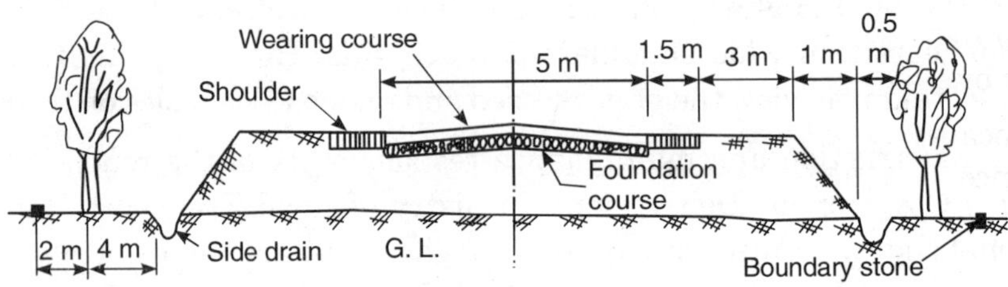

Figure 1.3 *Country road or minor district road*

4. Village Roads. These roads connect a village to a village or a village to a Railway Station or a District Road. The State Government and Local

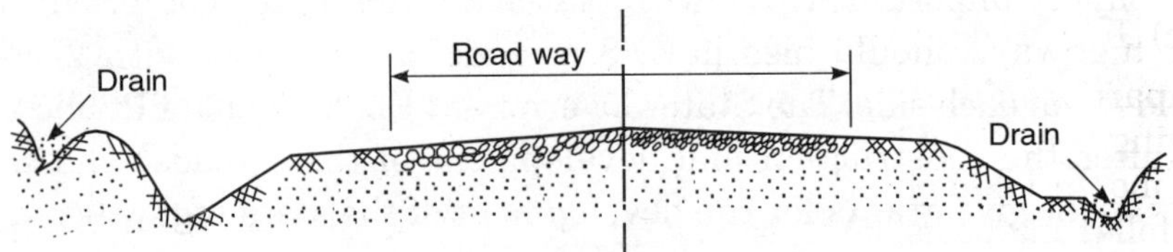

Figure 1.4 *Village road*

District Board Authorities are responsible for the construction and maintenance of these roads. These roads are generally unmetalled but are very important from the rural development of view.

1.6 ROAD FINANCING IN INDIA

In India roads are financed by the central or state governments or even local bodies, depending upon the class of road.

National Highways are financed by the central government. The construction, maintenance, improvement, construction of minor and major bridges, is the responsibility of the central government. The total length of the National Highways in this country, by the end of fifth Five-Year Plan, was 39,000 km.

The central government also help the state governments in the construction and improvement of State Highways, by giving grants in the following manner:

1. By allocating funds to the states from the Central Road Fund which is non-lapsable.

2. By giving special grants approved by the Planning Commission for the construction and improvement of roads of economic importance.

3. By giving funds for improving the geometric design standards of roads and bridges.

State governments finance the construction and maintenance of all roads other than the National Highways from the state revenues. Funds are allotted year by year and are lapsable.

Local bodies including district boards, municipalities, corporations etc., finance all roads coming under their purview.

1.6.1 Sources of Revenues

Roads are financed by realising taxes by the central and state governments.

The taxes levied by the central government are :

(a) Import or excise duty on motor vehicles, motor parts and accessories.

(b) Import or excise duty on motor tyre and tubes.

(c) Import or excise duty on petrol, mobile oil, diesel etc.

Apart from the above the central government has Central Road Fund by levying additional excise duty at the rate of 3 paise per litre of petrol, diesel and mobile oil to promote road development in India. This fund is divided on the following basis on a non-lapsable basis :

20 per cent of the annual revenue is reserved by the central government for the purpose of administration of road fund.

80 per cent is allotted to the state government on the basis of actual petrol consumption by the states.

A total of Rs. 600 million has so far been given by the central government.

State Governments also levy taxes under the State Motor Vehicles Act 1952, under the following heads. The rates are different in different states :

1. Registration fee on vehicles.

2. Issue and renewal of driving licence.
3. Passenger tax on vehicles carrying goods and passengers.
4. Sales tax on motor vehicles and their parts.

The local bodies like municipal boards, district boards, corporations etc. also levy some taxes in the form of Wheel tax, Toll tax etc. at various rates. These toll taxes are sometimes very irritating to the passengers particularly travelling in motor cars.

1.7 HIGHWAY DEVELOPMENT ECONOMICS

Improvement of existing highways sometimes become more important and economical than the construction of new roads to give more relief and facilities to the traffic using that highway.

Though the construction of new roads is also important and desirable and sometimes becomes essential for the development of trade, to reduce congestion, to segregate traffic, to meet up requirements of defence etc. but before launching project, a careful study of all features regarding improvement and future development of the existing roads and the original construction of a new road etc., should be taken into account. Public money to be spent on the construction of new roads or improvement of existing roads, must ensure, better utilization of funds, improving standard of living of the masses, improving existing traffic facilities, providing easy access to public utility services, assistance in the defence of the country, vehicle operation cost etc. A highway engineer has to determine cheapest of the alternatives and maximum facilities for the traffic in general Prof. W.M. Gillespie of USA and stated *"A minimum of expense is, of course, highly desirable, but the road which is truly the cheapest is not the one which has cost the least money, but the one which makes the most profitable return in proportion to the amount spent on it"*. Judicious selection of the alignment considering all present construction, future expansion programmes benefits and returns etc. will make the highway economical.

1.8 ADVANTAGES OF ROADS TO A COUNTRY

Roads are arteries of a country. Roads are a pre-requisite to speed, and speed is essential for progress. It is difficult to enumerate the advantages of roads. Without roads nothing is possible in a country. However, following are the main advantages of roads :

(1) A network of roads is an asset to national defence, during war as well as peace time.

(2) Roads facilitate the movement of men and material from one place to another.

(3) Better law and order can be maintained if there is a good network of roads in a country.

(4) Educational and cultural contacts can be maintained with each other, with the help of good roads.

(5) Roads help in the growth of trade, medical facility and other economic activities in and outside the villages and towns.

(6) Roads serve as a feeder for railways, airways and waterways.

(7) Natural resources of an area are easily tapped and improved with the help of good network of roads.

(8) They are essential for the economic prosperity and general development of a country. The country having comparatively more mileage of roads is said to be a more advanced and prosperous country.

1.9 IMPORTANCE OF ROADS IN INDIA

India is a vast country and to connect its different parts with a good network of roads is essential. India's deficiency in agriculture and economic progress is also due to the lack of good roads, especially in villages. The vast difference of culture between different parts of this country, for example, North and South, is also due to the lack of good roads connecting the two major portions. Some of the interior portions are absolutely cut off from the remaining country due to lack of roads. In villages mostly fair weather roads are there i.e., those roads which can be used only in fair weather, and disconnect the villages from towns and railway stations during rainy season. Hence for the uplift of villages and economic development of this country, good and uptodate roads are very essential. India lags behind many other countries as far as the mileage of roads is concerned. This stresses the urgent and dire necessity of the planning and development of adequate road system in this country.

1.10 RURAL ROADS DEVELOPMENT PLAN

The number of big and small villages in our country is nearly 5,75,856. 57% of the villages have population more than 1500. 36.3% villages are with population between 1000 and 1500 and the rest of villages with population less than 1000. Less than half of the villages are connected with all weather roads. Government have since realised the importance of village roads and are actively engaged in planning minimum need based road development in rural and semi-urban areas for an overall development of the country. With rural development and rural construction departments the rural road development will get boost up.

It is estimated that about Rs. 11,000 crore will be required for providing all weather roads to nearly 90% of the villages. In the seventh five year plan out of Rs. 3,800 crore for road development nearly Rs. 1,200 crore was earmarked for rural roads only. Various states are actively involved in planning rural roads in the eighth plan.

1.11 FEATURES OF ANCIENT ROADS

1.11.1 Roman Roads

During the period of Roman Civilisation many roads were built of thick stone blocks. Roman roads were constructed without the consideration of gradients. Total thickness of these roads used to be between 0.75 m to 1.2 m. For the construction of these roads an open trench was excavated by removing the loose soil. One or two layers of boulders with or without lime were laid. Over the boulders, a layer of lime concrete was laid. Over one or two layers of lime concrete dressed stones were laid for the wearing surface.

1.11.2 Metcalf Road

John Metcalf constructed a road in England in the year 1780–85. He was responsible for the construction of 290 km of roads in the northern region of England. Metcalf used stone boulders or big stones and laid them on edge. The cavities or voids in between the boulders were filled with smaller stones and spalls. The surface was then compacted by bigger pieces of stone to make the weaving surface.

1.11.3 Tresaguets Road

Piere Tresaguets of France developed a new technique of construction of road in the eighteenth century B.C. Tresaguets gave due consideration for sub-soil moisture, sub-grade strength and the drainage conditions. During the period of Nepoleon major development of road system in France took place.

In Tresaguets system of road construction, sub-grade was properly prepared. Large foundation stones were packed tightly on the sub-grade,

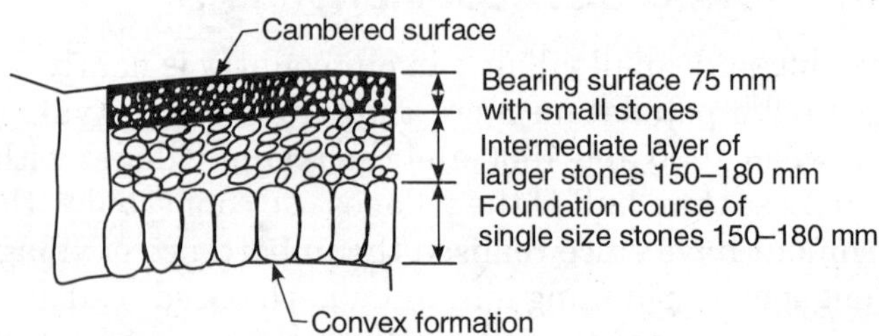

Figure 1.5 *Tresaguets road structure*

laid on edge. The voids on interstices of these stones were filled with smaller stones. For the first time camber was provided in the road for drainage. The shoulders were also made sloping towards outside to drain off rain water.

1.11.4 Telford Road

In England in the year 1786 or so, Thomas Telford started the work on road construction. He emphasized the work on road construction. He also pleaded for definite cross slope or camber in the surface of road, depending on the surface condition and intensity of rainfall. The constructional features of a Telford road are :

(a) The sub-grade was kept horizontal.

(b) Heavy foundation stones of size 17 cm were kept on the edges and bigger pieces of 22 cm size were kept in the middle of road.

(c) Two layers of broken stones were kept on the foundation layer. These layers of stones were properly compacted.

(d) A wearing course of nearly 4 cm thickness was then laid and completed.

(e) The total thickness of road varies from 35 cm at the edges to 42 cm at the centre.

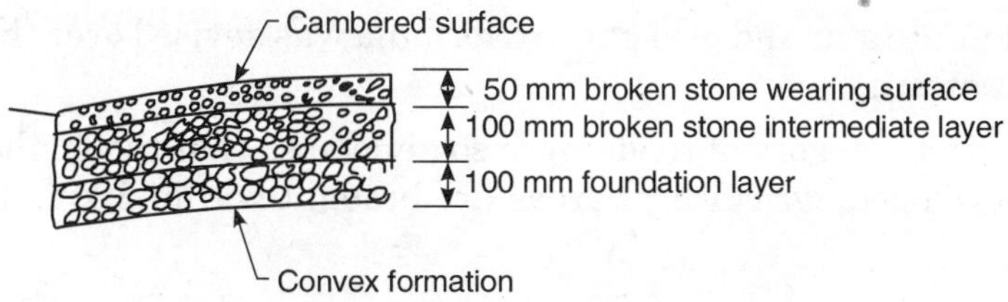

Figure 1.6 *Macadam road structure*

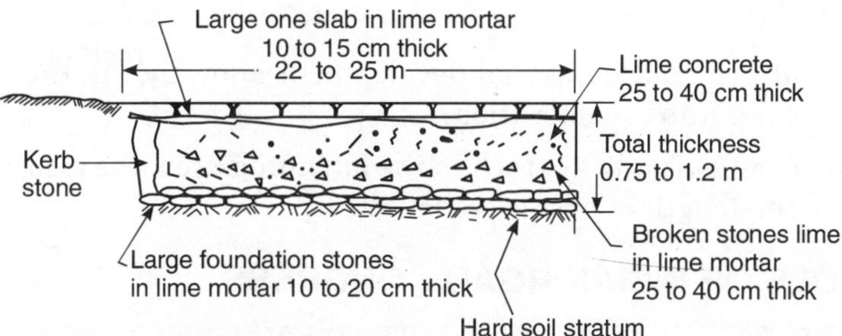

Figure 1.7 *Roman road structure*

1.12 INDIAN ROAD CONGRESS

In 1934 Indian Road Congress (I.R.C.) a semi-official technical body was formed by the Govt. of India to pool in experience, expertise and ideas for the planning and development of roads in the country to meet the demand of vehicular traffic. I.R.C. was also entrusted with the responsibility of formulating standards and specifications of different type of roads for their construction and maintenance. The I.R.C. has since become the main forum for exchange of ideas on the development and planning of road network. Now the Indian Road Congress has become an active body of Govt. of India controlling standardization of specifications and recommendations as regard materials, design and construction of roads and bridges. The I.R.C. publishes journals, research papers, standard specifications for different roads. I.R.C. has become the father of Highway Engineering.

1.13 SALIENT FEATURES OF NAGPUR PLAN

The following are the salient features of Nagpur Plan :

(i) The responsibility of construction and maintenance of various type of roads was assigned to central government, various state governments, departments, local bodies and other departments specifically.

(ii) It was a 20 year plan aimed at integrated road planning and development.

(iii) A definite star and grid pattern formula was devised over the existing irregular pattern.

(iv) The first category of roads were so divided so that the farthest points and the developed agricultural areas are brought within 8 km of metalled road.

(v) The second category of roads are meant to provide internal road system linking small villages with first category roads. The road length in the second category roads is worked out on the basis of villages of different population ranges.

(vi) Agricultural and industrial development allowance in the coming 20 years was taken into consideration.

(vii) The length of railway track in the adjoining area was also considered in deciding the length of first category roads.

1.14 ROLE OF INDIAN ROAD CONGRESS

History of Roads date back to Vedas. The Rig Veda speaks of mahapath on which royal chariots used to ride. Ramayan describes roads of varying nature used by Lord Rama in his journey to Lanka. There was a network of roads linking various parts of the country during the period of Ashoka. Upto World War II very little attention was paid to the construction of surface roads mainly due to paucity of funds. The Indian Road Congress (I.R.C) came into being with the principal objective of providing a national forum of pooling the experience and ideas of construction, maintenance and planning of the network of roads in the country. In more specific terms the following are the broad objectives of I.R.C :

1. To promote and encourage the science and practice of building and maintenance of roads.

2. To provide a channel to the expression of collective opinion on matters for the development of road in our country.

3. To promote use of standard specifications and to formulate newer specifications.

4. To advise regarding education, experience, research connected with roads.

5. To hold periodical meetings to discuss technical questions regarding roads.

6. To suggest legislation for the development improvement and protection of roads.

7. To suggest new methods of construction, administration, planning designing, testing of materials and maintenance of roads.

The vital role of road network in a country can be further emphasised, for the socio-economic development of a country. It is only roads which serve as a feeder for railway, airways and seaways and can penetrate deep into the rural areas. The importance of roads is increasing day by day and so its problems of construction and routine maintenance are to be taken care of. Due to financial constraints the problem of construction of new roads and their proper maintenance can be tackled through concerted efforts backed by experience and pooling of all available resources, knowledge and expertise. The I.R.C. has played a significant role in the promotion of these activities, through seminars, lectures and group discussions. Over the years the I.R.C. has touched upon and contributed to the improvement of the various facets of road development. Adaptation of standardized practice in design, construction and maintenance with regard to variabilities in terrain, soil and climate also comes under the purview of I.R.C. The Road Congress has been closely associated with road planning in the country which forms the forerunner to any development activity. Indian Road Congress also plays an important role for the development of rural roads. In 1958 the I.R.C. set up a committee which was then called the Community Project Road Maintenance Committee. The aim of this committee was to recommend measures and means for proper and adequate maintenance of rural roads which are otherwise fair weather roads. In 1969, the committee was renamed as Rural Roads Committee for bringing out a comprehensive report dealing with planning, specifications, financing, construction and maintenance of all rural roads.

REVIEW QUESTIONS

1.1. What are the different means of communication? Which of them, in your opinion, is the most important for general development of a country. Give reasons for your answer.

1.2. How are roads classified in India? What is the necessity of classifying roads in this manner. Describe in brief.

1.3. Define the following terms :

Right of way, carriageway, footpath, express way, urban roads.

1.4. Draw a neat sketch showing cross-section of the following roads :

(i) National Highway, (ii) Urban Road, (iii) Village Road.

1.5. What are the advantage of roads to a country?

1.6. Write short notes on :

(a) Shoulders.

(b) Footpaths.

(c) Ribbon development.

(d) Major and minor District Roads.

[**Hint**: (c) *Ribbon Development*.

Along the roadsides, construction of shops, hotels and other development of buildings, is known as *Ribbon Development.* The following are the disadvantages of ribbon development:

(a) It produces congestion on the road.

(b) Future widening of the road becomes very costly and sometimes impossible.

(c) Chances of accidents increase.

(d) It causes hindrance to free flow of fast moving vehicles]

1.7. Write an essay on the importance of roads in India.

2

Geometrics of Roads

GENERAL

In this chapter, we will study the techincal aspects of Roads. The design and standards of a road are the most important aspects of a road construction. The success and failure of road construction mainly depends on these vital factors. While designing a road all possible data regarding the type of traffic, its intensity etc., should be collected and sufficient thought should be given, to all these aspects, before finalising particular road project.

2.1 REQUISITES OF A GOOD ROAD

In order that a road surface may give a satisfactory service throughout the year, it must satisfy the following conditions :

1. It should remain dry free from sub-mergence.
2. It should have a good carriageway.
3. It should have mild gradients and large smooth curves.
4. It should have a good wearing surface.
5. It should be easy in construction and cheap in maintenance.
6. It should have an impervious surface.

2.2 ROAD STRUCTURE

Like other engineering structures a road has also a foundation and a super-structure. The top of the grounds on which the foundation of the road rests, is called *Sub-Grade.* The top of the sub-grade should be 60 cm above the highest flood level (H.F.L.) of that area. The foundation of the road is also called *Soling* or *Base* and the super-structure of the road is called *wearing layer, wearing course,* or *road surfacing.* In those places, where the bearing capacity of the soil is poor and the intensity of traffic is high, an additional layer between the soling and sub-grade is provided. This additional layer is called *Sub-base.*

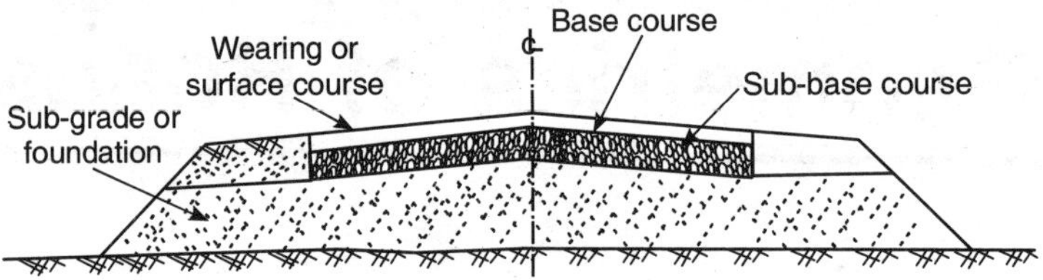

Figure 2.1 *Road structure*

The sub-grade soil is the most important part of a road structure. The strength and durability of a road depends on the type of soil provided. The sub-base consists of a layer of hard murum, hand packed boulders or rubble, bricks, compacted stone ballast etc. Lean cement concrete slab is also sometimes used as a soling for cement concrete roads. The type and thickness of soling depends on many factors, including bearing capacity of soil, intensity of traffic etc. The function of a road sub-base or soling is to transmit load of the traffic, from the road top to the sub-grade. Hence the strength and durability of a road depends upon its sub-grade, on which the load of the traffic is coming ultimately, and the base.

The base or soling can be of a flexible type stone boulders, brick-bats or brick ballast, stone ballast or a rigid type such as cement concrete slab. In case of flexible soling, it is necessary that the bearing capacity of the sub-grade should be uniform and the soft patches etc. should be refilled with good earth. In case of rigid base like cement concrete slab, it is not necessary as the cement concrete slab will act as a beam over the weak spots.

The thickness of a road structure can be calculated by the following empirical formula:

$$d = \sqrt{\frac{23w}{70p} + 0.7T^2 - 0.84T}$$

where

d = thickness of road structure, including surfacing and soling in cm

w = maximum wheel load including 50% impact. Hence w is 1½ times the maximum static wheel load. In India the value of static wheel load is taken as 2268 kg for the purpose of design

T = width of wheel in contact with road surfacing in cm. (The value of T is taken as 12 cm)

(For only bullock cart traffic, the value of static wheel load is taken as 1270 kg and T as 4.5 cm)

p = Safe bearing capacity of soil in kg/cm^2

The thickness of surfacing depends on the type of traffic and the type of material as well as the intensity of traffic. The thickness of base and sub-base will be equal to the thickness of road structure minus the thickness of road surfacing. The thickness of base in no case should be more than 30 cm. If the thickness comes out to be more than 30 cm, the sub-base and sub-grade should be improved so that the thickness of base should come within 30 cm.

The main function of road surfacing is to provide a smooth and stable running surface suitable for the type and intensity of traffic anticipated. The surfacing should have the following requisite properties :

1. It should be impervious so that it should protect the base, the sub-base and sub-grade from the action of weather and rain water.

2. It should be durable, so that the maintenance charges should be the minimum.

3. It should be stable and should transmit the load of the traffic to the base of sub-grade without undue deformation and should be sufficiently flexible to adjust slight unequal settlements.

4. It should be non-slippery especially in rainy season.

5. It should be dustless.

6. It should be economical in construction and maintenance.

2.3 ROAD CAMBER

The road surface has convexity upwards, with its highest point in the centre, in the straight portion of the road. The highest point on the surface is called *crown.* The word 'camber' is defined as the slope of the line joining the crown and the edge of the road surface. Thus a camber of 1 in 30 means that for a 30 metre wide road the crown of the road will be $\frac{1}{2}$ metre above the edge of the road or for a 60 metre wide road the crown will be 1 metre above the edge of the road. The camber is also sometimes referred to as *cross-fall* or *cross-slope.*

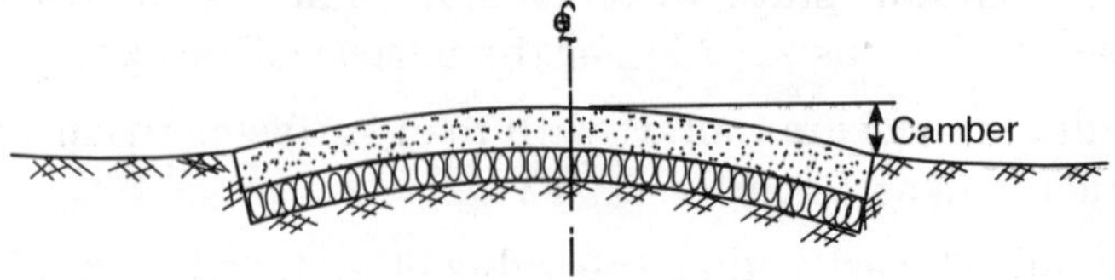

Figure 2.2 *Road camber*

The main object of providing a camber is to drain off rain water from the surface of the road, as quickly as possible. Hence a hard smooth surface will require less camber than a soft and rough surface. The amount of cross-fall or camber depends on the rainfall of that particular locality in which the road is to be constructed and the permeability of the road surfacing material. The steeper the camber, the more inconvenient it is for the traffic. In roads having steeper camber, the central portion of the road surface will deteriorate quickly as the traffic will tend to run on the central half portion. The following cambers have been recommended for different road surfaces:

TABLE 2.1 *Road cambers*

S. No.	*Type of Road*	*Recommended Camber*
1.	Earth roads and footpath	1 in 20 to 1 in 24
2.	Gravel road	1 in 24 to 1 in 30
3.	Murum road or Kankar Road	1 in 24 to 1 in 30
4.	Water Bound Macadam Road	1 in 30 to 1 in 48
5.	Bituminous Road	1 in 48 to 1 in 60
6.	Cement concrete roads	1 in 60 to 1 in 72

Types of cambers. In general three types of cambers are provided for road surfaces:

1. Barrel camber.
2. Sloped camber.
3. Composite camber.

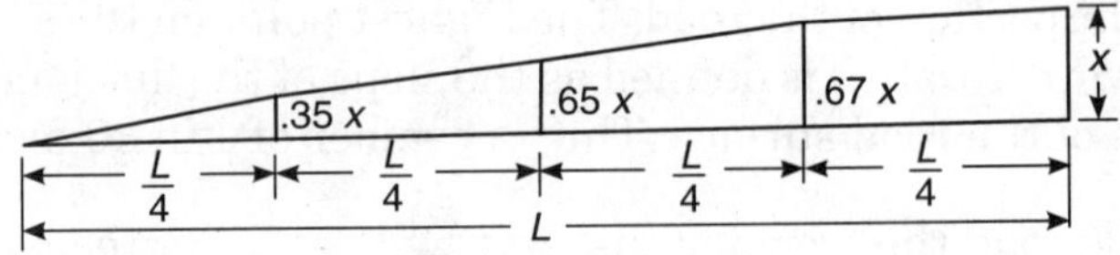

Figure 2.3 (a) *Elliptical*

Barrel camber consists of a continuous curve either parabolic or elliptical.

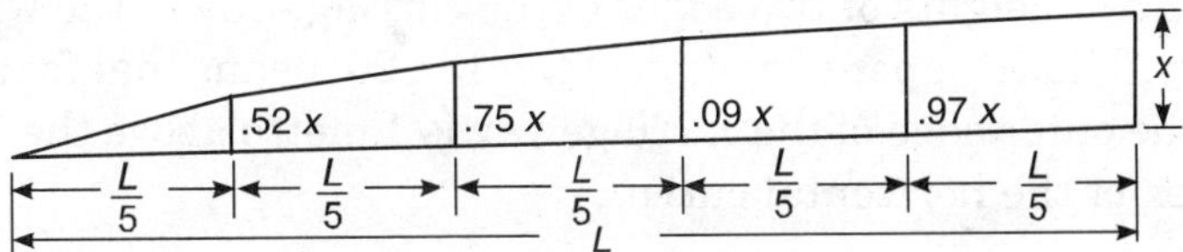

Figure 2.3 (b) *Parabolic*

Sloped camber consists of two straight slopes joining at the centre.

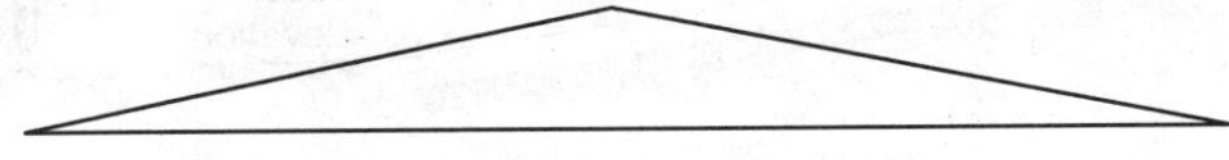

Figure 2.4 *Sloped camber*

Composite camber consists of two straight slopes with a parabolic crown in the centre.

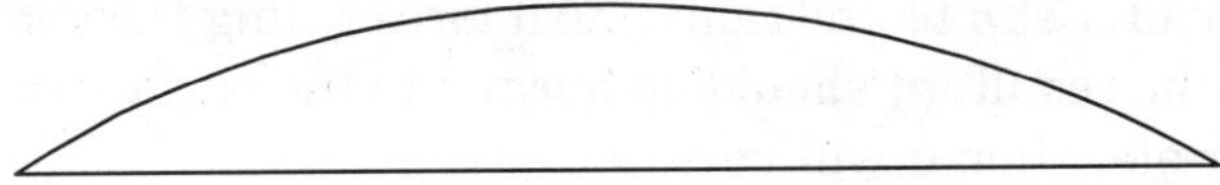

Figure 2.5 *Composite camber*

2.4 SUPER-ELEVATION

When a fast moving vehicle negotiates a horizontal curve, the centrifugal force acts on the vehicle and the stability of the vehicle is disturbed. This force is experienced by the wheels at right angle to the direction of motion. The frictional resistance between the wheel and road surface will act in the opposite direction. If the value of the centrifugal force exceeds the frictional resistance, the vehicle will have side slip and the outer wheels of the vehicle will be raised up from the road surface thereby causing instability to the vehicle. In passing from a straight to a curved path, a vehicle is under the influence of two forces namely, (i) the weight of the vehicle, and (ii) the centrifugal force, both of them acting through its centre of gravity. The centrifugal force always acts in the horizontal direction and its effect is to push the vehicle off the track. To balance this, it is necessary to make the road surface perpendicular to the resultant of the above two forces i.e., the outer edge of the road is raised above the inner edge. To avoid this, the outer edge of road at the horizontal curves is raised above the inner edge. *Super-elevation* is defined as the inward tilt or transverse inclination given to the cross-section of the road surface, throughout the length of the horizontal curve to reduce the effect of centrifugal force on the running wheels. It is also sometimes termed as *Cant* or *Banking*. It is expressed as the difference of heights of two edges of the carriageway to the width of the carriageway. Thus a super-elevation of the 1 in 20 means that for a 20 metre wide road the outer edge of the carriageway is 1 metre above the inner edge at the vertex of the horizontal curve.

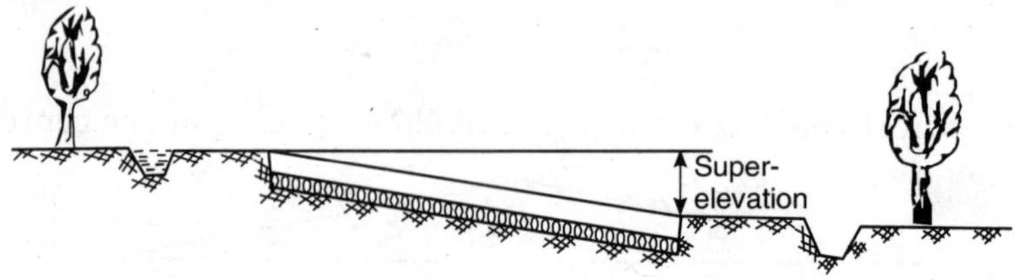

Figure 2.6 *Super-elevation*

Let a vehicle of weight W is negotiating a horizontal curve of radius R where outer edge has been raised by e over the inner edge. Let s be the total width of the road. Let p be the centrifugal force titling the vehicle outward. In order that the resultant should be normal to the surface of the road, the angle of super-elevation α will be

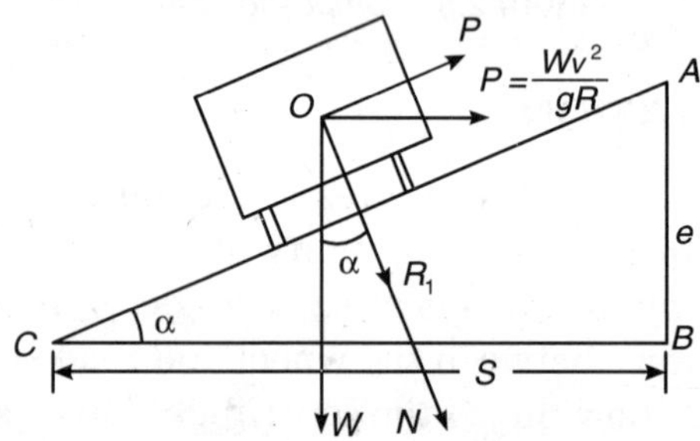

Figure 2.7 *Centrifugal force*

From Fig. 2.7

$$\frac{AB}{CB} = \frac{WN}{QW}$$

or

$$\frac{e}{s} = \frac{p}{w} = \frac{Wv^2}{gR}/W = \frac{v}{gR}$$

$\therefore$

$$e = \frac{v^2 s}{gR}$$

If v is taken as velocity of vehicles in km.p.h. and bringing R, g and s in the corresponding units the formula for super-elevation may approximately be taken as

$$e = \frac{v^2 s}{126R}$$

Here v = speed in kilometres per hour and R in radius of curvature in metres s = width of road in metres.

Example. Find the amount of super-elevation on a horizontal curve, having a radius of curvature 300 m. The average speed of vehicle can be taken as 50 km. p.h.

$$e = \frac{50 \times 50}{126 \times 300} = \frac{1}{15.12}$$

The super-elevation varies from $\frac{1}{14}$ to $\frac{1}{16}$ but the maximum super-elevation which can be provided in special cases is $\frac{1}{10}$. Greater the super-elevation, more the inconvenience to the slow moving traffic. Hence the super-elevation to be provided should be just minimum but it should not be less than the camber, prescribed for the remaining part of the road to ensure effective drainage of rain water.

Table 2.2 gives the recommended minimum radius of horizontal curves on the flat country at various speeds of the vehicles.

TABLE 2.2 *Minimum radii for horizontal curves for flat country*

S. No.	*Design speed of vehicle*	*Minimum Recommended Radii*		*Absolute minimum radii*
		For flat country	*For urban areas*	
	km/hr	*m*	*m*	*m*
1	100	500	—	370
2	80	300	—	250
3	60	250	—	155
4	50	170	125	100
5	40	130	100	60
6	30	100	60	50
7	25	—	50	30

Table 2.3 gives the minimum recommended radii for horizontal curves of hill roads on various types of roads.

TABLE 2.3 *Recommended minimum curve radii for hill roads*

S.No.	Classification road	Recommended minimum radius of curves			
		In steep terrain		In mountainous terrain	
		Snow bounded area	Areas unaffected by snow	Snow bounded areas	Areas unaffected by snow
		m	m	m	m
1.	State and National Highways	33	30	60	50
2.	Major District Roads	15	14	33	30
3.	Other District Roads	15	14	23	20
4.	Village Roads	15	14	15	14

2.5 METHODS OF PROVIDING SUPER-ELEVATION

The changing of the cambered surface of a road to a one way slope on a curved length is done progressively along a certain length of the road. This length of the road is equal to the *transition curve.* A transition curve is a curve so shaped as to follow the natural path and commences to move in a segment of a circle. A transition curve may be defined as *A curve whose radius gradually changes from infinity to a selected minimum or the converse for*

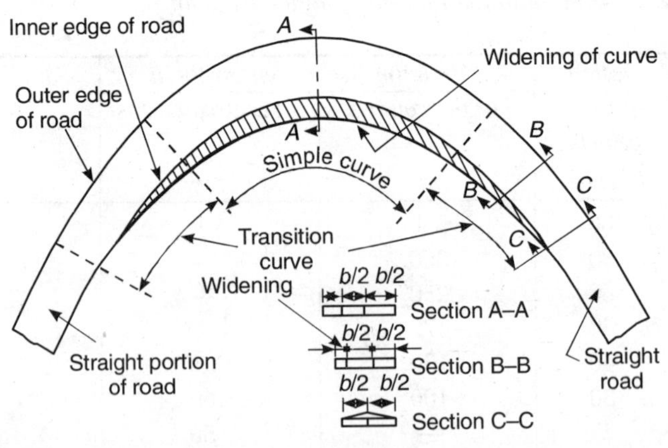

Figure 2.8 *Simple curve with transition curve and widening of road*

the purpose of easy changing of direction of a road. Hence a transition curve joins the straight portion of a road, with one end of a circular curve.

In cases when the transition curve is not provided, the length along with the super-elevation is provided, is equal to half the length of the circular curve, and the change from the cambered section to the required super-elevation starts at $\frac{1}{4}$th the length of the curve measured along the straight portion of the road from the tangent point and the full super-elevation is reached at $\frac{1}{4}$th the length of the curve measured along the curve from the tangent point.

Now the super-elevation is introduced gradually. The surface is assumed to be rotated about the crown that is the outer edge is raised at a uniform rate. The rate of rise of outer edge over the inner edge will be equal to the total super-elevation divided by the length in which the cambered section

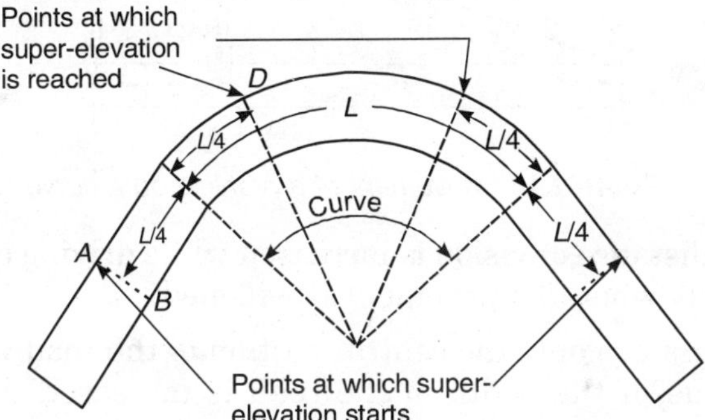

Figure 2.9 *Beginning of super-elevation*

Figure 2.9 shows the beginning of syper elevation is to be changed into super-elevated section. The rising of the outer edge is continued till it is in line with the inner edge.

Advantages of Super-elevation

The following are some of the advantages of super-elevation :

1. By the provision of super-elevation there is a decrease in the intensity of stresses on the road crust and sub-grade foundations.

2. As the whole width of the road is drained to one side where there is super-elevation, no gulleys are formed on the outer edge of the road.

3. Super-elevation increases the stability of the fast moving vehicles when they negotiate a horizontal curve.

4. On super-elevated curves the vehicles are not necessarily slowed down.

2.6 WIDENING OF ROADS

When a vehicle approached a horizontal curve, the steering wheel turn side-ways and occupy more width of the carriageway than on straight portion of the road. Hence the carriageway is increased on the entire portion of the curve on the inside. The widening is effected gradually, with a maximum on the central portion. When the radius of the curve is more than 460 m, this widening is not necessary. The following are the other *main reasons* of widening of carriageway on the curves :

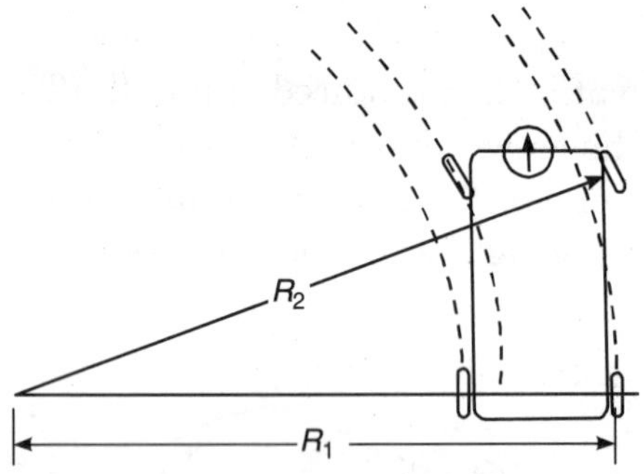

Figure 2.10 *Movement of a vehicle on a curve*

1. The sight distance or vision is increased while driving on the outer half of the curve, only when the pavement is widened.

2. Most drivers use only the central portion of the road while moving on the curves. Thus, if the width of the road is increased this will prevent accidents to a good extent.

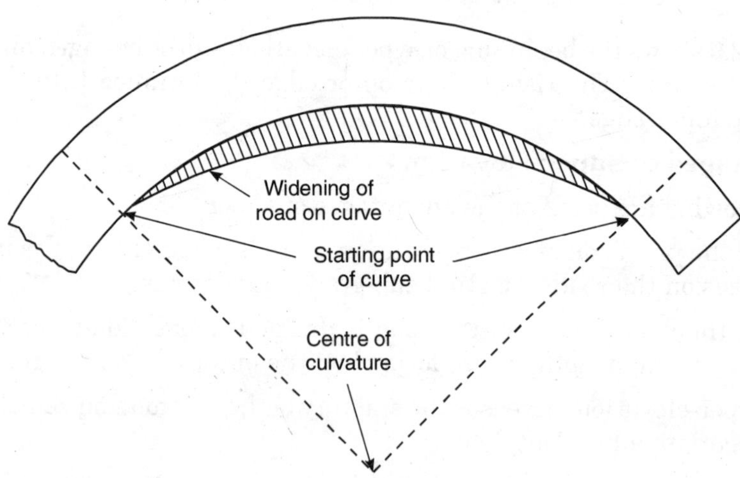

Figure 2.11 *Widening of road*

The value of uniform width added to the inside of the entire circular curve corresponding to the radii of curves is as follows :

TABLE 2.4 *Additional width of roads at curves*

Radius of circular curve	*305 m to 460 m*	*150 to 305 m*	*60 m to 150 m*	*Less than 60 m*
Value of uniform width for each traffic lane of carriageway having one or two traffic lanes	0.3 m	0.6 m	0.9 m	1.25 m

2.7 ROAD GRADIENT

The ground is never dead flat, and hence the road will also have rises and falls along the length of the road. The rate of rise or fall of the road surface along its length is called gradient. It is expressed as the ratio of the difference of heights between the two points and the distance between them.

Thus, if the difference of levels between two points A and B is 1 metre and their distance apart is 50 metres, the gradient is said to be 1 in 50. It is also sometimes expressed as a percentage, i.e., $\frac{1}{50} \times 100 = 2\%$. Gradient is necessary so that the surface water may be drained off easily through the side drains. Gradient of a road depends on the following factors :

1. Nature of Traffic. Steeper gradients are very inconvenient to the slow moving traffic. Hence while deciding upon the gradient to be provided the nature of traffic will have to be taken into account. Roads only meant for slow moving traffic such as bullock carts, etc., must not have very steep gradients.

2. Nature of Ground. The amount of gradient to be provided is directly related to the nature of the ground. Ground with steep undulations will have roads with steep gradient as it is sometimes not possible to provide unnecessary deep excavations or cuttings.

3. Rainfall of the Locality. The gradients in a road are mainly provided to drain off rain water from the side drains as quickly as possible. Hence more the rainfall, steeper gradients will have to be provided, in the side drains and the road.

2.7.1 Type of Gradients

The gradients which are provided in different portions of a road length are as follows :

1. Maximum Gradient.
2. Ruling Gradient.
3. Minimum Gradient.
4. Average Gradient.
5. Exceptional Gradient.
6. Floating Gradient.

1. Maximum Gardient. It is the maximum or steepest gradient which is to be permitted on the road, which on no account is to be exceeded, as steeper gradients are very incovenient to the traffic, specially to the slow moving traffic. This is also called *Limiting Gradient.* Its value has been fixed as 1 in 15 for *Hill Roads* and 1 in 20 for other roads. These steeper gradients are sometimes to be provided to avoid deep excavations and avoid long detours.

2. Ruling Gradient. It is the permissible gradient in the alignment of a road. This gradient is such that vehicles, whether they are animal driven or power driven, can overcome long distances of this gradient, without much fatigue or uneconomical fuel consumption. *Ruling Gradient,* is therefore, defined as the suitable gradient within which the Engineer must endeavour to design the road. The Indian Road Congress has recommended the value of ruling gradient as 1 in 20 in hills and 1 in 30 in plains.

3. Minimum Gradient. It has also been found that for efficient drainage of water from the road surface, a certain minimum gradient is essential. Such an essential gradient which has to be provided for the purpose of road drainage, is called *Minimum Gradient* and its value is usually fixed as 1 in 200. However, the minimum gradient will depend upon the nature of the ground as well as on the rainfall of that particular area. For cement concrete roads, a minimum gradient of 1 in 330 can be provided.

4. Average Gradient. It is defined as the total rise or fall between any two points chosen on the alignment divided by the horizontal distance between the two points. The determination of the average grade is useful in carrying out the first paper location or preliminary survey.

5. Exceptional Gradient. During the alignment of a road, there may come certain patches of length, where a gradient may have to be provided which may be either less than the minimum known as *Exceptional Gradient.* Such a gradient becomes necessary to avoid deep cuttings or excavations. Exceptional gradient should not be provided in a length more than 100 metres, in any case.

6. Floating Gradient. When a motor vehicle descends a gradient, some tractive effect is required to maintain it at uniform speed, but to the descent, there is the negative tractive force or resistance. This resistance will be equal to the tractive effort required to maintain the vehicle at a uniform speed with a particular gradient, such a gradient is called *Floating Gradient.*

2.8 SIGHT DISTANCE

When a fast-moving vehicle reached a horizontal or vertical curve, a certain distance of visibility, through which a driver can see the opposite vehicle, a pedestrian or some fixed object, is essential so that the driver may react and avoid any collision or accident. *Sight distance* or *visibility* is defined as the distance measured along the centre line of a road, over which a driver can see the opposite object on the road surface and the provision of this distance is necessary to avoid any accident. The distance should be such that drivers and pedestrians should be given sufficient time to react to an emergency and not only to avoid accident but extend road courtesy to each other. In short, a sight distance is the length of the road, which a driver can see, especially on the curves. Sight distance can be:

1. Crossing sight distance or safe sloping sight distance.
2. Non-passing or non-overtaking sight distance.
3. Passing or overtaking sight distance.
4. Lateral sight distance.

1. Crossing Sight Distance. On roads and highways, two vehicles coming in opposite direction, on seeing each other have to reduce their speed to enable each other to use the pavement edges or shoulders, as the case may be. This distance in such a case is taken as twice the distance required for a vehicle to come to stop, and is called crossing sight distance. So on horizontal and vertical curves, this minimum sight distance must be provided to avoid any collision of two vehicles coming from opposite directions.

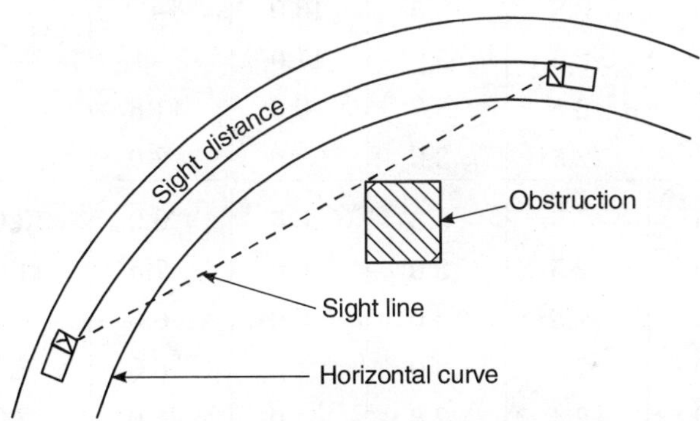

Figure 2.12 *Crossing sight distance*

Tables 2.5 and 2.6 give the minimum setback distance *s* required at horizontal curves on two lane roads for safe stoping sight distance.

2. Non-passing or Non-overtaking Sight Distance. It is defined as the longest distance at which a driver whose line of sight is 1.2 m above the road surface can see the top of an object 10 cm high on the surface of the road.

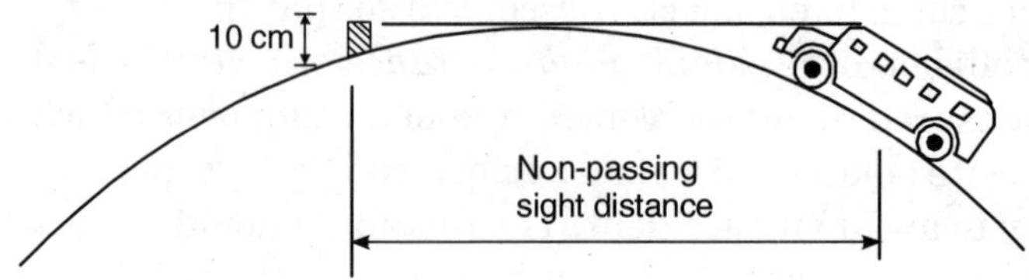

Figure 2.13 *Non-passing sight distance*

TABLE 2.5 *Minimum setback distance s at horizontal curves on two lane roads on flat country*

S. No.	Radius of curve along the centre line of the road	Setback distance 's' for various designed speed V and safe stopping sight distance D					
		D = 20 m V = 20 km/hr	D = 45 m V = 20 km/hr	D = 60 m V = 50 km/hr	D = 90 m V = 65 km/hr	D = 120 m V = 80 km/hr	D = 190 m V = 100 km/hr
	m	m	m	m	m	m	m
1	50	3.2	7.0	—	—	—	—
2	75	2.8	5.5	13.0	—	—	—
3	100	2.5	5.0	11.0	—	—	—
4	125	2.5	3.7	9.0	11.8	—	—
5	150	2.5	3.6	7.0	2.0	—	—
6	175	2.5	3.3	6.5	8.0	13.0	—
7	200	2.5	3.0	6.0	7.5	11.0	—
8	250	2.5	3.0	5.0	6.5	8.5	17.5
9	300	2.5	3.0	4.5	6.0	7.5	15.5
10	350	2.4	2.8	4.0	5.0	7.0	13.0

Contd.

Table 2.5 Contd.

	m	m	m	m	m	m	m
11	400	2.4	2.7	3.5	4.5	6.5	12.0
12	450	2.4	2.6	3.0	4.3	6.0	11.0
13	500	2.4	2.5	3.0	4.0	5.5	10.5
14	550	2.4	2.5	3.0	4.0	5.3	9.5
15	600	2.3	2.5	3.0	4.0	5.0	9.0

TABLE 2.6 *Minimum setback distance s at horizontal curves on hill roads for safe stopping sight distance*

S. No.	*Radius of curve along the centre line of the road*	*Setback distance s for various designed speed V and safe stopping sight distance D*				
		D = 20 m V = 20 km/hr	*D = 30 m V = 20 km/hr*	*D = 35 m V = 50 km/hr*	*D = 50 m V = 65 km/hr*	*D = 70 m V = 80 km/hr*
	m	*m*	*m*	*m*	*m*	*m*
1	15	3.2	—	—	—	—
2	20	2.4	5.3	—	—	—
3	23	2.0	4.7	—	—	—
4	30	1.7	3.7	5.0	—	—
5	33	1.5	3.4	4.5	—	—
6	50	1.0	2.3	3.0	6.0	—
7	60	—	1.9	2.4	5.0	—
8	80	—	1.4	1.8	3.8	7.5
9	93	—	1.3	1.6	3.4	6.5
10	150	—	—	1.2	2.3	4.0

3. Passing or Overtaking Sight Distance. On two-and three-lane highways, opportunities to pass a slow moving vehicle should be provided at intervals. On four-lane highways this facility is not required. *The passing or overtaking sight distance* is defined as the longest distance at which a driver whose line of sight is supposed to be 1.2 m above the road surface, can see the top of an object 1.2 m above the roadway. This will help in overtaking the slow moving vehicles easily and without any risk.

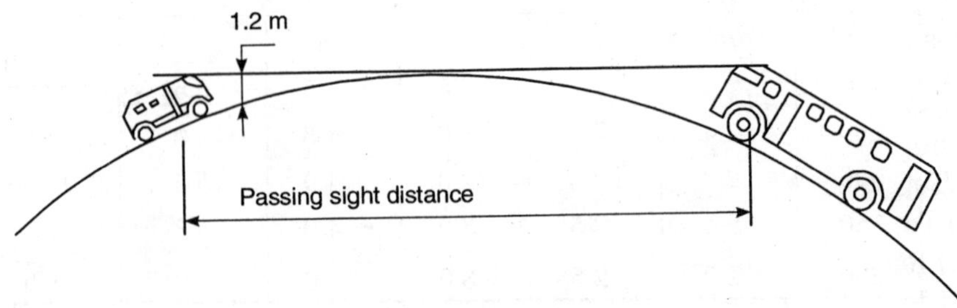

Figure 2.14 *Passing sight distance*

Table 2.7 gives the stopping sight distance and overtaking (or passing) sight distance for various speeds of the vehicles.

TABLE 2.7 *Stopping sight distance of overtaking sight distance for various speeds (Based on I .R .C : 66—1976)*

S. No.	*Design speed*	*Absolute minimum stopping distance*	*Minimum visibility distance along major roads at intersections*	*Minimum overtaking distance*	
				Single lane carriageway two-way traffic	*Two lane (undivided) carriageway two-way traffic*
	km/hr	*m*	*m*	*m*	*m*
1	20	20	—	40	—
2	25	25	—	50	—
3	30	30	—	60	—
4	40	45	—	90	165
5	50	60	100	120	235
6	60	80	135	160	300
7	65	90	145	180	340
8	80	120	180	240	470
9	100	180	220	360	640

4. Lateral Sight Distance. The corner sight distance of an intersection should be sufficient to allow an approaching driver unobstructed view of the entire intersection. The driver should be able to see the vehicle approaching intersection from the cross road sufficiently early, to provide for his reception and reaction time to change the speed of his vehicle. This distance, through which a driver reacts and applies brakes, is called *lateral sight distance*.

2.9 CURVES

Curves are provided on the highways in order that the change of direction at the intersection of straight alignments either in horizontal or vertical plane, shall be gradual. The necessity of providing curves arises due to the following reasons :

(i) Topography of the country and non-availability of land.
(ii) To provide access to a particular locality in straight length.
(iii) Restriction imposed by some unavoidable reasons of land etc.
(iv) Preservation of existing amenities.
(v) Avoidance of certain religious, monumental or some other structures.
(vi) Making use of existing sight of ways.

2.9.1 Factors Affecting the Design of Curves

The following factors will influence the design of curves :

(i) Design speed of the vehicle.
(ii) Allowable friction.
(iii) Maximum permissible super-elevation.
(iv) Permissible centrifugal ratio.

2.9.2 Types of Curves

Curves are of two types (1) horizontal and (2) vertical. The horizontal curves allow change in direction of the road while the vertical curves change in gradient. The curves used in the design of highways are :

(i) Circular curves ; and (ii) Transition.

Circular curves are of three types (a) Simple, (b) Compound, and (c) Reverse.

Transition curves can be divided into four groups: (1) True spiral or clothoid, (2) Cubic spiral, (3) Cubic parabola, and (4) Lamniscate.

2.10 CIRCULAR CURVES

Circular curves are defined as follows :

Simple curve. A simple circular curve consists of a single arc connecting two straights. In this country particularly a curve is expressed in terms of degrees subtended at the centre by an arc of 30 m radius (Fig. 2.15 a).

Compound curve. A compound curve consits of a series of two or more simple curves that turn in the same direction, and join at common tangent points. At each common tangent point, the adjacent curves have a common tangent and their centres are on the same side of the curve (fig. 2.15 b).

Reverse curve. A reverse curve consists of two simple curves of opposite direction that join at the common tangent point called the point of reverse curve. Their centres are on opposite sides of the curves.

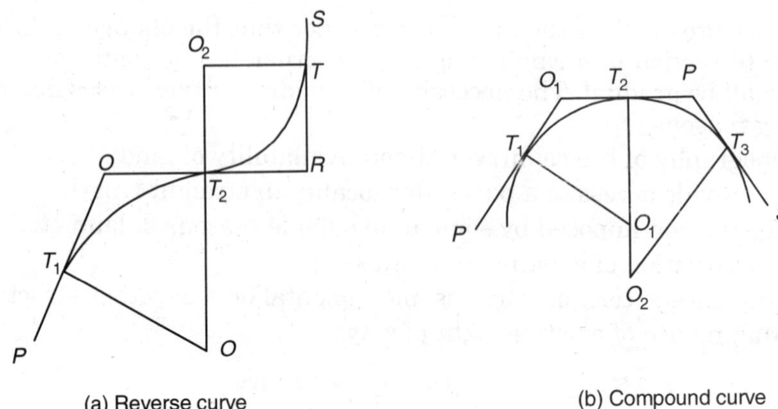

(a) Reverse curve

(b) Compound curve

Figure 2.15

2.11 TRANSITION CURVES

On a highway or railway, curves of varying radii are introduced. A transition or spiral curve or an easement curve as it is sometimes called, is one whose radius changes gradually from an infinite value to a finite value or vice versa for the purpose of giving easy change of direction. This tends to counteract the swaying outwards of a vehicle when subjected to sudden application of a centrifugal force at the instant of its entering or leaving of the curve.

2.11.1 Advantages of Providing Transition Curves

1. To have smooth work force entry from a straight road to a curved road and obtain *transition* from the tangent to the circular curve and from the circular to the tangent.

2. To obtain a gradual increase of curvature from a value of zero at the tangent to a maximum at the circular curve.

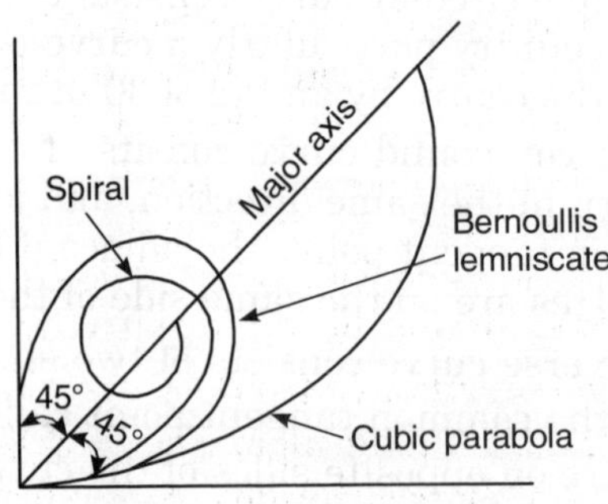

Figure 2.16 *Transition curves*

3. To have a gradual increase of super-elevation from zero at the tangent to a specific maximum at the circular curve.

4. To counrteact the form of centrifugal force which tends to sway outwards the vehicle.

2.11.2 Requirements of a Transition Curve

A transition curve should satisfy the following essential requirements:

1. It should meet the straight length of the road tangentially.
2. It should meet the circular curve smoothly the tangentially too.
3. The radius of curvature of the transition curve should be the same as that of the circular curve at the junction.
4. The rate of increase or decrease of the radius of curvature should be smooth and uniform.

2.11.3 Length of a Transition Curve

The length of a transition curve may be determined by any of the following methods:

1. By the rate of change of radial acceleration.
2. By time rate method.
3. By the rate of change of super-elevation.
4. By some empirical formula.

1. *By the rate of change of radial acceleration.* It has been observed that for maximum comfort to the passenger and the driver of the vehicle the rate of change of radial acceleration should be uniform and restricted to 0.3 m/sec^2 and the ratio $\frac{p}{W}$ should not exceed $\frac{1}{8}$ in any case, then

Length of transition curve = $4.491\sqrt{R}$

2. *By time rate method.* If the super-elevation is to be applied at the rate of x metres/sec and the velocity of the vehicle is v metres/sec, then the length of transition curve

$$= v \times \frac{h}{x} \text{ metres.}$$

3. *By an arbitrary rate of change of super-elevation.* For comfort condition, the super-elevation is changed at the rate of 400, then

$$L = 400 \times e \text{ where } e = \frac{sv^2}{126\,R}$$

4. *By empirical formula.* If v is the speed of the vehicle in km.p.h.

Length of transition curve can be expressed by the formula

$$L = \frac{v^3}{14R} \text{ metres}$$

2.12 MEDIANS

Medians or separators are provided between the two traffic lanes meant for traffic coming in opposite directions to prevent head-on collision. Medians also help in channelising the traffic into their respective lanes. Medians or separators can be provided by lane markings, mechanical separators embedded in the road surface, raised platforms etc.

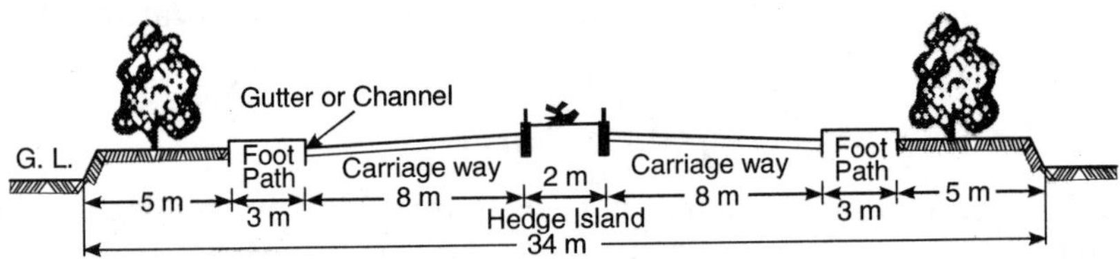

Figure 2.17 *Medians*

Pavement marking is the simplest of all. Ideally speaking the two lanes of opposite movement of traffic, should be separated by 5.0 m wide medians so that the glare of head-lights of two vehicles coming in opposite direction is minimised. The medians should be of uniform width on a particular road but if reduction in width is unavoidable a transition of 1 in 15 to 1 in 20 should be provided. For multilane highways medians must be provided. The width of a medians should not be less than 1.2 m.

2.13 ROAD KERBS

Road kerbs are indicators between the edge of a carriageway and the footpath, road islands, refuge islands, medians etc. The road kerbs may be of three types :

(i) Low kerbs (ii) Barrier type kerbs (iii) Semi-barrier type kerbs.

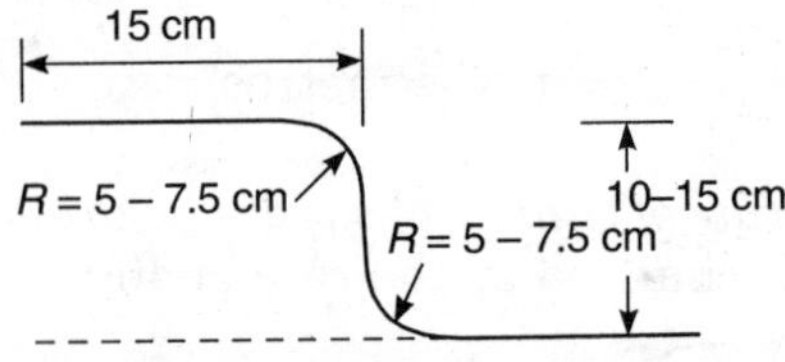

Figure 2.18 *Low kerb*

(i) *Low Kerbs.* They are also called mountable kerbs. These kerbs are indicators between the boundary of a road and the shoulders. The height of the kerb is such that the drivers find no difficulty in crossing these kerbs and use the shoulders in case of emergency. Normally the height of low road kerbs is only 10 cm above the payment edge. These kerbs channelise the traffic and also allow longitudinal drainage of road.

(ii) *Barrier type kerbs.* These type of road kerbs are provided in urban roads or city roads where the road passes through built up areas. The height of these kerbs is generally kept at 20 cm above the pavement surface.

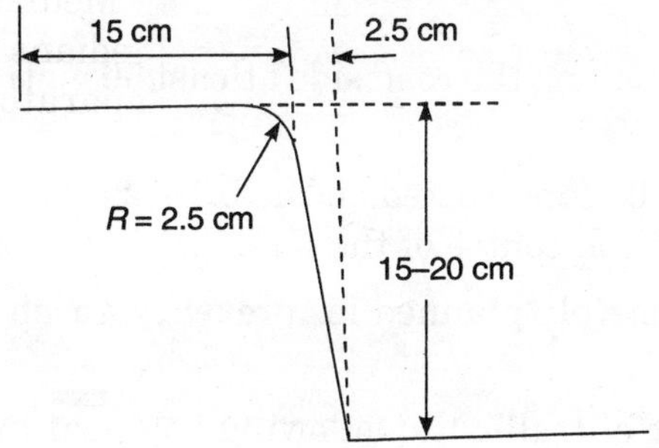

Figure 2.19 *Mountable kerb*

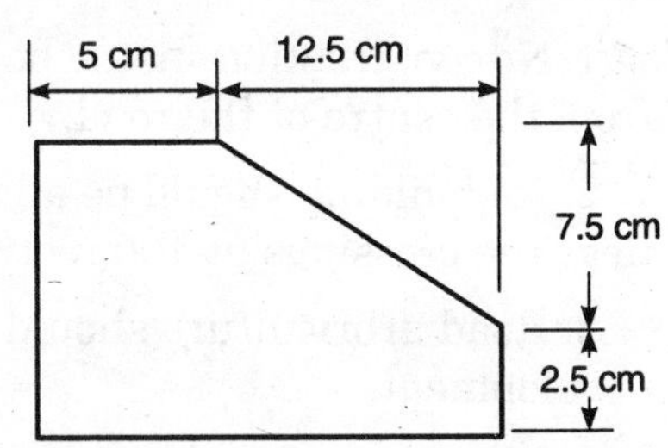

Figure 2.20 *Semi barrier type kerb*

(iii) *Semi-barrier type kerbs.* These type of road kerbs are used on the two ends of the highway separating the footpath and the pavement where pedestrian traffic is high. The kerb is generally 15 cm high with a batter of 1:1. This type of road kerbs prevent encroachment of parking spaces, footpaths, etc.

2.14 RIBBON DEVELOPMENT

This is the tendencies of most of population to live and have houses etc. nearest to a road, so when a new road is constructed near a city or in the sub-urban, construction of buildings, such as houses, shops, restaurants etc. starts. Sometimes the development of building is so haphazard that it becomes nuisanse for the traffic. Development of buildings in an un-planned manner is called *Ribbon Development.* Following are some of the disadvantages of ribbon development.

1. It produces congestion hinderances to traffic on the road.

2. Chances of accidents increases.

3. Future widening and development of road becomes very costly and very difficult.

4. Approach to cities and market places becomes hazardous and cumbersome.

5. It causes hinderance to the free flow of the thorough traffic.

6. It reduces the sight distances on crossings and disturbs road drainage.

7. Due to ribbon development, average travel speed decreases which causes inconvenience to the traffic.

2.14.1 Prevention of Ribbon Development

The following precautions should be taken for the prevention of ribbon development.

1. No construction should be allowed along the road side atleast 50 metres from the centre of the road.

2. No building should be allowed to be constructed on the bye-passes and near the crossings at 100 metres from the centre of the lane.

3. Road arbriculture should be effectively planned for preventing ribbon development.

For easy, comfortable and safe flow of traffic the following bold features should be adopted for modern highways :

1. Access to the main road should be completely controlled.
2. Sufficient number of traffic lanes should be provided.
3. Fast and slow moving traffic should be seggregated.
4. Parking spaces should be clearly indicated and provided.
5. Cross and opposite traffic should be separated by providing fly-overs, over and under bridges at convenient points.
6. Pedestrians should be segregated.
7. Comfortable and easy sight distances should be provided.
8. Separate by-passes should be provided for loading and un-loading of goods and passengers.
9. Modern street lighting should be provided in urban roads.

2.15 RIGHT OF WAY

It is defined as the land width acquired along the alignment of road. The *land width* of road will depend on its importance and future development. A minimum land width for different category of roads and different locations have been fixed by I.R.C. While acquiring land for the construction of a new road, care should be taken to see that sufficient land width is acquired for future development also as the cost of land invariably increases too much soon after the road is constructed particularly in sub-urban areas. The land width is governed by the following factors :

1. It primarily depends on the category of roads, e.g., national highway, state highway, district road and village road.

2. Height of embankment or depth of cutting will also matter in deciding the land width.

3. Side slopes and soil type.

4. Sight distance and nature of curves.

5. Drainage system and catchment areas.

6. Nature of surroundings.

Table 2.8 gives the minimum recommended curve radii and length of transition curves for various speeds as per I.R.C.

TABLE 2.8 *Minimum radi and transition length for various speeds (I.R.C.)*

S.No.	Curve radii	*Transition length of the curves*								
		Hill roads				*Plain and rolling country roads*				
		Speed of vehicle in km/hr				*Speed of vehicle in km/hr*				
		25	30	40	50	30	50	65	80	100
		m	*m*	*m*	*m*	*m*	*m*	*m*	*m*	*m*
1	45	30	30	–	–	75	–	–	–	–
2	60	15	15	30	–	60	–	–	–	–
3	75	15	15	30	–	45	–	–	–	–
4	90	15	15	30	45	30	45	–	–	–
5	120	15	15	15	30	25	60	–	–	–
6	150	15	15	15	15	25	45	75	–	–
7	200	15	15	15	15	15	45	60	–	–
8	250	–	–	–	–	15	30	45	90	–
9	300	–	–	–	–	15	25	26	75	–
10	400	–	–	–	–	–	15	25	45	100
11	600	–	–	–	–	–	–	25	45	80
12	750	–	–	–	–	–	–	25	30	65
13	900	–	–	–	–	–	–	25	30	50
14	1000*	–	–	–	–	–	–	–	–	45

* **Note.** It is usually not necessary to provide transition curve on the horizontal curves having radius greater than 1000 m.

REVIEW QUESTIONS

2.1. (a) What should be the qualities of a good surface?

(b) Draw a neat sketch, showing the different parts of a road structure.

2.2. What do you understand by the term 'camber'. Why camber is provided and on what factors the camber of a road surface depends. What are the disadvantages of a high camber. How much camber will you recommend for the following roads :

(a) Cement Concrete.

(b) Earth Road.

(c) Water-bound Macadam Road.

2.3. What is super-elevation, when and why it is provided? On what factors the super-elevation depends. Describe in brief the change over of cambered surface to that of super-elevated surface.

2.4. What is the maximum gradient and ruling gradient? Why the value of a gradient is different for roads in plain and roads in hilly places?

2.5. Describe in brief the meaning of 'Sight Distance'.

2.6. Describe the necessity of super-elevation, transition curve and widening of curves.

2.7. Write short notes on :

Transition curve, non-passing sight distance, introducing of super-elevation, composite camber, floating gradient, average gradient.

2.8. What are the requirements of a good road surface? How you will achieve to construct an ideal road ?

2.9. What do you understand by the term alignment of a road? What are the ideals which are kept in mind while aligning a road ?

2.10. Write short notes on :

(a) Super-elevation.

(b) Maximum gradient.

(c) Widening of road.

(d) Dual carriageway.

(e) Soling.

(f) Barrel camber.

2.11. What is super-elevation ? On what factors the super-elevation of a road depends ? What are the disadvantages of super-elevation ?

3

Road Project

GENERAL

Road project is a comprehensive term. Road project means planning, designing and construction of a new road, connecting two *terminus* points or a town with an existing road. The road project will consist of the following works :

1. Survey
2. Preparation of maps
3. Acquisition of land
4. Road alignment
5. Earth work
6. Traffic surveys
7. Estimates
8. Project report

In this chapter, we will underline the various stages of road project in brief.

3.1 SURVEY

To determine the location of a proposed road and to collect the necessary data, the following road surveys are undertaken :

1. Reconnaissance survey
2. Preliminary location survey
3. Final location survey
4. Construction survey

3.1.1 Reconnaissance Survey

This is a rapid examination of the ground and its adjacent natural features from approximate, preliminary details. This preliminary survey is done generally without the help of any instrument. The reconnaissance or *Recci* as it is sometimes called, is to determine the suitability of each alignment marked on the map during map study. *Recci* also helps in conducting the preliminary survey. During the *Recci* one has to examine a very wide belt on either side of the general location of the route and thus covering all possible alignments drawn during map study, has to improve upon the alignments, if possible. It is during the *Recci* that the visual inspection of the proposed alignment and the surrounding country is carried out. Special attention is paid to :

(i) River crossing and determination of suitable site for bridge or culvert construction.

(ii) Alignment which will entrail approximate quantity earth work or steep gradients, availability of land.

(iii) Sources of supply of water, and other construction materials, availability of labour and machinery required for the construction of road.

3.1.2 Preliminary Location Survey

It is a rough approximate type of survey, which is conducted to have a fair idea of the surrounding areas. The preliminary survey is done with the help of a plane-table, compass, Ghat tracer chain or steel tape. Rough levels are also taken at different points so as to have an idea of the earth work.

After the preliminary survey, the necessary plans i.e., drawings are prepared corresponding to the survey work and the approximate, preliminary estimate of the cost of the road and its ancillary works for each proposed route is prepared. The most economical and the best of these alignments is selected.

3.1.3 Final Location Survey

Before the final location survey is started, the centre line of the finally selected route is marked on the ground. Now leves are taken along this centre line. Cross-sectional levelling is also done at the required points of centre line. Actually speaking the final location survey is the detailed levelling along the centre line and at right angles to it. From this survey work, final plans, designs and estimates are prepared.

3.1.4 Construction Survey

The construction survey is done for the setting out of the road. It consists of the following steps :

(a) Clearing of jungles, bushes, grass and any other objectionable materials.

(b) Setting out the centre line and edges of the road. The centre line is defined by the alignment pegs which are driven at regular distances. On the curves pegs will be driven close together. The line between these pegs is clearly demarcated by cutting a narrow V-shaped cut in the ground along the centre line. This cut is called *lockspit* or *dagbel.*

(c) Setting out areas for *barrow pits* and *banks.*

(d) Fixing bamboo and string profiles at regular distances for earth work.

3.2 PREPARATION OF MAPS

The following drawings are generally required :

(i) *Topographical maps.* These maps show the general details of the area such as existing metalled and unmetalled road, cart tracks, position of wells or other sources of water, hospitals, schools, police stations or outposts etc. Contours are also shown. The topographical maps also show villages and towns having a population more than 500. These maps are generally prepared by the Survey of India maps to a scale of 1″ = 1 mile or 1 cm = 0.66 km.

(ii) *Population maps.* These maps show the distribution of population in a particular area. These maps show the important topographical features of villages, towns etc. and their population.

(iii) *Agricultural and industrial maps.* These maps are prepared on the basis of agricultural and industrial productivity. The maps are divided into following three groups :

(a) Those villages and towns which are highly developed agriculturally or industrially. Sometimes industries are also divided according to their importance, (b) Those which are particularly developed, and (c) Those which can be developed.

(iv) *Proposed plans.* These maps show the proposed alignment, and detailed sections of the proposed road, along with the earth work details such as cuttings and embankments. The details of ancillary works of roads, such as bridges, culverts, and railway crossings etc., are also drawn on separate sheets. The alignment maps should clearly show the position of these ancillary works etc.

3.3 ACQUISITION OF LAND

For the construction of a road and its ancillary works, some or more land is to be acquired. It is always advisable not to pass the alignment through agricultural land or through built-up areas, because it becomes difficult to acquire the required land. There are two types of lands viz., permanent land and temporary land. Permanent land is the land-width required permanently for the road and its ancillary works. The following are permanent land-widths recommended by the Indian Road Congress: (Table 3.1).

TABLE 3.1 *Permanent land widths*

Class of road	*Normal width*	*Minimum width*
National Highway	60 m	45 m
State Highway	45 m	30 m
Major District Roads	30 m	20 m
Minor District Roads	25 m	15 m
Village Roads	20 m	15 m

Apart from this permanent land, some more land is also acquired for borrow-pits and soil banks etc. This land, is called *Temporary Land.* This temporary land, after the construction of road is given to the farmers on lease, only for agricultural purposes.

3.4 ROAD ALIGNMENT

The course or route along which the centre line of a road is located in the plan, is called *Road Alignment.* Before starting the actual construction, the centre line of the road is first marked on the plan and then on the site. The following points should be kept in mind while aligning a particular road :

(i) The alignment should be as short and straight as possible.

(ii) The straight alignment should be deviated when it is required to give the benefit of the road to an intermediate town, an important village, a railway station, a market place or some other important highway.

(iii) The alignment of a road should cross another road, a railway line or a stream, preferably at right angles.

(iv) The alignment should cross a river or a stream at such points where the width of river or stream is minimum and where good and durable foundations are possible for the construction of a bridge or culvert.

(v) There should be minimum number of crossings, bridges or culverts in the alignment of the proposed road.

(vi) As far as possible, alignment should not pass through thick forests, built up areas, or agricultural, garden land.

(vii) There should be minimum of cutting and minimum of banking. Practically speaking the amount of cutting and of banking i.e., filling should be equal.

(viii) The alignment should ensure easy gradients and smooth curves.

(ix) The alignment should pass through such points where the material for the construction and maintenance of road is easily available.

(x) Unnecessary zig-zags in the alignment should be avoided.

3.5 FACTORS CONTROLLING ALIGNMENT OF ROADS

The various factors which govern the alignment of roads are :

(i) *Obligatory points.* For an alignment to be shortest, it must be straight. For connecting the obligatory points which the road alignment has to pass, the alignment has to be deviated. The various obligatory points are bridge, railway track, intermediate town, high ridges, mountain pass, tunnel etc. Various alternatives are to be worked out for a suitable alternative. For a crossing a hillock or a ridge, sometimes a tunnel is more suitable than going around a ridge. Sometimes alignment is deviated to touch some important points. The various obligatory points and the deviation in the alignment are shown in Fig. 3.1.

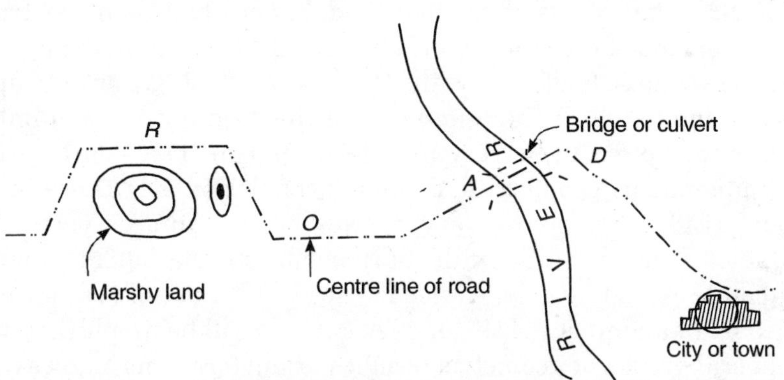

Figure 3.1 *Obligatory points*

There are certain obligatory points which should be avoided such as religious places, costly structures, thick jungles, marshy lands, grave yards, military protected areas etc. Alignment of such areas will have to be avoided otherwise either the alignment will be too costly in construction and maintenance or it will create other problems of law and order, acquisition of land etc.

(ii) *Traffic.* The volume and trend of traffic also affect the alignment of road, so while aligning a road desire line should be drawn before taking a final decision.

(iii) *Geometric design.* Road curves, gradients, sight distance and such other factors also affect the alignment of roads. If the straight alignment causes steep gradients, then longer route to get ruling gradient is planned. Similarly if straight alignment does not provide passing or crossing sight distance, then also the alignment will have to be deviated. Similarly to avoid steep curves even we have to deviate from the shortest alignment. The gradient should be within the ruling gradient.

(iv) *Economic Factors.* Economic considerations play a very important role in the alignment of road. The economic considerations should be for easy and cheap acquisition of land, travel cost or running cost. Longer the route, more the construction, maintenance and running cost. Deep cuttings high embankments, tunnelling, railway crossings etc. are the other economic considerations of alignment. Drainage, hydrological factors, political considerations, monotomy, military strategic points also demand considerations in the alignment of roads.

3.6 RE-ALIGNMENT

Most of the Indian roads of pre-independence era had been constructed to meet the demand of the traffic which was much slower than the present day traffic. Most of these roads were either constructed in stages or were upgraded in phases due to which the originality of the road have been diminished. They are now found to be very much deficient in their characteristics of design, alignment, geometrics, drainage conditions etc. There are many National Highways having very narrow bridges and culverts that too unsuitable for heavy traffic, with sharp bends, steep gradients, inadequate sight distances and poor lane width. These defects of highways are to be rectified by re-alignment of the present road. It will be worthwhile to adopt more liberal values of geometric design parameters where excessive costs are not involved. It may not be always possible to re-align the road for the sake of making improvements and rectifying defects. At the national level it has been decided that National Highways should as far as possible, be able to cater to the traffic demands moving at the design speed, fulfilling the safety requirements and comforts, both for the present and future needs. For achieving this object, it is imperative on the part of the traffic engineer to plan improvements and remove deficiency of geometrics etc. within the means. The following situations may warrant the necessity of re-aligning a road :

1. For improving geometric designs such as horizontal and vertical curves, super elevation, sight distance etc.

2. For improving design elements like steep gradients, undulations, hair curves or zig-zag curves etc.

3. For raising the level of the road on account of change in H.F.L. and water logging.

4. For re-constructing or widening and strengthening of weak culverts, narrow bridges etc.

5. For constructing over bridges or railway crossings, grade separation etc.

6. For constructing a bye-pass so as to avoid congested localities.

7. For connecting some strategic point such as a new railway station, some tourist spot or some defence requirements.

During the process of re-alignment, special care should be taken to improve the geometric design of the highway such as smooth curve, super-elevation, sight distance etc. This is possible only when the alignment is considered as a whole and not in piece meals. Improvement in transition curves is not a very costly process and hence can be improved only by improving the horizontal curves. Sight distances will automatically improve. While alignment attempts are made to improve overtaking sight distances on summit curves. Stopping sight distances should also be improved on strategic points. Water logged area and sub-merging of road or a portion of road should be avoided.

The following procedure should be adopted for the re-alignment of a project :

1. Extensive survey should be conducted for collecting various datas like traffic, soil conditions, drainage and drainage structures, right of way and its cost etc.

2. Detailed drawings such as topographical plans, detailed cross-sections at various points, longitudinal section, contour maps etc. should be prepared for the existing and the proposed alignment/alignments for the purpose of comparison.

3. Compare the economics of the proposed alignment to see whether the project is worthwhile (see chapter on Highway Economics).

4. Preparation of design of pavement, detailed working drawings of culverts, bridges, fly-overs etc. and their estimates.

5. The construction work along the new alignment should be started by first marking the centre-line, dagbelling, earth work etc.

3.7 EARTH WORK

Earth work in its widest sense includes excavation in rock as well as ordinary clayey soil and formation of banking.

(a) *Cutting.* It denotes excavation, under the surface of the earth and the removal of the excavated material. When the depth of cutting is less than 3 m it is called shallow-cutting and when it is more than 3 m, it is called deep-cutting. Cutting more than 16 m should be avoided.

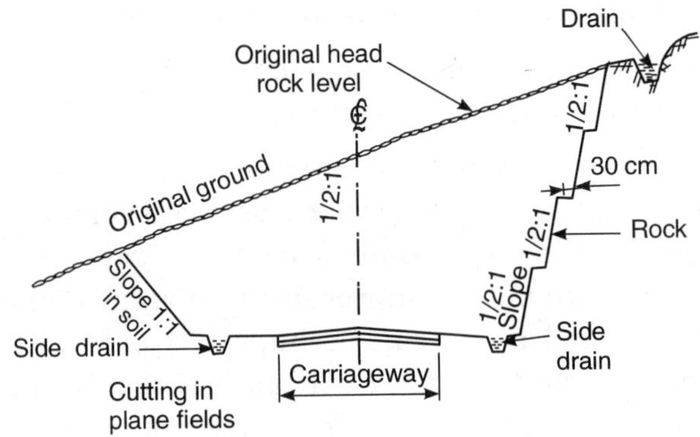

Figure 3.2 *Cutting in rocks*

(b) *Embankment.* It denotes lump of earth, collected together from the excavation and built to a definite shape to carry a roadway (or a railway) over it, above the ground level. In a low lying area embankment more than 3 m is called high embankment.

Excavated earth, when broken down, increases in volume. This loose earth when deposited in banks will settle down under its own weight and through the action of atmosphere. The decrease of height of embankment is called settlement and the amount by which the new embankment is to be made higher than the formation level is called *allowance for settlement.*

3.8 SIDE SLOPES

Earth embankment of usual height will stand safely with side slopes of 2 in 1. In cutting a side slope of 1 : 1 will serve the purpose well. Rocks when cut are even stable at 1 in or $\frac{1}{4}$ vertical even overhanging depending on rock properties and rock joints. Such slopes were adopted in as they involve minimum of earth work, but the tendency these days is to provide flatter slopes as they result in the safe operation of vehicles and the maintenance cost is also reduced. Steep slopes in embankments have the following disadvantages:

1. Steep slopes result in accident hazards. The vehicle can overturn if the wheel goes over the edge.

2. Steep slopes on gutter ditches may also result in accidents.

3. Steep slopes erode badly, hence the maintenance cost is increased much.

4. Tree plantation or grass growing is difficult.

Indian Road Congress has recommended the following side slopes in embankments:

(a) Upto 60 cm height—1 in 4.

(b) Over 60 cm height—natural slope or 1 in 2 whichever is flatter.

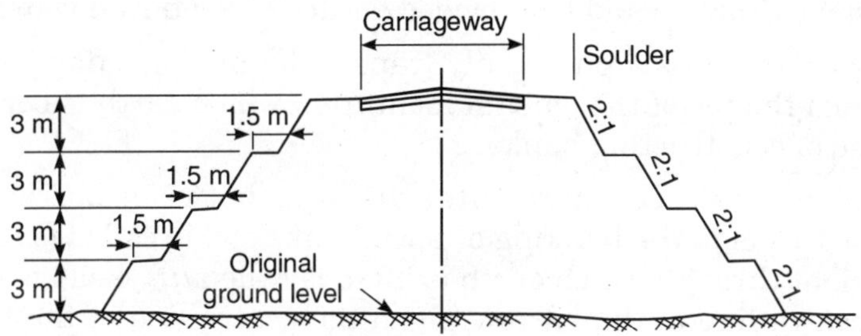

Figure 3.3 *Embankment*

3.9 SOME TECHNICAL TERMS IN EARTH WORK

(1) *Balancing of earth work.* The excavated earth from cutting should be utilized in forming embankments, when the cutting and filling both are required at not very long lead. This cutting and formation of banks simultaneously along the alignment of a road is generally known as *balancing of earth work.*

(2) *Berms.* A strip of land is always left in between the toe of the embankment and the inner edge of the barrow-pit and the boundary line to prevent damage of banks by erosion. In case of high embankment and deep cuttings, the long side slopes are stepped to avoid speedy flow of rain water and thereby causing *rain-cuts.* These horizontal strips are called *berms.*

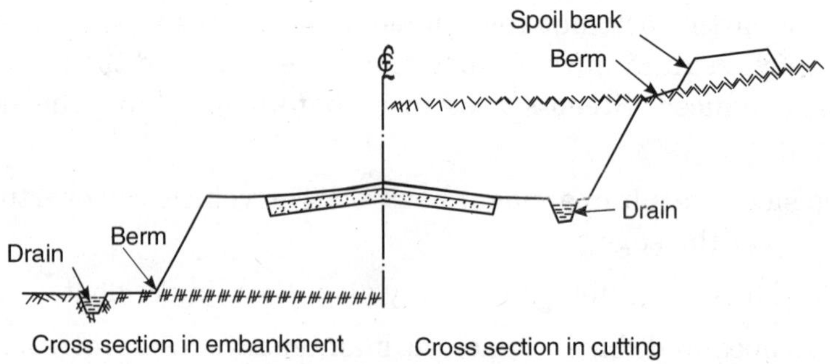

Figure 3.4 *Berms*

(3) *Spoil banks.* Earth from excavation, if not utilized in banking or to fill up the low lying areas nearby, is deposited in heaps, parallel to the length of the cutting in the form of a bank at some distance away from the top edge of the cutting. These banks are called *spoil banks.* The height of spoil banks should be nearly 125 m. Between the toe of the spoil bank and top edge of cutting a small drain should be provided for draining off rain water.

(4) *Barrow pits.* It is a pit usually trapazoidal in shape dug out at some distance from the toe of the embankment from which earth is borrowed for the purpose of constructing banks.

(5) *Lead and lift.* The horizontal distance between borrow pit i.e., excavated earth is to the banking or spoil banks etc. is called *lead,* and the vertical height through which earth is lifted is called *lift.* Lead is measured from the centre line of the barrow pit to the centre line of the bank. The initial lead and lifts are taken as 100 ft (30 m) and 5 ft (1.5 m) respectively.

(6) *Angle of repose.* Loose earth or any granular solid when heaped together, slides down and finally takes up a shape of equilibrium. The maximum angle, at which the surface of the earth stands up by itself, with the horizontal, is called the natural angle or *angle of repose.*

(7) *Benching.* Embankments on ground which have a considerable slope or cross fall, say of more than 1 in 10, the ground is benched to prevent slipping down of earth. *Benching* means cutting of series of steps at right angles to the line of slope along the length of the bank.

(8) *Formation.* It is the finished surface of sub-grade of a roadway (or railway) in cutting, or embankments. The formation width is defined as the finished top width of an embankment or a cutting. The portion of the formation width which is left unmetalled on either side is called *shoulder.* The shoulders should be nearly 1.5 m wide. The following are the formation widths recommended for various classes of roads by the Indian Road Congress (I.R.C.) :

TABLE 3.2 *Road formation widths*

Class of Road	*Minimum formation width in plane country*	*Minimum formation width in hilly country*
National Highways	12 m	8 m
State Highways	9.75 m	8 m
Major District Roads	7.25 m	6.75 m
Minor District Roads	7.25 m	6.75 m
Village Roads	5.5 m	4.25 m

3.9.1 Procedure of Earth Work in Embankment

For a roadway (or railway) construction, earth work is essential throughout the length of the road. It may be either in the form of embankment or cutting. The following procedure should be adopted for earth work in embankment:

(1) If suitable earth is available from cutting, banking should be done by the excavated earth from the cutting portion only.

(2) In case of high embankments, earth should be taken from cuttings and borrow pits.

(3) Banks should not be constructed entirely with sand but if it is unavoidable they should be covered with a layer of good earth. That is, sand layers should be covered with a layer of good quality of earth. Otherwise, if possible the earth to be used for banking should contain two parts of sand and one part of clay.

(4) When the ground has a considerable cross fall, it should be benched to prevent slipping down of soil.

(5) After dogbelling, the usual demarcations of pegs line, on both sides of the central line, should be made. Bamboo profiles are set up at an interval of 100 m.

(6) Embankments are raised in regular layers of 20 cm to 25 cm thickness. The layers should be horizontal having a uniform slope. Each layer should be compacted and consolidated properly in camber.

(7) Allowance for shrinkage is to be given for each particular class of soil. Earth work should not be done in rainy season.

(8) The formation should be finished only after two successive monsoons. The gradients and camber should then be checked.

3.9.2 Procedure for Earth Work in Cutting

1. The cutting should be started from the extreme dogbelled edges and from cross-section to cross-section.

2. In shallow cuttings the central portion in first excavated with vertical faces but deep cuttings are taken in steps, then the slopes are finally formed.

3. During cutting, check measurements should be taken at intervals from the guard pillars on each side of cutting, so that excavation should not be carried below the formation level.

4. Suitable catch-water drains are to be provided on either side or on one side of the cutting.

5. The earth obtained from cutting should be utilized for making embankments adjacent to it, otherwise it should be deposited in the form of spoil banks on one or both sides as may be necessary. The spoil banks are formed in the same way as the embankment.

3.10 TRAFFIC SURVEYS

Before deciding width of formation and the type of surfacing to be provided, it is essential to have a detailed study of the nature of the traffic anticipated over the proposed road. The following information should be collected in addition to data relating to existing trades, industries and agricultural products etc.:

(1) Nature and intensity of traffic anticipated.

(2) Possibilities of trade and industrial development resulting from the proposed road.

(3) General resources of the locality.

From the above information the width of formation and the type of surfacing to be provided is decided. Traffic survey also helps in the proper alignment of the proposed road.

3.11 ESTIMATES

The following items of estimates are to be prepared. Actually speaking the items of estimates depend upon the nature of the project. In some cases, where the project is minor, a few items are required :

(i) Acquisition of land.

(ii) Clearing of jungles from the site.

(iii) Earth work.

(iv) Materials to be used for the construction.

(v) Estimates (which are prepared separately) for ancillary works such as bridge, culvert, railway crossing, etc.

(vi) Supplying and fixing of kilometre, 200 metre and boundary stones.

(vii) Arboriculture.

(viii) Construction of temporary site buildings etc.

3.12 PROJECT REPORT

Before the construction is actually started, a project report is prepared and submitted for approval. A project report is a concise statement, containing important information regarding the project. It should also include the possibilities of future developments and the benefits to the general masses. The following detailed information should be included in the project report.

(1) Details of rainfall, floods and other climatic conditions.

(2) Justifications for the selection of the particular alignment of the proposed road.

(3) History of the existing road system and other means of communications.

(4) Earth work details.

(5) Design and justification for the type of surfacing, proposed.

(6) Benefits served to the people in general, by the construction of this road.

(7) Details of bridges, culverts, railway crossings, etc.

(8) Availability of constructional materials and labour near the alignment.

REVIEW QUESTIONS

3.1. What are the essential constituents parts of a project ? Describe in brief how you will proceed to start a road project.

3.2. What is meant by earth work, what is the procedure for earth work in embankment ?

3.3. For a road project, how many types of engineering survey are there. What is the difference between traffic survey and engineering survey ?

3.4. Write short notes on :

(i) Reconnaissance (ii) Topographical maps

(iii) Construction survey (iv) Spoil banks

(v) Borrow pits (vi) Dagbell

3.5. Write down specifications for earth work in cutting.

3.6. What is meant by alignment of roads ? What are the points to be kept in mind while aligning a road ?

Highway Drainage

GENERAL

Most of the damages happens to highway during rainy season only, hence adequate and efficient drainage is an most essential factor in the design, construction and maintenance of a road. With efficient drainage system the efficiency and life of a road is considerably increased. The water enters the road structure from the following sources:

(i) From the top of pavement through crack, patches and voids of pavement materials.

(ii) Surface water from the sides of pavements.

(iii) From the under-side of the pavements by sub-soil water.

Road or highway *drainage* means removal of surface water and sub-soil water. The removal of rain water from the surface of road is called *surface drainage* and the removal of sub-soil water from the sub-grade is called *subsurface-drainage.*

4.1 SURFACE DRAINAGE

During the rainy season, water gets collected at the top of the road surface and percolates into the road surface and the sub-grade and thereby weakens the sub-grade. To prevent this percolation of water into the sub-grade from the surface of the pavement, the following remedies have been suggested:

(1) Providing a water-tight or impervious type of road surfacing at the top most layer i.e., wearing surface.

(2) Providing sufficient cross and longitudinal slopes in the road surface.

(3) The top surface of the berms should be of an impervious material and should have proper slopes towards side drains.

(4) Providing side drains on both sides of the road. The drains should have a proper slope or gradient so that the water from the road surface should be drained off easily. The side drains should not be less than 1.85 m or 6 ft from the edge of the formation of the road. In case of cutting, the surface drains should be just after the edge of formation.

(5) In those places where the rainfall is heavy, the side drains should be trapezoidal in section and in those places where there is less rainfall the section of the drain is triangular or sauccer [(Figs 4.1 (a), (b) and (c)].

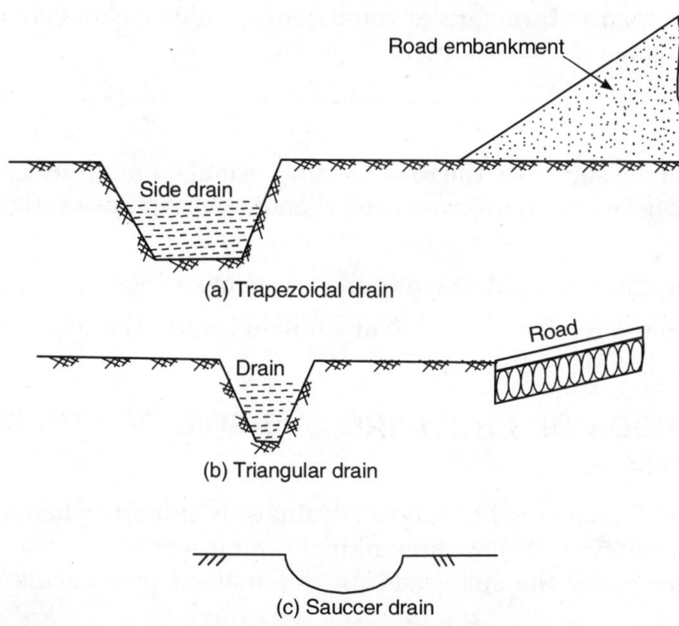

Figure 4.1 *Surface drainage*

(6) The height of the embankment should be nearly 60 cm or 2 ft. above the highest flood level (H.F.L.) of the area.

4.2 SUB-SOIL OR SUB-SURFACE DRAINAGE

Water which penetrates into the ground will continue to flow underground until it meets with some impermeable material, then the flow ceases and the water accumulates. The top surface of this underground water is called *Water Table.* The ground above this level will remain unsaturated and below this level, saturated. The water table generally tends to be parallel to the ground level, but the depth of water table below the ground level may vary and it depends upon the geological conditions.

If this water table is very near to the sub-grade of the road, the consistency of the soil will change from dry to plastic state and the bearing capacity will consequently decrease. Another adverse effect which results from the increased moisture content is the volumetric increase which will ultimately cause cracks in the road surfacing. Hence it is essential to drain off this water to prevent the sub-grade, which is the foundation of the road from these adverse effects of sub-soil moisture. The method of removal of the sub-soil moisture is called *Sub-soil drainage.*

Following are the conditions under which sub-soil drainage should be provided :

(1) If the road is through flat country and water stagnates on adjacent lands which makes the road bed soft and unstable.

(2) When the road is in cutting and there is considerable seepage in the slopes.

(3) When the surface of the road has the normal underground water table sufficiently below the road crust even then due to capillary action moisture may rise to the surface of the road or the sub-grade.

(4) When the soil is subjected to the action of springs.

(5) When the road is at the foot of a hill and water therefrom flow on the road and it damages the road.

4.3 METHODS OF PROVIDING SUB-SOIL OR SUB-SURFACE DRAINAGE

Pipe drains. This method of sub-soil drainage is suitable when a road runs in a flat country with low embankments and where the sub-soil water accumulates below the sub-grade. In this method pipe drains are placed

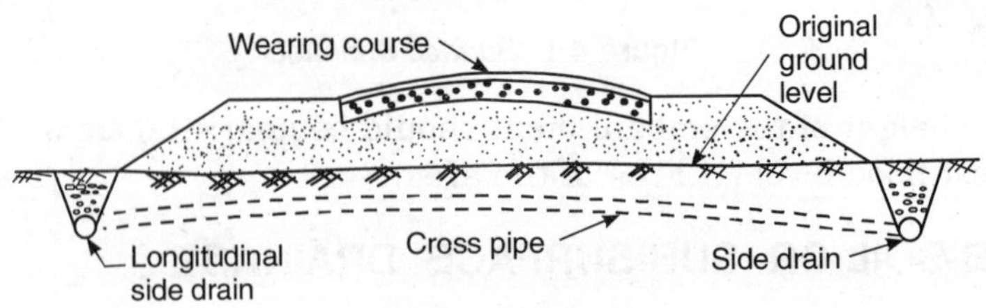

Figure 4.2 *Cross pipe*

below the surface of the ground in the permeable saturated stratum. The pipes are usually made of vitrified clay and are placed on a bed of sand, crushed stone, or clay 15 cm thick, with open joints butting against each other. To prevent the earth from above entering the pipes, through open joints, a tarred paper is sometimes provided to cover the joints. These pipes are placed in the trench with proper slopes both cross and longitudinal. The pipes are 15 to 20 cm in diameter. Cross or transverse pipes which are 6 to 10 cm in diameter are also laid at a distance of 6 m to 20 m apart. The trench is then filled with hard porous materials as shown in Fig. 4.2.

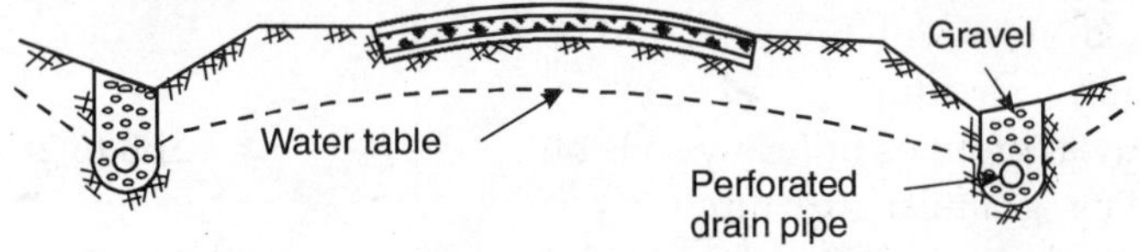

Figure 4.3 *Longitudinal pipe of sides of raods*

The main longitudinal pipes may be laid in the centre or on both sides of the road depending upon the moisture conditions. The main pipe discharge their water into the surface drain.

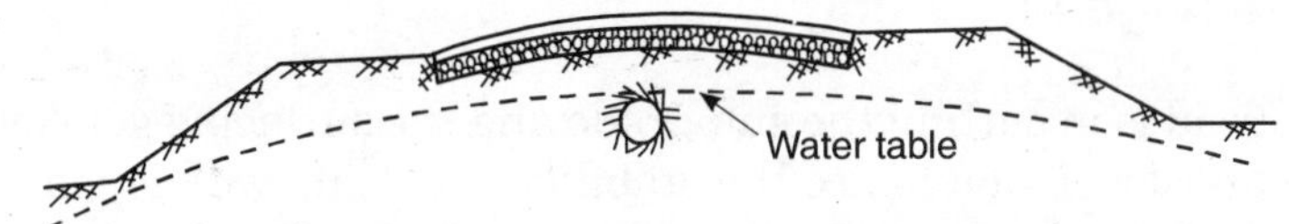

Figure 4.4 *Longitudinal pipes in the centre of road*

Table 4.1 gives the recommended spacing of sub-soil drains, which are used for draining off the sub-soil moisture.

TABLE 4.1 *Centre to centre distance of sub-soil drain pipes*

S.No.	*Nature of sub-soil*	*Depth of the invert of the sub-soil drain from the ground level*	
		0 .6 m to 0.9 m	*0.9 m to 1.2 m*
1	Sand	45.5 to 60.0 m	60.0 to 91.5 m
2	Sandy loam	26.0 to 30.0 m	30.0 to 46.0 m
3	Loam	18.0 to 27.5 m	27.5 to 30.5 m
4	Clay loam	13.5 to 17.0 m	17.0 to 20.0 m
5	Sandy clay	10.0 to 12.0 m	12.0 to 14.0 m
6	Clay	7.5 to 9.0 m	9.0 to 11.0 m

As a thumb rule a drain line draws the moisture from the sub-soil of 120 metres from each side for each metre of depth.

C.G. Elliot gives the following (Table 4.2) spacing of the sub-soil drains for guidance.

TABLE 4.2 *Spacing of sub-soil drains as per C.G. Elliot*

S. No.	*Nature of sub-soil*	*c/c spacing of the sub-soil drains*
1	Close dense soil, having largely clay	9.0 to 12.0 m
2	Coastal plain lands composed of mixed clays with fine sand and of uniform structure	18.0 m
3	Alluvial grounds or heavy soils but having granular structure	21.0 to 24.0 m
4	Alluvial glacial drift and sandy loam silts with clayey sub-soil	38.0 m
5	Sandy lands and soils containing considerable quantity of vegetable matter	45.0–60.0 m

4.3.1 Drainage in Cutting

When the road is in cutting the sub-grade and the surfacing get closer to the sub-soil water level, and hence the stability of cutting will depend upon the efficient removal of water from the slopes.

To remove this water, burnt clay pipes having a diameter of 5 cm to 10 cm or ordinary drains having suitable cross-sections are provided on the slopes at an interval of 10 to 12 m, which discharge their water into the central drain.

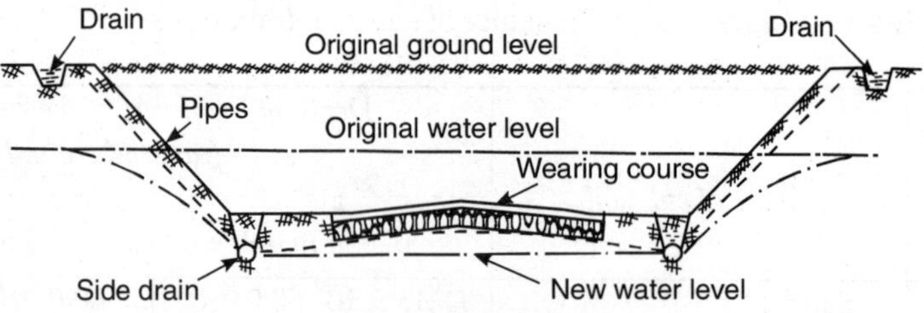

Figure 4.5 *Drainage in cutting*

4.3.2 Drainage in Side-long Ground

In embankment the stability of the side slopes is usually more important than the bearing power of the sub-grade or formation. If the embankment is built on ground with considerable crossfall or the road has a section partly in embankment and partly in cutting, the drainage problem should be surface water or sub-soil water, is not drained off immediately, it will reduce the bearing power of the soil and will almost invariably cause an

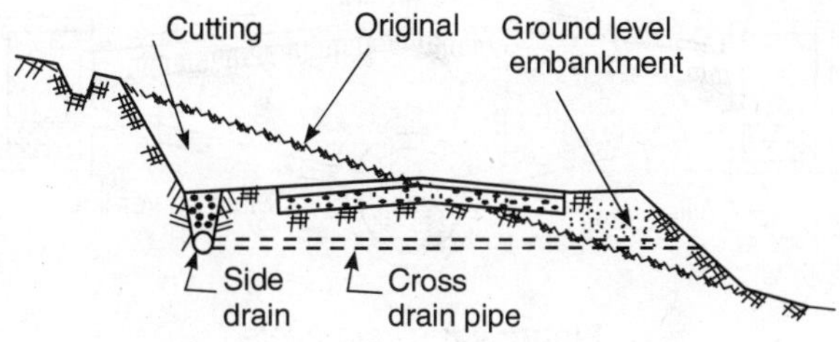

Figure 4.6 *Drainage in slopy ground*

embankment to slip. The arrangement of providing drains in a road section in partial cutting and partial embankment is shown in Fig. 4.6.

4.4 DRAINAGE IN CITY OR URBAN ROADS

The drainage of road in urban or built-up areas is different from that of other places. Open drains in built-up areas cannot be provided as they are unsightly, and occupy a lot of space ; and are a source of danger to the traffic. The drainage is, therefore, provided through vents and gratings. The surface water enters through those gratings and vents (through kerbs) into an underground pipe system.

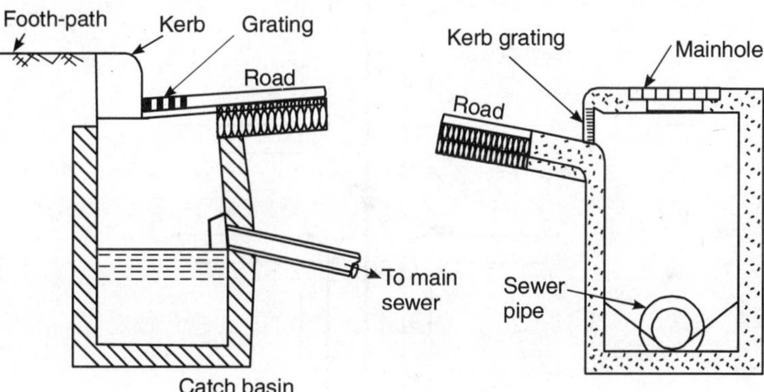

Figure 4.7 (a) *Catch basin* **Figure 4.7 (b)** *Kerb opening storm water inlet*

Pipes are laid underground and water is led to them through transverse pipes. Water is admitted through the gullies by means of hollow kerbs known as *Kerb inlet* or by means of gratings which are fixed at the edge of the road.

Kerb inlets are generally preferred to the gratings as the gratings reduce the carriage width and sometimes the gratings are broken due to heavy impact of vehicles. Surface drains are generally provided on one side only as

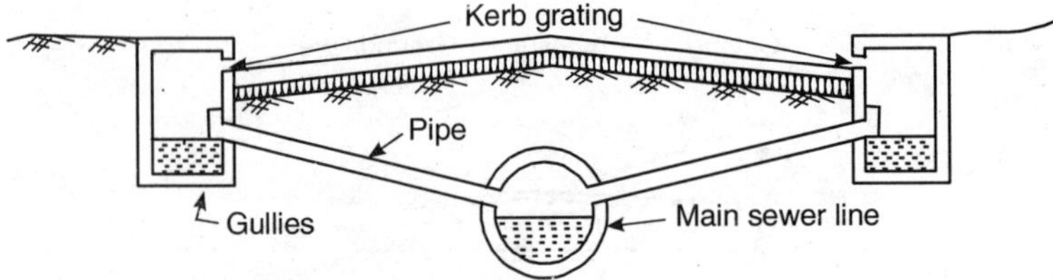

Figure 4.8 *Road Inlets*

it is uneconomical to run drains on both sides of the road unless the house drains are to be pitched up.

4.5 DRAINAGE IN CROSS-ROADS

Surface drainage in cross-roads, that is, at the road junctions or intersections requires careful planning. When two roads are crossing each other, the grade of underground drains, or channels and the cross slope of roads is so adjusted that the low spot is displaced a few metres away from the turn out.

Figure 4.9 shows the arrangement of gullies for road junction where both carriageways have the normal arched camber. It should be noted that

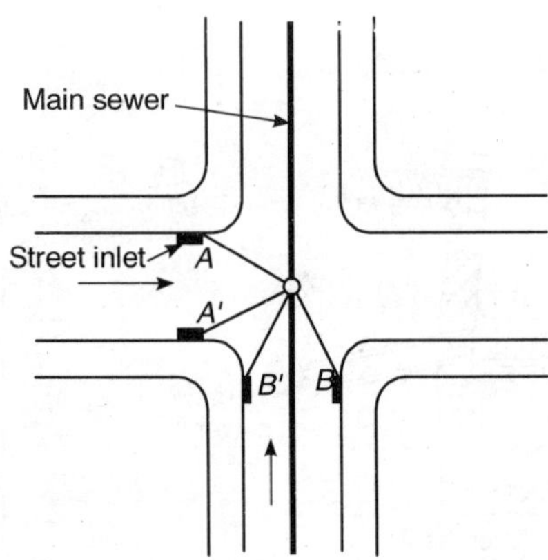

Figure 4.9 *Position of road-side gullies*

the water should not be allowed to collect at the crossing but should be intercepted at points *A*, *A´*, *B*, *B´* etc., before it reaches the crossing.

To remove this water, burnt clay pipes having a diameter of 5 cm to 10 cm or ordinary drains having suitable cross-sections are provided on the slopes at an interval of 10 to 12 m, which discharge their water into the central drain.

4.6 WATER TABLE

Water table is responsible for the stability and strength of a highway. Lower the water table more the stability. The level of water table should be 1 to 1.2 m below the sub-grade of the road and pavement layers are not subjected to moisture from below. In places where water table is high, sometimes as high as the ground level, the alignment should be diverted to avoid such a site. But sometimes it is unavoidable and the alignment has to be taken through such places. In that case, the water table has to be lowered. *Lowering of water table* is comparatively very easy in permeable soils but in the case

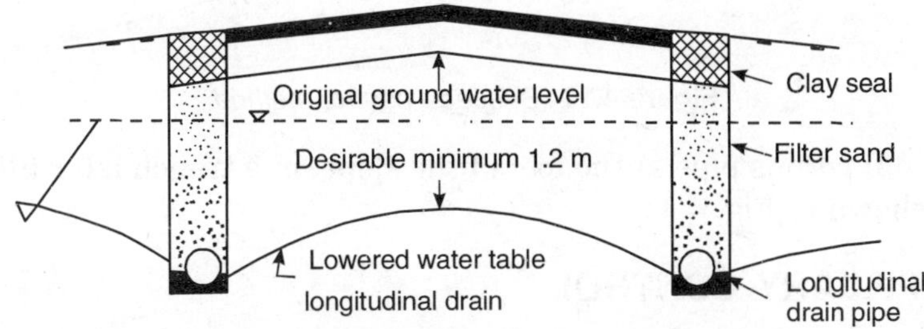

Figure 4.10 *Lowering of high water table*

of impermeable soil it is difficult. For permeable soils, porous pipes are laid along the edges of the road longitudinally and over the pipes the trenches are filled with filter sand. The pipes are placed with longitudinal grades parallel to the road gradient. The porous pipe suck the water from the sub-soil thereby lower the water table to the desired level. Sometimes transverse pipes are also placed to remove the excess water from the longitudinal pipes. To drain out water from longitudinal drain pipes are discharged in some culverts, rivers etc. whose HFL are lower than these longitudinal pipes.

4.7 SEEPAGE

When the general ground level is slopy and even the strata of soil is sloping, the problem of seepage does occur. Due to continuous seepage the sub-grade gets weaken and the road structure is exposed to failure. When the seepage zone is at a depth of less than 0.8 m or 0.9 m, seepage control measures are to be adopted. Seepage zone is ascertained by the general level of water table in the adjoining area. The usual method of seepage control is to provide

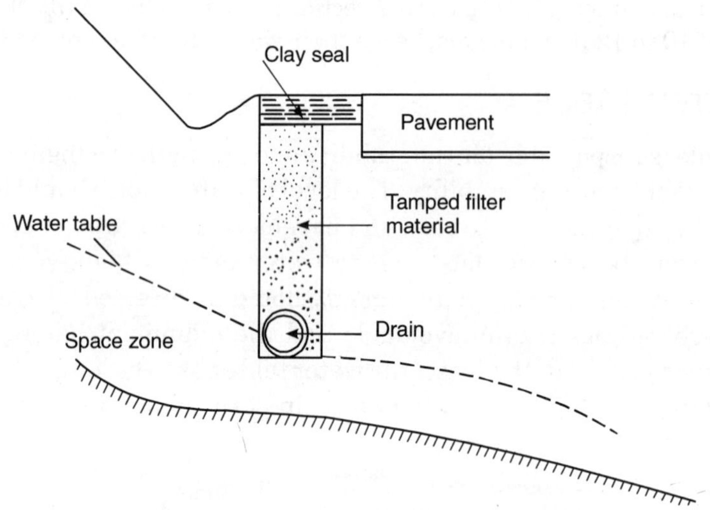

Figure 4.11 *Seepage control method*

longitudinal porous pipe at the foot of the slope, in a trench later filled with sand as shown in Fig. 4.11.

4.8 CAPILLARY CONTROL

The water uses from the moist sub-soil due to capillary action which reduces the bearing capacity of the sub-grade. This problem of capillary rise occurs when the road is in bank and the water table is close to the ground level.

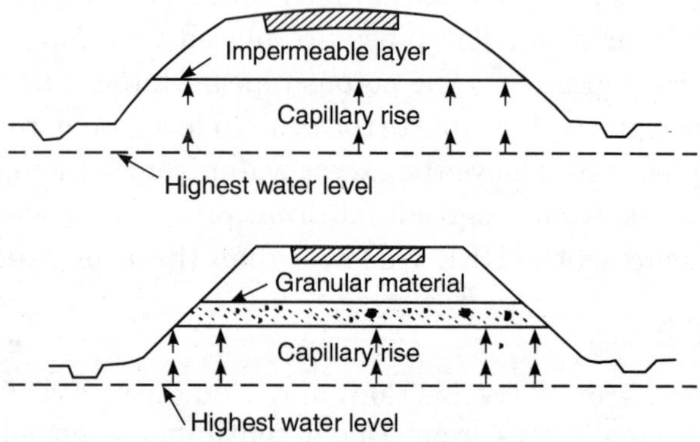

Figure 4.12 *Capillary control*

This can be prevented by providing a layer of granular material of a suitable thickness or by providing a layer of some impermeable material or a bitumen layer between the sub-grade and the ground level. This layer of granular material or impermeable material layer will act as a cut off layer beyond which the water can not rise due to capillary action.

4.9 CROSS DRAINAGE WORKS

A road alignment has to cross a river, a nallah or a stream or a sheet flow of rain water in a low lying area. For crossing the stream or nallah or a river the following cross drainage works are constructed along the alignment of road:

1. Bridge
2. Culvert
3. Irish bridge or causeway

The detailed study of these works is beyond the scope of this book. The readers are advised to refer some book on Bridges. Some important definitions will be useful for ready reference.

4.9.1 Bridge

It is defined as a structure that affords over low ground, water or other obstructions to cross a roadway, railway, footpath etc.

4.9.2 Culvert

It is defined as a small structure, having a maximum span of 6 m. Structurally there is very little difference between a bridge and a culvert.

4.9.3 Causeway

It is defined as a dip which allows flood water to pass over the road. Traffic get stopped on causeways when flow depth exceeds more than 0.6 m.

4.9.4 Submersible Bridge

It is slightly different from a causeway. It is defined as a normal bridge which allows normal flood water to pass below it through the vents and, heavy flood water to pass over it. The submersible bridges are designed in such a way that during heavy floods when the flood water passes over the road, the traffic is not held up for more than 3 days at a stretch and not more than six times in a year.

REIVEW QUESTIONS

4.1. Give neat sketches in plan and section, illustrating fully the surface drainage arrangements that you would propose for a hill road half in cutting and half in embankment.

4.2. Why is surface drainage of roads necessary? Outline in brief the circumstances under which the necessity of sub-soil drainage occurs while setting out a new road.

4.3. What methods do you suggest for making such a road which should remain dry throughout the year. Illustrate your answer with neat sketches where necessary and possible.

[**Hint**. *The road surface should remain dry, means that the water from the surface of road should be drained off easily, and immediately*.]

4.4. Describe in brief the methods of providing surface drains in city roads. Illustrate your answer with neat sketches.

4.5. Under what circumstances sub-surface drainage is provided? What are the different methods of providing surface drainage?

4.6. Draw a neat sketch showing the arrangement of drainage system in city roads crossing each other. Describe the method in brief.

4.7. Discuss in brief the damages caused by water to highway subgrades and pavements. What precautions would you suggest to safeguard the road against these damages?

4.8. Distinguish between the rural and urban highway drainage systems. State the factors which control the design in each case.

4.9. What are the causes of changes in moisture contents in the sub-grade? How can you improve the drainage of a road and control higher water level?

4.10. Explain the need for efficient drainage of the road surface. Describe with sketches the methods used for draining a road on a steep hill slope.

5

Earth Roads and Soil Stabilization

GENERAL

An ordinary earth road is one whose foundations and wearing surface are composed of natural soil, available along the alignments of the road. Earth roads are used for very low volume of traffic. In India, earth roads are called *kachcha* roads. They are cheap, easy in construction and maintenance. They provide a link between villages and villages, a village with district roads or a railway station. These roads are fair weather roads as they become muddy in rainy season and dusty in dry weather.

There are two types of earth roads, viz., (i) ordinary earth roads which are constructed from the local material available at the site, and (ii) stabilized earth roads. The stabilized earth roads are better in wearing conditions. They afford less resistance and give neat appearance. Soil stabilisation improves soil properties.

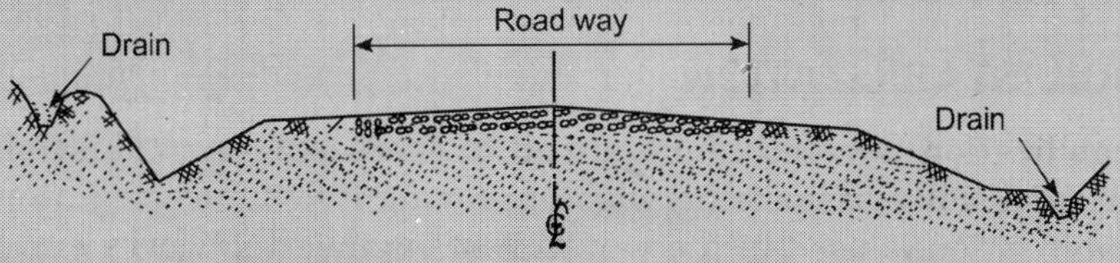

Figure 5.1 *Village road*

5.1 CONSTRUCTION OF EARTH ROADS

The first stage in the construction of earth roads is the preparation of formation. All grass and weeds should be removed from the site. As per the configuration of the country, these roads may be taken through a flat country, in cutting or in embankment as the case may be. If the road is to be in cutting the excavated earth should be deposited in the form of spoil banks. On the other hand if the road is going in embankment, earth is obtained from the borrow pits. In any case the formation level should be kept minimum 0.6 m higher than that of the surrounding land for easy drainage. The width of formation varies from 6 m to 9 m according to the requirement.

The *sub grade* is given a camber of 1 in 24 and the surface is rolled and watered. A layer of 10 cm graded soil is spread evenly and rolled at optimum

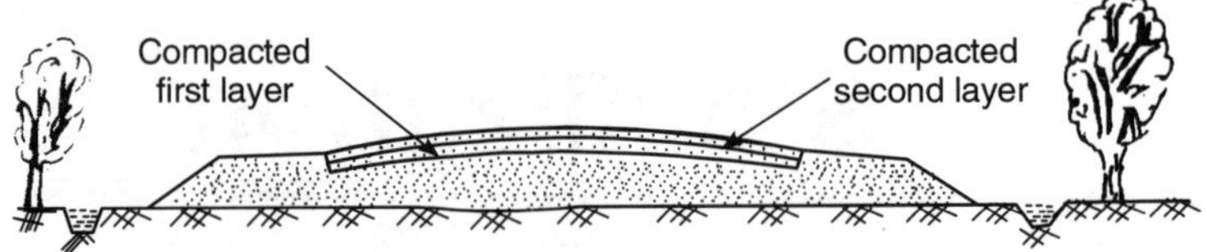

Figure 5.2 *Earth road*

moisture content (O.M.C.) with steep foot rollers and finally finished with a lighter roller. Sometimes another layer of nearly 10 cm thick is also spread and rolled properly to act as *wearing course.*

The surface is then watered for 4 to 5 days for curing and no traffic is allowed to pass on the road during this period. Then for a period of 10 to 15 days the surface is sprinkled with water until completely cured or set, but the traffic is allowed to pass over it during this period.

Maintenance. These roads require constant attention. If these roads are not maintained properly, they will wear out and will become unserviceable very soon. Hence periodic repairs of the pot-holes and ruts is very essential. The ruts and pot-holes are filled with earth and compacted by hand rammers. Side drains should also be repaired properly. Actually speaking, the life and efficiency of earth roads mainly depends upon the efficient drainage system.

5.2 SOIL STABILIZATION

Soil stabilization is a technique aimed at increasing or maintaining the stability of a soil mass or otherwise improving its engineering properties. Soil stabilization is used in the varieties of engineering works e.g., for the construction of earth road or other cheap roads, for making an area easily trafficable within a short period of time for military use, for reducing the permeability and compressibility of soil and so on.

Earth roads, constructed as described above, will wear out very soon and will become unserviceable within no time. Constant repair of earth roads in villages is also not practicable. Hence to reduce the headache of maintenance and to keep the road in serviceable condition for the major part of the year, *stabilized* earth roads should be constructed. Soil stabilization is defined as the process of treating a soil in such a manner so to improve or alter its physical conditions or properties so that it may become more stable and durable. The soil so stabilized may form the sub-grade or wearing layer of a road.

5.2.1 Objects of Stabilization

1. To increase compressive strength irrespective of moisture content.

2. To increase resistance to softening action of water.

3. To reduce shrinkage due to withdrawal of moisture, and swelling due to wetting.

(4) To increase flexibility to take the wheel load without deformation and cracking.

(5) To increase shear-strength and resistance to punching.

5.3 PRINCIPLES OF SOIL STABILIZATION

As pointed out earlier, soil stabilization means improving the bearing capacity and stability of soil by the use of controlled compaction, proportioning or addition of suitable stabilizers. The soil stabilization should precede the following:

(i) Evaluating the properties of the soil.

(ii) Deciding the method of supplementing the desired characteristics.

(iii) Method of designing the stabilized soil mix to achieve the desired characteristics or qualities of the soil.

(iv) Deciding upon the adoptation of the appropriate method of construction.

The following changes are expected after the stabilization :

(i) Stability will increase and change in the properties like density, shrinkage and other physical characteristics.

(ii) Changes will occur in the chemical properties.

(iii) Bearing capacity will increase and the soil will become more resistant to water.

5.4 METHODS OF SOIL STABILIZATION

Following are the different methods of soil stabilization:

1. Mechanical stabilization
2. Cement stabilization
3. Lime stabilization
4. Bituminous stabilization
5. Chemical stabilization
6. Grouting
7. Electrical stabilization
8. Complex stabilization

5.4.1 Mechanical Stabilization

This is the simplest method of stabilization. It has been observed that if a soil consists of coarse grains and fine grains in the ratio of two and one or if a soil consists of two parts of sand and one part of clay, and is compacted at optimum moisture content (O.M.C.), it will have better wearing qualities and will not deform easily under normal wheel load. So in this method the soil is tested. If the soil is coarse-grained, fine-grained soil is so added that the proportion if coarse and fine grains is 2 and 1. Similarly, if the soil is sandy, requisite quantity of clay is added to adjust the proportion.

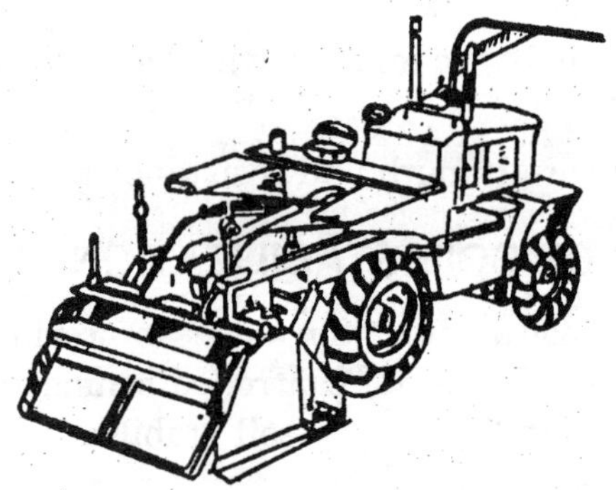

Figure 5.3 *Road stabilizer*

The soil is ploughed to a depth of nearly 15 cm, and pulverized ; and then the required quantity of fine or coarse grains is added. After sprinkling water the surface is compacted by light rollers and then left to be cured for about 4 to 5 days.

5.4.2 Cement Stabilization

Cement is a binding material. When mixed with soil, it forms a sort of low strength concrete in which the soil acts as aggregate and cement as matrix. So the soil is excavated to a depth of nearly 15 cm and 8 to 12% of cement is mixed. Sufficient quantity of water is then added and the soil-cement mixture is compacted properly by road stabilizers. After it has been compacted it is then cured for about 7 to 8 days by simply sprinkling water over it.

The primary object of soil cement stabilization is to improve the wearing conditions of the soil and reduce the deformation either in the road surfacing or its sub-grade due to wheel load irrespective of moisture content. Because, if the soil deforms, it will not distribute the load on sub-soil and due to repeated deformations, the road surface is likely to crack or break. However the objects of stabilization may be enumerated as follows.

The cement binds the particles of clay and forms a flexible dustless and durable road surface. The cement stabilized roads are water-tight, require less maintenance and are very suitable for light traffic. It is suitable for sand predominant soils.

5.4.3 Lime Stabilization

In this case the process of stabilization is similar to that of cement stabilization. The soil is loosened, pulverised, sieved and mixed with 5 to 10% by weight of hydrated lime. The two are thoroughly mixed. Sufficient quantity of water is added and the surface is compacted. The lime helps in reducing the shrinkage and swelling of soil. It is suitable for clay predominant soils.

5.4,4 Bitumen Stabilization

In this method the soil is treated with about 8 to 10% of road oils, cut backs or emulsions, according to the nature of the soil. Their object is to glue together the soil particles and fill up the voids.

The soil is graded and then compacted properly at O.M.C. About 5.5 litres of oil is applied on 1 sq. metre of the surface so that the oil may penetrate nearly 1.25 to 2.5 cm in the soil.

This method of soil stabilization is unsuitable for soils containing more than 50% of fine grained particles. Before applying this method it is essential that the soil should be well graded and compacted properly, otherwise more quantity of bitumen will be required. It is suitable for sand predominant soils.

5.4.5 Chemical Stabilization

Hygroscopic materials such as calcium chloride, sodium etc. are mixed with the soil at the rate of 1 kg per 5 sq. metres of the surface and the soil is thoroughly compacted. These hygroscopic materials help in retaining proper amount of moisture in the soil and add to its stability. This method senes more as dust palliative than soil stabilization. However, dampness in the surface reduces shrinkage and cracking. It is suitable for clay predominant soils.

5.4.6 Grouting

Grouting or injecting is a process of introducing a stabilizer of fluid consistency into soil and rock formations. The stabilizer used is known as *grout.* Grouting of soil formations is usually done (i) to reduce the permeability, (ii) to increase the strength for the purpose of increasing the bearing capacity, and (iii) to reduce the compressibility.

The usual grouting materials are cement, soil, bitumen and chemicals. Holes are driven at regular intervals and of desired depth and the grouting material of fluid consistency is injected under heavy pressure with the help of a grouting pump. The grout having the cementing properties will bind the soil particles. It is suitable for all types of soils.

5.4.7 Electrical Stabilization

Electrical stabilization is a method of drawing out the fine-grained soil by passing *direct current* through them. It is also sometimes called *Electro-osmosis.* With the damage of the fine particles the volume of the soil decreases i.e., the soil is consolidated and the shear strength is increased. During the *electro-osmosis,* electro-chemical decomposition of the electrodes takes place and the metal salts are deposited in the soil pores. There may also be an *ion-exchange,* and an alteration of soil particle arrangement. This will ultimately lead to hardening of soil and process is sometimes known as *electro-chemical hardening.*

5.4.8 Complex Stabilization

Complex stabilisation is defined as the method of stabilization with more than one stabilizer. Difficult soils such as organic soils, highly plastic for clays and soils with *easy-soluble* salts, require more than one stabilizer for their effective treatment. Complex stabilization involves the use of binding material and surface acting additives or electrolytes. At present the following combinations are considered to be the best:

(i) Cement + calcium chloride + lime

(ii) Cement + bituminous emulsions

(iii) Cut-back + lime

(iv) Cement + naphtha soap

5.5 MOORUM ROADS

Moorum is a disintegrated rock, gritty and silicious in nature, having grains not exceeding 18 mm in size. It is generally mixed with natural clay of calcarious origin.

Moorum roads are also cheap type of roads and are mostly used in villages. In this type of roads, the traffic-way has the surface of moorum which is consolidated in layers to get the required finished surface. The surfacing material is laid directly on the sub-grade. No foundation of soling is required. The surface is properly compacted by rolling. Proper camber and gradient is maintained during the process of rolling. Generally 1 in 24 camber is provided.

5.6 STABILIZED MOORUM

Stabilized moorum may be used for sub-base of a road. This will fulfill the requirement of a material to be usefully used as a sub-base material. Moorum is abundantly available in many parts of our country. Various tests are conducted on different samples of moorum and were classified as per I.S. 1948–1970 classification and was found that most samples belonged to 'SC' group of I.S. classification. Procter's compaction test and soaked CBR tests are conducted and was found that the test results are quite encouraging. Moorum samples passing through 20 mm sieve and retained on 0.425 mm sieve, it is found there is appreciable increase in the CBR value. The CBR values so obtained varies from 16.5% to 40.4%. Although sufficient improvement in the CBR values had resulted by excluding fractions passing 0.425 mm sieve, but is not a practical solution of improving CBR value. The next alternative of improving CBR value is treating the moorum with lime because it is found that the fines of moorum are reasonably plastic. It is found through extensive, experimentation that moorum treated with 3–5% lime and cured for seven days give 27.6% – 75% CBR values which are really encouraging and appreciable. So lime stabilized moorum can be used as a sub-base material or as a surfacing for moorum roads. The studies are conducted by Uttar Pradesh (U.P.) P.W.D. Research Institute.

5.7 INVESTIGATIONS OF SOIL STABILIZATION

Undisturbed and disturbed soil samples are collected from shallow pits after removing the top layer and organic matter along the alignment and from the sides of the alignment at regular intervals. These samples are then tested in the laboratory for physical chemical and engineering properties. Identification and classification tests such as plastic limit, liquid limit etc. are carried out. CBR tests are also conducted, the investigations are carried out in the following manner:

(i) Investigation of route and alignment.

(ii) Survey for the availability of materials to be used for stabilization.

(iii) Soil surveys and field investigation of soil.

(iv) Soil classification.

REVIEW QUESTIONS

5.1. Outline the importance of earth and other cheap roads in India. Where are these roads used in India ?

5.2. What is meant by soil stabilization ? What are the different methods of soil stabilization ? Describe in brief.

5.3. Describe in detail the construction of an earth road. Describe the procedure of construction step by step.

5.4. What is the difference between stabilized and unstabilized road surface ? What is the object of stabilization ?

5.5. On what factors the stabilization of soil depends. What is the role of moisture in stabilization ?

5.6. Discuss the method of bitumen stabilization of soils and state also the limitation of the method.

5.7. Elucidate in brief the method of Moorum Stabilization for Road construction.

5.8. Elucidate the principles of soil stabilization.

5.9. State the procedure and compaction of earthen embankments.

5.10. Differentiate between compaction and consolidation. How soil can be stabilized by compaction.

5.11. State the principles and procedure of soil-cement stabilization.

5.12. Differentiate between soil-cement, cement modified cement and soil plastic cement.

5.13. State the process of grouting for soil stabilization. Name the various methods generally used.

5.14. State the advantages of soil stabilization and discuss bitumen stabilization in brief.

5.15. Describe in brief the construction of stabilized moorum roads.

6

Gravel and Waterbound Macadam Roads

GENERAL

Gravel and Waterbound Macadam (W.B.M.) roads are cheap quality roads and are mostly used in Indian villages. The construction of gravel roads is very much similar to that of moorum roads. The waterbound macadam road is a superior road which is used in many cases as a foundation layer for other superior roads. The W.B.M. road, as it is known, is the mother road for all types of road construction.

6.1 GRAVEL ROADS

It is a cheap type of road and is mostly used in Indian villages. Gravel can be obtained from the river beds or by crushing the stone. The surface layer of gravel roads consists of 6 to 36 mm size gravel mixed with sand and clay. It works as a binder. The function of binder is to fill up the voids and bind the particles of gravel.

The sub-grade is prepared and properly compacted. Proper camber and gradient is given to the sub-grade surface. The mixture of gravel, earth and sand is spread on the sub-grade, in a thickness which varies from 15 cm to 30 cm according to the requirement of the traffic. The layer of gravel is rolled by light rollers. During the process of rolling proper camber and gradient is maintained. The gravel is generally spread in two layers. During rolling some water is sprinkled on the surface. Gravel roads may be constructed either *Trench* type or *Feather* type as shown in Fig. 6.1 (a) and (b).

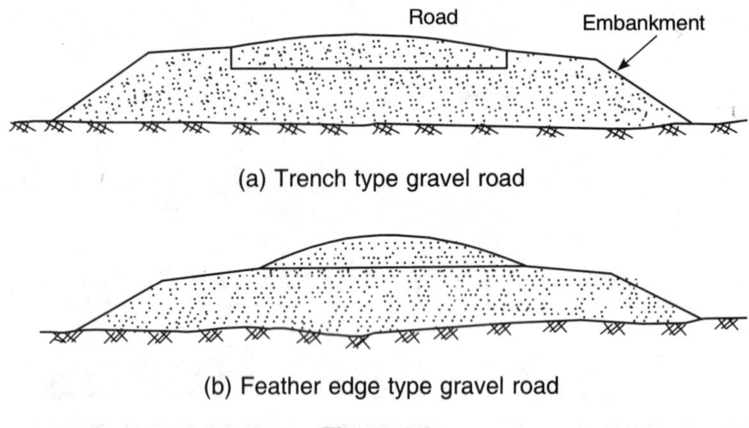

(a) Trench type gravel road

(b) Feather edge type gravel road

Figure 6.1

Trench type gravel roads are more common in this country. In the feather type gravel roads, the gravel is spread on the formation directly without excavating a trench.

The surface drainage in the gravel roads should be very efficient, so that the surface must not be worn out by water. Periodic repairs are necessary to maintain proper camber and gradient.

6.2 WATERBOUND MACADAM ROAD

The waterbound macadam road is one of the oldest methods of road construction known to mankind. The road metals used for the construction of this road are known as *macadam.* The W.B.M. road is used as village road, or as a base for the important modern roads such as surface painted roads, premixcd carpet, bituminous macadam and cement concrete roads. This type of road consists of road metals bound together with water only. The smaller pieces of the metals are used to fill up the voids.

6.3 SIZE OF AGGREGATE FOR W.B.M. ROAD

Crushed or broken stones, crushed slag, overburnt brick aggregate or other types of naturally occurring aggregates such as kankar, laterite etc. are used for the construction of the waterbound macadam roads. The size of the aggregate mainly depends on the intensity of traffic, weight of the vehicles to pass over the road, bearing capacity of the soil and the hardness of the aggregate to be used. As a general rule the maximum size of the aggregate may not exceed three-quarters of the total consolidated thickness of any one layer of construction. Some engineers recommend that the maximum size of the coarse aggregate may be equal to the consolidated thickness of each layer, the reason being that larger stones engage the base course and the surface has no tendency to 'roll' or wave.

Following nominal sizes of coarse aggregate are recommended for W.B.M. roads:

(a) For limestone, flint, kankar, laterite and vitrified brick ballast aggregates 40–63 mm.

(b) For granite, trap, basalt, diorite etc. hard stone aggregates 20–50 mm.

The size of the aggregate should be reduced with its hardness. Coarse-grained rocks are not suitable for use in roads in small sizes.

6.4 GRADING OF COARSE AGGREGATE

Table 6.1 gives the various recommendations for the grading of the road metal for W.B.M. roads.

TABLE 6.1 *Recommended metal grading*

Nominal size of Aggregate	*% of Aggregates*					
	1st Recommendation	*2nd Recommendation*	*3rd Recommendation*	*4th Recommendation*	*5th Recommendation*	*6th Recommendation*
63 mm	—	—				
50 mm	60%	—		50%	50%	60%
40 mm	30%	60%	50%		25%	
32 mm	—	—			—	
25 mm	—	30%			—	
20 mm	10%	—	30%		15%	30%
12 mm	—	10%		50%		
6 mm	—	—	20%		10%	
Stone Dust	—	—				10%

Table 6.2 gives the size and grading of coarse aggregate for waterbound-macadam roads as per I.R.C. 19–1972.

TABLE 6.2 *Size and grading of coarse aggregate for W.B.M. road (as per I.R.C. 1971–1972)*

Grading No.	*Nominal size of Aggregate and material*	*Sieve Designation in mm*	*Percentage weight of Aggregate passing the Sieve*
1.	90 mm to 40 mm	100	100
	Jhama Brick	90	65 – 85
	Aggregate	63	25 – 60
		40	0 – 15
		20	0 – 5
2.	63 mm to 40 mm	80	100
	Soft-Stone:	63	90 – 100
	Lime stone, flint, kankar,	50	35 – 70
	quartzite, laterite and	40	0 – 15
	vitrified brick ballast etc.	20	0 – 5
3.	50 mm to 20 mm	63	100
	Hard Stone:	50	95 – 100
	Gramite, trap,	40	35 – 70
	basalt, diorite etc.	20	0 – 10
		10	0 – 5

Grading No. 1 is more suitable for sub-base course, but it should not be used for courses having compacted thickness less than 90 mm.

6.5 THICKNESS OF LAYERS ON COURSES

The thickness of the consolidated metal layer should not normally be less than 11.5 cm, except where the road in future is to be treated with tar or bitumen. If hard stone metal is used the thickness of the consolidated layer may be 7.5 cm for light traffic. I.R.C. has recommended that the thickness of a single compacted layer in case of hard metal should not exceed 10 cm.

As the pressure intensity of a loaded iron-tyred cart is more than due to a 12 tonne road roller, therefore for such traffic the thickness of the course can be increased.

For all important roads on medium sub-grades two courses of stone metals (7.5 cm thick each) over a sub-base of 7.5 cm of gravel or clinker should be constructed. In case of light traffic one layer of 11.5 cm consolidated thickness should be provided instead of two layer of 7.5 cm thickness.

The thickness of one consolidated layer for hard metal should not be more than 7.5 cm and 11.5 cm for soft metal. If the designed pavement thickness is more it should be laid in two or more layers. Care should be taken that lower course should not be fully consolidated but nearly consolidated, so that while compacting the next layer above it, it can be fully consolidated. No binding material should be used on the first layer.

6.6 CONSTRUCTION OF W.B.M. ROAD

The stages of construction of a W.B.M. road are as follows:

(i) *Subgrade and Kerbs.* After the construction of embankments or cutting is completed, the subgrade is prepared in the shape of required camber and gradient. It is properly compacted by light rollers or hand rammers. Earthen kerbs of 15 cm × 15 cm section are prepared along the edges of the proposed road to hold the materials.

(ii) *Base or foundation layer.* The base or foundation course consists of 12 to 18 cm size boulders or broken pieces of stone, large size kankar, overburnt brick-bats or flat bricks. Stone boulders or bricks are each placed by hand, firmly bedded in position with as little voids as possible. In the case of bricks the larger edge of bricks is placed transverse to the direction of the roadway. In the case of stone base the voids if any should be properly filled with stone pieces and the surface by lightly rolled with a 10 tonne roller. The width of the foundation should be 60 cm wider than the pavement, projecting 30 cm on each side.

(iii) *Wearing course.* This is also known as metalling course and can be laid in one or two layers depending upon the total thickness required. The thickness of the loose metal in each layer should not exceed 15 cm. It is because of the reason that ordinary rollers will not be able to roll more than 15 cm thick layer. If very heavy rollers are used, the metal will be crushed. The wearing course is prepared in the following order:

(a) First of all stone metal is evenly spread over the proposed surface in the required thickness. After spreading, the metal is hand packed properly to the required camber and gradient. Generally 1 in 30 to 1 in 40 camber is provided.

(b) The layer of stone metal is then rolled by 8 to 10 ton rollers. Rolling should first be started from the edges and gradually shifted towards the centre. This dry rolling will provide interlocking of aggregates.

(c) After dry rolling, stone screening or 12 mm grit is evenly sprinkled on the surface at the rate of 1 cu metre per 100 sq m. It is known as bindage layer. This surface is now rolled dry, and then sufficient quantity of water, to wet the aggregates, is sprinkled, and the surface is again rolled. During the process of rolling proper camber and gradient should be maintained.

(d) The next day final coating of sandy soil containing clay and moorum and 75% sand is spread in a thickness of 5 mm and the surface is wetted by cupious quantity of water. This layer of sandy soil is known as bindage layer. Nearly 40 to 60 litres of water is used per sq. metre of the surface. The surface is then rolled. During rolling water is sprinkled on the roller wheels so that the metal should not stick to the roller surface. The object of spreading so much of water is that the material such as sand, clay etc., should reach interstices of stone.

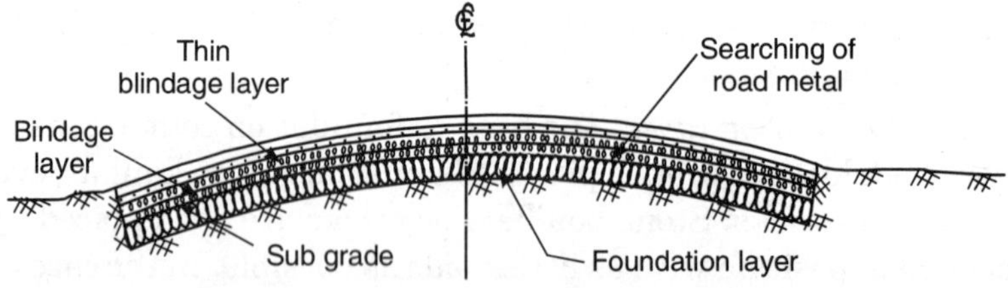

Figure 6.2 *Waterbound macadam road*

The rolling is continued on the next day also. Some quantity of sand at the rate of 1 cu metre per 160 to 180 sq m of road surface is spread. The surface is then kept wet for about seven days and then it is opened for traffic.

6.7 CAUSES OF DISINTEGRATION OF WATERBOUND MACADAM ROADS

Following are the main causes of disintegration of W.B.M. roads.

(a) In the W.B.M. road the stone aggregates are binded together by means of grit, sand and clay. The rain water wash away the soil binder and the stone aggregates protrude out or get loose on the surface layer, thus it forms the pot-holes and ruts.

(b) The high temperature of the sun during day and cool temperature during night causes disintegration.

(c) The heavy mixed traffic and adverse weather conditions damage the water-bound macadam road in no time.

(d) Steel-tyred bullock carts crush the road metal and the fast moving pneumatic-tyred vehicles create a partial vacuum and loosen the surface metal. The pulverized metal is blown by the wind and the fast moving vehicles.

6.8 CORRUGATIONS ON W.B.M. ROADS

Corrugations are the wave-like deformations formed in the surface of the roads. The corrugations make the travel uncomfortable and cause great damage to the vehicles. Vehicles cannot move with high speed on the corrugated road surfaces, due to heavy vibrations.

Following are the main causes of formation of corrugations on W.B.M. roads:

(a) The wheels of the vehicles due to spun action throw up the loose surface blindage and cause pots and depressions in the surface.

(b) During the construction of the road, the defective rolling of thick layers of the metal, causes corrugations. While rolling by road roller, due to the effect of the metal creeping in front of the roller, a hard ridge is formed followed by a depression.

To prevent the formation of the initial waves while rolling, the road roller gearing should be in good condition to prevent jerky rolling. Similarly long runs should be taken to minimize stopping and reversing the roller, which also cause corrugations. While rolling the longitudinal evenness should be checked with string.

(c) Imperfect grading of coarse aggregates also cause uneven displacement of stones resulting into corrugations under the traffic load.

(d) Improper consolidation of the embankments also causes corrugation.

(e) Heavy loaded vehicles than the designed load also cause corrugations.

(f) The braking and starting action of the vehicles also causes corruption.

(g) During rains when the bearing capacity of the road reduces, the movement of traffic causes sinking of road at various places.

6.9 MAINTENANCE

To keep the road surface in a serviceable conditions, periodical maintenance of W.B.M. road is very essential. The blindage is sucked up by fast moving traffic and the road surface without bindage disintegrates in no time. Also the surfacing gives way in isolated patches known as *pot holes.* These *ruts* and *pot holes* should be repaired as soon as they appear on the surface. The *pot holes* and *ruts are* cleaned with a wire bruch and filled with fresh metal, properly hand packed and compacted by hand rammers. Bindage is now spread to fill up the voids and the surface is again rammed by sprinkling water. If, however, the *pot holes* occur on more than one-third area of a road surfacing in a particular reach, this area should be repaired by scarifying the surface, rolling the old road metal, spreading the required fresh layer of loose metal, rolling and preparing the surface in conformity with the old road. This sort of repairing is called *partial resurfacing.*

6.10 IMPROVEMENTS ON W.B.M. ROAD SERVICING

The waterbound macadam roads (and other cheap type of roads such as earth, moorum and gravel roads), produce dust when in service and produce more dust when they are badly worn out. Therefore to prevent this disturbance the following improvements are suggested:

1. Water should be sprinkled on the surface periodically. (Too much of water will make the road muddy and wet road surface will get worn out quickly).

2. A coat of road oil may be given to the road surface. This coat will help in keeping the dust down and will increase the load carrying capacity of the road.

3. Some hygroscopic material such as calcium chloride is sprinkled on the surface of W.B.M. road. This powder will absorb moisture from the atmosphere and thus the bindage will also remain moist and intact

6.11 I.S.I TESTS FOR AGGREGATES

The properties of the mineral aggregates which form the pavement structure are of great importance in the highway construction.

The following laboratory tests are carried out on the mineral aggregates to determine their essential properties important to highway engineer.

1. Sieve Analysis Test. This test determines the grading of aggregate particles. A weighed quantity of mineral aggregates is placed over a set of

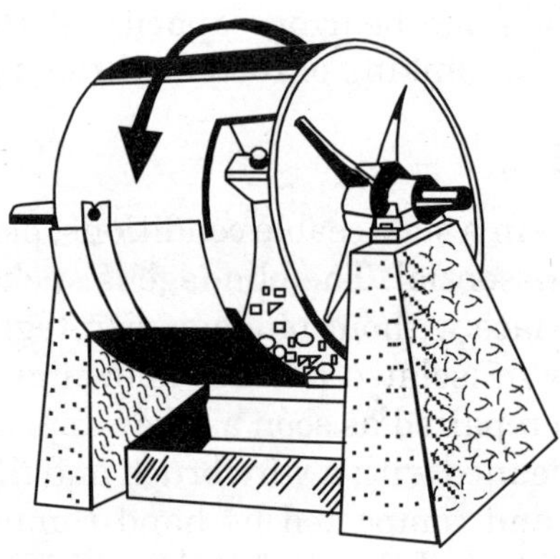

Figure 6.3 *Loss Angel's abrasion testing machine*

standard sieves of square opening and being shaked by a standard sieve shaker for a specified time. The aggregates of each sieve is weighed and the percentage retained on each sieve is calculated.

2. Particle Shape Test. With the help of this test, the shape of particles is determined as *flaky* or elongate. Those particles whose least dimension is less than 0.6 times of their mean size are termed *as flaky* and those particles whose greatest dimension is more than 1.8 times their mean size, are termed as elongated.

3. Abrasion Test. This test is conducted in a Loss Angel's abrasion testing machine. With the help of this test, it is found as to how much percentage of the aggregate will wear out in a specified time. The Loss Angel's abrasion testing machine consists of a hollow steel cylinder having an internal diameter of 700 mm and inside length 500 mm. It is fixed on strong steel axils. It is fitted with a dust proof opening cover. Stone aggregates are heated in an oven for 4 hours at a temperature of 100°C to 110°C. The weight of aggregates which are fed in the cylinder is 5 kg or 10 kg. Alongwith the stone aggregates, steel balls of 40 mm diameter are also fed and the drum is rotated at the rate of 30 to 33 revolutions per minute for 500 to 1000 revolutions. After the test the aggregates are sieved on Indian Standard Sieve No. 170 and the percentage loss on wear is calculated.

4. Weathering Test. This test is performed on cylinders of diameter 6 cm and length 7.5 cm cubes of 6 cm side to find out the effect of atmospheric action. The test specimens are heated for 24 hours at a temperature of 105 ± 2°C. The specimen is weighed. Let this weight be W_1. After this the specimen is kept in a desicator for 24 hours at a temperature 20–30° C. After cooling the specimen is again weighed. Let this weight be W_2. After that the specimens of stone are kept in a saturated solution of sodium or magnesium sulphate at a temperature of 105 ± 2° C and heated till all water is evaporated. The process is repeated 29 times and each time the specimen is weighed. Let the average weight of the specimen is W_3.

Now the specimens are kept in pure water for 24 hours. Let the weight of the specimens before and after immersion in water is W_4 and W_5 respectively.

∴ Percentage water absorption

$$J_1 = \frac{(W_2 - W_1)100}{W_1}$$

Volume of specimen

$$V_1 = \frac{W_3 - W_3}{D}$$

where

D = sp. gravity of the stone.

Percentage water absorption after 30 times immersion in saturated solution of sodium or magnesium sulphate $\quad J_2 = \dfrac{(W_4 - W_1)100}{W_1}$

Now volume $\quad V_2 = \dfrac{W_4 - W_5}{D}$

Increase in the percentage of water absorption $= \dfrac{100(J_2 - J_1)}{J}$

Percentage increase in volume

$$= \frac{V_2 - V_1}{V_1} \times 100$$

From the above experiment various varieties of stones are compared and the best suited variety is selected.

REVIEW QUESTIONS

6.1. Describe in brief the construction of gravel road. Where this type of road is used ? What improvements can you suggest to improve the working conditions of this road ?

[**Hint.** See Sec. 6.4. The importance which have been suggested for W.B.M. road can also be applied to gravel roads.]

6.2. Describe in detail step by step the construction of a waterbound macadam road. What improvements do you suggest to reduce the dust nuisance of W.B.M. road ?

6.3. Write down detailed specifications for road metalling, bindage and bindage layers to be used in W.B.M. road.

[**Hint.** See Sec. 6.2. Detailed specifications means constructional details to be followed, including thickness of each layer or course, size and quantity of metals to be used etc.]

6.4. What do you understand by the waterbound macadam road ? Name the different metals which are generally used for the construction of such a road. Draw a neat sketch of W.B.M. road having a carriageway of 4 m.

6.5. Give in brief the circumstances under which a waterbound macadam road is used.

6.6. Write short note on the grading of the coarse aggregate for W.B.M. roads.

6.7. Write short note on following:

(i) Size of aggregate for W.B.M. roads.

(ii) Thickness of W.B.M. road layers.

(iii) Causes of disintegration of W.B.M. roads.

(iv) Construction of W.B.M. roads.

Bituminous Road

GENERAL

Bituminous roads are those roads having binding material such as bitumen, coal-tar or asphalt in surfacing. Such roads are also known as *black top* roads. These roads are common and popular in cities and towns.

7.1 TYPES OF BITUMINOUS ROADS

There are various types of bituminous roads which are useful for different intensities of pressures of wheel loads and different types of traffic. The following are the various types of bituminous surfaces :

(1) Surface painting or surface dressing or premix carpet as wearing coat.

(2) Bituminous macadam or dense bituminous macadam as crust coarse.

(3) Bituminous concrete or mix seal surfacing or semi dense bituminous concrete.

(4) Sheet asphalt or asphaltic mat.

7.2 BITUMEN, COAL-TAR AND ASPHALT

It will not be out of place to give some idea of bitumen, coal-tar and asphalt, before proceeding further.

7.2.1 Bitumen

There are various definitions of bitumen. The *international definition of bitumen* is that the bitumen is a mixture of natural pyrogenous hydrocarbons and their non-metallic derivatives which may be gaseous, liquid, viscous or solids but must be completely soluble in carbon disulphide. Bitumen, in other words, is a hydro-carbon compound in a solid or semi-solid state. It contains 85 parts of carbon, 12 parts of hydrogen and 3 parts of oxygen. The properties of bitumen depend upon its source and the method of preparation. However, it is blackish in colour, sticky in nature, semi-solid and melts when heated.

Bitumen is obtained by the partial distillation of crude petroleum. The distillation may take place over centuries as Nature Work, or the distillation may be carried out in refineries, where the lighter oil fractions, which hold the bitumen in solution, are distilled off. Thus bitumen is an important by-product of the fractional distillation of crude petroleum.

7.2.2 Asphalt

Asphalt is a substance containing a high percentage of *bitumen* and some mineral substances. Bitumen is the basic constituent of asphalt. When crude petroleum is fractionally distilled in nature, the residual product is called asphalt. (If distillation of crude petroleum is done in refineries the product is called Bitumen).

It is blackish brown in colour, non-inflammable but burns with a smoky flame at a temperature of 250° C.

Asphalt can be classified as (1) Rock asphalt, and (ii) Lake-asphalt.

(i) *Rock asphalt.* It is a bituminous lime stone. It is a compound containing 30 to 90 parts of lime stone and 20 to 10 parts of bitumen. It is used for pavements.

(ii) *Lake asphalt.* It is a natural asphalt found in lakes and contains 40 to 70 per cent of bitumen and about 30 per cent of water, lime, clay and sand. Lake asphalt is refined by boiling it in a tank when water evaporates and the impurities float on the surface and are removed. Refine lake asphalt is the best material for road making.

7.2.3 Coaltar

Coaltar is a jet black, viscous liquid. It is the residual product obtained by the destructive distillation of coal. Coaltar is usually obtained as a by-product in the manufacture of *coal gas.* It is used as a preservative for timber and for making roads.

Table 7.1 gives the application temperatures for different grades of bitumen and tars.

TABLE 7.1 *Application temperatures for bitumen and tars*

Sl. No.	*Grade of the bitumen tar*	*Application temperature for spraying purpose*	*Application temperature for mixing with aggregate*
	(A) Straight run or paving Bitumen		
1	30/40 penetration	170 –185°C	170 – 185°C
2	60/70	165 – 180°C	155 – 165°C
3	80/100	160 – 175°C	150 – 165°C
4	180/200	150 – 165°C	135 – 150°C
5	A–90 and S–90 grades	175 – 161°C	150 – 175°C
6	(B) Cut back Bitumen **Note :** Some cut backs are for cold application and do not require heating	161 – 161°C	150 – 165°C
	(C) Road tars		
7	Grade RT –293– 104°C	93 –104°C	
8	RT –3	104 – 116°C	116°C
9	RT –4	132°C	132°C
10	RT –5	138°C	138°C

7.3 CUT BACKS AND EMULSIONS

7.3.1 Cut Backs

They can be defined as the solution of bitumen (asphalt or coaltar) in solvents. The solvent is the distillate of the petroleum or coaltar.

The function of cut backs* is to soften the heavy product and convert it into a liquid form so that they can be used without heating or require only a light heating. When the cut backs are mixed with aggregates or spread on road surface, the volatile oil evaporates leaving behind the thick oil. Cut backs contain nearly 80 per cent asphalt and 20 per cent solvent. The type of solvent to be used, depends upon the rate of evaporation required of the cut backs, to suit different types of road construction.

7.3.2 Emulsions

Emulsions are defined as a mixture comprising two unmixable liquids, the

* In refinery, the volatiles which are distilled off, are known as cuts. To manufacture *cut backs*, the *cut* is put *back* into the prepared residue and hence the name.

one dispersed in the other in the form of fine globules or droplets. Oil and water when mixed and vigorously shaken, the oil is broken up into small globules floating in water. This is a water oil emulsion. Bitumen emulsion consists of asphalt or tar particles of size about 3 microns $\frac{3}{1000}$mm dispersed in water in the presence of an emulsifying agent which is added nearly 1 per cent. The emulsifying agent is absorbed on the surface of each globule forming a film on its surface. Upon the permanency of the film and its degree of homogeneity, it depends on the stability of the emulsion. Generally soap and resinous substances are used as emulsifying agents. Emulsions contains nearly 55 to 65 per cent of bitumen by weight.

Bitumen emulsions can be applied at normal temperature without heating. When emulsion is spread on the road, it changes from brown to black colour while the water soaks in or evaporates, allowing the bitumen particles to reunite. Emulsion is useful for damp surface. Before applying emulsions, the road surface shall be thoroughly cleaned and slightly damped with water.

7.4 I.S.I. TESTS FOR BITUMEN

Bitumen is available in a variety of types and grades. Different grades of bitumen are used for various types of road construction. Following tests have been prescribed by I.S.I. to test the workability of various grades of bitumen .

7.4.1 Penetration Test

This test indicates the hardness of bitumen used in road construction by increasing the distance in *tenths of a millimetre* to which a standard needle

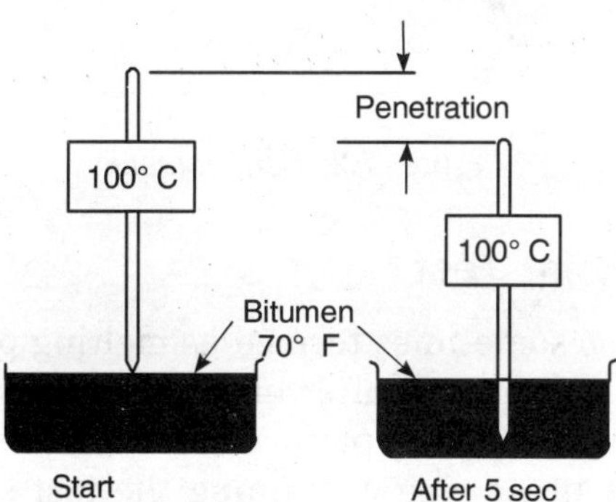

Figure 7.1 *Penetration test*

will penetrate vertically into a sample of bitumen under stipulated condition of temperature, loading and time. (I.S. : 1203 –1958) unless otherwise stated, the bitumen sample is penetrated for 5 seconds at a temperature of 25 °C and the load on the needle is 100 gm. The softer the bitumen the greater will be its number of penetration units.

The test is being performed by an instrument known as *penetrometer*. The sample of bitumen is placed in the bottom pan and the needle is lowered down and left for 5 seconds. The temperature of the sample is kept at 25°C. The weight of the needle is 100 gm.

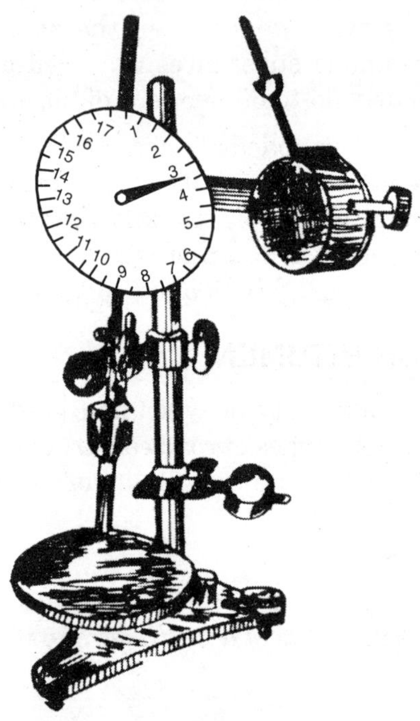

Figure 7.2 *Penetrometer*

7.4.2 Softening Point Test

Softening point is also sometimes termed as melting point. As the bitumen is heated it becomes softer and softer until it flows readily with the rise of temperature. This test is usually performed by the ring and ball apparatus (I.S. : 1205–1958). A brass ring containing the test sample of bitumen is suspended in water at a given temperature. The brass ring has an internal diameter of 15 mm and thickness 8 mm. The sample is poured in the ring and 10 mm dia. ball is placed over it. The water is then heated at the rate of 5 degrees per minute, till the bitumen is soft enough to allow the ball to pass through it and touch the bottom plate placed at a specified distance of

2.45 cm. If the melting point of bitumen is higher than that of water, glycerine should be used in place of water.

7.4.3 Ductility Test

The ductility of bitumen is defined as the length or distance in centimetres to which a standard briquette of bitumen can be stretched before the thread breaks, at a temperature of 25°C, when the ends of the briquette are stretched at the rate 5 cm per minute (IS : 1208–1958).

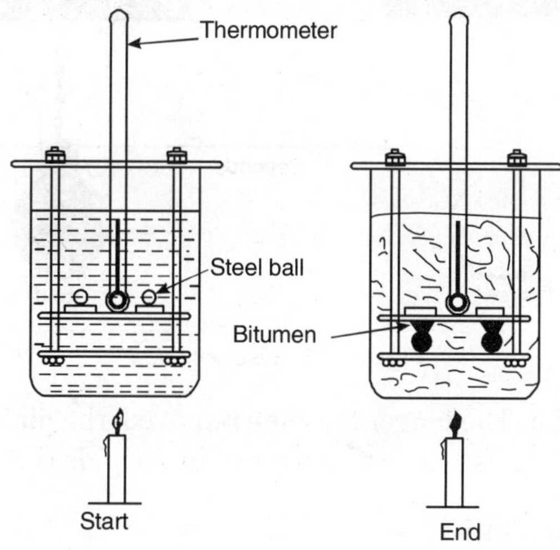

Figure 7.3 *Softening point test*

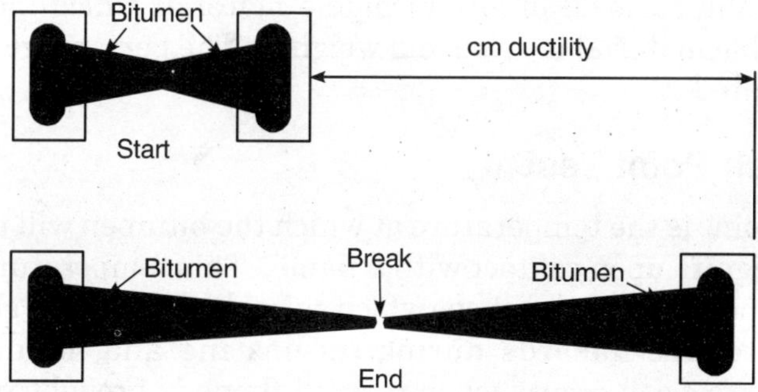

Figure 7.4 *Ductility test*

7.4.4 Viscosity Test

Viscosity is defined as the resistance to the flow of liquid bitumen i.e., cut-back or emulsion. Viscosity at any specified temperature is measured by recording the time in seconds for a given quantity of the product at the same temperature to flow through an orifice of standard dimensions into a receiver

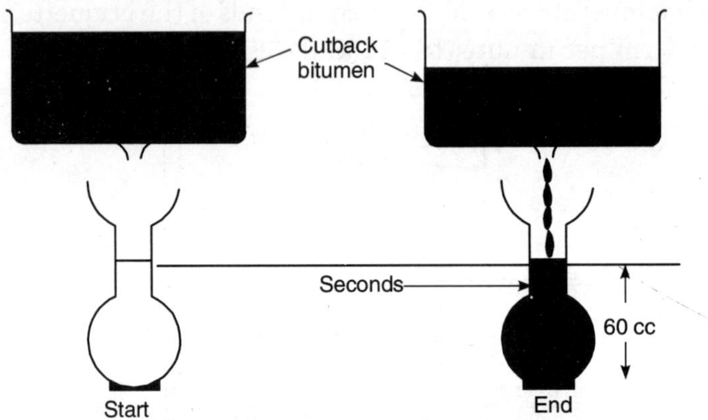

Figure 7.5 *Viscosity test*

as shown in Fig. 7.5. The longer the time required, the higher is the viscosity of the material. Test values are expressed in seconds (I.S.: 1206–1958).

7.4.5 Solubility Test

This test measures the extent to which bitumen goes into solution either in carbon disulphite (CS_2) or carbon tetrachloride (CCl_4). A weighed sample of the material is treated with an excess of the solvent. The solution is then filtered for the removal of any insoluble material which may be present. This insoluble material is dried and weighed. The percentage of solubility is then calculated.

7.4.6 Flash Point Test

The flash point is the temperature at which the bitumen will evolve vapours which will ignite upon contact with a flame. This temperature signifies the critical temperature at and above which suitable precautions should be taken to eliminate free hazards during its heating and handling. At this temperature a flash occurs when a small flame is brought in contact with the vapours of a bituminous product. The test is performed by heating a sample material in an open cut at a specified rate and determining the

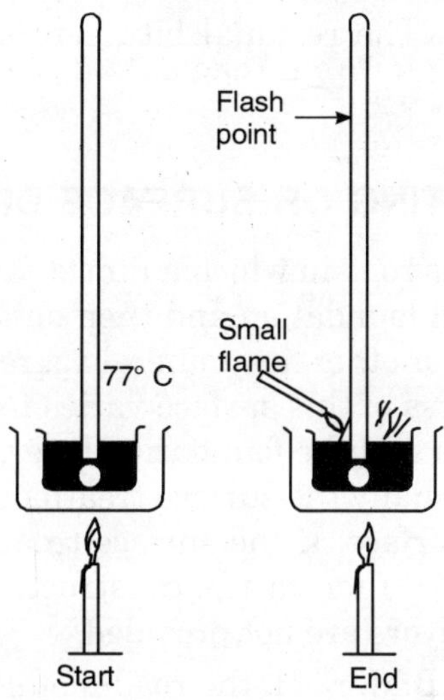

Figure 7.6 *Flash point test*

temperature at which a small flame passed over the surface will cause the vapours to ignite or flash. (IS : 1209–1958).

7.4.7 Distillation Test

It indicates the rate which a cut-back will cure after application. This test determines the relative proportions of bitumen and the volatile flux present in

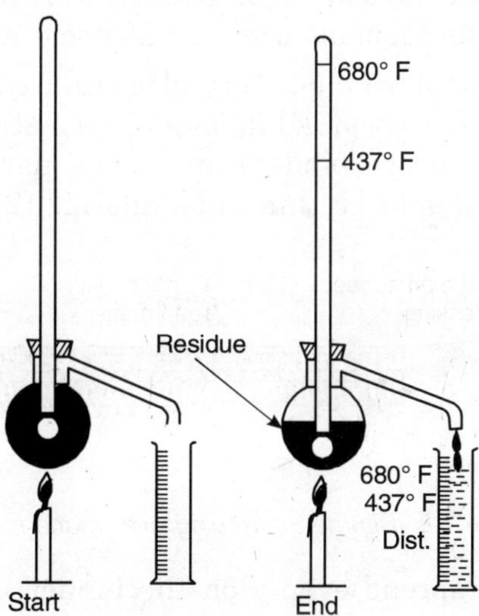

Figure 7.7 *Distillation test*

cut-back. It also measures the amount of volatiles driven off at certain specified temperatures. The residual bitumen is tested for penetration, ductility and solubility. I.S. : 123–1958 describes the standard test method of this test.

7.5 SURFACE PAINTING OR SURFACE DRESSING

It is a kind of bituminous road in which a film of tar or asphalt is applied on the prepared top of road foundation and then on this film is spread a thin layer of stone chippings or other fine mineral aggregate and the surfacing is then rolled. The thickness of this surface varies from 1 cm to 2 cm and the purpose of this coat is to seal the foundation layer, which may be a W.B.M. surface, gravel layer. Usually the surface treatment is given to a new or an existing W.B.M. road surface. If the surface treatment is to be given to a new W.B.M. road surface then in the construction of W.B.M. road itself, bindage and blindage layers are not provided.

In the case of new W.B.M. road, the road metals are cleaned of dust etc. and are properly rolled when road metals are properly interlocked, the surface dressing is applied.

In the case of an existing W.B.M. road, the surface is brought to the required camber and gradient i.e., the road is reconditioned. It is then cleaned by wire brushes and surface dressing is applied.

7.5.1 Methods of Applying Surface Dressing

The surface dressing can be applied either by cold process or hot process. In the hot process the tar or asphalt is heated. Tar, if used, is heated to a temperature of 120°C and asphalt, if used, is heated to a temperature of 180°C.

A uniform thin layer of hot binder is laid over the *clean and dry* surface of W.B.M. road. The binder is spread in longitudinal strips, starting from the edges of road. Nearly 1.8 kg of binder is used per square metre of the surface. The layer of binder should be thin but uniform. 12 to 20 mm size stone

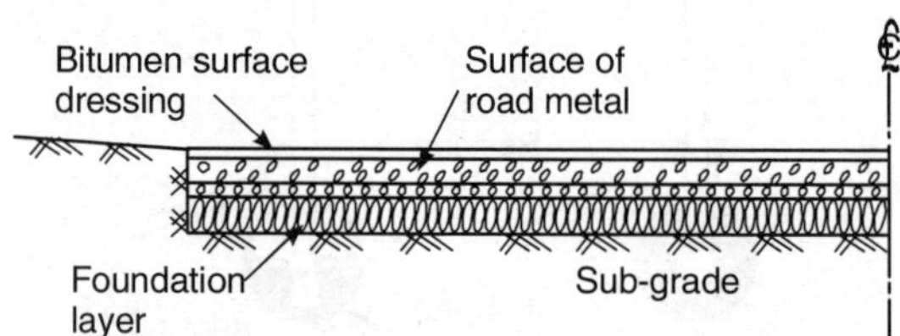

Figure 7.8 *Bitumen surfacing over macadam road*

chippings should be spread evenly on the binder. About 1.8 to 2 m^3 of chippings will be required per 100 m^2. After *broadcasting* the stone chippings, the surface is lightly rolled with light roller. Usual precautions of rolling (i.e., the rolling should be started first from the edges and the roller should move to and fro), should be taken.

In case of cold process, the rapid setting emulsions are spread on a clean and moist surface at the rate of 2 kg/m^2 of the surface and then gritted with 6 mm size stone chippings before the emulsion breaks. (When the emulsions changes its colour from brown to black the emulsion is said to have been broken). The water of the emulsion will evaporate as time passes and the binder will become thicker and thicker. Cut-backs are applied at the rate of 1.5 kg/m^2. Cut-backs are used when surface is damp.

7.6 TWO COAT SURFACE DRESSING

The method of surface dressing or surface treatment explained above is called *single coat surface dressing*. For heavy traffic of mixed type i.e., consisting of fast moving traffic and bullock carts, two coat surface dressing will be required. The thickness of two coat dressing is nearly 3 cm. There are two methods of laying two coats. In one method the two coats are applied simultaneously as explained above. In the other method, the first coat is applied and the road is opened to traffic for one to two weeks. The surface is then cleaned thoroughly, and the second coat is applied. In the hot process the second coat will require bitumen at the rate of 1.0 kg/m^2 and stone chippings of 9 mm size at the rate of 1.5 m^2/100 m^2 of road surface. Similarly surface dressing may be given with emulsions or cut-backs in a similar manner.

In some cases bitumen and stone chippings are first mixed in a hot state and the mixture is spread and rolled to the desired thickness. This is known as *premixed surface dressing*. In this case 64 kg of bitumen is taken for 1 cu. m of stone chippings.

7.7 RENEWAL COAT OF SURFACE DRESSING

Due to constant wear and tear, the surface is worn out under the mixed traffic. If pot holes and ruts appear on more than one-third of road surface in a particular portion of the road, a renewal coat is necessary. The surface is cleaned and worn out portion scrapped if necessary. After cleaning the surface bitumen is applied in a thin uniform layer. When the bitumen has become tacky, stone chippings are broadcast at the rate of 1.0 m^3/100 m^2. The size of chippings should not be more than 9 mm. The bitumen to be applied should be 1.0 kg/m^2 of the road surface. The surface is then properly rolled. During rolling proper camber and gradient should be maintained.

7.8 BITUMINOUS MACADAM

The surface painted roads are only suitable for light traffic. Under heavy and mixed traffic, such type of road surfaces are worn out very easily. For heavy traffic Bituminous Macadam surfaces are very suitable. There are two types of bituminous macadam surfaces viz., (i) grouted macadam, and (ii) premixed macadam.

Grouted Macadam. The grouted macadam surfaces may be semi-grouted or full-grouted.

When the binder is allowed to penetrate to a small depth of 2 to 2.5 cm below the surface of macadam, it is called *semi-grouted macadam.* In this method cold process can also be used. When the penetration is substantial i.e., upto 5 to 7.5 cm, it is called full grouted macadam. Cold process is generally not used as the bituminous macadam roads constructed by hot process are superior to that constructed by cold process.

7.8.1 Construction of Full-grouted Macadam

The base consists of a W.B.M. or W.M.M. of sufficient thickness to withstand the traffic load. Generally 15 to 20 cm thickness is recommended.

Earthen kerbs are constructed along the edges of the base course. The base surface is cleaned and the aggregate is spread evenly to the required profile. The thickness of loose metal varies from 9 to 6 cm. The surface is now lightly rolled dry. After the rolling is complete, hot bitumen or asphalt is sprayed with a spraying machine or spout can. The rate of bitumen to be sprayed varies from 8 to 13.5 kg/m^2 of the surface, depending upon the gradation of aggregate used and the desired depth of penetration. After spraying the binder, neat grit 10 mm size are immediately broadcast over the surface at the rate of 1.5 to 2 m^3 per 100 sq. metres of the surface. The surface is now rolled by 10–12 tonne rollers. The rolling should be first started from the edges and should be shifted towards the centre. After the surface has been rolled to compaction, it is opened to the traffic for a week. After this, the premix coat and seal coat is provided.

7.9 BITUMINOUS CONCRETE

These are the surfacing of premix of fine and coarse aggregates mixed together with tar or bitumen in a mixer. This is a costly and superior type of surfacing for heavier and mixed traffic.

The old surface of the waterbound macadam road is primed by bitmant emulsion or WMM. The surface is dried and cleaned. Side kerbs of earth of bricks are formed to support the bituminous surfacing. The thickness of the W.B.M. base should not be less than 15 cm. Well graded road metal of 6 to 30 mm size is heated to a temperature of 180°C in the drier of the mixing plant. Sand of 1 mm to 3 mm size well graded is also heated and mixed with the road metal in the ratio of one and two. Tar or bitumen is also heated to a temperature of 180°C and mixed with the hot aggregate. The binder is taken at the rate of 48 kg/m^3 of road metal and 128 kg/m^3 of sand. For proper mixing $\frac{2}{3}$rd (of the binder is mixed with the road metal and the remaining $\frac{1}{3}$rd) is mixed when the sand is added. The whole thing is thoroughly mixed. The bituminous concrete is spread on the prepared base of W.B.M. road, between the kerbs to a thickness which varies from 5 to 10 cm. The surface is then rolled by 6–8 ton rollers maintaining proper profile. The rolling should be continued for a few days till the surface is thoroughly compacted. After this premixed seal, coat consisting of coarse sand and bitumen is applied as usual. The surface is again rolled and is finally opened to the traffic.

This type of surfacing is better than grouted macadam and can carry heavier traffic. It is also suitable for iron wheeled traffic. (*Sheicrete mix is a patented type bituminous concrete*).

7.9.1 Sheet Asphalt

Sheet asphalt surfacing is used on the W.B.M. base or cement concrete base. Sheet asphalt is elastic having less friction. It gives a better type of wearing surface. It consists of 6 cm thick layer of asphalt concrete over which 2 to 4 cm thick layer of asphalt mortar consisting of coarse dry sand mixed with asphalt in a hot state. The method of laying and mixing is the same as explained for Bituminous Concrete.

7.10 MIX DEFECTS

Following are the common possible defects in the asphaltic concrete mix:

(a) Lack of Bitumen. If the quantity of the bitumen is less, the mixes will have dry lean appearance. They will lack in the black lustre shining of the well coated mix. Such mixes will be difficult to spread and will have transverse cracks while rolling under road roller, and will not readily heal up.

(b) Mix too Cold. When the bitumen is not properly heated, such mixes shall be stiff and difficult to handle. The large size particles cannot be coated properly. The spreading will be difficult and there will be a tendency for the mix to tear under the screed.

(c) Excess Bitumen. The mix containing excess bitumen will have slopy mix, which will slump too readily even in the lorry or the container.

(d) Poor Mixing. When the mixing is not thoroughly done, after laying some areas will appear lean and brown and others as shiny and black having excessive bitumen.

(e) Poor Grading. If the grading of the aggregate is not properly done, the workability will be poor. The appearance after laying will be rough open textured if the quantity of coarse aggregate is more. If fine aggregate is more, the surface will show lean and too finely graded texture.

(f) Over-heated Mix. During heating in the boiler, if excess of blue smoke is seen coming, it indicates the over heating of the bitumen. Over heating of the bitumen should not be allowed, because after laying, such mix will have dull burnt appearance. The binding strength of such mix is also poor.

(g) Excess Moisture. If the aggregate used for mixing has excess of moisture, the appearance of the surface will be similar to those with excess of bitumen. The excess moisture will evaporate while heating and the stearn rising from the mix indicate its presence.

7.11 AGGREGATE GRADATION FOR BITUMINOUS ROADS

Table 7.2 gives the aggregate gradation for premixed bituminous carpets of various thickness.

TABLE 7.2 *Aggregate gradation for premixed bituminous carpets*

Sieve designation	*Semi-dense carpets of asphaltic concrete*			
	Aggregate percent by weight passing the sieve			
	20 mm compacted thickness	*25 mm compacted thickness*	*25 to 40 mm compacted thickness*	
			Grading 1	*Grading 2*
20 mm	–	100	–	100
12.5 mm	100	75–100	100	80–100
10.0 mm	75–100	60–85	80–100	70–90
4.75 mm	35–55	35–55	55–75	50–70
2.36 mm	20–35	20–35	35–50	35–50
600 micron	6–16	6–16	13–23	13–23
150 micron	4–12	4–12	8–16	8–16
75 micron	2–8	2–8	4–10	4–10
Binder content	The binder bitumen should be 5 .5% by weight of the total mix.		The binder bitumen should be 5% to 7.5% of the total mix.	

7.12 CAUSES OF DISINTEGRATION OF BITUMINOUS ROADS

Practically it has been seen that there are many failures with surface painting even with light traffic. Most of such road failures are not due to wrong use of the surface treatment but due to the improper preparation of the sub-grade and the foundations. The base of the road is usually the weak point, which causes failure. The water or dampness is the main enemy of the surface treatment. If the dampness reaches the road structure due to rain, due to capillary action, from side drains etc., it will weaken the sub-grade and will cause disintegration of the road.

The sub-grade is the main element of the road structure, which carries the load of the traffic through wearing course and distribute it to the sub-soil over wider area.

If due to any reason the water reaches the underneath into the sub-grade and the sub-soil, it will reduce its bearing capacity, this will be the cause of the road failure.

Following are some of the main causes of failure of the treated surfaces :

7.12.1 Use of Incorrect Quantity of Binder

If the quantity of the binding material is in excess, it will cause smoothing and softness of road. On the other hand use of less quantity of binder material will make the surface brittle having the tendency to easy disintegration off grit-particles, ravelling of surface or crumbling.

7.12.2 Incorrect Aggregate Proportion

It will result in too dense or too open mix, which may fail.

7.12.3 Overheating

During mixing if the bitumen is overheated, the volatile oils will be lost and the binder material will loose cohesion and strength of binding and will become brittle.

7.12.4 Waving

The waving in the road surface is caused due to excess of binder which act as lubricant or it may be due to the sub-grade being smooth and allowing the road crust to slide on it. Sometimes unsuitable metal, incapable of interlocking also becomes the cause of waving.

7.12.5 Sub-soil Movement

If the sub-soil under the road moves or slides due to any reason the road cracks and fails.

7.12.6 Bleeding or Flushing

During hot weather, due to excessive heat the binding material comes out at the surface of the road from the surface course. It is generally caused due to (i) use of excess of binder, (ii) insufficient quantity of binding materials, (iii) loss of cover aggregate. To stop bleeding, extra quantity of coarse sand and fine aggregate is spread on the surface.

7.12.7 Streaking or Striations

This defect is caused due to (i) inexperience of the sprayer man, (ii) improper adjustment or careless operation of the binder distributor, and (iii) mechanical defects in the mixer.

7.12.8 Scabbing

It is also called dislodgement of cover aggregate and is caused due to (i) use of too hard grade of binder, (ii) lack of adhesion of the aggregate with binder due to moisture, dust etc. on the surface of the aggregate, (iii) insufficient binder to hold the aggregate properly and firmly in position, (iv) loosing and whip-off of aggregate by fast-traffic, (v) raining immediately during or after the construction, (vi) undue delay in spraying of the aggregate after spraying at the binder material.

7.13 MAINTENANCE AND REPAIR OF BITUMINOUS ROAD

A bitumen-grouted macadam road, laid on sound sub-grade can carry upto 1000 or more medium light weight vehicles per lane per day. The life of the bituminous treated roads is about 12–15 years traffic upto 1200 tonnes per day. These roads should have routine maintenance of surface dressing etc. at every 2–3 years.

While doing the repairs and maintenance of the roads, the preventive maintenance should be carried out at the first indication of the failure, so that major defects may not develop.

The source of the trouble or reasons of the road failure should first of all be determined before starting the repair works. There is no use of doing surface repairs on the defective sub-grade or base. Before starting any repair it is better to first check the condition of the base. The patch repairing should be done properly. Undue strengthening of the weak spots should not be done, because it will create difference in traffic wear and impact, which will further damage the adjoining surface areas.

While doing repairing works every effort should be made to reduce the traffic interference. Warning signs and barriers should be placed at proper places. It is better to stack the maintenance materials along the road-side at suitable places, from where it can be taken.

All the pot holes and ruts should be repaired as early as possible, otherwise they will quickly cause further damage to the surface. The main cause of large disintegration area is untimely repair of the pot holes. The rain water is collected and get way through the pot holes into the sub-grade and causes failure of the road.

The old surface is loosened by pick axes to a depth of about 4 cm for hard stone old aggregate and upto 6.3 to 7.5 cm for soft stones. The loosened metal is raked over to bring the metal from 20 mm gauge upwards to the surface. New metal of 40 mm gauge is added at the top of the old surface, to make a total thickness of atleast 11.5 cm, and is given the required shape. The aggregate so spread is consolidated and surface dressed.

On the roads which have thin crust of hard metal and it is proposed to lay the new metalling over it, without digging the old one, the old surface should be scored for the full width of the road with diagonal criss-cross lines about 40 mm deep and 30 cm apart. In case of soft metal top, the V-shaped trenches can be made at 60 cm intervals instead of scored lines for the purpose of keying between the old surface and the new surface.

As the main purpose of the scarifying is to get the proper bond between the old and the new layer, therefore the old surface should be disturbed only, what is necessary. In case the scarifying is done for the full depth of the metal, all the scarified metal should be removed and screened. Aggregate from 3 mm to 20 mm gauge are stacked separately. The mixture of the dust and fine grit are stacked properly. Now the extra required materials such as aggregate, earth etc. are added, and are spread along with the chipping materials on the sub-grade. The foundation so prepared is thoroughly watered. The old big size aggregate is now spread over it and combed through so that big stuff come on the top. Now new metal of 50 mm to 63 mm size are added according to the requirements and the whole surface is consolidated by road roller.

As the metal is properly graded and large percentage of old metal is used, the stabilization of the surface is done in less time. The wet soil underneath also help in the stabilization. While doing wet rolling, the water should be gradually added so that first the complete mechanical lacking of the metal is done, before the underneath earth cushion begin to be forced to the surface through the interstices. After doing the full consolidation, the fine material obtained from the screening is spread over the surface, thoroughly washed with copious quantity of water and allowed to stand for 24 hours. After its partial drying the surface is rolled with light roller. The road should be cured with water for 7 days.

The first coat of surface dressing should not be done before 14 days after the proper drying of the surface. Before applying top bitumenous coat, the surface should be cleaned out for atleast 10 mm depth.

REVIEW QUESTIONS

7.1. Describe the process of laying two coats of surface dressing with stone chips and bitumen on a new surface of stone metal. Calculate the quantities of materials with specific mention of size and grade required for two coats of surface dressing for 100 sq. m of a road surface.

[**Hint :** Stone chipping of 12 to 20 mm size 1.8 to 2 m^3 will be required nearly for 1st coat and 9 m size chippings will be required 1.5 m^3 for the second coat. For 1st coat 200 kg of binder will be needed while for the second coat 150 kg of binder.]

7.2. Describe fully the process of laying the single coat of paint with bitumen (by hot process) on a water bound macadam road as a surface dressing. Give quantities of grit and paint (i.e., binder) you would require for one km of road.

7.3. What are emulsions and cut-backs ? Compare the merits and demerits of emulsions and bitumen.

7.4. Under what circumstances would you provide a surface painted road ? What are the advantages of a surface painted road over the waterbound macadam road ?

7.5. Write down the detailed specification for the first coat of surface painting. Give quantities of materials required for 4 m wide carriage way 1 km long.

7.6. Describe in brief the construction of premixed carpet road. Give detailed specification for a 4 cm thick premixed carpet.

7.7. Describe in detail, how you would lay a bitumen semi-grouted road surface 4 m wide over an old macadam pavement carrying light mixed traffic. Work out the quantities of materials required for laying 1 km long road.

7.8. Describe in detail the construction of a bituminous macadam full grouted road.

7.9. Explain the various works involved in the maintenance of bituminous roads.

7.10. Write a short note on the possible defects in the asphaltic concrete mix.

7.11. What do you understand by the causes of disintegration of bituminous roads ? Describe them.

7.12. Write short note on the aggregate gradation for bituminous roads.

8

Cement Concrete Roads

GENERAL

Cement concrete roads or concrete roads as they are generally called, had become very popular in this country and being used on a large scale. They are superior to many other roads especially bituminous roads. Their advantages are manifold.

8.1 ADVANTAGES AND DISADVANTAGES OF CONCRETE ROADS

Advantages

1. The concrete roads, if constructed properly, last longer and the maintenance charges are very low.

2. They provide a safe and easy riding surface under all weather conditions.

3. They are dustless.

4. They can be laid on any sub-grade.

5. They do not develop corrugations.

6. They can be easily reinforced when it is so desired to resist high stresses.

7. The tractive resistance of such roads is very low due to non-slipperiness.

Disadvantages

1. The initial cost of construction is very high.
2. The construction of such roads require skilled labour and skilled supervision.
3. They develop cracks due to variation of temperature in the atmosphere.
4. They cause glare due to reflected sunlight.
5. Resiliency is less than bituminous roads.

8.2 KINDS OF CONCRETE ROADS

The concrete roads may be divided into two parts:

1. Premixed concrete roads.
2. Cement bound macadam roads.

The premixed concrete roads are constructed by mixing the various ingredients of concrete viz., cement, sand, coarse aggregates (stone metals) and water in proper proportions in a concrete mixer or central mixing plant and laid on the prepared bed of sub-grade or base as the case may be. It is consolidated and finished to a slab of desired thickness. In this case rolling is not required.

In the case of cement bound macadam roads, the road metal is laid on the prepared bed of subgrade or base and rolled to a desired thickness. Over the rolled surface, thin cement mortar or cement slurry is spread and the surface is finished.

8.3 CONSTRUCTION OF PREMIXED CONCRETE ROADS

The construction of these roads is carried in two different methods :

(i) Alternate Bay Method, and

(ii) Continuous Construction Method.

8.3.1 Alternate Bay Method

In this method of construction, concrete is placed in bays of 4 to 5 m length and width equal to the width of the carriage way. The whole length of road is divided into rectangles or bays and concrete is filled in alternate bays and finished. The intermediate bays are filled after at least a week of the construction of adjoining bays. This method is only suitable when the carriage way is less than 4.5 m in width. This method is generally unsuitable for hot countries.

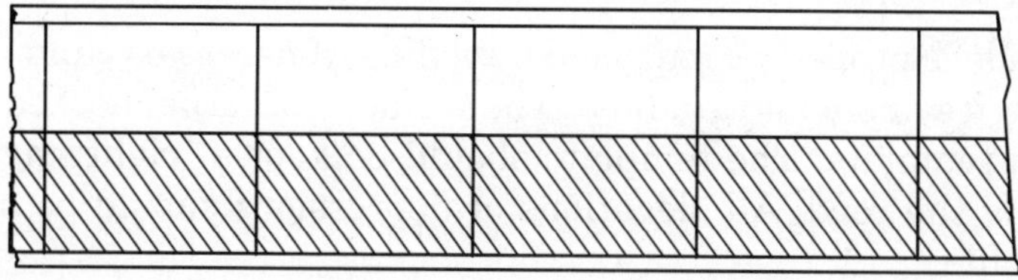

(a) Successive bay method

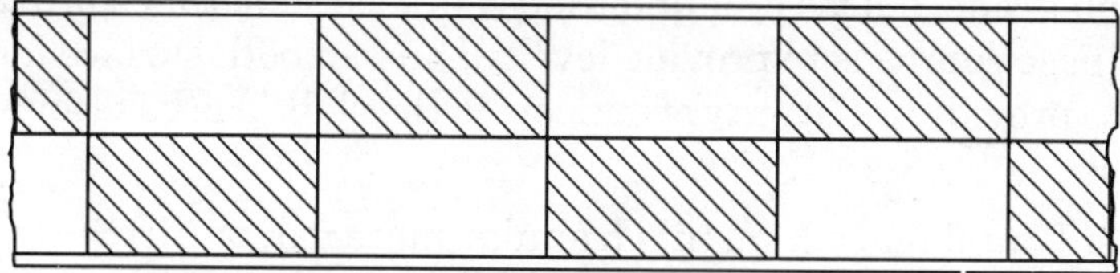

(b) Alternate bay method

Figure 8.1

8.3.2 Continuous Construction Method

When the width of traffic way is more than 4.5 m the slab is laid in longitudinal strips, having a width of 2.5 m to 3.5 m (This is also called strip method.) The slab along each strip may be laid in successive bays or in alternate bays as is shown in Fig. 8.1 (a) and (b). Usually the construction in alternate bays is preferred.

8.4 CONSTRUCTION DETAILS

The construction of concrete roads involves the following constructional stages :

1. Preparing the sub-grade.
2. Preparing the base course.
3. Placing of form work.
4. Watering the prepared base.
5. Mixing and placing of concrete.
6. Compaction and floating.
7. Belting.
8. Brooming.
9. Checking the finished surface.
10. Curing.
11. Edging.

8.4.1 Preparing the Sub-grade

The sub-grade is properly compacted by rolling, if necessary, and is brought to the required camber and gradient. The surface should be checked by means of a scratch template. The template should be kept at right angles to the centre line of the road. All elevations or depressions should be brought to the true profile.

8.4.2 Preparing the Base Course

Over the sub-grade, a base or foundation course is sometimes provided. The function of base course is to provide levelled and smooth surface for the slab and hence to provide uniformity of support to the slab. The base for a concrete road may be :

(i) Water bound macadam, (ii) Granular material, or (iii) Stabilized soil.

(i) In the case of W.B.M. base the thickness should not be less than 15 cm. The base width should be at least 50 cm more than the carriage way, projecting 25 cm on each side. The surface of the W.B.M. should be thoroughly compacted by rolling. The surface should not be made smooth but it should be left rough so that there should be proper anchorage between W.B.M. surface and concrete slab.

(ii) Granular material base provides high intrinsic strength and should be so graded that all particles will not knit together into a tightly packed mesh when compacted. The reason for this is, that there must be proper anchorage between the base and the cement concrete surfacing and hence it is desirable that the top of the base course should remain rough.

(iii) Stabilized earth base is also sometimes provided. The type of stabilization will depend upon the type of material available at the site and the type of soil.

(iv) Dry lean concrete base is also provided as base.

8.4.3 Placing of Form Work

After the base course is completed, form work for the concrete slab is laid. The forms, which are set on the edges, can be of steel or timber. If of timber, they should be at least 5 cm thick capped with 5 cm angle-iron along the upper edge from the inside. The depth of the form work should be equal to the thickness of the slab. Channels are sometimes used for steel-form-works. The timber or steel-form-work should be fixed rigidly well in advance of concrete work to be started. It should be oiled and checked to line and grade.

8.4.4 Watering

After the form work has been laid, the surface of base or subgrade, as the case may be, is wetted with as much water as it can readily absorb. Water should not be allowed to stand on the surface. The surface should be kept wet for about 12 hours before concreting. The object of watering is to saturate the sub-grade or base so that it should not absorb any water from the concrete when it is laid over it.

8.4.5 Mixing and Placing of Concrete

The ingredients of concrete are mixed in proper proportion in a dry state. Generally 1 : 2 : 4 mix is adopted. Mixing should preferably be done in mechanical mixers. Measured quantity of water should be added to have the designed *water cement ratio.*

The concrete is placed in the form works by manual labour, and it should be laid in layers of thickness not more than 5 to 8 cm or two or three times the size of aggregates. The concrete should be laid in the entire width of the form work and proceeded lengthwise. Each layer should be sliced and spaced with special tools, immediately it is laid. The top-most layer should be laid 6 to 12 mm higher than the actual profile for further tamping. The top layer should be laid to the required camber and gradient. Necessary transverse and longitudinal joints should be provided.

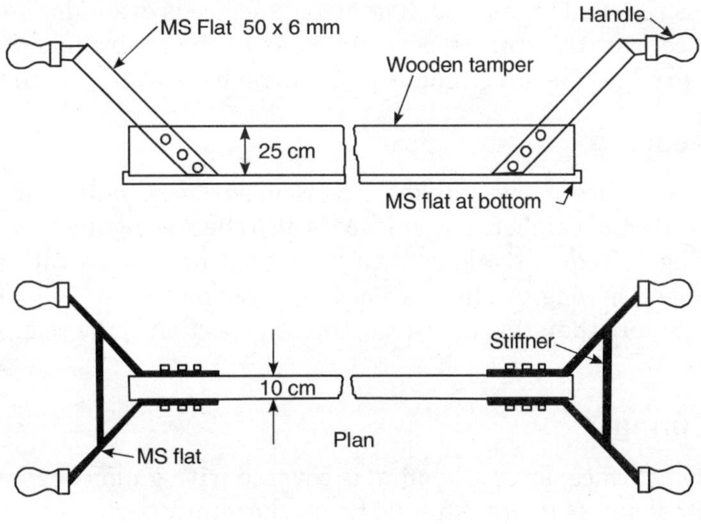

Figure 8.2 *Hand tamper*

8.4.6 Compaction and Floating

The compaction and consolidation of concrete is done by manual labour or by means of hand tampers as shown in Fig. 8.3. Consolidation may also be done by mechanical vibrators. The hand tamper is a wooden beam 10 cm thick and 25 cm deep, having a length equal to the width of the bay plus 30 cm. To the under-side of this beam a metal plate 5 mm thick is fixed. The tamper is used across the bay and tamping is done along the length of the bay. The surface is then tamped longitudinally. The irregularities caused due to tamping are rectified. The surface is then finished by hand floats.

The purpose of floating is to produce a uniform and even surface of concrete, free from transverse waves or corrugations. Floats are made of wooden boards 20 cm wide and 5 cm thick provided with suitable handles. The floating is carried out in longitudinal direction. The float is drawn in slow back and forth motion and slowly shifted towards the other edge. This will produce even surface free from corrugations.

8.4.7 Belting

The purpose of belting is to finish the surface of the concrete. The belt is 15 to 30 cm wide strip of canvas or rubber, fitted with handles at both ends. The belt is moved longitudinally by two persons. The process is rarely used in this country.

8.4.8 Brooming

Brooming is resorted to, if a rough or gritty surface is desired. The broom is gently pulled over the surface perpendicular to the centre line of the road from edge to edge. The brooming is done immediately after belting.

8.4.9 Checking

The finished surface is now checked. The road surface should conform to the required grade and camber. The surface is also checked against unevenness. The checking is frequently done during the floating process with a wooden straight edge 3 m long. A tolerance of 2 mm per metre can be given. If the difference is more than this tolerance, then the portion of the road should be corrected.

8.4.10 Curing

The finished surface, after 12 hours, is covered with gunny bags which are kept wet for about 24 hours. After 24 hours the gunny bags are removed and the surface is covered with a layer of sand which is kept wet for about 14 days. Sometimes the whole surface is divided into small bays by forming earthen banks or ridges 5 cm high. The bays are now filled with water to a depth of nearly 4 cm for 14 days. After curing, the surface is cleaned and washed.

8.4.11 Edging

Before the road is opened to traffic, brick edging is constructed to protect the slab. Earth is then spread on the berms upto the top of the brick edging.

8.5 JOINTS IN CONCRETE ROADS

The main difficulty with the concrete pavements is, that due to temperature variations, cracks are developed in the slab. If sufficient care is not taken at the time of construction, these cracks will make the road unserviceable as these cracks can never be repaired satisfactorily and will become a perpetual source of headache. To overcome this difficulty *joints* are provided. If the joints are properly spaced, cracks are reduced and if at all, only minor cracks will develop.

Joints can broadly be divided into two types :

(1) Expansion joints.

(2) Contraction joints.

The first provides for the expansion of concrete and the second to prevent irregular cracking due to warping or contraction. The expansion and contraction joints can also be grouped as :

(a) Longitudinal, and

(b) Transverse or cross joints.

(a) Longitudinal Joints. The joints are provided to prevent longitudinal cracks and enable the road to be constructed in convenient widths.

There are two main causes of longitudinal cracks. They are (i) high warping stresses, and (ii) variation of sub-soil moisture under the slab. The longitudinal joints which are also called *construction* joints are provided

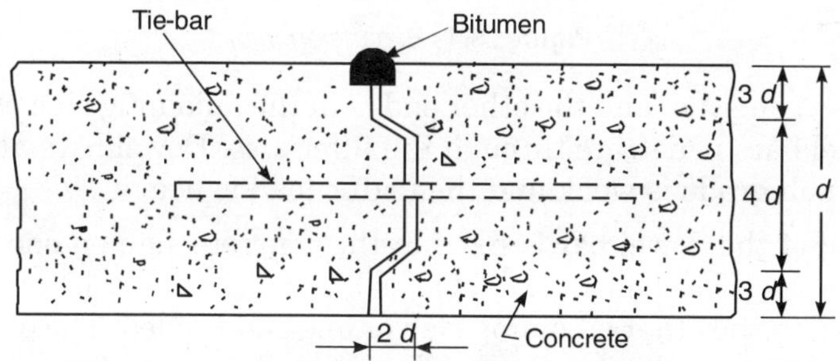

Figure 8.3 *Longitudinal joint*

between the two strips. When the width of the traffic way is more than 5 m, the slab is laid in two or more strips. Thus a 6 m wide carriage way will be laid into two strips of 3 m width each. There will be longitudinal joint between these two strips. The longitudinal joint consists of a tie bar inserted between the two strips with a key as shown in Fig. 8.4. The top of the joint is sealed with bitumen.

(b) **Traverse Joints.** There are four types of transverse joints. They are :

(i) Expansion joint.

(ii) Contraction joint.

(iii) Warping joint.

(iv) Construction joint.

(i) *Expansion joint.* Expansion joints form a complete separation between the adjoining slabs and permit the concrete to expand, as the temperature increases. The joints should be provided at right angles to the centre line of the road and should extend to the full width and depth of slab. The expansion joints are provided at a distance which varies from 18 to 27 m. To provide for the transfer of wheel load from one slab to the other, dowel bars are provided. The dowel bars are 30 to 60 cm long, 15 to 18 mm diameter and are provided at a distance of 30 to 50 cm centre to centre. One end of

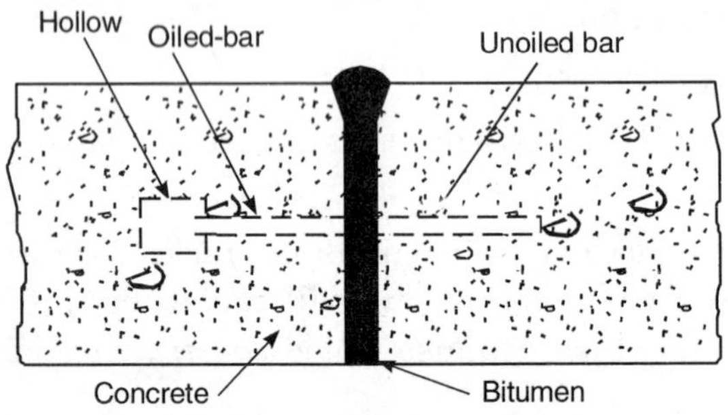

Figure 8.4 *Expansion joint*

this bar is in one slab and the other end is in the adjoining slab on the other side. Dowel bars penetrate through the filler also. Dowel bars are provided when the sub-grade is weak and the traffic load is heavy.

One end of the dowel bar is fixed in the concrete slab on one side of the joint while the other end of dowel bar is free to move in a 10 cm long metal sleeve fixed in the other adjoining slab or one end is oiled. The oiled end will work as a free and whereas the unoiled end will act as a fixed end.

(ii) *Contraction joint.* They are used to allow for contraction or shrinkage of concrete slab and prevent or control cracking. The contraction joints are of two types :

(a) Plain, vertical but joints extending to the full depth of the slab.

(b) Dummy joints extending to about one-third of the depth of the slab.

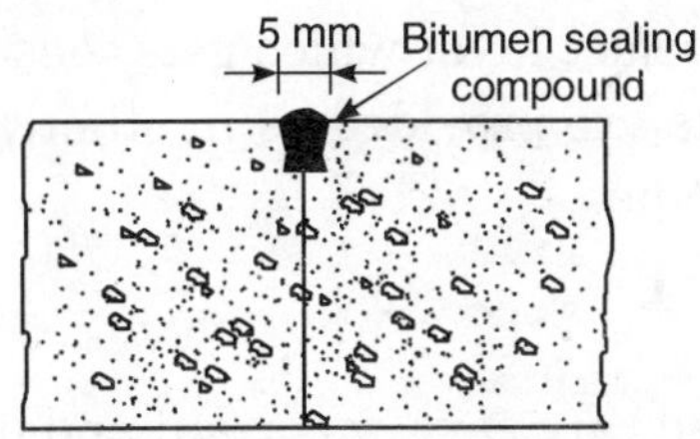

(a) Plain butt type contraction joint

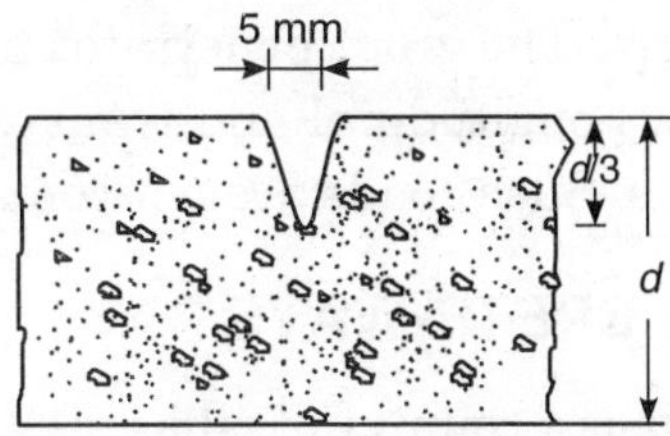

(b) Dummy contraction joint

Figure 8.5

The contraction joints are provided along the length of the road at a distance of 4 to 5 m. The plain butt joints are cuts in the full thickness of the slab. The top of the joint is filled with bitumen seal. The dummy contraction joints are cut in the upper one-third or quarter depth of the slab and are filled with bitumen.

(iii) *Warping joints.* These joints like the contraction joints are simple breaks in the continuity of the concrete slab. The opening of these joints is prevented by tie bars.

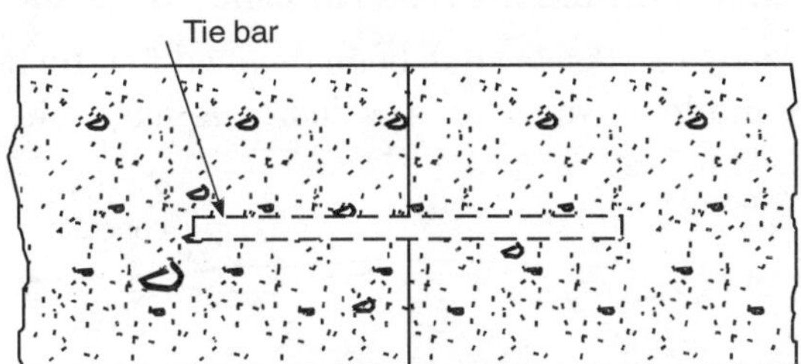

Figure 8.6 *Warping joint*

(iv) *Construction joint.* The object of construction joint is to facilitate the construction. They are provided when it is necessary to stop the work of concreting at any point other than the expansion or contraction joint. Tie bars of 10 to 12 mm diameter and 1 m long are provided at 60 cm centre to centre, at a depth of nearly 3 to 5 cm from the top of the concrete slab.

8.6 CEMENT BOUND MACADAM ROADS

On the reconditioned surface of W.B.M. road, a 12 cm thick layer of road metal of 36 to 48 mm size is spread and rolled tightly with 8 tons roller to get a partially consolidated layer of 10 cm thickness. Cement and sand are taken in the ratio of 1 : 2. About 5 bags of cement per cubic metre of road metal and 34 litres of water per bag of cement is used to form a fluid cement mortar or grout. This grout is poured on the prepared surface of macadam. After grouting, stone chippings of 15 to 18 mm size are spread on the surface and forced into the mortar on the surface by longitudinal tamping or light rollers. The grout penetrates into the surface layer which gets consolidated also. Longitudinal and transverse joints are provided as in other concrete roads. Such roads are not common in India.

8.7 CRETE-WAYS

India is a country of villages. Village roads are very important and it should be tried to have durable roads under iron-wheeled traffic and at the same time they should be cheap in construction and in maintenance. Generally the following types of roads are constructed in villages :

(1) Earth roads (2) Moorum roads

(3) Gravel roads (4) W.B.M. roads

These roads give way easily under the bullock cart traffic and hence it becomes necessary to repair them periodically or to recondition the whole surface. This becomes a very costly affair in the long run. The present economic condition of the country do not afford to construct other durable type of roads such as bituminous roads or concrete roads.

A type of *trackway* or *wheeler* has been devised for the bullock carts. This type of trackway made of concrete is called *crete-way*. The trackway can be

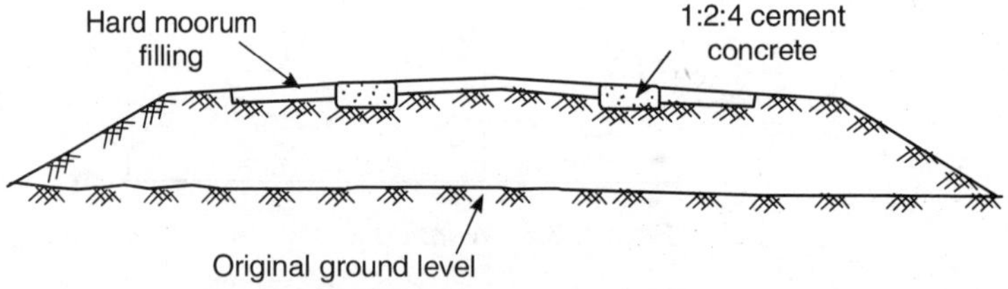

Figure 8.7 *Crete-ways*

made of stone pavings or slabs or bituminous surfacings. These trackways are quite durable and cheap and are becoming common and popular especially in those places from where heavy bullock cart loads move to nearby industrial areas or sugar mills, etc.

Crete-way consists of two narrow longitudinal strips of road surface on which the two wheels of the bullock cart or any other vehicle can move. The strips are generally 60 cm to 75 cm wide, 10 to 15 cm thick and are 1.30 m to 1.5 m centre to centre apart.

The crete-way slabs may be constructed in situ or precast. Generally precast slabs of 2 m length and 10 cm thickness are placed in the trenches excavated for the purpose. The beds of trenches are properly watered and compacted. Tongue and groove joint is provided between the two adjoining slabs.

REVIEW QUESTIONS

8.1. Under what circumstances you would recommend the construction of a cement concrete road ? What are its advantages over the other types of road surfaces ?

8.2. What are the stages of construction of a cement concrete road ? Describe in brief.

8.3. What do you understand by the term 'crete-way' ? Where they are used and what are their advantages ? Give specifications for a crete-way and describe in brief the construction of a precast crete-way.

8.4. Why are joints provided in a cement concrete pavement ? What are the various types of joints ? Describe the construction of any one joint.

8.5. A water bound macadam road in a city area is to be improved either by adopting a bituminous pavement or a cement concrete pavement. Which of the two, do you prefer ? Give reasons for your choice.

Hill Road Construction

GENERAL

Hill roads are also called *Ghat* roads in India. Hill roads present more difficulty in their design, construction and maintenance. These roads are very dangerous and sometimes fatal accidents occur on such roads. Hence very special care is taken in the design and construction of such roads. Surface drainage is very important in hill road construction. Hill roads are classified as :

(i) Motor roads
(ii) Bridle paths
(iii) Village paths or tracks

Motor paths are important from the engineering point of view. They are generally meant for fast moving traffic.

Bridle paths are meant for pedestrians and other transports such as horses, ponies etc. They serve as feeders to the motor roads, thus connecting interior places with the main communication system. The width of the bridle paths generally varies from 2 to 2.5 m. The ruling gradients for bridle paths is 1 in 10. The maximum permissible gradient can be 1 in 7½. The bridle paths can have either a W.B.M. surface or earth or natural surface brought to the required camber and gradient. Side drains should be provided for bridle paths.

Village tracks or paths connect small villages and other working places over which cattle and people move. The width of village paths varies from 1 m to 1.5 m.

9.1 ALIGNMENT OF HILL ROADS

The success and utility of a ghat or hill road depends on the proper alignment of the road. During the alignment of these roads care should be taken to see that the road should be as short as possible and should have easy gradients. In the alignment of the road, a number of sharp curves such as hair pin bends and corner bends will have to be provided. The geometrics of various hill roads are to be provided and designed as per IRC 52–1981. As far as possible the minimum radius of the curves should be 15 m. The desirable radius is of course 30 m. The aim of the alignment should be to establish the

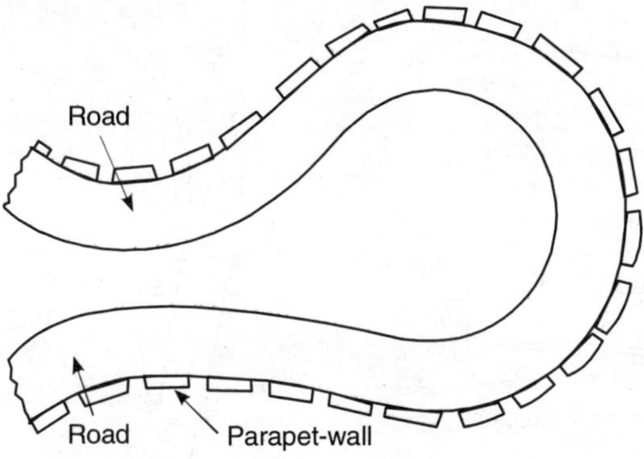

Figure 9.1 *Hair-pin curve*

easiest, shortest and the most economical line of communication between the obligatory points in consideration, physical features of the country and

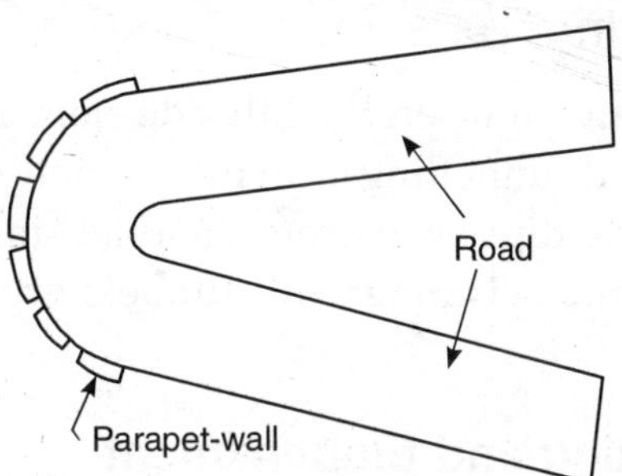

Figure 9.2 *Corner-Bend*

the traffic need of the area served. The principle which the engineer should bear in mind is to align the road so that the vehicles can travel with ease and in safety and the expenditure of motive power and wear and tear of the vehicles should be reduced to a minimum, as also the cost of construction and maintenance of the road should be as low as possible.

9.2 FORMATION

The formation of a hill or ghat road may be :

(i) Wholly in cutting.

(ii) Partly in cutting and partly in embankment.

(iii) Wholly in embankment.

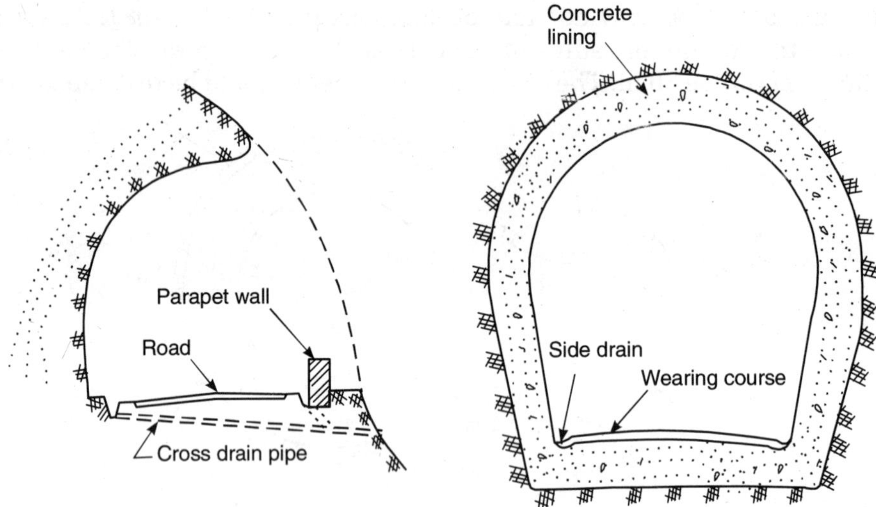

Figure 9.3 Road in half tunnelling

Figure 9.4 Road in tunnelling

9.2.1 Road in Cutting

The road in cutting is taken when the hill side slope is very steep. The road in cutting may be in half tunnelling or in full tunnelling. Half tunnel is used where the strata of rock, dips away from the road side and when the rock is sound and hard. The roads through full tunnels are rarely constructed as they are costly.

9.2.2 Road in Cutting and Embankment

Hill roads are generally constructed partly in cutting and partly in embankment. Road sections may be $\frac{2}{3}$rd in cutting and $\frac{1}{3}$rd in embankment.

Road in cutting and embankment is constructed when the side slope is not very steep and the cost of cutting through rock is too much.

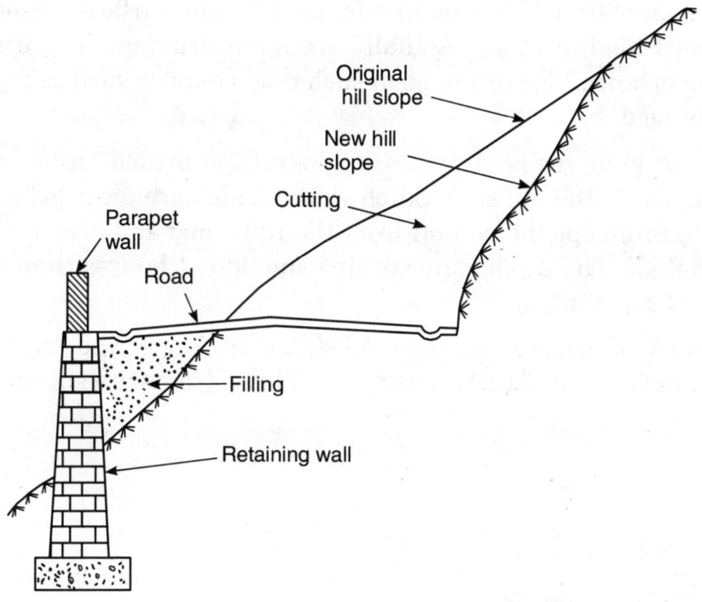

Figure 9.5 *Road in embankment and cutting*

9.2.3 Road in Banking

Sometimes the entire road will have to be taken in embankment. This happens when a depression comes in the alignment. In this case two retaining walls one on each side of the road will have to be constructed. This type of construction is also very costly and hence avoided as far as possible.

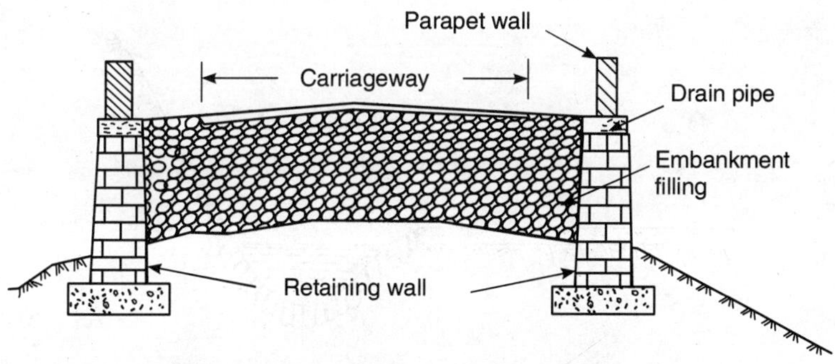

Figure 9.6 *Road in banking*

9.3 DRAINAGE IN HILL ROADS

For better service and low maintenance, efficient surface drainage should be provided for hill roads. Actually speaking drainage in hill roads is of prime importance. The drainage in such roads may be surface drainage and cross drainage.

In hill or ghat roads, drains are generally provided on one side of the road, usually on the hill side. Catch water drains are provided higher up in the hill to intercept the run off from the hill slope and divert it towards a nearby nallah. The catch water drains should not be less than 4.5 m from the edge of the road.

The shape of drains near the road side is (i) Angle type drains, (ii) Kerb and Channel type, and (iii) Saucer type. These drains are shown in Fig. 9.7.

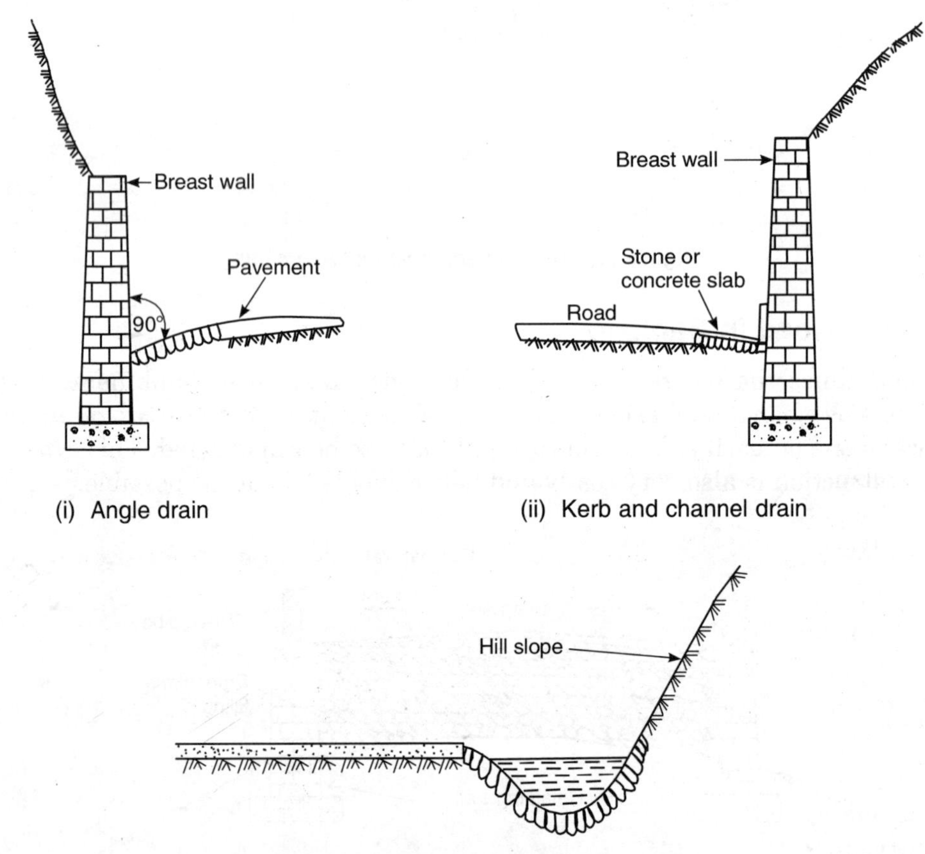

(i) Angle drain

(ii) Kerb and channel drain

(iii) Saucer drain

Figure 9.7

To prevent the side drains from overloading and thereby causing the road surface flooding, cross drainage should be provided at frequent intervals. Cross drains also help in reducing the size of side drains. Cross drains are provided by means of small *under-drains, scuppers, Irish bridges,* etc.

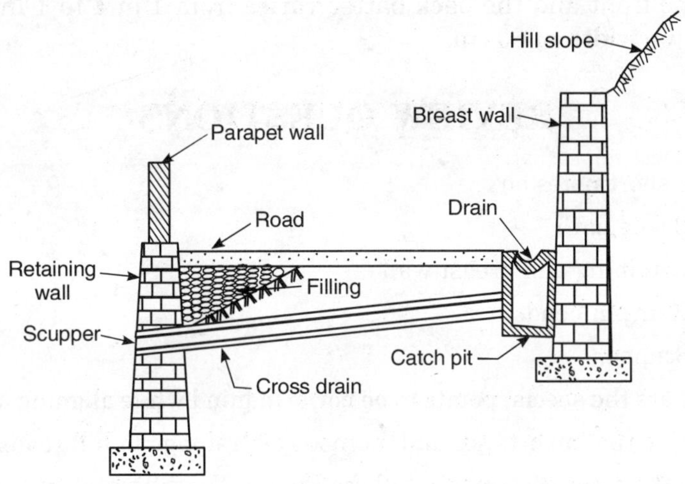

Figure 9.8 *Section through cross drain and catch pit*

9.4 RETAINING AND BREAST WALLS

9.4.1 Retaining Wall

It may be defined as a wall built to resist the earth pressure of filling and the traffic load of the road. It is commonly used in hill roads when the road goes in embankment, partly cuttings and partly embankment. Retaining walls can be constructed in random rubble dry stone masonry, stone masonry, brick masonry or cement concrete. These are designed as a gravity retaining wall.

Random rubble dry stone masonry retaining walls are the simplest and are mostly used in hill road construction. The stability of such walls does not depend upon the quality of materials used in the masonry but on the workmanship, that is, the arrangement of stones in the wall. The stones to be used should be of a large size and should be roughly hammered dressed. The wall should have a minimum top width of 60 cm and the front face should have a batter varying from 1 in 4 to 1 in 3. In principle the height of dry stone retaining wall should be restricted to 6 m only. For higher walls, the upper 6 m portion of the wall should be made in dry masonry and the lower portion in mortar.

9.4.2 Breast Walls

These are stone walls provided to protect the slope of hill cutting in natural ground from the action of weather. The section of a wall to be adopted depends upon the height of the wall, the nature of the backing, and the slope of cutting. The front and the back batter varies from 1 in 4 to 1 in 2, with a minimum top width of 60 cm.

REVIEW QUESTIONS

9.1. Write short notes on :

(a) Bridle path

(b) Retaining and breast wall

(c) Hair pin bends

(d) Scupper

9.2. What are the special points to be borne in mind while aligning a hill road?

9.3. Describe the importance and methods of drainage in hill roads.

9.4. Draw the cross-section of a hill road under the following circumstances :

(i) Full tunnelling

(ii) Partial cutting and partial embankment

(iii) Full embankment

9.5. Write an essay on the construction and alignment of Ghat roads.

10

Highway Machinery

GENERAL

Today is the age of machines. With the help of machines, many complicated and laborious works have been simplified. For the construction of roads also, an engineer has to depend on machines. A highway engineer must be familiar with the use of modern road machinery. It will be beyond the scope of this little volume to describe each of this machinery in detail. However the particular use of important road machinery and plants have been outlined in brief, so that the reader must have a fair idea about the use of all such machinery and plants used in road construction.

10.1 ROAD MACHINERY

Road making plants and machinery is used in various jobs. The various works included in the road construction are :

(a) Clearing trees, shrubs, jungles etc.

(b) Forming the soil, to level by cutting ridges or filling hollows.

(c) Spreading the metal and rolling.

(d) Surfacing or applying wearing course.

A. The following tools and plants are generally used for clearing the jungles, shrubs etc.

(1) Bull dozer

(2) Rooter

(3) Tractor

(4) Towed scraper

(5) Shovel units with trucks

B. For cutting, filling and preparing the formation or subgrade for a road, the following machinery is generally used :

(1) Tractor
(2) Dozer
(3) Grader
(4) Shovel
(5) Clamshell
(6) Trucks
(7) Trench cutter
(8) Plough
(9) Rollers

C. For spreading the road metal or aggregates on the prepared sub-grade and rolling the same, the following tools and plants are used :

(1) Crusher

(2) Trucks

(3) Aggregate distributor and

(4) Rollers

D. The surfacing generally consists of (a) Bituminous surfacing, and (b) concrete surfacing.

(a) For Bituminous pavements following tools and plants are used :

(i) Bitumen Boiler

(ii) Sprayer

(iii) Aggregate spreader

(iv) Bitumen mix spreading machine paver

(v) Grouting machine

(vi) Rollers

(b) For cement concrete pavements the following tools and plants are used:

(i) Central batching and mixing plant

(ii) Concrete mixers

(iii) Concrete pavers

(iv) Concrete spreader and finisher

(v) Concrete vibrators

In the following articles some of the important plants and machinery will be briefly described.

10.2 TRACTOR

It is an important machinery which is used for different type of works. The tractor works with diesel engine having horse power varying from 20 to 200. There are two types of tractors:

(i) Track type tractor (ii) Crawler type tractor

The track or wheel type tractors move on pneumatic tyres. This type of tractor is used for even and smooth country. The crawler type tractor moves on

Figure 10.1 *Crawler-type tractor*

an endless chain and is used for uneven and rough country. Tractors are intended primarily to exert a powerful tractive force for towing other heavy equipments. Tractors are generally provided with various attachments such as a dozer, scraper, harrow, plough etc. Tractors with various attachments

Figure 10.2 *Track-type tractor*

are used for clearing the shrubs, roots etc. and for earth work in formations. The tractor is primarily an earth moving machinery.

10.3 DOZERS

Dozers are attachments of a tractor which are used for various works. In particular, dozers are important machines which are used for various operations in road construction. There are three types of dozers viz., bull dozers, angle dozer and tree dozer.

Figure 10.3 *Bull dozer*

A *bull dozer* consists of a blade, attached to the front side of the tractor. The bull dozer is used for excavating the material and pushing the material in a forward direction. It is mainly used for levelling a heap of excavated material.

Angle dozer consists of a blade attached to the front side of the tractor, which can be set obliquely at any angle to the direction of motion of the tractor. The maximum possible angle of inclination of the blade to the front face of the tractor is 30°. Angle dozer is used for pushing the material to the side, right or left.

A *tree dozer,* which is also sometimes called *a stumper,* consists of a slightly curve blade, with its concavity in the forward direction of motion. The blade is specially designed for uprooting the trees, shrubs etc.

10.4 ROOTER OR RIPPER

This is a very useful instrument used to remove stiff clay, soft rock or other hard soils. It is mounted on wheels and is towed to a tractor. It has one or more tyres or teeth which are driven through the ground.

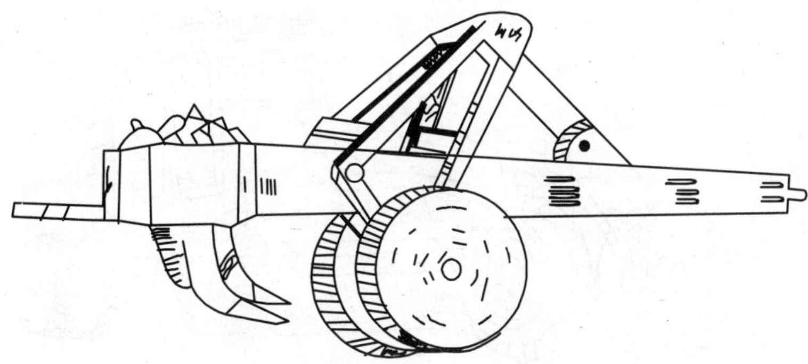

Figure 10.4 *Rooter or ripper*

10.5 SCRAPER

A scraper is an earth moving equipment generally drawn by a tractor. It is used for earth work of a light nature and where the load is also very small. It provides one of the most economical machines used during earth operations of a road. It is a self-sufficient machine, as it can do all the operations necessary for light soil viz., it can dig, load, carry, dump, and spread the earth or soil. It consists of :

(a) A cutting edge which is lowered into the ground to make the cut.

(b) An apron in front of the bowl, which is required to take in or throw out the stuff.

(c) The bowl–which is the main carrier.

(d) It is mounted on four or two pneumatic wheels.

Figure 10.5 *Scraper*

10.6 GRADER

Graders are earth-moving machines which are either self propelled or towed by a tractor. Graders are used for light and medium works. They are used

Figure 10.6 *Grader*

for shaping the sub-grade, constructing earth roads, and for spreading the loose material.

Grader mainly consists of an angled blade 3 to 4 m long supported on a frame-work which is mounted on wheels. The blade can be turned, lowered or raised as may be desired. The self-propelled grader is called *motor grader* and the one which is towed to a tractor is called *elevating grader*.

10.7 ROAD ROLLERS

It is one of the most essential road making machinery. It was the only highway equipment used in India some few years back. Rollers are generally self-propelled but some of the rollers are towed by tractors etc. The following are the common types of rollers which are generally used:

1. Cylindrical roller.
2. Flat-wheeled rollers or smooth wheeled rollers.
3. Pneumatic tyred rollers.
4. Sheep foot rollers.

As has been said earlier, a roller is an essential tool for the construction of a road. Its main function is compacting the surface such as sub-grade, base or soling ; and wearing surface etc.

10.7.1 Cylindrical Roller

It is a light roller of iron, concrete or stone, which is drawn by hand or bullocks. These are available in various sizes and weights. Usual size is 100 cm in diameter and 150 cm long. This roller exerts about 7.0 to 10.0 kg/cm^2 pressure on the ground. These rollers are used for very light work—for rolling village roads, parks, game fields etc. These may be vibratory and non-vibratory type.

10.7.2 Flat-wheeled Roller

This is a very common type of roller and is commonly used. It can be driven by steam or diesel oil. The steam rollers have generally three wheels, a steering wheel, and two rear compaction wheels. The oil rollers have generally 2 wheels. A two-wheel roller is also called *Tandem roller*. The rollers are built in sizes varying from 3 to 20 tonnes. It is mainly used for compacting or consolidating thick layers of macadam or road metal.

Figure 10.7 *Flat-wheeled roller*

10.7.3 Pneumatic Roller

This roller consists of a small truck with two or more axles with a number of pneumatic-tyred wheels. The truck may be loaded so that the required

Figure 10.8 *Pneumatic roller or wheeled rubber tyred roller*

intensity of pressure may be attained. This type of roller is capable of consolidating the surface slowly and uniformally, without causing damage to the aggregate.

These rollers are used for compacting cold laid bituminous pavements, soft base course and layers of loose soil. These are also suitable for compacting closely graded sands, and fine-grained cohesive soils at nearly plastic limits. These are useful for finishing the embankments compacted by sheep foot rollers or on loose sandy soil.

The pneumatic rollers have got the following specialities :

(i) They are able to knead the aggregate in the binder without breaking or crushing the aggregate pieces, which is very important.

(ii) When the aggregate and the binders are spread on the road surface, the aggregate binder joint is very weak for the time being and the pneumatic wheel rollers do not have the tendency to dislodge the aggregate.

(iii) On steep gradients and sharp curves, they work very efficiently and have been found very useful in the construction of ghat roads.

10.7.4 Sheep-Foot Roller

It consists of heavy metal cylinders usually of 1 to 1.5 m diameter and 1.2 to 1.5 m long. Each of these drums have 16 to 18 cm projections on their surface and each drum weighs about 3 to $4\frac{1}{2}$ tonnes. The cylindrical drums are

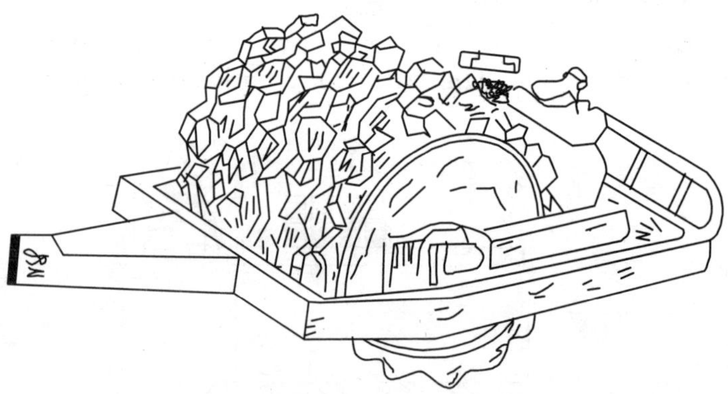

Figure 10.9 *Sheep-foot roller*

hollow from inside and they can be filled with water or sand to increase their weight. Two rollers form one set to provide flexibility of movement and control of larger area. The cylinders are generally towed to a tractor.

With this roller the surface is compacted from bottom upwards. The feet of the roller penetrate into the surface and knead it and thereby ensuring the proper compaction of soil from the bottom itself. These rollers are used for compacting filled up earth in banking.

10.8 ROLLING

The weight of the road roller to be used for compaction of road depends on the size and hardness of the stone aggregate and the thickness of the layers of the materials to be compacted. For ordinary soft stone aggregate 6 to 8 tonne roller is sufficient. For brick aggregate, 6 tonne road roller is used. For roads and highways near towns, 10 to 12 tonne road rollers are used. The actual depth of compaction of any particular road roller should be determined by the actual field tests. Usually the thickness of the loose soil for compaction should be between 15–23 cm and that of stone aggregate upto 10 cm. If more depth is to be rolled, it must be done in layers.

The operation of the rolling should be started from the road edges and it should progress gradually towards the centre from both the sides. The rolling is done parallel to the centre line of the road. In case of curves where super-elevation is to be provided, the rolling should be started from the inner edge to the outer edge of the curve. While rolling care should be taken that each pass of the roller should uniformally overlap not less than one-third of the track rolled in the preceding pass.

The actual number of passes of the roller for obtaining the desired compaction, should also be determined by the actual field test. Usually 50 passes are sufficient for compaction of the dry material. The rolling should be done at the minimum speed of the roller which is usually 3 km/hour for smooth wheeled roller and 5 to 6 km/hr with pneumatic-tyred and sheep-foot rollers. In case of light work the speed can be increased.

Table 10.1 gives the weight of the roller and maximum layer thickness for various types of rollers.

TABLE 10.1

S. No.	*Weight of road rooler in tonnes*	*Description of material*	*Maximum thickness of each layer which should be compacted time in cm*	*Approx. moisture content in the material in per cent*
1	6 – 8	Clayey sand	20 – 23	10 – 18
2	6 – 8	Lean mix concrete	20 – 23	–
3	8	Ashes, clinker etc.	13 – 15	1 – 20
4	8	Clay gravel (stable)	2 – 23	6 – 10
5	10 or more	Crushed rock or hard brick-bats	10 – 12	–

10.9 OUTPUT OF ROAD ROLLER (APPROX.)

Table 10.2 gives the output of a 8–10 tonne road roller with various type of materials for 8 hour/day work.

TABLE 10.2

S.No.	*Material/work*	*Output*
1.	Rolling of 6 mm chips with bitumen/ asphalt spread at about 1 cu. m per 100 sq. m	2000 – 2200 sq. m per day
2.	Rolling of 12 mm chips spread at about 1.5 cu. m per 100 sq. m	1600 – 1800 sq. m per day
3.	Consolidation of the premixed carpet	500 – 600 sq. m per day
4.	Rolling of two coats of surface dressing done together	1000 sq. m per day
5.	Rolling of 3.8 m wide road soling	1000 m per day
6.	Rolling of premixed metal (base-course)	1100 – 1400 sq. m per day
7.	Rolling of hard road metal to full compaction	20 – 25 cu. m per day
8.	Rolling of soft metal to full compaction	50 – 60 cu. m per day

10.10 DUMPERS

The dumpers are used for conveying the materials to the spot and lay them in their proper places. The dumpers are trollies fitted on trucks which can be

Figure 10.10 *Dumper*

tilted in one or the other direction. The materials are loaded, conveyed and dumped in places by dumpers.

10.11 BITUMEN MIXING MACHINERY

The machinery, which is used for laying and mixing bitumen and aggregate and applying bituminous surfacing on the prepared surface, consists of the following :

1. Bitumen Boiler. *Bitumen boiler* is used for hot process, where the bitumen is required to be heated. The bitumen or tar boiler, as the name suggests, is used to heat the bitumen to the required temperature. The boiler consists of a continuous system, capable of supplying hot bitumen at the required temperature.

2. Bitumen Sprayer. *Bitumen sprayer* is a bitumen boiler, fitted with a pumping and spraying equipment. In this case the boiler is fitted with a double acting pump and is so designed that one man can easily obtain and maintain the necessary pressure at the delivery point. The bitumen passing out of the pump flow through a long flexible tube at the end of which a straight metal pipe is fitted with a nozzel. Through this nozzel, the bitumen is sprayed uniformly on the surface.

3. Hot mix plant i.e., Bitumen Mixer and Spreader. *Bitumen mixer and spreader* is a road making machinery in which the road metal or grit are mixed with the required quantity of bitumen at the desired temperature and spread on the surface. This machine is used in the case of premixed carpet and bituminous concrete roads.

4. Gritting Machine. *Gritting machine* is used for broadcasting grit of the required size uniformly over the surface. The machine is fitted with a sieve of the required size through which the grit passes and is spread on the prepared surface in uniform thickness. These machines are seldom used in India.

5. Aggregate Heater or Drier. *Aggregate heater* is an ordinary sheet metal pan in which the aggregate is heated to drive off moisture. The aggregates can also be heated in revolving drum.

10.12 CONCRETE MIXING MACHINERY

For the construction of cement concrete roads, the following machinery is generally used:

(1) Concrete mixers

(2) Concrete pavers

(3) Concrete finishers

(4) Concrete vibrators

10.12.1 Concrete Mixers

The ingredients of concrete such as fine and coarse aggregates and cement are mixed in concrete mixers along with the requisite quantity of water. The mixers can be stationary or portable, tilting or non-tilting type. The tilting type portable mixers are generally used. The tilting mixers usually have a conical or bowl-shaped drum,which is rotated either electrically or manually. They are charged by means of a feed hopper. Mixers are of various sizes which is indicated by a number equal to its rated capacity in cubic metre of mixed concrete.

10.12.2 Concrete Pavers

Concrete pavers are used when the mixing is done on the site. The concrete pavers do the mixing and laying of concrete on the sub-grade. They consists of a bucket and boom arrangement which carry the concrete and discharge it on the required position. Concrete pavers are rarely used in India.

10.12.3 Concrete Finishers

After spreading the concrete in a uniform layer, the finisher does the job of finishing. The finishers have two screeds, which work in transverse direction like a saw. The machine moves on the slides of form work on its wheels. The front screed acts as a strike off bar and brings the surface to its final level. The rear screed gives the exact shape contour in the finished pavement.

10.12.4 Concrete Vibrators

Vibrators are used to compact and consolidate the concrete. There are three types of vibrators which are commonly used for different concrete works.

(a) Surface Vibrators

(b) Internal Vibrators

(c) Poker or Needle Vibrators

REVIEW QUESTIONS

10.1. Name the various types of road machinery for the construction of a W.B.M. road.

10.2. How many types of dozers are there ? What are their main functions ? Describe in brief.

10.3. You are the incharge of a big road project. List the various types of road machinery you will require for the completion of the various types of works involved in the construction of the road.

10.4. What type of road machinery is generally used for the construction of a cement concrete road ? Describe in brief.

10.5. Describe the various types of road rollers. Enumerate the special features of each of these rollers.

10.6. Write short notes on :

(i) Ripper

(ii) Bull dozer

(iii) Scraper

(iv) Concrete mixer

(v) Bitumen sprayer

10.7. What do you understand by rolling ? How it is done ? What are the main purposes of rolling ?

10.8. Write short note on the output of the road rollers.

Road Arboriculture

GENERAL

The highway engineer with his inherent ingenuity and patience can achieve wonderful results with only small cost to beautify the road. Road arboriculture is one of the architectural effects, which adds to the general or overall appearance of the road. *Arboriculture* means tree culture, that is, care and planting of trees. In this small chapter, some thought will be given to the planting and care of trees along the road-side.

11.1 OBJECTS OF ROAD ARBORICULTURE

The following are some of the objects of planting trees along the road-side :

(1) Trees merge with the natural landscape of the surrounding area and hence provide an attractive appearance to the road user.

(2) They provide shade to the traffic.

(3) By planting useful timber trees and fruit bearing trees, it becomes a source of supply of fruit and timber.

(4) They break the monotony of the road, specially in country roads or in un-built-up areas.

(5) Trees also act as wind breakers and reduce the temperature variations on the surface of the road.

11.2 SELECTION OF TREES

The selection of trees mainly depends upon the nature of the soil and the climatic conditions of the locality. However the following should be the characteristics of road-side trees :

(1) Trees should have long life.

(2) They should yield either fruit or timber.

(3) They should not be very clumsy and should grow centrally so that they should not cover the roadway.

(4) They should be dense and should yield shade all the year long.

Table 11.1 shows the type of trees suitable for a particular type of soil.

TABLE 11.1

S.No.	*Type of soil*	*Trees which can be planted*
1.	Loamy soil	Shisham, mango, jamun, kathal, imli etc.
2.	Clayey soil	Gold mohar, jamun, mango, malina etc.
3.	Sandy soil	Shisham, safedsin, arroo, kanjee etc.
4.	Usar	Neem, imli, babool, dhak etc.

11.3 PLANTING OPERATIONS

The tree planting along the road-side involves the following operations :

1. Excavation of pits
2. Preparation of seedlings
3. Transplanting
4. Watering and care of trees

11.3.1 Excavation of Pits

Pits are excavated of 1.2 m × 1.2 m × 1.2 m size or less and the excavated earth should be dumped along the edges of the pit in the shape of a bund. Good soil mixed with some manure is filled into the pit upto a depth of 15 cm below the ground level and the soil is allowed to settle for some time.

11.3.2 Preparation of Seedlings

Seedlings are prepared in a nursery, located in a nearby place. (A nursery is a place where plants are grown in a most favourable condition.) A nursery should be as near the road line as possible so that during transplantation the plants may not be injured.

11.3.3 Transplantation

When the seedlings are grown to a sufficient height, they are first transferred to pots and kept there till they are nearly 1.15 m high and then transferred to pits. Before the plants are received in pits, the soil in the pits is made loose and mixed with some measure. The best season for transplantation is when the rainy season has commenced.

11.3.4 Watering and Care of Trees

After the transplantation, the plants require regular attention and care including watering for about 3 to 5 years. The plants should be protected by tree guards made of steel, wood, earth, or manure. They are also to be protected against the decaying agents such as fungus, white ants etc. by spraying D.D.T. or tobacco water. The plants should also be watered at regular intervals.

11.4 LOCATION OF TREES

Planting of trees along the road-side can be either on one side or on both sides. The tree should be planted in a position where it does not interfere with the traffic or damage the road surface. Generally trees are planted on both sides of the road. The minimum distance from the edge of the road and line of trees should be 2 to 3 m. The distance between two trees depends on the

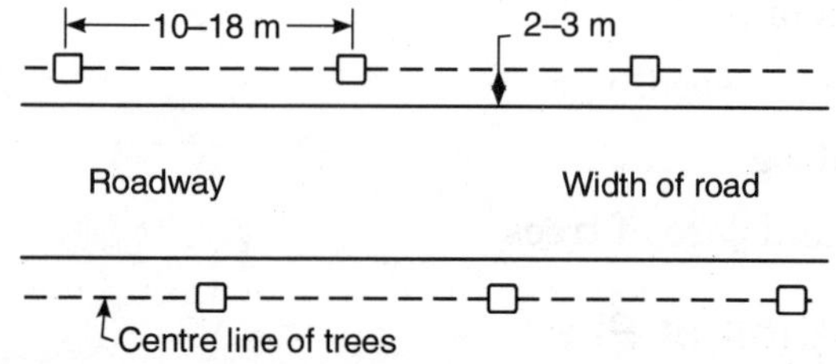

Figure 11.1

nature of the trees and it varies from 10 m to 18 m. The trees should be planted in a staggered way, that is, a tree in one row should be in between the two trees in the other row, on the other side of the road as shown in Fig. 11.1.

When the trees have grown up, care should be taken to see that the branches of the trees should not interfere with traffic otherwise all branches of the trees within 4.5 m height from the road surface should be trimmed

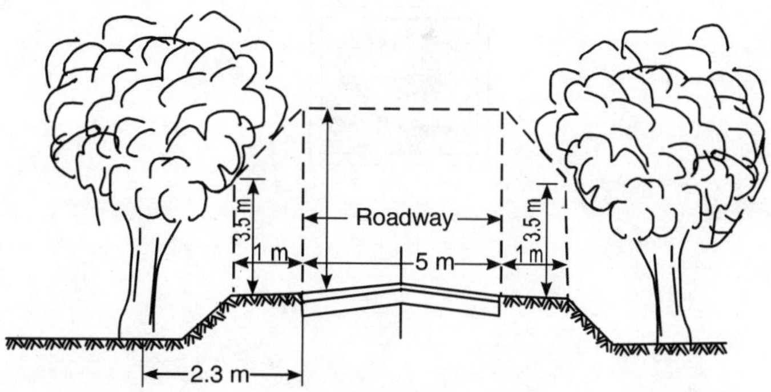

Figure 11.2

off. It should also be seen that branches of the trees do not interfere with the telegraph and telephone wires.

REVIEW QUESTIONS

11.1. What do you understand by the term road arboriculture ? Why are trees planted along the roadside and what are their advantages ?

11.2. What type of trees would you recommend in loamy soil and clayey soil ? Draw a typical cross-section and a plan of road showing position of trees.

11.3. Discuss the factors that influence the selection of roadside trees.

[Hint. In India, rainfall, temperature, topography, soil etc. are different from one place to another. Trees which grow well in one place may not even grow in other type of soil. Hence according to the nature of the soil and the climatic conditions of a particular locality, the trees are selected. (*See* Article 11.2)

12

Traffic Engineering

GENERAL

Traffic engineering may be defined as that phase of engineering which deals with the planning and geometric design of roads and adjoining land. It also deals with the traffic problems which are necessary for the safe, convenient, rapid and economic transport of men and material from one place to another.

Road is meant for safe transport of traffic. The road engineer has to devise certain methods which may ensure safe and free use of road by the traffic. Traffic engineering includes the analysis of traffic characteristics, the planning of regularity measures, the design and control application devices and the geometric design and junctional planning of routes, terminals etc.

12.1 ACCIDENTS

Accidents are the results of collision of one highway user with the other highway user. The accidents cause death, serious injuries and damage to vehicle property. A traffic engineer is not concerned with the legal cause of accidents but to prevent and avoid such accidents, he has to give some thought to the geometric design of pavements in order to educate and provide such facilities to the traffic by suitable signals and road regulations, by means of which it may be possible to reduce or minimize the chances of fatal accidents.

12.1.1 Causes of Accidents

The main causes of accidents on the road are as following:

1. Want of segregation between fast and slow moving traffic.
2. Lack of sufficient sight distance at the corners and junctions.
3. Road junctions not controlled by proper road signals.
4. Constant speed variations of vehicle due to congestion and the traffic jam.
5. Insufficient *weaving length,* between the intersection of roads.

12.1.2 Methods of Prevention of Accidents

To prevent or minimise accidents, the road engineer, the road user and the legal authorities will have to go hand in hand. The road engineer will design the pavement in such a way so as to minimise all possible chances of accidents by providing the following facilities to pedestrians and vehicular traffic :

1. Providing traffic signal for pedestrians.
2. Installation of pedestrian islands.
3. Improving street lights.
4. Construction of under and over bridges where essential.
5. Proper design and marking of cross walks.
6. Efficient layout of crossings, junctions and corners.
7. Providing proper cross-section of road.
8. Providing suitable super-elevation and transition curves at all curves.
9. Provision of non-skid and uniform surface.
10. Providing traffic segregation where possible.

The road users should realize their responsibilities towards each other. A road user must realize the large difference in speed between the pedestrians and vehicular traffic especially pneumatic traffic. A driver and a pedestrian must realize the distance required by the driver to react and apply brakes to stop his vehicle, especially under unfavourable conditions such as slippery road surface.

The legal authorities should get the laws of road enforced. It is the duty of the traffic police to guide and force the traffic to obey the rules of the road, keeping in view the facilities of the road users.

12.2 ROAD JUNCTIONS

Road junctions are places where two or more roads meet or cross each other at different angles. Safety of vehicular traffic and pedestrians is very essential at such places. They must be well-planned and properly signalled. Road junctions are the places where accidents generally occur, unless proper precautions are taken in the design or layout. As the traffic has to take different routes, it has to cross each other, and so proper segregation and control of the traffic is essential.

The following principles should be kept in mind while designing and planning the road junctions :

1. The intersections must, as far as possible, be at right angles and acute-angled crossing should be avoided.

2. The main road traffic should be independent of the traffic from branch roads and for this purpose, if possible, either an over-bridge or under-bridge should be provided.

3. Sufficient space should be provided for proper visibility as also for any future extension and improvement.

4. Proper provision should be made for pedestrians crossing the road.

5. As far as possible, change of gradients at the junctions should be avoided.

6. The camber usually allowed is avoided and as far as possible the whole area of the road crossings should be in one level.

7. *Round-abouts* should be provided especially when more than three roads cross.

8. While designing the crossings consideration of the minimum turning radii at junctions as given in Table 12.1 should also be kept in view.

TABLE 12.1 *Minimum turning radii at junctions (Turning speed 30 km/hour)*

S.No.	*Angle of the road junction*	*Minimum radius of curve for cars uplo 6.1 m length*	*Minimum radius of curve for truck/bus*
1	90°	9.1 m to 10.7 m	15.2 m to 16.8 m
2	105°	10.7 m	18.3 m
3	120°	13.7 m	21.3 m
4	135°	18.3 m	27.4 m
5	150°	36.6 m	45.7 m

12.2.1 Types of Road Junctions

There are many different types of road crossings having their own advantages and disadvantages and are planned as per the local requirements. The following are the important types of junctions :

1. Square Junctions. These are junctions, where two roads of equal importance or two main roads of nearly the same width, cross each other at right angles. Square junctions are of the following types :

(a) All paved type
(b) Circular round-about type
(c) Separate turning lane type
(d) Turbine type

(e) Elliptical round-about type

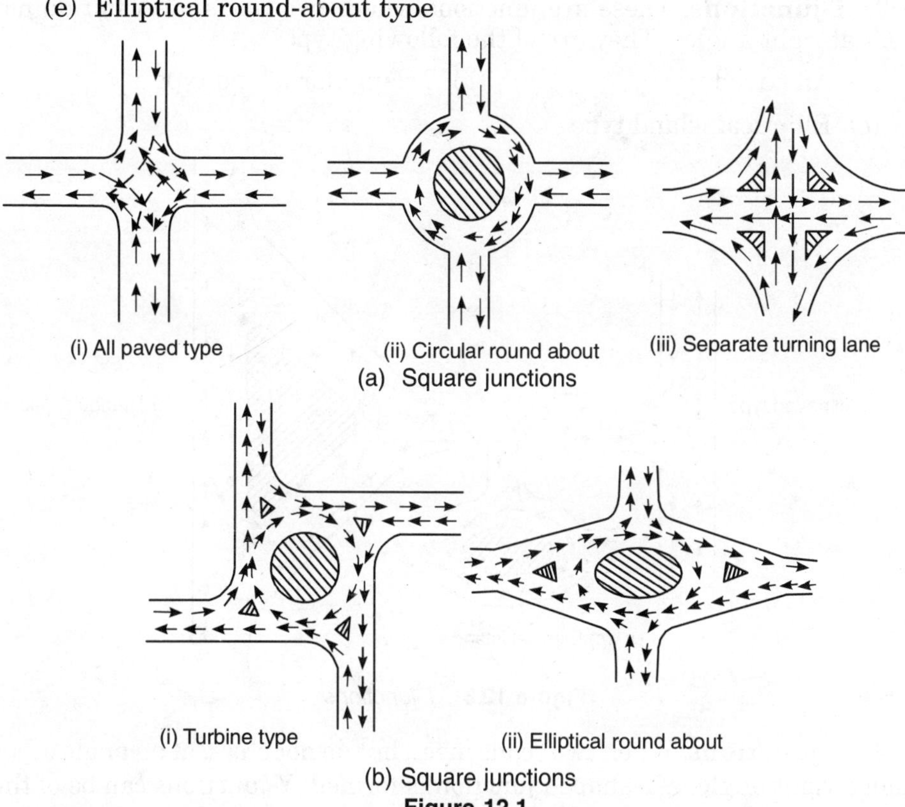

Figure 12.1

2. Acute Junctions. These are junctions where two main roads meet or cross each other at an angle other than right angle. The following are the different types of acute crossings or junctions :

(a) All paved type (b) Island type, and (c) Slip round type

(a) All paved type

(b) Island type

(c) Slip round type

Figure 12.2 *Acute crossings*

3. T-junctions. These are junctions where two roads meet but do not cross at right angles. They are of the following types :

(a) All paved type (b) Triangular island type

(c) Elliptical island type

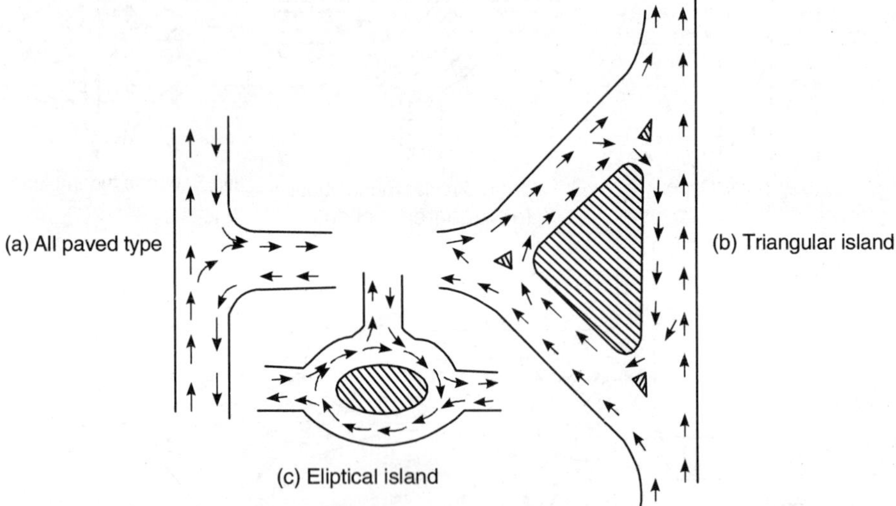

Figure 12.3 *T-junctions*

4. Y-junctions. When two roads meet, but do not cross, at an angle other than a right angle, a Y-shaped junction is formed. Y-junctions can be of the following types :

(a) All paved type (b) Triangular island type

(c) Circular island type (d) Channelised Y

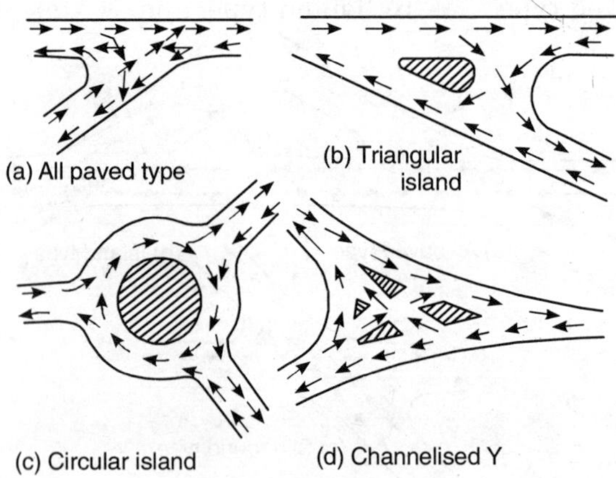

Figure 12.4 *Y-junctions*

5. Staggered Junctions. In this type of junctions the branch roads meet the main road at a good distance apart and at right angles.

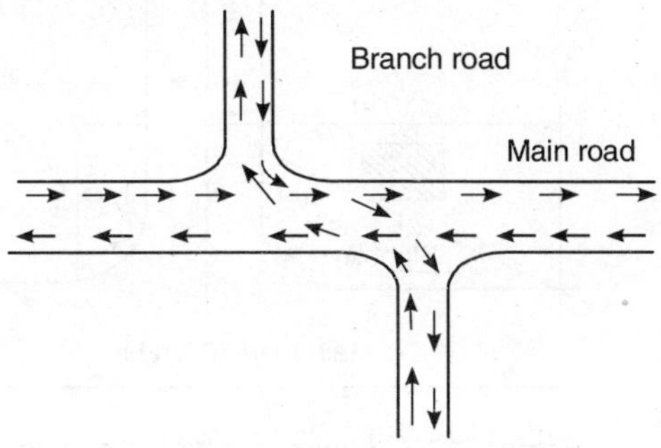

Figure 12.5 *Staggered junctions*

6. Multiple Junctions. This is a junction where more than two roads meet. This type of junctions are highly dangerous and special precautions should be taken in their planning and layout.

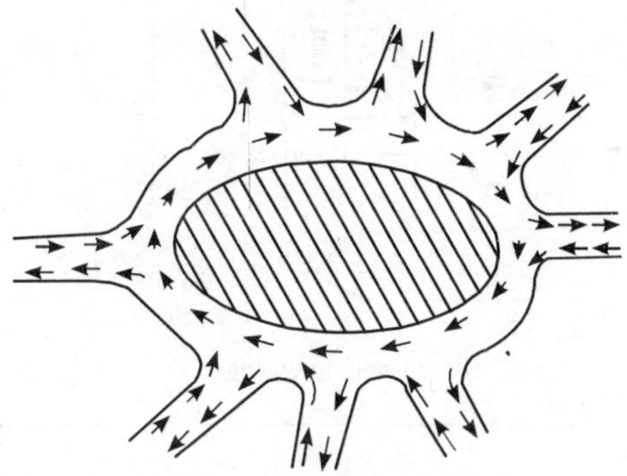

Figure 12.6 *Multiple junctions*

12.3 VISIBILITY OF CORNERS, BENDS AND JUNCTIONS

In the interest of road safety it is of primary importance that vehicle drivers should have full view of the approaching vehicles at all corners, bends and junction. Figure 12.7 shows the minimum sight distance which is required in the towns and the cities. Appropriate traffic signs are provided in the towns and the driver have sufficient warning of the presence of the junctions, corners or bends. At such places the speed of the vehicle is normally below 30 km/hour.

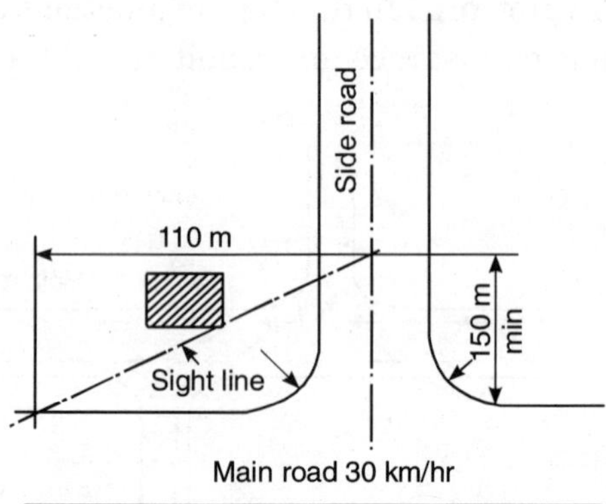

Figure 12.7 *Minimum sight distance*

Figure 12.8 shows the method of providing sight distance at corner by cutting a corner of the building by 4.6 m on each side. The figure of 4.6 m is taken by

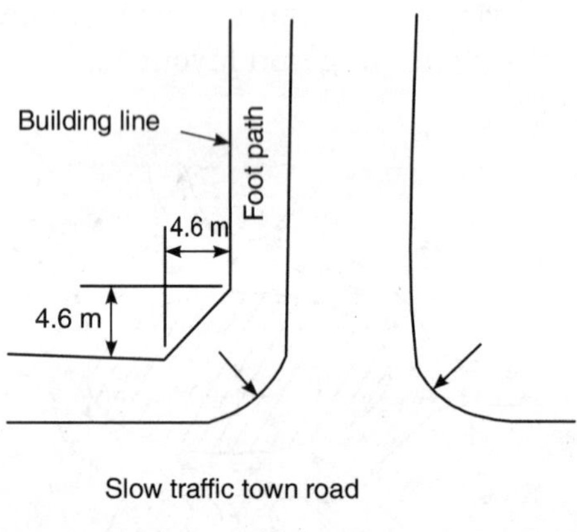

Figure 12.8 *Method of providing sight distance at corner by cutting corner of a building*

assuming that the width of the footpath on both sides of the road is 4.6 m. In the town, adjustments and changes in this 4.6 m can be made according to the need of the crossings.

Road crossings are mainly classified under two groups, depending on the importance of the road as follows:

Group 1. Priority intersections, where the minor road meet with the major road. At such a crossing the driver moving on the major road has virtual precedence over the other driver moving on the minor road.

Group 2. Uncontrolled intersection, or the crossings where both the roads are of the same category or equal importance.

Table 12.2 gives the minimum visibility distance along major roads at priority intersection for Rural Roads as per I.R.C.

TABLE 12.2

Design speed of major road in km/hour	*50 km/hr*	*65 km/hr*	*90 km/hr*	*100 km/hr*
Minimum visibility distance along major roads at priority intersection for rural roads	110 m	145 m	190 m	220 m

Minimum visibility distance of 15.0 m along the minor road is taken. On this basis the sight triangle at priority intersections should be formed by measuring 15.0 m along the minor road, and the distance given in Table 12.1 along the major road.

The minimum turning radius R of a vehicle is the radius of the sharpest curve that can be traced by its outer front wheel as shown in Fig. 12.9. On the

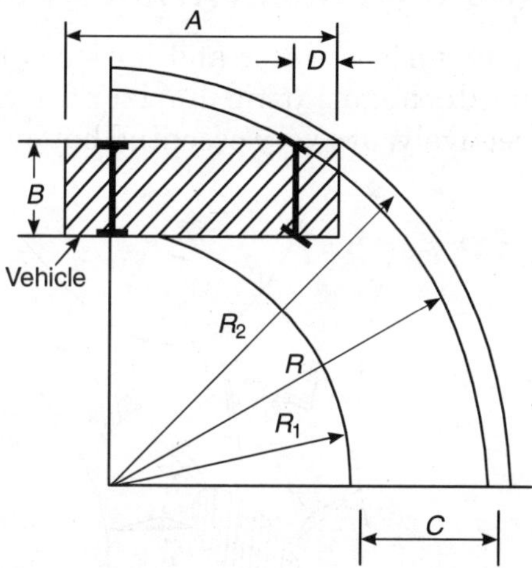

Figure 12.9 *Minimum turning radii of a vehicle*

curves when a vehicle is turning its rear wheels do not follow the tracks of the front wheel. The transverse position of the road kerb is usually determined by the radius R_1 of the inner rear wheels of the vehicle as given in Fig. 12.9.

In congested town areas which are used by light four wheels vehicles, a curve radius of 4.6 m should be absolute minimum radius having 4.6 m road width. The curve radius for buses, trucks, tractors and trailers should be 17.0 m minimum. When the turning crossing is more than 90°, greater radius will be required.

Table 12.3 gives the minimum turning radii for different classes of vehicles with reference to Fig. 12.9.

TABLE 12.3 *Minimum turning radii for different classes of vehicles (Fig. 12.9)*

Type of vehicle	Vehicle dimensions		Width of the path C	Vehicle overhang D	Minimum desirable radius			Minimum right angle turn radius		
	Length A	Width B			R	R_1	R_2	R	R_1	R_2
	m	m	m	m	m	m	m	m	m	m
Car	6.1	2.1	2.6	0.40	11.7	9.1	12.3	8.5	5.9	8.9
Bus or Truck	10.67	2.44	3.9	0.61	18.7	16.8	19.3	13.7	11.4	14.3

12.4 CAMBER AND SUPERELEVATION AT JUNCTION

For proper drainage of surface water and for smooth running of the fast moving vehicles while negotiating horizontal curves, camber and superelevation, respectively, are very essential but on junctions they create

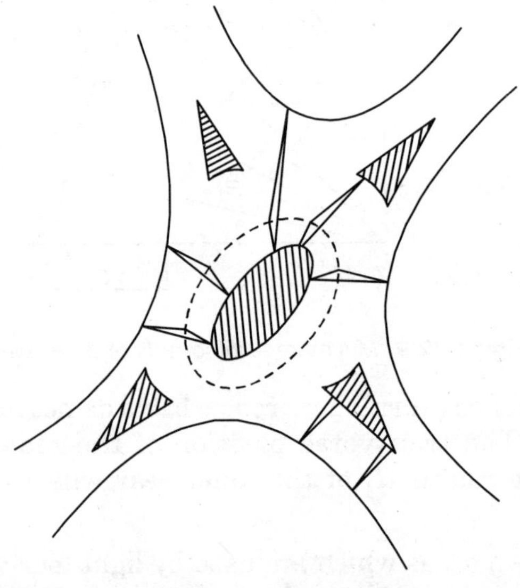

Figure 12.10 *Superelevation and Camber at junctions*

problems in their fixing. The crown of the roads should be adjusted in such

a way that if line on the point of intersection of the paths of the vehicle leaving and entering the junction or rotary.

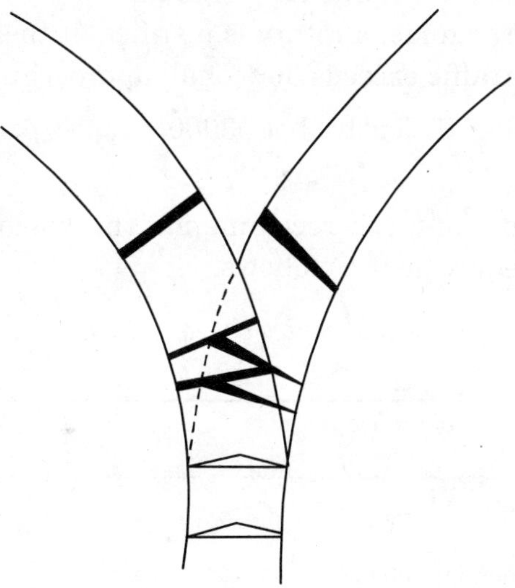

Figure 12.11 *Negative superelevation*

At Y-junctions where the two curved roads are meeting, both the roads will have superelevation or cant (as it is called in railways) the outer edge of one road will become the inner edge again the junction road and similarly inner edge of the other road will become the outer edge of the junction road. The edges being at different grade will pore a problem for which on one of the roads (whichever is unimportant) a *negative superelevation* is provided so as to bring the edges of the two roads at the same level.

12.5 TRAFFIC ISLANDS

Traffic islands or rotary is an enlarged intersection where all converging vehicles are compelled to move round an island in one direction before they can weave out of the road, into their respective direction radiating from the island. They also act as shelter for pedestrians, while crossing the road. The efficiency of a rotary depends upon the diameter of the circle. The traffic circles should be free from all obstructions, as the drivers should be able to see ahead and take necessary precautions. They are marked with special colour lights, kerbs of special types and painted with broad strips of blacks and white, so that they may be conspicuous even at night.

Roundabouts are usually provided where the traffic density exceeds 500 vehicles (cars) per hour on all the intersecting roads. Some engineers recommend that rotaries should be provided at the crossings where right hand turn traffic is more than 50% or where the straight cross traffic is not more than 30% of the total traffic having density of more than 400 vehicles per hour. As a general guide, a rotary is justified at the crossing where the right hand turning traffic exceeds 30% of all approaching traffic.

A rotary can handle efficiently about 3000 vehicles per hour under Indian traffic conditions.

The Indian Road Congress recommends the various vehicles to the passenger car unit equivalent as follows.

TABLE 12.4 *Passenger car unit equivalents of various vehicles (I.R.C)*

S.No.	*Description of vehicle*	*Passenger car unit equivalents*
1.	Cars, Jeeps	1.0
2.	Light commercial vehicles	1.0
3.	Two-wheelers motor-cycles and scooters	0.75
4.	Pedal cycles	0.5
5.	Animal drawn vehicles	4 to 6
6.	Buses	2.9
7.	Trucks	2.8

12.6 SHAPES OF TRAFFIC ISLANDS

The islands may be circular, elliptical, rhombus, turbine shape, or tangent shape.

1. Circular Island. This is the best type of island and is suited where the two roads intersect ; which are of equal importance and width. As it has the shortest perimeter for a given area, the extra distance which the traffic has to travel is the minimum. It avoids the turns of short radius which hinder traffic. (See Fig. 12.1 b).

2. Elliptical Island. It is used where the traffic on one road is more than that on the other. The elongation is in the direction of the greater flow, to facilitate thorough traffic. (See Fig. 12.1 e).

3. Rhombus Island. This is similar to elliptical island but instead of providing an ellipse, a rhombus type island is constructed on the same basis.

4. Turbine Island. In this type, the traffic is forced to slow down, when it enters the rotary, as it has to take a left hand turn and while leaving the rotary, it has a tangential exit. (See Fig. 12.1 d).

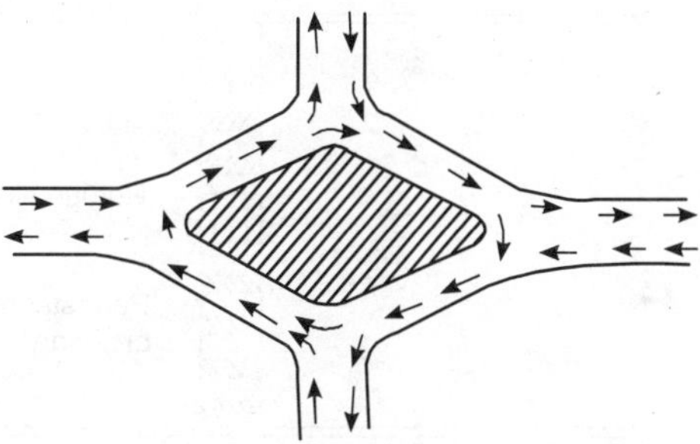

Figure 12.12 *Diamond or rhombus island*

5. Tangent Island. In this case more than two roads enter a rotary and leave it tangentially. In this type of island the thorough traffic is not obstructed and moves with a high speed. Such type of islands are very dangerous as there are chances of accidents.

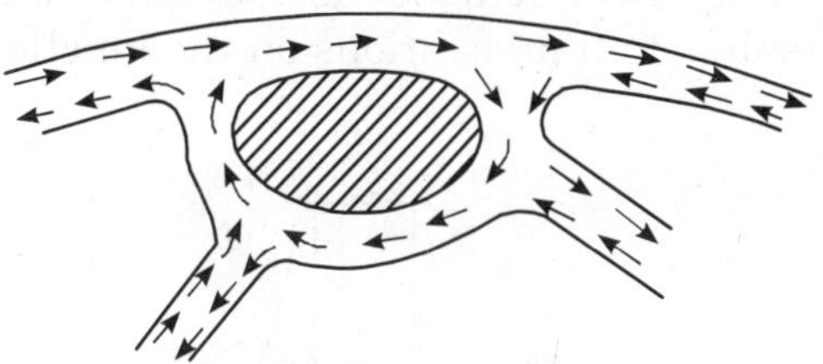

Figure 12.13 *Tangent island*

12.7 REFUGE ISLANDS

Refuge islands are safety islands and serve the dual purpose of affording protection to the pedestrians and segregating the traffic into its proper channels. These refuge islands are constructed in the centre of the wide roads, carrying a heavy volume of traffic. They form sanctuaries of pedestrians who are caught between the opposing lines of traffic and are a special boon for the aged and infants.

Refuge islands are 15 to 20 cm above the road level and are in the form of two triangular strips with a square one in between. The width of the island depends upon the width of carriageway.

The following are the advantages of refuge islands :

1. It is a protection to the pedestrians.
2. It acts as a guidance to the drivers in negotiating on junctions.

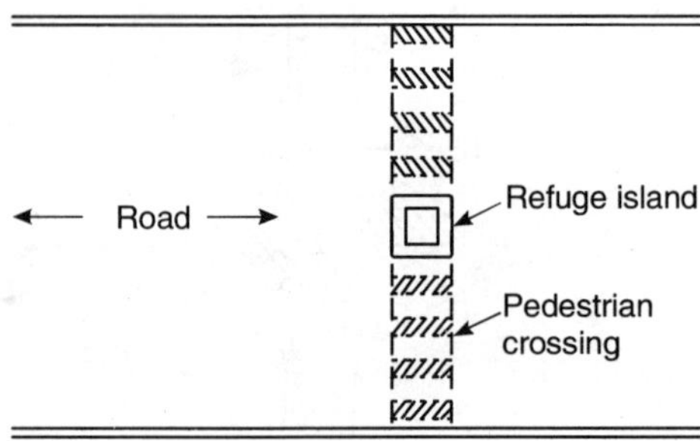

Figure 12.14 *Refuge island*

3. It increases the efficiency of traffic.
4. It segregates the traffic into proper channels.

12.8 FLYOVER JUNCTION

It has been observed that about 30–35% of the road accidents occur on junctions only. Inspite of all precautions on the junctions, mixing of traffic

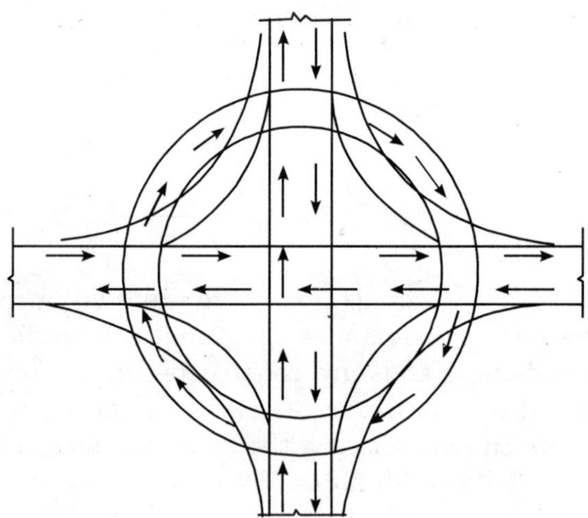

Figure 12.15 *Clove leaf junction*

coming from different directions is unavailable but can be considerably reduced by separating a changing the grads of the cross roads. Such type of junctions in which the two cross roads meet at different levels are known as

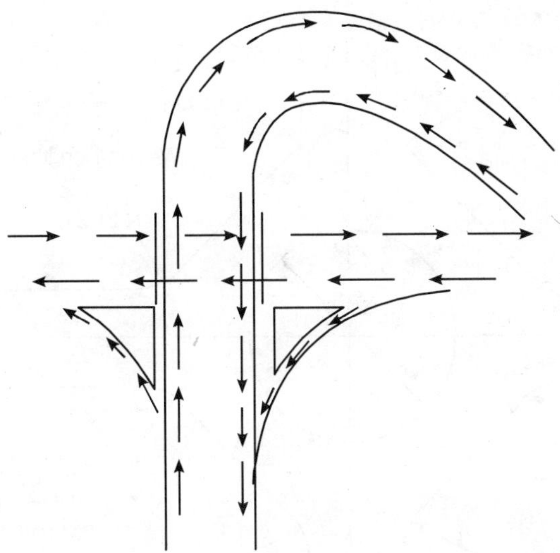

Figure 12.16 *Trumpet flyover*

flyover. The most important type of flyover or flying junctions are Clover Leaf Junction trumpet type fly-over.

12.9 DESIGN OF TRAFFIC ISLANDS

The common shapes of the traffic islands are circular or elliptical (elongated). The choice of shape mainly depends upon the site conditions, angle of intersections of the roads and the traffic density.

The circular shape is most common and economical where roads carry nearly equal volumes of traffic and the roads intersect at nearly equal angles. The elliptical or elongated shape is adopted at intersections where the cross traffic is small. The elongation of such islands should be kept in the direction of the greater flow and it should be so designed that the radius of curvature at the sharpest point is not less than 15.2 metres.

The exact shape of the island is determined by connecting the sections of the road round the central island to form a closed figure giving at least the minimum weaving length between the adjacent radial roads.

At the intersection of two or more roads the basis of the design of the rotary should be on the highest design speed of any of the highway, irrespective of the greatest volume of the traffic it carries.

Practically it has been seen that a vehicle can travel at 25 km/hr speed round an island of 30.0 m diameter and at a speed of 40 km/hr round an island of 46 m diameter. As far as possible the diameter of the central island should not be less than 30.0 m.

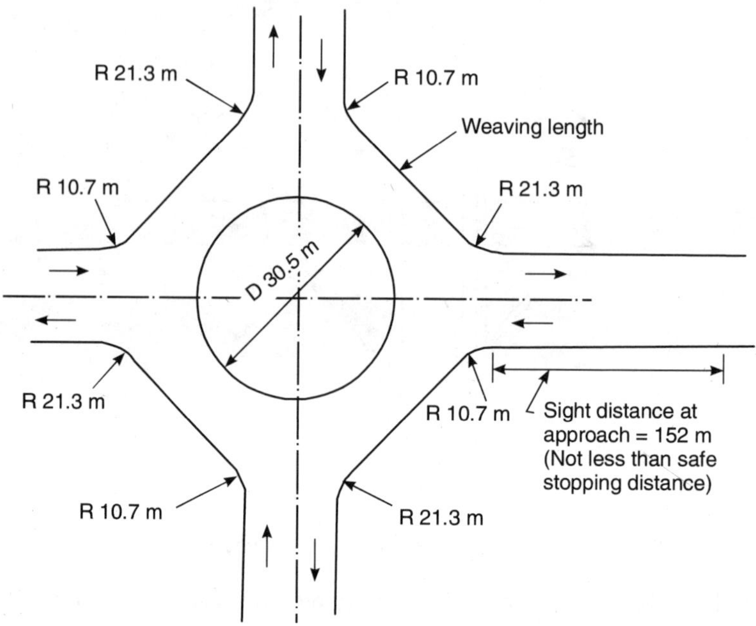

Figure 12.17 *Round about*

Indian Roads Congress has recommended a radius of 25 m to 35 m for design speed of 40 km/hr and 15 m to 25 m for design speed of 30 km/hr for radius of entry of rotary. Similarly the radius of the exit curve has been recommended to be 1.5 to 2.0 times that of the entry radius.

Indian Road Congress has recommended that the radius of the central island should be 1.33 times that of entry radius. The radius should be as large as possible and practicable.

Table 12.5 gives the width of that carriage way at the entry and the exit and radius of curve at the intersections having rotary.

TABLE 12.5 *Width of carriageway at entry and exit* (*Based on IRC 65–1976*)

S. No.	*Width of the approach road or carriageway*	*Curve radius at entry*	*Width of carriageway and entry and exit to rotary*	
1	7.0 m (Two Lane)	25 m – 35 m	6.5 m	Inclusive of widening required on account of curvature
2	10.5 (Three Lane)	25 m – 35 m	7.0 m	
3	7.0 m (Two Lane)	15 m – 25 m	7.0 m	
4	10.5 m (Three Lane)	15 m – 25 m	7.5 m	

Indian Road Congress has also recommended that minimum width of 5.0 m with necessary widening to account for the curvature of the road should be provided.

Width of the carriageway around the island. This should not be less than one-fourth the total width of all the radial roads and not less than one-half the width of the widest roads plus width of one lane, whichever is more.

Weaving length and width. The minimum weaving length of 4.5 m for 40 km/hr speed and 30 m for 30 km/hr designed speed should be provided. The maximum weaving length should not be more than twice the above lengths. The width of the weaving section of the rotary should be one traffic lane of 3.5 m wider than the mean entry width thereto.

Width of the carriageway around island. This should not be less than the weaving width. It should be equal to the widest single entry road into the rotary.

12.10 PEDESTRIAN CROSSINGS

The pedestrian accidents are more frequent and fatal and hence special attention is to be given to the pedestrian crossings. The pedestrian crossing may be at the same level, above or below the road level. Crossings at the road level are cheapest and are mostly used. The safety of the pedestrians depends upon the pedestrians themselves and the drivers. Strict supervision at such places is very essential and the pedestrians should be allowed to cross the road when the traffic is held up temporarily.

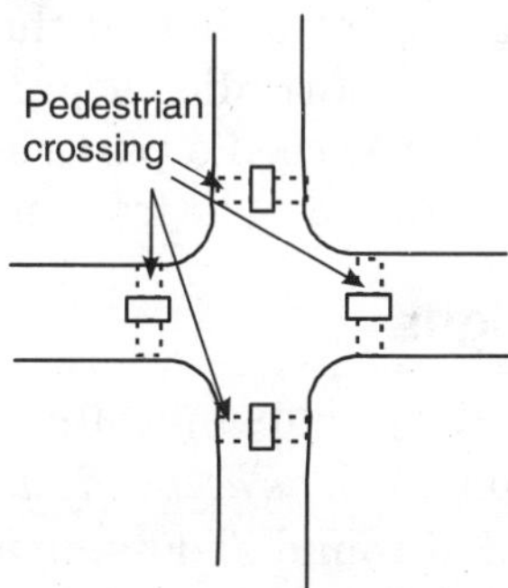

Figure 12.18 *Pedestrian crossing*

The crossing is marked on the road by means of rubber or steel studs embedded in the road. The studs should be flush with the road surface or at the most of 5 mm projecting out from the road surface and are placed at a distance of nearly 35 cm centre to centre.

The pedestrian crossing should be located just near the intersection of roads where the vehicle reduces the speed (generally). The minimum width of crossing should be 2.5 m. The crossing should be clearly visible from pedestrian crossing a distance to the drivers of the fast moving vehicle.

12.11 ROAD SIGNS

Road signs are used to prevent accidents. Properly designed road or traffic signs placed at appropriate positions are a must for safe and efficient movements of traffic. They should be so located as to be clearly visible from a distance. They have been standardised by the Indian Road Congress and are divided into four main groups :

(1) Warning signs
(2) Prohibitory signs
(3) Mandatory signs
(4) Informatory signs

12.11.1 Warning Signs

Warning or cautionary singns indicate to traffic by suitable standardised symbols the approach of a place where some precaution is required for the safety of the traffic. The warning or cautionary signs are provided at a distance of 100 m on an up grade 110 m on a level road and 130 m on a down grade, from the specified danger point. These signs are meant for hair pin bends, level crossing, steep hill, round-about, school, cross road, T-junctions, narrow road bridge etc.

12.11.2 Prohibitory Signs

Prohibitory signs indicate to the traffic that the use of a particular road is prohibited by a specified class of traffic, or prohibits the use of horns in a particular area of the road, or exceed a particular speed limit as on sharp curves, or parking at a certain place, or from entering into a road.

12.11.3 Mandatory Signs

Mandatory signs or regulatory signs indicate to the traffic to comply with certain statutory regulations such as *keep left*, *turn right* etc. The mandatory signs are usually provided at round-abouts, flyovers etc.

12.11.4 Informatory Signs

Informatory signs convey certain information and guidance to the traffic such as road directions, parking places, direction to a town or city, along with distance etc.

The different road signs are shown in Figs. 12.19 (a), (b) and (c).

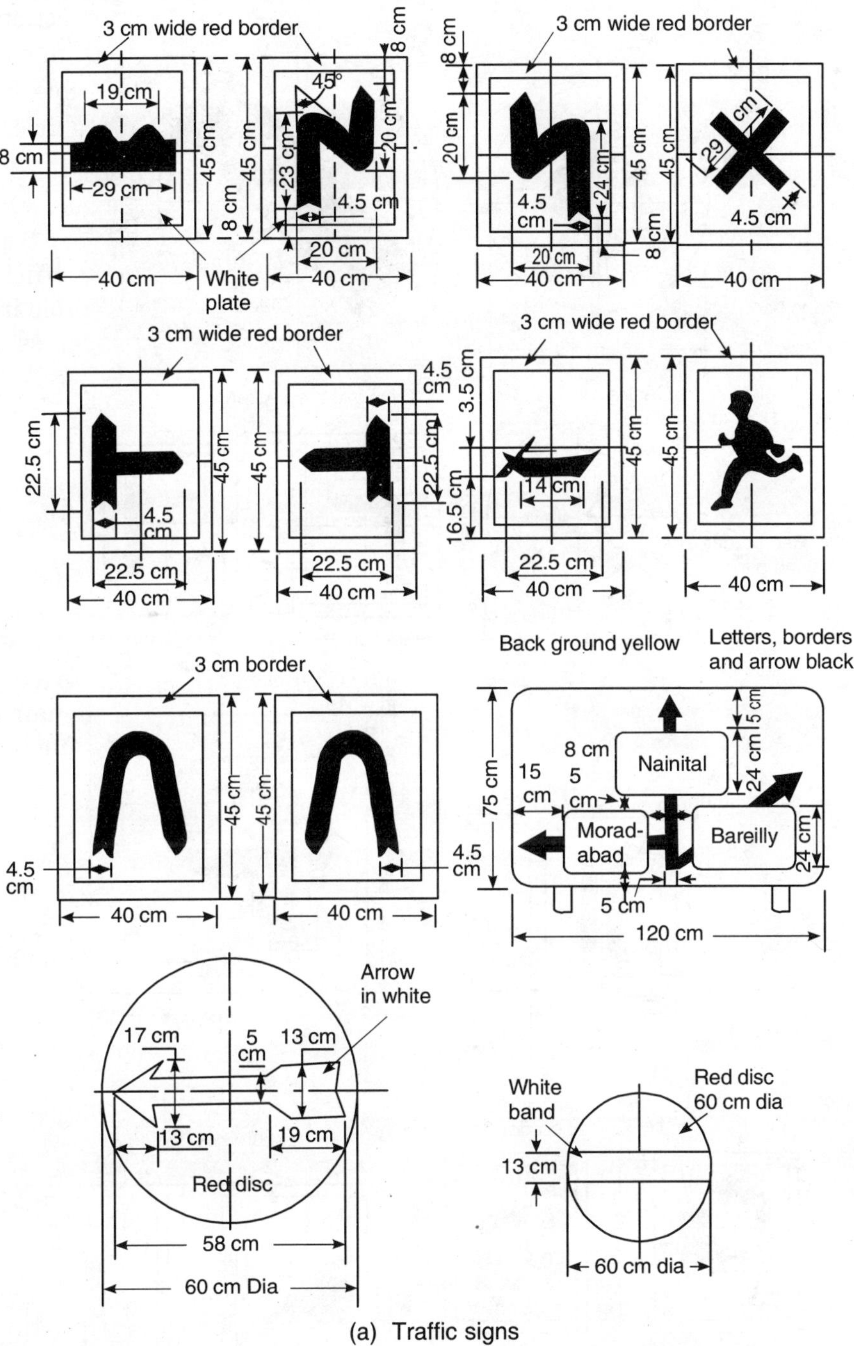

(a) Traffic signs

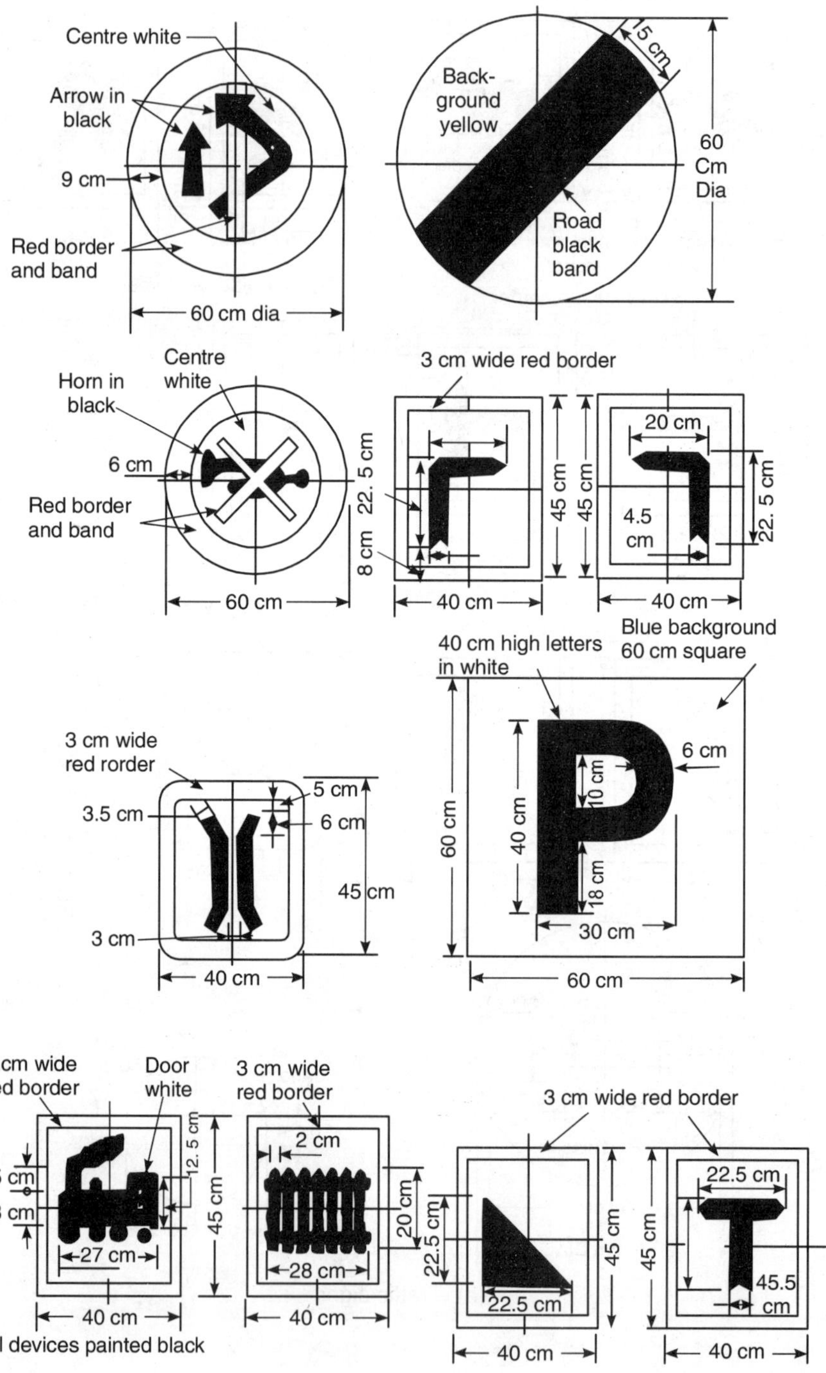

(b) Traffic signs

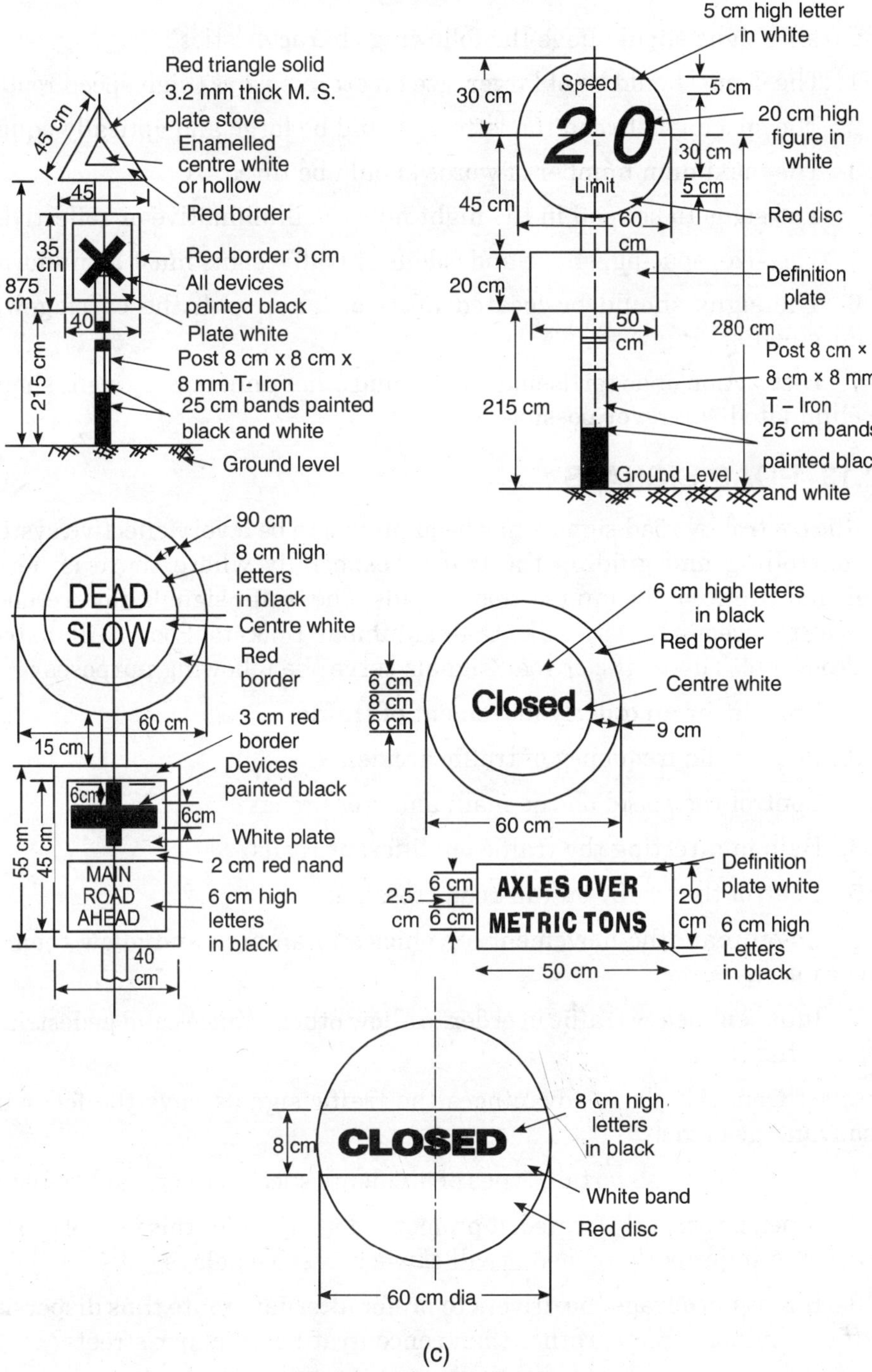

(c)

Figure 12.19 *Traffic signs*

12.12 Characteristics of Traffic Signs

The traffic signs should have the following characteristics :

1. The signs should be of larger size on *express ways* (high speed roads).
2. The spacing between the letters should be large and optically equal.
3. The maximum number of words should be three.
4. The signs to be read in the night must be illuminative or reflective.
5. The size, spacing, shape and colour of traffic signs must be uniform.
6. The signs should be located in accordance with the carriageway requirements.
7. Distraction or advertisement signs and other unnecessary signs should be eliminated wherever possible.

12.13 ROAD SIGNALS

Traffic control by road signals has been proved to be a very effective system of controlling and guiding the traffic, especially where there is heavy vehicular traffic on the intersection of roads. The traffic signals have recently been introduced in big cities like Delhi, Mumbai, Kolkata, Poona, Allahabad, Lucknow etc. The traffic or road signals serve the following purposes :

1. Provide for an orderly movement of traffic.
2. Reduce the frequency of traffic accidents.
3. Control the speed on the main and cross roads.
4. Help in directing the traffic on different roads.
5. Control the traffic on rail-road crossings.
6. Co-ordinate the movement of vehicles in an area and allow them to flow cautiously.
7. Intercept heavy traffic in order to allow other vehicles and pedestrians to cross the road.

Apart from the above advantages the traffic signals have the following disadvantages too :

1. It may increase certain types of accidents such as re-entrant collision.
2. When improperly located, it promotes disrespect for this type of control device and improperly turned it will cause excessive delays.
3. It may encourage the drivers to prefer alternate route thus dispersing traffic on minor streets, rather than concentrating on major streets.

12.14 HIGHWAY AND STREET LIGHTING

In the olden days, when there was not much of the fast moving vehicles, oil-lamps were used on the road-side. This practice is still in use in Indian villages. These oil lamps are only meant to show the path to the pedestrians and the bullock cart traffic. This was actually more a psychological than the actual benefit. With the advancement of fast moving traffic, hazards of travel for pedestrians and the automobiles have increased considerably due to dusk, darkness and impaired vision. In modern road development, efficient street lighting has become an important feature for the reduction of crime expedition of traffic.

12.14.1 Factors Affecting Street Lighting

For efficient and proper street lighting the following factors should be kept in mind:

(i) The electric posts poles should be located at proper places.

(ii) The light from the source should illuminate the entire road surface uniformly.

(iii)The posts should be spaced uniformly.

(iv) The light should not produce glare.

(v) The height of lamps should be adjusted in such a way that it should produce such a shadow which may not darken the entire line.

12.14.2 Design of Highway and Street Lighting

Lighting from one side of the road only usually is unsatisfactory except on bends and narrow roads. Due to economy single side lighting is usually adopted even for two lane roads. The maximum distance between two opposite rows of lights should not be more than 9.0 m. The roads upto 9.0 m width can be effectively lighted by lights mounted on the kerb-line in vertical line above the kerb.

When the width of the road is increased, brackets are used for overhanging the lights above the road. The maximum overhang of brackets over the kerb should be 1.8 m to avoid undue dark patches on the kerb and footways.

Following are the I.R.C. recommendations for horizontal clearance for lighting poles.

(i) For roads with raised kerbs (as in urban roads).	0.3 m minimum desirable 0.6 m from the edge of the raised kerb.

(ii) For roads without raised kerbs (as in rural roads).	1.5 m minimum from the edge of the carriageway, subject to a minimum of 4.0 m from the centre line of the carriageway.

The side mounted lamp posts can be arranged opposite to each other on both sides of the road or arranged staggered. The staggered arrangement is more efficient.

The usual mounting height of lamp post is 6.1 to 7.6 m in 4.6 m (minimum) wide streets. The mounting height is between 7.6 to 9.0 m in highways with 35 to 45 m centre to centre spacing. The ratio of spacing of poles to the mounting height is kept between 8 : 1 in congested areas and 10 : 1 in thickly populated areas. High mounting greatly reduces the blinding effect of direct glare.

The spacing of the light poles is reduced on the curves, bridges, level crossings etc. for better light requirements.

I.R.C. has specified the minimum vertical clearance required for electric power lines upto 650 volts as 6.0 m. All the poles carrying overhead power and telecommunication lines except the urban areas, should be erected at least 10.0 m away from the nearest edge of the roadway. The distance of this line should be 5.0 m from the nearest line of avenue trees.

If the trees are interfering with the side lamps, they should be centrally suspended. On the S-curves the lamp posts should be changed from one side of the road to the other at some point in the middle of the bend.

At all the road junction and roundabouts, it is most important for safety to ensure that sources of light are so placed that a driver cannot only see from some distance away that he is approaching a junction but he can also see the route, he has to follow. At the small roundabouts upto 18.0 m diameter, a single light centrally mounted at a height of 9 to 10.6 m above the carriageway will be suitable. For bigger roundabouts, the light should be provided along its circumstances.

At the mouth of the crossing, a bright patch must be provided. On the intersection of two roads, four light lamps at not more than 12.0 m along each road should be provided.

12.15 PARKING PLACES AND LAY-BYS

For parking the cars 23 sq. m for each car is required including allowance for access roads, irregularities in manoeuvring and for opening of doors. For parking of cars parallel to the road kerb 2.5 m wide strip is required. For parking the cars at right angle or diagonally to the kerb, 6.0 m wide strip is required.

Indian Road Congress has recommended the following minimum parking spaces :

(a) For cars :

(i) 3 m × 6 m when individual parking space is required.

(ii) 2.5 × 5 m when parking is done in lots for community parking.

(b) For trucks :

(i) 3.75 × 7.5 m parking space is required.

12.16 LAY-BYS

It is the space provided on all the important roads at suitable intervals to enable the vehicles to draw-off the road for temporary parking during break-downs and repairs. In the cities lay-bys are also provided for bus-stops. 30 m × 2.5 m space is usually sufficient for lay-bys.

REVIEW QUESTIONS

12.1. State the causes of road accidents. What remedies do you suggest for avoiding very frequent accidents ?

12.2. What are the main advantages of road signals ?

12.3. Why are traffic islands provided ? What are the different kinds of traffic islands, which are generally provided ?

12.4. How many types of road junctions are there ? Describe with sketches.

12.5. What are the different types of road signs ? Give example of each with appropriate sketches of each type ?

12.6. Write short note on the visibility at corners, bends and crossings.

12.7. Write short note on the design of highway and street lighting.

12.8. Write short note on the parking places and lay-bys.

13

Road Specifications

GENERAL

Engineering structures are constructed according to certain standards, which are called specifications. While preparing a project, detailed specifications or particulars for carrying out the work, are drawn, so that the work should be carried out according to these standards or specifications, as they are called. Detailed specifications are of great importance to the engineer and the contractor (or the person executing the work). Unless clear and concise specifications for a particular work are drawn up, it becomes difficult to get the things executed properly in the desired manner. The specifications should include the method of carrying out the work, precautions to be taken and the quantity of material to be used.

13.1 SPECIFICATIONS FOR THE CONSOLIDATION OF STONE BALLAST

1. Before spreading stone metals the old metalled surface shall be thoroughly cleaned of all dirt etc. The holes etc. should be the roughly filled. The whole of the old surface may be scarified if required and the surface may be brought true to template.

2. Two parallel mud walls 20 cm wide and 15 cm high, made of well primed earth or clay shall be formed along the outer edges of the metalled surface to confine and prevent the new metal from spreading under the pressure of roller. The mud wall shall be properly aligned with flags and the clear width between them shall be the exact width of the new metalling.

3. The stone metal shall then be spread evenly over the surface true to the template. The metal shall be carefully hand-packed, the bigger size metals shall be placed below the smaller size metals in the interstices of the bigger pieces. The templates shall be placed at a distance not exceeding 12 metres in any case, and shall be fixed truly horizontal so as to ensure that both sides of the roads are at desired level. No organic material whatsoever, shall be mixed or spread over before, during or after consolidation.

4. The metal shall then be rolled with a road roller, commencing at the edges and working towards the centre. The metal shall be dry rolled until no lines of the roller are left on the surface i.e., until the metal has been thoroughly compacted together. The surface shall then be sprinkled with water and re-rolled until thoroughly compacted and no marks are left on the surface by the roller or any chart which may pass over it.

After this, the blinding material be spread evenly over the whole surface, and be of such thickness as well, when watered and rolled, will only just fill all interstices. The blinding materials shall not contain clay or any organic matter. After the spreading of the blinding material, the surface shall be brushed slightly with *besom brushes,* backward and forward in front of the working roller and the rolling and brushing shall continue until the surface has become smooth and have been passed by the engineer incharge.

13.2 SPECIFICATIONS FOR LAYING SOLING

Soling shall always be provided except where the road is founded on a very hard natural soil such as rock etc. It shall consist of the hardest suitable material locally available or easily procurable. Stone boulders, kankar, or over-burnt bricks are suitable for the purpose.

The width of soling shall always be 30 cm more than the proposed width of carriageways. The thickness of soling should be as under :

(i) 15 cm if stone boulders are used.

(ii) 10 cm if over-burnt bricks are used.

(iii) 10 cm if kankar is used.

The size of soling material should not be less than 12 cm in any dimension if boulders are used, and 10 cm if kankar or overburnt bricks are used.

Before laying the soling coat, the sub-grade should be thoroughly levelled. Care shall be taken to see that the sub-grade is hard and well compacted and no soft patches or depressions are there while laying stone boulders or over-burnt bricks, and these too, should be properly hand packed so that no interstices are left. The soling shall then be dry rolled with a road roller, depressions etc. being filled up with shingle, gravel or kankar bajri, sufficiently to give a hard true surface, sloped to the correct camber. Earth shall on no account be used in filling depressions.

13.3 SPECIFICATIONS FOR SURFACE PAINTING

Following is the detailed specification for surface painting :

Surface painting shall never be applied on a water bound kankar surface or of soft stone. Old stone or brick surface shall duly be painted when in good condition and free from patches pot holes and ruts.

When consolidation is done with the intention of painting the surface, no blinding or binding coat shall be used.

All new surfaces which have not been previously painted, shall be painted in two coats. The first coat may be bitumen cut-backs or bitumen and should be applied hot.

The grit shall be either of broken stones or coarse river shingle, bajri. The hardest material available at a reasonable price shall be used, irrespective of its shape. For example, broken stones may be of angular shape whereas coarse river shingle, bajri may be round or of any shape. For the first coat the grit shall be screened on 2 cm mesh and all the particles must be retained on 6 mm mesh. The grit for the second coat or for repainting an old painted surface, should be screened on a 12 mm mesh and 100% retained on 6 mm sieve.

The period which must elapse between the completion of compaction and the commencement of painting depends on the material to be used for a time of one year and the nature of traffic. No painting shall be done in December and January. No hot painting shall be done during the rainy season. Similarly painting with cut-backs or emulsions, not to be done while it is raining or rain appears imminent.

In no case painting be delayed for more than one month from the date of completion of compaction or consolidation, and if it is anticipated that more than this period will elapse before the road is fit for hot paint, emulsion should be applied as soon as possible after the consolidation is complete.

13.3.1 First Coat with Emulsion

When the first coat is to be done with emulsion, either by choice or for the reasons stated above, this should be applied within ten days of completion of consolidation, during which time, the surface shall be protected with a blindage layer.

13.3.2 First Coat with Cut-back

The road must be reasonably dry but need not to be bone dry right through.

Two days dry weather, even during the monsoon will usually suffice to produce sufficiently dry conditions.

Before the application of paint, whether it is tar, cutback or emulsion, the road surface must be cleaned of all caked mud, crowdung, with stiff brooms or wire brushes. The surface is then swept clean with ordinary brushes. Finally and immediately prior to painting, it shall be blown free of all dust by means of gunny bags or if available with mechanical blowers. The dirt and dust removed from the surface shall be collected and deposited at a distant place from the road and away from the prevailing winds. Only that much area of the road shall be cleaned and broomed which can be painted in that day. The final cleaning by blowing shall not precede the application of paint by more than half an hour.

A hot paint shall be heated to a temperature not higher than the specified by the manufacturer and in boilers fitted with thermometers. Heating of the paint must be gradual and only that much quantity of tar or bitumen shall be heated which may be used for nearly 30 sq. m of the road surface. Boilers must be cleaned thoroughly every month.

The paint shall be applied to the road surface with specially made pouring canes with wide mouths and of known capacity. To obtain correct and even distribution of paint, the road surface shall be divided into rectangles of known area, each suitable for the contents of one pouring can and then the paint is poured longitudinally and brushed evenly over the surface with bass brooms. Brushing shall always be done from she sides towards the crown. Brooms must be cleaned at the end of the day's work. Nearly 2 kg of binder should be used per square metre of the road surface.

As soon as the paint has been spread, in the case of hot paint or as soon as the paint commences to break down, in the case of an emulsion or cut-back, the grit shall be broadcast on to the surface. Before doing so, the grit should be screened into two portions, one containing the larger pieces and the other smaller pieces. The larger grit shall be broadcast first, made into an even layer by hand and then the smaller grit spread over it. About 1.8 to 2 m^3 of grit will be required per 100 m^2 of road surface.

After the grit has been spread evenly, the surface shall be rolled with the lightest roller available. Rolling shall be continued only sufficiently to press the grit into the painting material and to fill the interstices of the metal. Excessive rolling crushes the grit and hence should be avoided. During the rolling coolies should follow the roller, fill the lean surface with extra grit and brush all the extra grit from those places which are above the general level. Close and even packing of grit is essential for a good result. The roller must be driven at the lowest speed and care must be taken not to start or stop the roller with a jerk. The roller shall never be permitted to stop on a part of the surface which has not been finished and should never stop at the same stop twice.

As a general rule, the *edges* of the painted road shall be protected by bricks or water bound macadam surface, at least 30 cm wide and 10 cm thick.

The second coat of paint shall be applied as soon as the first coat has reached its *optimum condition.* This will be done when the paint has hardened, the surface has become smooth, exhibits a mosaic and all loose grit has been absorbed. The period of time interval, which will elapse between the first coat and second coat of paint will depend on the volume of traffic and on the material used for painting. It is essential that tar should be allowed to harden before being repainting with bitumen as otherwise it remains as a soft layer under the relatively hard bitumen and the road surface becomes uneven. In majority of cases, the time interval between the first coat and the second coat will not be less than one month.

13.4 SPECIFICATIONS FOR CEMENT CONCRETE ROADS

A cement concrete road consists of a cement slab of a specified mixture of suitable materials, laid on a suitable sub-grade. Slabs less than 8 cm in thickness are called thin slabs and those of 8 cm and more are called thick slabs. The top of the sub-grade shall not be less than 30 cm above the anticipated water level in the side drains or the highest flood level of the locality. The following are the general specifications for a Cement Concrete Road :

1. All slabs shall be anchored to the sub-grade ; the surface of the sub-grade shall be brushed so as to remove all fine and loose material such as stone, kankar or moorum, and then the concrete shall be deposited on the surface which has thus been cleaned. If the upper surface is of a water bound macadam, the surface shall be given a wash of mix of cement and water just prior to depositing the concrete.

2. All cements to be used must comply with the Indian Standard Institution Specifications. A certificate of test shall be obtained from the manufacturers. Cement shall be ordered in quantities which can be utilized within one month and can be stored in a dry place close to the work.

3. The water to be used for mixing of concrete shall be clean, free from oils, acids and alkalies. Moreover, it should be *drinkable.* The water to be mixed must be measured properly and only measured quantity of water shall be added, depending upon the Water-Cement Ratio specified for that particular mix. (In general, the water-cement ratio varies between 0.46 to 0.5).

4. The fine aggregates shall be of such size that they pass through a 5 mm sieve and shall consist of sand or stone screening and shall be clean, hard, sharp and durable. Before using the sand, it must be washed and then dried. River sand need not be washed. It should be of specified fineness modules.

5. Coarse aggregate consists of crushed stone or river shingle and shall be clean, hard, durable, free from thin elongated or laminated pieces. *Limestone shall not be used.* The coarse aggregate shall pass through a screen having square meshes whose linear dimensions are not greater than half the least thickness of the slab.

6. The concrete shall be mixed in the specified proportions. Hand mixing shall be resorted to only in emergency and petty works otherwise it is preferable to postpone the work until a mechanical mixer is available.

Specified quantity of sand and cement is measured accurately and mixed first. It is always preferable to measure the quantity of sand in terms of bags of cement. All dry cement and sand shall be turned over by shovels at least three times until the mixture is of uniform colour. The specified quantity of coarse aggregate shall now be added and the whole mixture turned over again for three times. The specified quantity of water shall next be added slowly through a hose pipe attached to a watering can, while the process of turning over with shovels is being carried on. The mixing shall be continued until the whole batch has reached an even consistency and the mortar is spread evenly through the batch.

As far as possible, machine mixing of concrete should be adopted. The various ingredients of concrete shall be dry mixed as explained above and then fed into the concrete mix. The ingredients of concrete shall be mixed in the mixer for about 2 to 3 minutes. Measured quantity of water shall then be added. Concrete shall be mixed for not less than $2\frac{1}{2}$ minutes after water has been added and until there is a uniform distribution of the materials and the mass is of uniform colour and consistency. The mixer shall rotate at a speed of 15 to 20 revolutions a minute.

7. Before placing the concrete, the sub-grade shall be sprinkled with as much water as it will readily absorb but there shall be no pools of water standing on it. It is preferable to have the sub-grade sprinkled with water or thoroughly wetted 12 hours in advance of placing concrete, where such a procedure seems necessary.

8. Concrete shall then be laid and vibrated by concrete vibrator in its final position before setting has commenced and shall not subsequently be disturbed. Concrete shall be laid in two layers if the thickness is more than 15 cm and in one layer if it is to be 10 cm or less. The time to elapse between the laying of two layers shall not exceed 20 minutes.

Concrete shall be laid over the entire width of the slab and between joints in one continuous operation. While being laid it shall be sliced and spaded with suitable tools and shall then be brought to correct camber by means of heavy creed or tamper of solid wood about 25 cm deep and not less than 10 cm wide. The bottom of the tamper shall be shod with sheet iron. During tampering the surface should be carefully inspected for high and low spots and any correction necessary be made by adding or removing concrete.

9. After tampering, the entire surface shall be floated longitudinally with a wooden board not less than 5 metres in length and 20 cm wide with handles at each end. The float shall be operated by a man at each end standing on a suitable platform spanning the road. The float is drawn backward and forward with strokes of about 60 cm and advancing slowly from one side of the road to the other side. When the platforms or bridges are shifted forward, they should be so placed that the next floating overlaps the first one by nearly 1 metre.

10. The surface shall then be finished by using a strong canvas belt not less than 15 cm in width and at least 60 cm more than the width of the slab. The belt is to be applied with a combined cross-wise and longitudinal motion, the strokes being not more than 30 cm but these should be reduced to 10 cm as soon as the *water sheen* has disappeared from the surface of concrete.

11. There shall be kept at the site of work, a number of strips of hessian or burlap sufficient to cover at least two bays of the road. These strips shall have bamboo or light wood strips sawn at each end and as soon as the belting and edge finishing is complete, one of these strips previously damped, shall be placed on the surface of concrete. As the work proceed more strips shall be brought forward and similarly placed so that the moment the work on any portion is finished, the concrete is protected from drying. These strips should be kept moist by light spraying. When the concrete so laid is nearly two hours' old, the strips may be replaced by damp (empty) cement bags.

On the following day, the bags should be removed and the whole surface of concrete shall be covered with earth to a depth of 15 m. The earth shall be thoroughly wetted as soon as it is laid and the whole surface shall then be divided into squares of nearly 2 m length ; which shall be kept saturated with water until 21st day of laying. The earth shall then be removed and the road opened to traffic after 28th day.

12. Immediately prior to opening the road, the joints must be filled and protected. The joints should be filled up with bitumen which should overlap about 5 cm on either side of the joint.

13. After the completion of the concrete slab, the edges shall be protected by soldering of at least 30 cm width and 10 cm thick, water-bound macadam or brick, kankar or similar material.

REVIEW QUESTIONS

13.1. What do you understand by the term specifications ? What is the importance of specifications in engineering construction ? Explain briefly.

13.2. Write down specifications for laying the soling coat.

13.3. Describe briefly the constructional specifications for a cement concrete road.

Earthwork

GENERAL

For the construction of any road either in plains or hills, the irregularity of the natural ground surface must have to be removed. Earth work in its widest sense includes excavations in the soil and rock and formation of banking. Earthwork means cutting the high ground and filling the low areas, cutting of drains etc. (see Chapter 3 on Road Project).

14.1 EARTHWORK COMPUTATIONS

For the construction of a road or a railway track some amount of cutting or filling will have to be done. The amount or quantity of cutting or filling is called total earth work. Finding out the total quantity of earth work is called Earth Work Computations. Cross-secting of a road or railway track in cutting or banking is generally trapezoidal. The total quantity of earth work can be computed by taking the total length along the centre line multiplied by the average cross-sectional area for that particular length of the road or track.

Quantity or volume of Earthwork

$$= \text{Average cross-sectional area} \times \text{Length}$$

$$= \frac{1}{2}\,(\text{Top width} + \text{Bottom width}) \times \text{Average depth} \times \text{Length}$$

or (Area of central rectangle + Area of two side triangles) × Length

$$= B \times d + 2\left(\frac{1}{2} Sd \times d\right)$$

$$= Bd + Sd^2$$

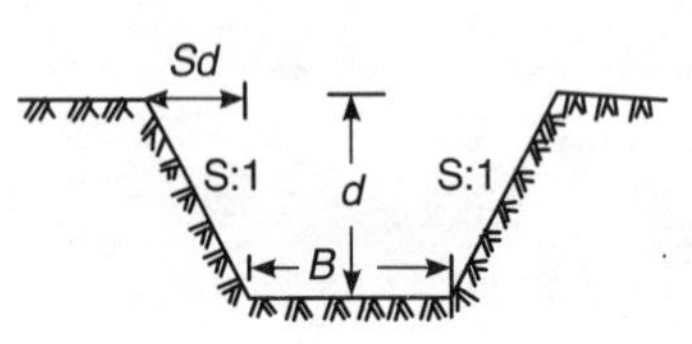

Figure 14.1

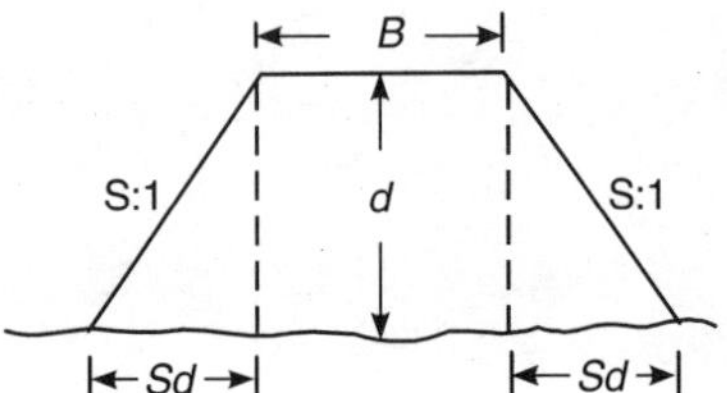

Figure 14.2

For the calculation of earthwork, the longitudinal section of the length of the road on the centre line is taken, and on this longitudinal section, a number of cross-sections are taken at right angles to the centre line. Across-section should be taken at regular intervals and at such points, when the slope of ground changes sufficiently to effect the volume of earth work appreciably. On the longitudinal section, the line of formation of the road should be shown and on every cross-section, the width of formation, the side slope and the depth of cutting or filling should be shown.

14.2 METHODS OF EARTHWORK COMPUTATIONS

There are three methods of computing earthwork :

(i) Mid-Sectional Area Method. Volume of earth by this method is calculated by taking the area of cross-section of the middle section by taking mean of the end sections multiplied by the length. Let d_1 and d_2 be the

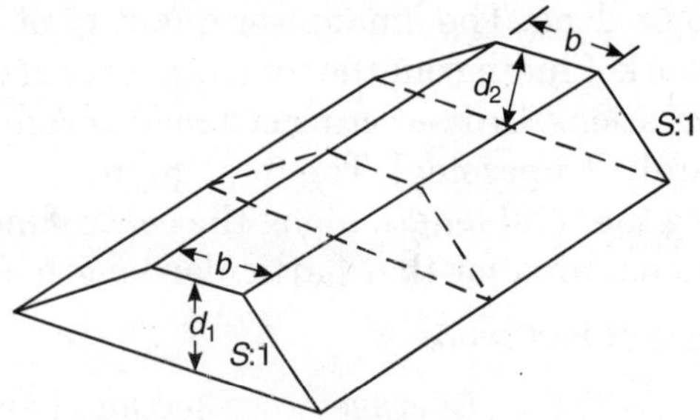

Figure 14.3

heights of the banks at the two ends of the embankment, b the formation width and $S : 1$ the side slope ($S : 1$ means, s horizontal and 1 vertical).

Mean height of embankment

$$= \frac{d_2 + d_2}{2} = d_m$$

Area of middle section = Area of rectangular portion + Area of the two side triangles

$$= b.d_m + \frac{1}{2}\, sd_m^2 + \frac{1}{2}\, sd_m^2$$

$$= bd_m + sd_m^2$$

Volume or quantity of earthwork

$$= (bd_m + sd_m^2) \times L$$

For calculating the earthwork, the following tabular form should be used :

Station or chainage	Depth or height d	Mean depth or height d_m	Area of Central rectangle $b.d_m$	Area of side traingles sd_m^2	Total Area $bd_m + sd_m^2$	Length L	Quantity $(bd_m + sd_m^2) \times L$	
							Fill-ing	Cutt-ing

(ii) Average Cross-sectional Area Method. According to this method, the area of cross-section of the two ends is calculated and average or mean is calculated and is multiplied by the length to find out the volume of earth.

Sectional area at one end $A_1 = bd_1 + sd_1^2$

Sectional area at the other end $A_2 = bd^2 + sd_1^2$

Mean or average sectional area

$$A = \frac{A_1 + A_2}{2}$$

Volume or quantity of earthwork

$$= \frac{A_1 + A_2}{2} \times L$$

The following table is used for calculating work :

Station or chainage	Height or depth	Area of central portion	Area of side triangles	Total sectional area	Mean sectio-nal area	Length	Quantity	
							Fill-ing	Cutt-ing

(iii) Presmoidal Formula Method. In this method the area of cross-section at the two ends and mid-sectional area is obtained and mean depth of cutting or banking is also calculated, then according to the prismoidal formula, the quantity of earthwork will be

$$= \frac{d_m}{6}(A_1 + A_2 + 4A_m)$$

where

d_m = mean depth of cutting or filling

A_1 and A_2 = Area of cross-section at the two ends

A_m = Mid-section area

This method is the most accurate method of calculating earthwork.

14.3 TYPICAL SECTIONAL AREAS IN HILL ROADS

In hilly terrains, when there is a considerable longitudinal slope, the cross-sections can be wholly in cutting, wholly in banking and partly in cutting and partly in banking.

14.3.1 Wholly in Cutting

(i) *Assuming the ground to be level and the section is wholly in cutting.* Let b is the formation width, $s : 1$ the side slope of the cutting and d is the depth of cutting at the centre, d_1 and d_2 be depths of cutting at the two ends, then

Area of cross-section = $b \times d + Sd.d$

$= d(b + sd)$ (i)

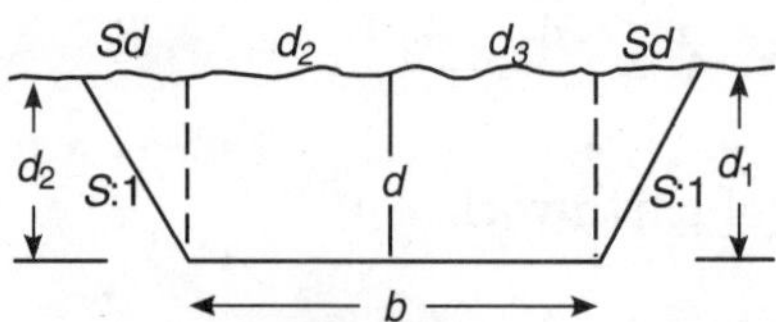

Figure 14.4

When the ground is horizontal

$$d_1 = d_2 = d$$

This formula will also hold good when the formation is in full banking and the ground is horizontal.

(ii) *Assuming the ground to be slopy and have a side long slope of m : 1 (m horizontal and 1 vertical).* From Fig. 14.5, we have

A = Area 1 + Area 2 + Area 3 + Area 4 + Area 5 + Area 6

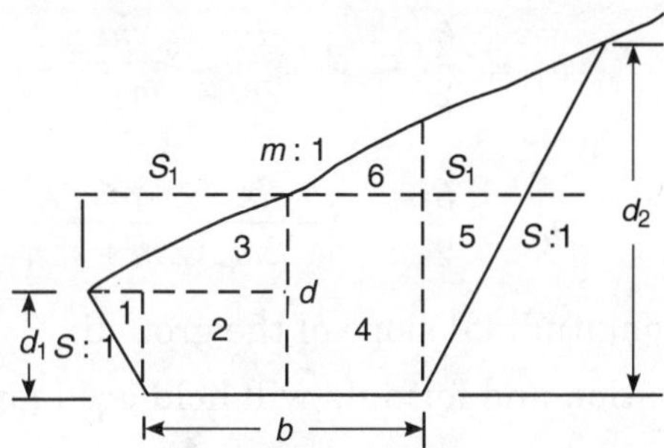

Figure 14.5

$$\text{Area 1} = \frac{1}{2} \times d_1 \times sd_1$$

$$\text{Area 2} = \frac{b}{2} \times d_1$$

$$\text{Area 3} = \frac{1}{2} \times (d - d_1) \;\; \frac{b}{2} + sd_1$$

$$\text{Area 4} = d \times \frac{b}{2}$$

$$\text{Area 5} = \frac{1}{2} \; d_2 \times sd_2$$

$$\text{Area 6} = \frac{1}{2} \times \frac{b}{2} \times (d_2 - d)$$

$$\text{Total area} = \frac{1}{2} \; sd_1^2 + \frac{1}{2} \; bd_1 + \frac{1}{2} \; (d - d_1) \;\; \frac{b}{2} + sd_1$$

$$+ \frac{1}{2} \; bd + \frac{1}{2} \; sd_2^2 + \frac{1}{2} \times \frac{b}{2} \; (d - d_1)$$

On simplification, we have

$$A = \frac{1}{2} \quad (s_1 + s_2) \quad d + \frac{d}{25} \quad - \frac{b^2}{25}$$

or

$$A = \frac{1}{2} \quad \frac{b}{2}(d_1 + d_2) + d(s_1 + s_2)$$

where

$$s_1 = \frac{b}{2} + \quad d + \frac{b}{2m} \quad \frac{ms}{m - s}$$

and

$$s_2 = \frac{b}{2} - \quad d - \frac{b}{2m} \quad \frac{ms}{m + s}$$

where m : 1 is the longitudinal slope of the ground.

The same explanation and formula will hold good for banking also.

14.3.2 Partial Cutting and Partial Banking

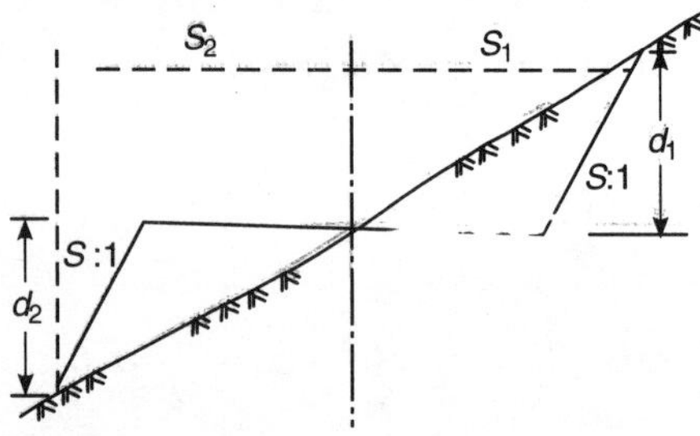

Figure 14.6

The ground has a longitudinal slope of m : 1 and the formation is such that a part of it is in cutting and a part in banking. Let d_1 be the depth of cutting and d_2 be the height of banking and d be the depth of cutting at the centre of formation. From Fig. 14.6

$$\text{Area of banking} = \frac{b}{25} - d \quad (s_2 - md) - \frac{b}{5} \quad (b - md)$$

$$= \frac{1}{2} \quad \frac{b}{2} - md \quad d_2$$

$$\text{Area of cutting} = \frac{1}{2}\ \frac{b}{2} - d\ (s_1 + md_1)\ - \frac{b}{2s} + \frac{b}{2} - md$$

$$= \frac{1}{2}\ \frac{b}{2} + md\ d_1$$

where $$s_1 = \frac{b}{2} + \ d + \frac{b}{2m}\ \ \frac{ms}{m-s}$$

and $$s_2 = \frac{b}{2} + \ \frac{b}{2m} - d\ \ \frac{ms}{m+s}$$

14.4 METHODS OF CALCULATING EARTHWORK

The earthwork quantities can be calculated by either of three formula e.g., mid-sectional area method, mean sectional area method or presmoidal formula method. The following illustrative examples will be found useful.

SOLVED EXAMPLES

Example 14.1 *The formation width of a road is 10 m, which is to be constructed on a hill having a slope of 9 :1. The depth of cutting at the centre is 5 m and side slopes of the cutting and banking is 1:1. Calculate the quantity of earth work in horizontal length of 30 m.*

Solution

From Fig. 14.7, we have

$$b = 10 \text{ m},\ d = 5\text{m},\ m = 9,\ s = 1$$

$$L = 30 \text{ m}$$

$$ms = \frac{b}{2} + \ d + \frac{b}{2m}\ \ \frac{ms}{m-s}$$

$$= \frac{10}{2} + \ 5 + \frac{10}{18}\ \ \frac{9 \times 1}{9-1}$$

$$= 5 + \ 5 + \frac{5}{9}\ \ \frac{9}{8}$$

$$= 5 + \frac{50}{9} \times \frac{9}{8}$$

$$= 5 + \frac{25}{4} = \frac{45}{4} = 11.25 \text{ m}$$

$$s_2 = \frac{b}{2} + \frac{b}{2\,m} - d_1 \quad \frac{ms}{m-s}$$

or

$$s_2 = \frac{b}{2} + d - \frac{b}{2\,m} \quad \frac{ms}{m-s}$$

$$= \frac{10}{2} + 5 - \frac{10}{18} \quad \frac{9 \times 1}{9+1}$$

$$= 5 + \frac{40}{9} \times \frac{9}{10} = 9 \text{ m}$$

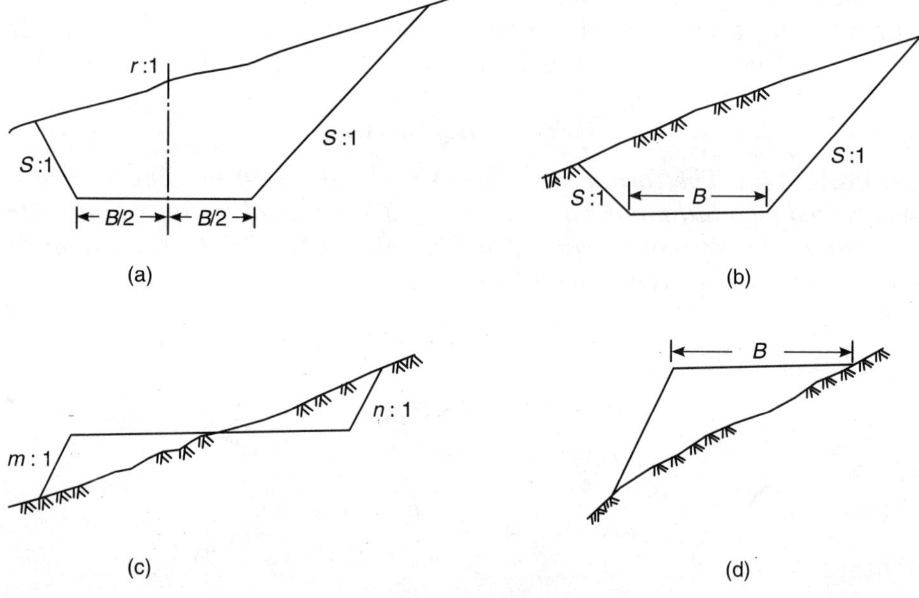

Figure 14.7

Area of banking

$$= \frac{10}{2 \times 1} - 5 \;(9 - 9 \times 5) - \frac{10}{2 \times 1}\;(10 - 9 \times 5)$$

$$= (5 - 5)\,(-36) - \frac{10}{2} \times -35$$

$$= 165 \text{ sq. m}$$

Earthwork in banking = 165 × 30 = 4950 cu. m.

Example 14.2 *Calculate the quantity of earthwork in embankment, by the data given below:*

Height of bank at the near end = 3 m

Upward gradient of formation = 1 in 150

Downward gradient of the ground = 1 in 30

Horizontal length of embankment = 100 m

Width of formation = 10 m

Side slope of bank = 1 in 2

Solution

Here $b = 10$ m

$s = 2$

$m = 3$

Height of embankment at the near end,

$d = 30$

Height of embankment at the far end

$$d_1 = 3 + \frac{100}{150} + \frac{100}{30} = 5.33 \text{ m}$$

Area at the near end $= (b + sd)\,d$

$= (10 + 2 \times 3) \times 3$

$= 16 \times 3 = 48$ sq. m

Area at the far end $= (b + sd_1)\,d_1$

$= (10 + 2 \times 5.33) \times 5.33$

$= 110$ sq. m

By the average end area method volume of earthwork

$$= \frac{(A_1 + A_2)}{2} \times L$$

$$= \frac{(48 + 110)}{2} \times 100$$

$= 79 \times 100 = 7900$ cu.m

Example 14.3 *Calculate the quantity of earthwork for a road in banking having a length of 100 metres, on a uniform ground. The width of formation is 10 m and the side slopes are 2 :1. The height of banks at the two ends is 1.00 m and 1.6 m.*

Solution

Volume of earthwork $= (bd + sd^2)\,L$

Here $b = 10$ m

(Mean depth) $d = \dfrac{1.0+1.6}{2} = 1.3$

$s = 2$

$L = 100$ m

$= (10 \times 1.3 + 2 \times 1.3^2) \times 100$

$= (13 + 3.38) \times 100$

$= 1638$ cu m.

Example 14.4 *Find out the area of the banks of an embankment 10 m wide and 100 m long having a side slope of 2.11. The height of embankment at the two ends is 2.5 m and 3.5 m.*

Solution

Mean height of the embankment

$$d = \frac{2.5+3.5}{2} = 3 \text{ m}$$

Width of slope at the mid-section

$$= \sqrt{2^2+1^2 \times d}$$

$$= \sqrt{5} \times 3 = 3\sqrt{5} \text{ m}$$

Area of two slopes $= 100 \times 3\sqrt{5} \times 2$

$= 600\sqrt{5}$ sq. m

$= 1342$ sq.m.

Example 14.5 *Calculate the quantity of earthwork for a road in embankment given the following data :*

Length of embankment is 300 m

Width of embankment is 10 m

Side slopes of embankment are 2 :1

The formation level has a downward gradient of 1 in 150 upto 120 m and then the gradient changes downward to 1 in 100 from 120 m to 300 m. The R.L. of formation at the start is 107. The R.L. of ground are :

0	30 m	60 m	90 m	120 m	150 m	180 m	210 m	240 m	270 m	300 m
105.0	105. 6	105.44	105.90	105.42	104.3	105. 0	104.10	104.62	104. 62	103.3

Solution

Here $b = 10$ m and $s = 2$

R.L. of formation at the different points can be calculated from the given gradient of the formation.

The R.L. at the start = 107.00 m

The R.L. at 30 m = 106.80 m

at 60 m = 106.60 m

at 90 m = 106.40 m

at 120 m = 106.20 m

Now the gradient changes

The R.L. at 150 m = 105.90 m

at 180 m = 105.60 m

at 210 m = 105.30 m

at 240 m = 105.00 m

at 270 m = 104.70 m

at 300 m = 104.40 m

Calculations of the earthwork will be computed in the following tabular form.

Length of bank	*Height of banking m*	*Mean height d of banking*	*Area of Central portion m^2*	*Area of triangles m^2*	*Total sectional area m^2*	*Length L m*	*Quantity of earth work cu. m.*
–	2.0	–	–	–	–	–	–
30 m	1.2	1.60	16.00	5.12	21.12	30	633.6
60 m	1.16	1.18	11.80	2.78	14.58	30	437.4
90 m	0.50	0.83	8.30	1.38	9.68	30	290.4
120 m	0.78	0.64	6.40	0.82	7.22	30	216.6
160 m	1.60	1.19	11.90	2.83	14.73	30	441.9
180 m	0.60	1.10	I1.00	2.42	13.42	30	402.6
210 m	1.20	0.90	9.00	1.62	10.62	30	318.6
240 m	0.38	0.79	7.90	1.25	9.15	30	274.4
270 m	0.70	0.54	5.40	0.58	5.98	30	179.4
300 m	1.10	0.9	9.00	1.62	10.62	30	318.6

Total Earthwork – 3,513.6 cu.m

14.5 MASS HAUL DIAGRAM

A mass haul diagram is a plot between the cumulative volume of earthwork and its distance from the site. The ordinate of the curve represents the algebraic sum of the volumes of cutting and filling and the abscissa represents the distance. While computing the volume the cutting is assumed as positive and the filling as negative. Allowance should for settlement of embankments. Generally the excavated earth increases in volume when removed from the ground the decrease or increase in volume varies from 10 to 20% depending on the nature of soil. The mass haul diagram is used in the following manner:

1. Divide the length of the road into several dimensions.
2. Compute the volume of earthwork by any of the methods for each section.
3. Plot a longitudinal section of each section.
4. Calculate the accumulated volume at various points.
5. Plot the cumulative volume as ordinate and distances as abscissa for the mass haul diagram by plotting positive cumulative volume above the baseline and negative below the baseline.
6. Join the ends of the ordinates with a smooth curve.
7. The mass haul diagram should always be constructed below the L-section.

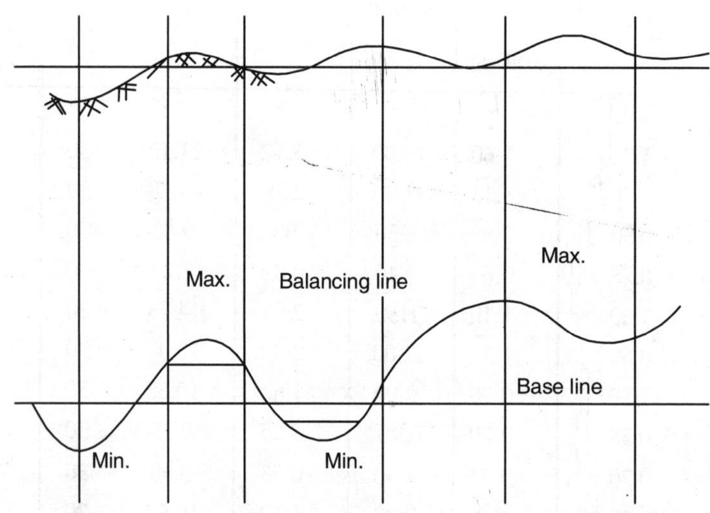

Figure 14.8

14.5.1 Observations

1. The mass haul diagram uses in case of cutting and falls in case of filling.

2. The difference between the ordinates at two points is equal to the volume of cut or fill provided there is no maximum or minimum point between them.

3. The vertical distance between a maximum point and next minimum point is equal to the volume of filling.

4. The vertical distance between a minimum point and the next maximum point is equal to the volume of cutting.

5. If the diagram cuts the baseline at any two points in succession, the volume of cutting between the two points is equal to the volume of filling as the algebraic sum of the ordinates (volume) is zero.

6. Any horizontal line drawn parallel to the baseline and intersecting the curve at two points is a balancing line. It indicates a length over which the volume of cutting is equal to volume of filling.

14.6 DEFINITIONS

1. **Haul distance.** It is the distance between the working face of an excavation along the centre and the tip of the formation from the excavated material.

2. **Average haul distance.** It is the distance between the centre line of excavation and centre of embankment.

3. **Haul.** It is defined as the product of excavated earth and average haul distance. The unit of haul is station metre. One station metre is equal to 1 m^3 of excavated material moved through 100 m.

4. **Free haul distance.** It is the distance through which the excavated material is moved by the contractor free of charge.

5. **Overhaul.** It is the distance beyond the free haul distance for which extra payment is made to the contractor.

6. **Spoil.** It is the extra excavated material which is not used for embankment and is stacked along the alignment. These dumped material is properly arranged in the form of banks which are called **Spoil banks**.

7. **Limit of economical haul.** It is the maximum limit of the haul distance beyond which it is not economical to haul and use the material the limit of economical haul is obtained by the following formula

Limit of economical haul (LEH) = Free haul distance (F.H.D.)

$$+ \frac{\text{cost of excavation}}{\text{cost of overhaul}}$$

8. **Lead.** It is the horizontal distance through which the excavated material is moved for embankment. The average haul distance and lead have the same meaning.

9. **Lift.** It is the vertical distance through which the excavated material is moved. Generally 1.5 m lift is free.

14.7 CALCULATIONS OF OVERHAUL

Following procedure is used to calculate overhaul.

Draw the longitudinal and mass haul diagram.

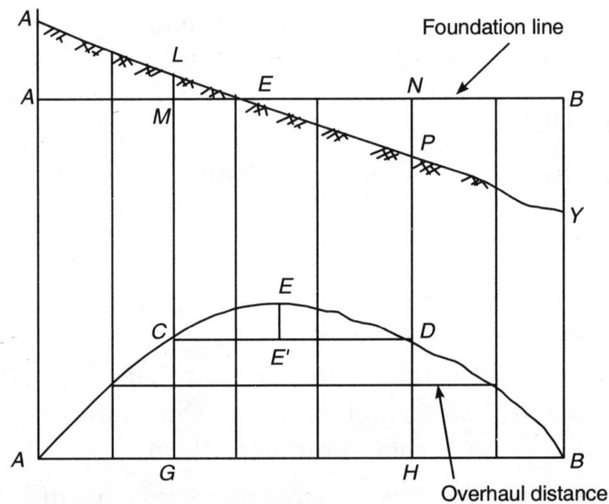

Figure 14.9 *Mass haul diagram*

Mark a distance of 100 m on a scale and move the scale up and down on the mass haul diagram, keeping it parallel to the base line, so that it meets the mass haul diagram at two points *CD* and the project these two points on to the *L*-section. This is called balancing line and it indicates that the amount of cut is approximately equal to the amount of fill. The volume indicated by the vertical line *EE´* indicates the free haul distance. The area enclosed by the balancing line and the mass haul diagram curve is equal to the haul in that reach.

Hence Free haul = Area *CED*

The remaining area *XLMA* which is indicated by the ordinate *CG* and the area *NBYP* which is indicated by the ordinate *DH*, are to be balanced. The overhaul volume *XLMA* has to be shifted to *NBYP*. The average overhaul distance will be the centroid of the area *XLMA* and the centroid of area *NBYP*.

$$\text{Overhaul} = \text{Area } A'GC \times \text{Area } B'HD$$

14.8 MOVEMENT OF EXCAVATED MATERIAL

In a mass haul diagram, if the loop curve is cut off by a balancing line above the base line, the excavated material is to be moved forward. However when the loop is below the balancing line, the material must be moved backward.

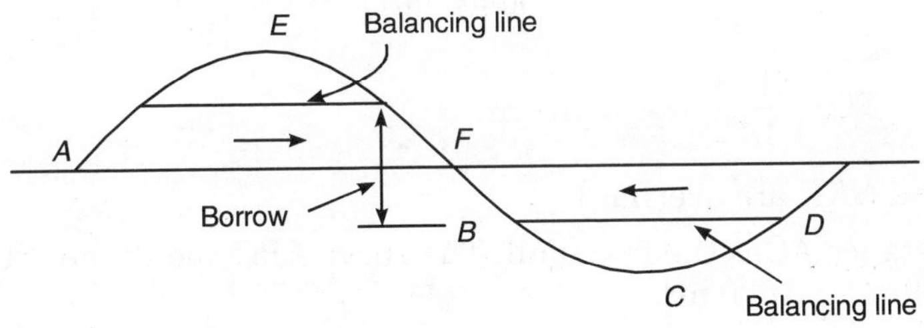

Figure 14.10 *Movement of material (Mass-haul diagram)*

14.8.1 Observations

1. The length of a balancing line intercepted by a loop of the curve indicates the maximum distance of movement of material.

2. The area enclosed by a loop and the balancing line is equal to the haul.

3. The vertical distance between the two successive balancing lines indicates the waste material or the material to be borrowed, depending upon the position of the balancing line.

4. To minimise the haul is to reduce the area between the loop and the balancing line, but it should be noted that the aim is to reduce the cost and not to reduce the excavation.

14.9 OVERALL COST CALCULATION

Suppose the station distance is 100 m, free haul is 200 m, cost of excavation is Rs. 30/- per cubic metre and cost of overhaul is Rs. 15 per m^3 per station then Limit of economical haul (L.E.H.) will be

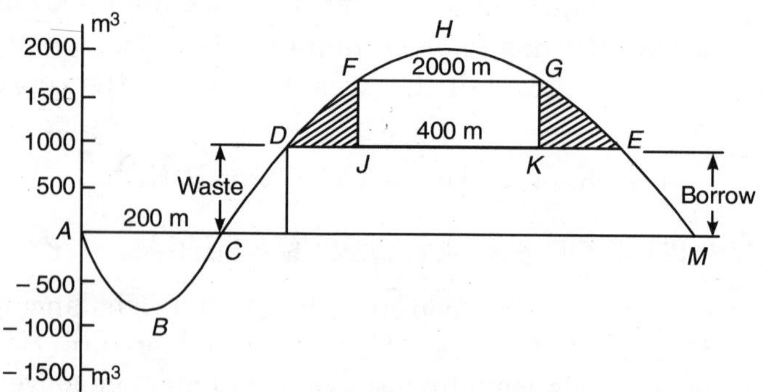

Figure 14.11

$$200 + \frac{30}{15} \times 100 = 400 \text{ m}$$

DFJ and *GKE* are overhaul.

The distance *AC* is the free haul. The curve *ABC* shows the fill and *CHM* is cut. Volume = 1000 m^3

From *C* to *D*, there is a waste of 1000 m^3

From *E* to *J*, the borrow = 1000 m^3

Excavation

$$B \text{ to } C = 1000 \text{ m}^3$$

$$C \text{ to } H = \frac{2000 \text{ m}^3}{3000 \text{ m}^3}$$

Borrow

$$E \text{ to } J = 1000 \text{ m}^3$$

Total excavation = 4000 m^3

Excavation cost = 4000 × 30 = Rs. 1,20,000

Overhaul Area *DFJ* + Area *GKE*

= 500 station metre

Cost of overhaul = 500 × 15 = 7500/-

Total cost of earthwork = 1,20,000 + 7500 = Rs. 1,27,500

14.10 USES AND CHARACTERISTICS OF A MASS HAUL DIAGRAM

1. The mass haul diagram can be used to compare the alternatives of the utilization of excavated material.

2. The mass haul diagram indicates the locations from where earth can be borrowed or can be dumped economically.

3. Most economical solutions can be found by trial and error method by changing the balancing line or shifting it.

4. With the help of mass haul diagram, the alignment and design can be modified.

Additional Examples

Example 1. *Calculate the volume of cut and fill by the end area method of the following alignment. If all the excavated earth is to be used for filling and the distance of transportation to be kept minimum. Calculate the chainage at which the excavated material is to be dumped.*

Chainage m	*0*	*50*	*100*	*150*	*200*	*250*	*300*	*350*	*500*	*450*	*500*	*550*	*600*
Area of cut	*0*	*20*	*44*	*42*	*26*	*10*	*2*						
Area of fill								*4*	*32*	*54*	*52*	*46*	*20*

Solution

Chaingae	*Area*		*Volume*		*Accumulated*
	Cut	*Fill*	*Cut*	*Fill*	*Volume*
0	—		—	—	—
50	20		500		500
100	44		1600		2100
150	42		2150		4250
200	26		1700		5950
250	10		900		6850
300	2		300		7150
350		4		50	7100
400		32		900	6200
450		54		2150	4050
500		52		2650	1400
550		46		2450	– 1050
600		20		1650	– 2750

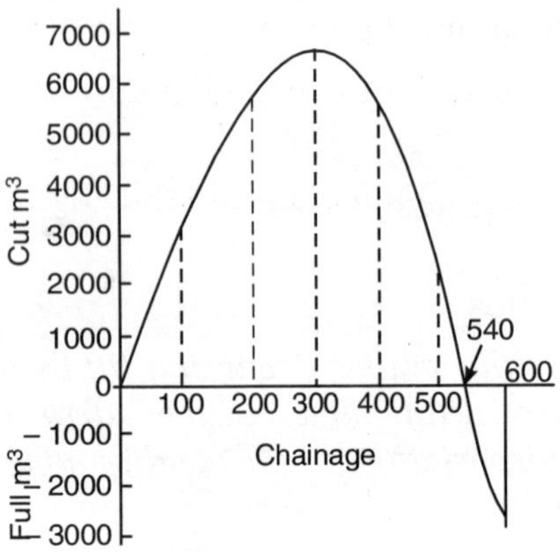

Figure 14.12 *Mass haul diagram*

Total volume of cut = 7150 m^3

Total volume of fill = **9850 m^3**

Example 2. *Calculate the length of road, volume of fill in the embankment of width 40 m and average height 15 m and average gradient 1 in 40 from contour 150 m to 590 m. Transverse slope is 1 in 10.*

Solution

$$\text{Length of road} = \frac{590-150}{1/40} = 400 \times 40 = 17{,}600 \text{ m}$$

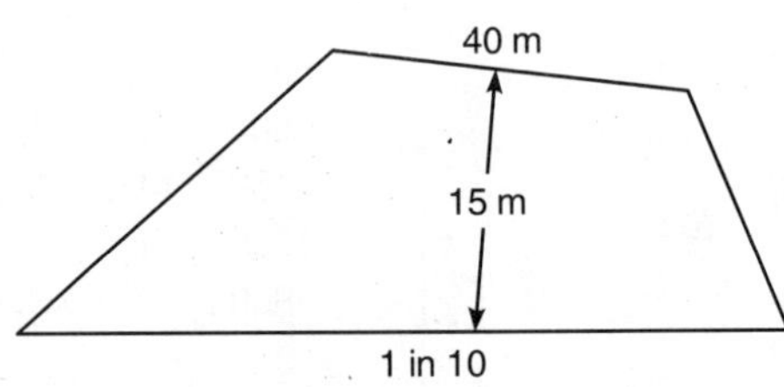

Figure 14.13

$$\text{Arear of cross section} = \frac{m^2 n}{m^2 - n^2} \; h + \frac{b}{2n} \; - \frac{b^2}{4n}$$

wnere $m = 40$, $n = 1$, $b = 40$, $h = 15$

$$= \frac{40 \times 40 \times 1}{40 \times 40 - 1 \times 1} \; 15 + \frac{40}{2 \times 1} \; - \frac{40^2}{4 \times 1}$$

$$= 837.4 \text{ m}^3$$

$$\text{Volume} = 837.4 \times 17600 = \mathbf{14{,}73{,}800 \text{ m}^3}$$

Example 3. *An embankment is 500 m long and is 9 m wide at the formation with side slope 2 in 1. The R.L. of the ground at 100 m intervals are*

Chainage	*0*	*100*	*200*	*300*	*400*	*500*
R.L.	*107.8*	*106.3*	*110.5*	*111.0*	*110.7*	*112.2*

The formation level at zero chainage is 110.5 and it has a rising gradient 1.2 m for 100 m. Calculate the earthwork.

Solution

Chainage	*0*	*100*	*200*	*300*	*400*	*500*
R.L.	*107.8*	*106.3*	*110.5*	*111.0*	*110.7*	*112.2*
Formation Level	*110.5*	*111.7*	*112.9*	*114.1*	*115.3*	*116.5*
Height	*2.7*	*5.4*	*2.4*	*3.1*	*4.6*	*4.3*
Area of cross section $(9 + 2h)\,h$	*3888*	*106.92*	*33.12*	*47.12*	*83.72*	*75.68*

Volume by prismoidal formula

$$= 500 \quad \frac{38.88 + 75.68}{2} + 106.92 + 33.12 + 47.12 + 83.72$$

= 32816 cu.m

N.B. Area of cross section at different distances from the origin or chainage are obtained by substituting the values of h as indicated above.

Example 4. *Calculate the earthwork of the pit.*

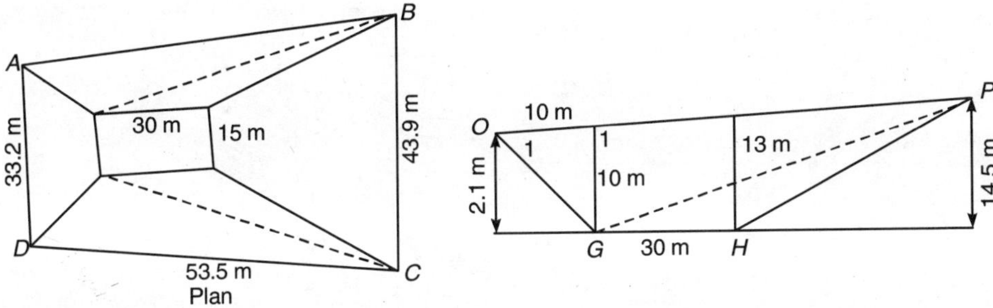

Figure 14.14

Solution

Divide the pit into two wedges as shown by the dotted line. Volume of wedge *GHP* is

$$= \frac{L}{6}(a + b + c) \times h$$

Here $L = 30, a = 15 \text{ m}, b = 15 \text{ m}, c = 43.9 \text{ m}$ and $h = 14.5 \text{ m}$

$$= \frac{30}{6}(15 + 15 + 43.9) \times 14.5$$

$$= 5334.34 \text{ m}^3$$

Similarly Volume of *OGP* is

$$= \frac{L_1}{6}(a_1 + b_1 + c_1) \times h_1$$

Here $L_1 = 53.5 \text{ m}, a_1 = 15, b_1 = 33.2\ c_1 = 43.9$ and $h_1 = 14.5$

$$= \frac{53.5}{6}(15 + 33.2 + 43.9) \times 14.5$$

$$= 8226.32 \text{ m}^3$$

Total volume of pit $= 5334.34 + 8226.32 = 13660.66 \text{ m}^3$

15

Waterlogging of Roads

GENERAL

Waterlogging is the effect or damage caused to a sub-soil due to excessive sub-soil moisture. The problem of waterlogging has taken a serious turn in our country on account of increase in the irrigated areas and shift on the emphasis from protective irrigation to intensive irrigation for increasing the agricultural production. A large area of roads passing through agricultural fields is greatly affected by waterlogging. Waterlogging is mostly due to the creation of irrigation channels in the locality of roads without considering adequate and proper drainage of seepage water. *A land is said to be waterlogged when the soil pores are effectively saturated due to lack of aeration of the soil.*

15.1 CAUSES OF WATERLOGGING

Waterlogging is the result of rising of water table due to the following reasons:

1. Due to increased inflow of water in the sub-soil of embankment or sub-grade of the road surface from the adjoining areas.

2. Due to constant inflow of seepage of irrigation channels from the adjacent upper region.

3. Due to inflow of seepage water from the adjacent tanks, water reservoirs etc.

4. Due to percolation of water from improper surface drainage or sub-surface drainage along the road-side.

5. Due to lateral flow of sub-soil water raising the water-table on the upstream side due to impervious obstruction.

6. Due to poor natural drainage i.e., flat terrain with depression leads to detention of storm water which adds to raising the water-table and hence waterlogging.

7. Due to underground ridges and barriers perched water-table acquifers.

15.2 HARMFULNESS DUE TO WATERLOGGING

Waterlogging raises the water-table of the area which in turn helps the moisture to rise due to capillary action and thereby reducing the bearing capacity of soil. As a result of this the sub-grade is deformed and unequal or excessive settlements take place which make the road surface undulated and unfit for traffic movement. Waterlogging produces some deterimental salts such as sodium sulphates, magnesium sulphates and sodium carbonates etc. which are very dangerous to the embankment of roads.

15.3 ANTI-WATER LOGGING MEASURES

The anti-waterlogging measures may be preventive as well as curative.

15.3.1 Preventive Measures for Waterlogging

Waterlogging may be prevented by the following methods :

1. Reducing percolation from canals, by lining the canals, lowering the full supply level (F.S.L.), providing intercepting seepage or parallel drains etc.

2. Reducing the sub-soil flow of the ground water by providing artificial seepage drains, providing natural seepage drains, pumping sub-soil water by *well point system.*

3. Reducing percolation of irrigation water by providing intercepting drains and by improving surface drainage of roads and embankments.

15.3.2 Curative Measures for Waterlogging

Curative measures constitute extraction of excessive water from the sub-soil of roads and embankments. Actually speaking when the land adjoining a roadway or railway or an embankment has become waterlogged, it becomes really difficult to eradicate water and prevent waterlogging. Water is mainly extracted from the waterlogged areas by well point system. Under this system vertical perforated pipes are driven at short distances along the road-side or embankment. The top of these vertical pipes is connected by a horizontal pipe, through which the water is sucked up by pumps. This will help in lowering the water-table and hence waterlogging.

15.4 PRECAUTIONS TO BE TAKEN WHILE CONSTRUCTING ROADS ON WATER-LOGGED AREAS

While constructing roads on waterlogged areas the following precautions should be taken to prevent any damage to the road :

1. The sub-grade should be thoroughly compacted.

2. The thickness of crust should be designed properly so as to rectify unequal settlements of sub-soil.

3. Roads in cutting should be avoided as far as possible.

4. Deep and narrow pits should be constructed and should be joined with some adjacent stream or nallah. These barrow pits will suck moisture from beneath the road and help in lowering the water-table and hence water-logging.

5. For the construction of bitumen roads, a thin layer of bitumen should be sprayed over the sub-grade to seal the pores of soil and prevent rising of moisture by capillary action. Due to this moisture, the binder gets stripped off, thereby reducing the strength of the road.

15.5 EMBANKMENT ON MARSHY LAND

Construction of embankment on marshy land pose a real big problem. The reason being that the bearing capacity of the soil is almost negligible and to get a dry compact embankment is real task. Before starting the construction of an embankment on a marshy land, the pre-requisite is that the sub-soil water must be drained off completely either by well point system or by constructing lateral drains. After the water has been drained off to a reasonable limit, the construction of embankment is started.

A fascine consisting of bundles of green wood of diameter 2.5 cm to 4.0 m tied at intervals into bundles of 15 to 18 cm diameter and placed at a distance of 1 to 1.2 m apart, is placed over the base of the embankment. The green wood and the rope should be such that it is not easily spoiled in the wet soil. The fascine which is just like a raft is filled with sand covered with the branches of trees. Now over this green wood raft or fascine, filling for embankment is started. For the stability of this bank, the water-table should be kept as low as possible.

15.6 CONSTRUCTION OF ROADS IN WATER LOGGED AREAS

When the problem is only of water logging with flooding or detrimental salts, the following procedure of road construction may be adopted:

1. The sub-surface water level or water table should be lowered by suitable means or drainage system.

2. The road should be constructed much above highest flood level.

3. Capillary cut off should be provided below the sub-grade to arrest capillary rise.

4. The sub-grade and pavement thickness should be sufficient to match the soil conditions.

5. Vertical sand drains should be provided to ensure rapid drainage of water below the sub-grade.

Sometimes detrimental salts are present in the water, in such cases superior pavement construction methods may be adopted to withstand the effects of detrimental salts. An effective capillary cut off below the subgrade and a water proofing layer on the pavement surface and on shoulders will minimise the effects of detrimental salt.

Sometimes in water logged areas, prolonged floods also add miseries to road structures. In such cases, cement concrete and bituminous concrete surface pavement should be preferred.

REVIEW QUESTIONS

15.1. What is waterlogging and what are its ill-effects?

15.2. What are the protective and creative measures of waterlogging?

15.3. Elucidate the various causes of waterlogging.

16

Highway Maintenance

GENERAL

Maintenance is defined as preserving and keeping the serviceable conditions of a structure as normal as possible, so as to maintain its original condition as constructed or as subsequently improved. Maintenance is an important activity which helps in providing better service facilities, longer life and better appearance. Highway location, design and construction have a bearing on the maintenance cost. At the time of alignment studies, if proper consideration is given to drainage problems, soil conditions, directness of route, landslide problems etc., it will generally reduce maintenance cost and maintenance problems. The maintenance of a highway will include (i) maintenance of surfacing, (ii) maintenance of shoulders and footpaths, (iii) maintenance of underground and surface drainage, and (iv) maintenance of bridges, culverts and other structures.

16.1 MAINTENANCE OF SURFACING

Whatever type of road surface it may be, the maintenance should be a preventive maintenance. For important highways and busy lanes, greater is the importance of preventive maintenance. Defects in the road surface should be corrected as soon as they appear so that they should not become too serious to be remedied quickly. This will reduce the maintenance cost and keep the road surface in good running condition. Surface maintenance is more important than any other maintenance because it is only the surface which is being used by the traffic. About 50% or more of the maintenance budget is spent on the maintenance of road surface only. Most of the maintenance work is done departmentally by employing work-charge staff on muster-roll as the maintenance work is so diverse, so subject to variation that it does not lend itself to competitive bidding.

16.2 MAINTENANCE OF BITUMINOUS SURFACES

Maintenance of bituminous surfaces consists in repairing patches, pot holes and ruts, surface treatment, resurfacing etc.

16.2.1 Repair of Pot Holes, Ruts and Patches

Patch repair may be needed for the damaged portions of a road, due to patches, pot holes etc. Sometimes patching is also required to remove inequalities in shape and surface and removing waviness in order to make the surface smooth riding. Pot holes are formed due to poor gradation of surface materials aging of bitumen, stripping of bituminous mix etc. Corrugation or waviness is caused due to incorrect gradation of aggregates, deficiency of bitumen, or bad quality of bitumen and excessive bitumen in the mix, traffic excessiveness or overloads. Corrugations are caused in the bitumen mix due to unequal settlement of sub-soil moisture.

Patches and pot holes are cut in square or rectangular shape and the loose material is removed. These are then cleaned by a wire brush. Heated bitumen or emulsion is applied on the cut portion of the pot holes. After applying bitumen, patch mix, consisting of bitumen and grit or stone ballast, is filled and thoroughly compacted by iron runner or rollers. The surface of the patch should be a little higher than the normal surface of road so as to allow for compaction by the traffic.

Sometimes pot holes occur on more than one-third of the surface ; it is preferable to resurfacing the entire road surface than to repair the patches. (The method of repainting or resurfacing has been discussed in the chapter on Construction of Roads.)

16.2.2 Repair of Base-course

Flexible pavements may cause structural failure due to inadequate thickness of base, excessive soil in WBM, WMM weak sub-grade and sub-base. Excessive moisture or unequal settlement may also result in the failure of base. Poor consolidation of the base, sub-base or sub-grade may cause failure too. Constant percolation of surface water may also be responsible for the failure of base. Before undertaking the repair of base course, investigations by trenching or visual inspection should be done. Repair work should not be started unless the cause of the failure is determined and remedy done.

If the failure is due to improper surface drainage which helps in the percolation of water to the base course, proper drainage should be provided so as to facilitate easy removal of surface water. This will help in avoiding unnecessary expenditure.

However, if it is decided to increase the base thickness, the whole surface is loosened by scarifying to the full depth. All loose pulverised material, soil are removed, sub-grade is rolled and a new base or sub-base of the required thickness is laid and rolled. Now the new base is covered with suitable surface treatment or a premixed carpet.

16.2.3 Repair of Bleeding Surface

If the surfacing consists of excessive bitumen, the surface will become slippery during rainy season and will bleed during hot weather. Bleeding will cause unnecessary inconvenience to the traffic as the bitumen will stick to the tyres, corrugation or rutting in the surface of the road shall be formed. Bleeding normally occurs just after the construction of the road. Repair of bleeding surfaces is also called *Surface Treatment.* Bleeding can easily be corrected by spreading a layer of dry coarse sand in a thickness varying from 5 mm to 10 mm and rolling the surface.

16.3 REPAIR OF CEMENT CONCRETE ROADS

The maintenance of rigid pavements consists (i) in filling and sealing joints and cracks in pavement surfaces, (ii) patching of damaged areas, (iii) repairing of blow-ups.

(i) *Sealing of joints and cracks.* Joints and cracks will allow water to percolate to the base and sub-grade. Prevention of infiltration of surface water to reach sub-grade under pavement and collection of water in cracks and joints, will help in improving the life of the road, its serviceability and reducing the maintenance cost. It is for this reason that the joints and cracks should be properly sealed and filled. Before filling the joints, the joints should be cleaned. Sometimes compressed air is used to clean the joints before filling or sealing. For better and proper sealing, the old sealing material should be removed. The sealing material consists of bitumen and sand, cork, impregnated fibre board or a ready-made sealing compound.

(ii) *Patching of damaged areas.* Patching of damaged concrete surfaces is a very difficult job and cannot be done satisfactorily. Small patches and holes are usually filled with bituminous materials or synthetic resins, grit mix like Araldite etc. The method of filling and repairing is the same as already explained.

(iii) *Repairing of blow-ups.* Due to longitudinal expansion, the transverse joints are buckled up i.e., the ends of the slab are raised up. This is known as blow-up. These blow-ups are sometimes too severe and the pavement remains in a raised position unless broken down. Blow-ups are a traffic hazard and should be repaired as soon as they appear on the surface. The blow-ups are repaired by removing the damaged portion of the pavement and replaced with a patch of concrete or bituminous materials. Sometimes the blow-up is temporarily repaired by road-side soil which has the advantage of allowing the pavement to settle down to its original position under the traffic load itself. After that the permanent patch is applied.

16.4 REPAIR OF SHOULDERS AND FOOT-PATHS

Shoulders and foot-paths constitute a very important part of a highway for the following three reasons, viz., (i) provide additional space on the carriage-way for the traffic to be used in emergency, (ii) protects the edges of the road thereby providing stability to the road surface, and (iii) provide walk for the pedestrians, cyclists etc. thereby reducing congestion on the road. Unless the importance of shoulders is understood properly the importance of its maintenance will remain half-way. It is desirable that there should be a colour contrast between the shoulder and the road surface. While maintaining shoulders the function of shoulders should be kept in mind. Shoulder and berms should be repaired as soon as they are damaged.

Foot-paths are raised platforms on one side or both sides of the road, for the use of pedestrians. Foot-paths are important feature of the urban roads. Foot-paths reduce congestion on the carriageways and thereby reducing chances of accidents. Therefore it is very important that foot-paths should be properly maintained and repaired.

16.5 REPAIR OF HIGHWAY DRAINAGE

Proper maintenance of highway drainage system means removal of water from highway surface and appurtenances as quickly as possible so as to keep the highway in good running condition. For efficient surface drainage the road surface should be smooth and should have proper gradients and camber. The water should be drained off smoothly to the side drains through the shoulders. The shoulders should be flush with the pavement so that water while flowing from the road surface should not enter the joints and thereby harming the sub-grade or base.

The side drains should be cleaned and properly maintained. The drains should have sufficient capacity to take the run-off effectively.

In urban roads sub-surface drainage (underground drains) also contribute a lot to the maintenance problems. The maintenance of sub-surface drains consists of maintaining and cleaning of manholes, inspection chambers, lamp holes etc. Periodic inspection of these appurtenances of the sub-surface drainage will reduce maintenance problems and hazards.

REVIEW QUESTIONS

16.1. What is maintenance and repair of highways? Also elucidate the importance of maintenance.

16.2. What are the essential features of maintenance of roads? Elucidate in brief.

16.3. How will you justify that the maintenance of shoulders directly affects the functioning of highway ?

16.4. What are the causes of pot-holes and ruts and how are they repaired in a bituminous road ?

16.5. What are the points to be considered while maintaining surface and sub-surface drainage of highways ?

Pavement Design

GENERAL

For economic and efficient construction of highways, correct design of the thickness of pavements for different conditions of traffic and sub-grades is essential. The science of pavement design is relatively new. In India, previously road crust was designed on some rational datas but more on the experience of the road engineer. Some arbitrary thickness of pavements were used which lead to costly failures and wastage as in some cases the thickness of pavements was insufficient and in other cases expensive. As there was no proper design criteria, the construction of roads was more or less uneconomical in almost all the cases. Hence judicious method of designing and calculating the crust thickness on the basis of estimation of traffic loads and bearing capacity of subgrade etc., will lead to economical construction of roads.

17.1 TYPE OF PAVEMENTS

Pavements can be classified as flexible pavements and rigid pavements. Flexible pavements are those pavements which can resist tensile stresses and having very little resistance to deformation under the wheel loads. The example of the flexible pavements are earth roads—stabilized and unstabilized, bituminous roads of all types such as surface painted, bituminous concrete premixed carpets etc. and water-bound macadam roads. The rigid pavements are those which provide great resistance to deformation under the wheel load. Plain and reinforced cement concrete roads are the examples of rigid pavements.

The essential differences between the rigid pavements and flexible pavements are:

(i) Temperature variations in different weathers do not produce stresses in the flexible pavements, but heavy and dangerous stresses are developed in the rigid pavements due to temperature variations.

(ii) The flexible pavements yield to occasional excessive wheel loads for which they are not designed but rigid pavements at once rupture due to excessive stresses caused due to occasional wheel loads.

(iii) If the sub-grade is of varying strength, the rigid pavement will not adjust to the differential settlements and will act as a beam or slab at the points of support whereas the flexible pavements will not adjust to the irregularities due to differential settlement.

17.2 DESIGN CRITERIA

As have been said earlier the pavement design is relatively a new science and in the initial stage some stress-strain assumptions were made on the characteristics of soil and distribution of the wheel load through the thickness of pavement. Most of the design methods are based on past experience. Some are empirical methods based on soil classification tests, soil strength tests etc. Some of the design methods are based partly on theory and partly on experience such as C.B.R. (California Bearing Ratio) method and Waterguard method for rigid pavements. The recently developed method is Group Index Method of USA, which gives approximate thickness of pavement.

Early methods of design of the pavements were based on the assumption that the load through the pavement is spread on the sub-grade through a cone. The pressure on the sub-grade thus calculated was compared with the actual bearing capacity of the soil and the thickness of the pavement thus determined. On the basis of theoretical and practical assumptions, the following factors have been vitally considered for a rational design :

1. The characteristics of the natural sub-grade which underlies the road.
2. The intensity and nature of traffic (to be assumed).
3. The amount of moisture present in the sub-soil and its drainage conditions.
4. The climatic conditions of the locality in respect of rains, snow, landslide etc.

17.3 FLEXIBLE PAVEMENT DESIGN METHODS

There are several methods of designing flexible pavements but the following two methods are commonly used in our country.

17.3.1 Group Index Method

Group Index is a number which characterises the nature of soil. The Group Index Method was first introduced by D.J. Steele. The Group Index is the

Group Index = Sum of readings on vertical scale of charts

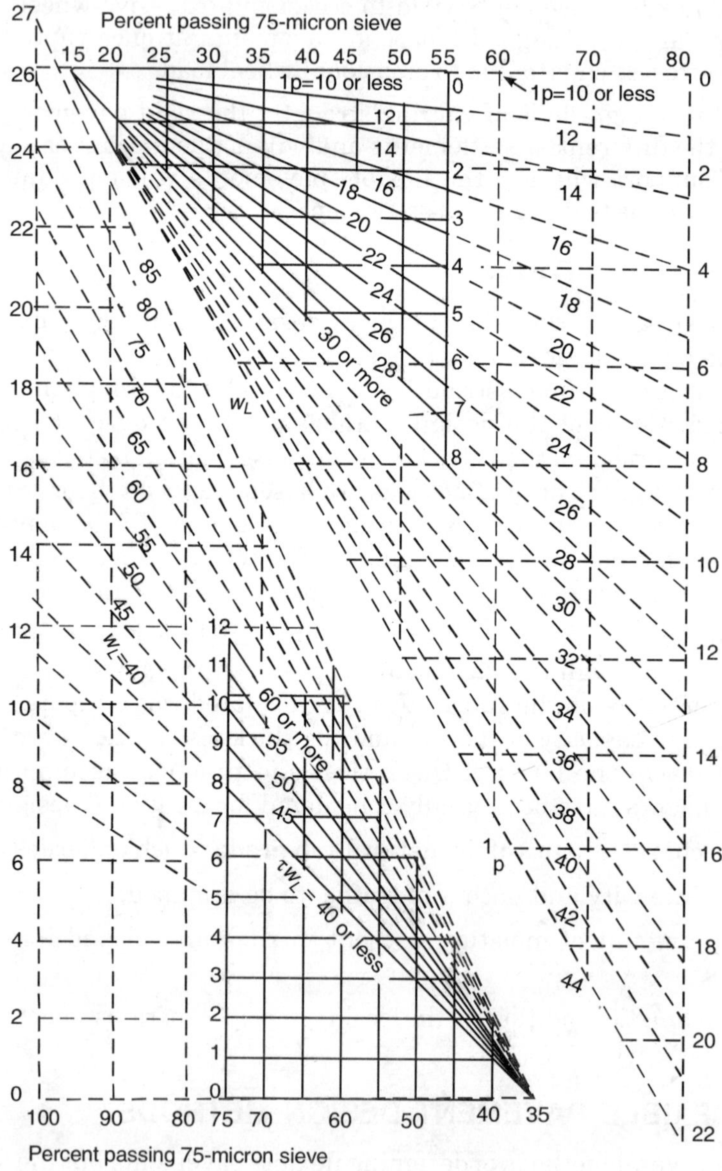

Figure 17.1 *Group index*

inverse measure of the qualities of a subgrade. More the Group Index number, lesser the bearing capacity of soil i.e., more the G.I. of a soil, the poorer the soil is, and hence more the thickness of the pavement and sub-base. The actual thickness of the base and the surfacing will depend on the intensity of traffic. The group index is a function of the amount of material passing the 75-micron I.S. Sieve, the liquid limit and the plastic limit is given by the following equation:

Group Index (G.I.) = $0.2\,a + 0.005\,ac + 0.1\,bd$

where

- a that portion of percentage passing 75-micron sieve greater than 35 and not exceeding 7.5 expressed as a positive whole number between 0 and 40.
- b that portion of percentage passing 75 micron sieve greater than 15 and not exceeding 55 expressed as a positive whole number (0–40)
- c that portion of the numerical liquid limit greater than 40 and not exceeding 60 expressed as positive whole number between 0 and 20.
- d that portion of the numerical plasticity index greater than 20 and not exceeding 30 expressed as a positive whole number between 0 and 20.

Figure 17.1 shows a chart of Group Index numbers from which G.I. can be conveniently obtained as the sum of readings on the vertical scale of the chart.

As has been said earlier, higher the group index the poorer is the sub-grade and greater is the thickness of pavement. The thickness of the sub-base will increase with the increase in the group index.

The thickness of surfacing and the base will depend on the traffic volume and intensity. If the number of commercial vehicles passing on the road per day is less than 50, it is termed as light traffic, if the number of commercial vehicles is between 50 and 300, it is termed as medium traffic and if the number of commercial vehicles passing on the road is more than 300 per day, it is termed as heavy traffic.

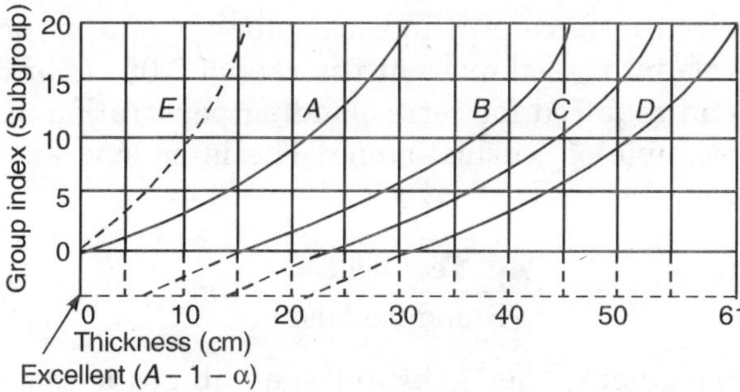

Figure 17.2

Curve *A* gives the thickness of sub-base to be provided, curve *B* gives the overall thickness of the road crust for light traffic, curve *C* for medium traffic and curve *D* for heavy traffic.

17.3.2 C.B.R. Method

The California Bearing Ratio (C.B.R.) method is the most commonly used method of designing flexible pavements. The C.B.R. is defined as the ratio

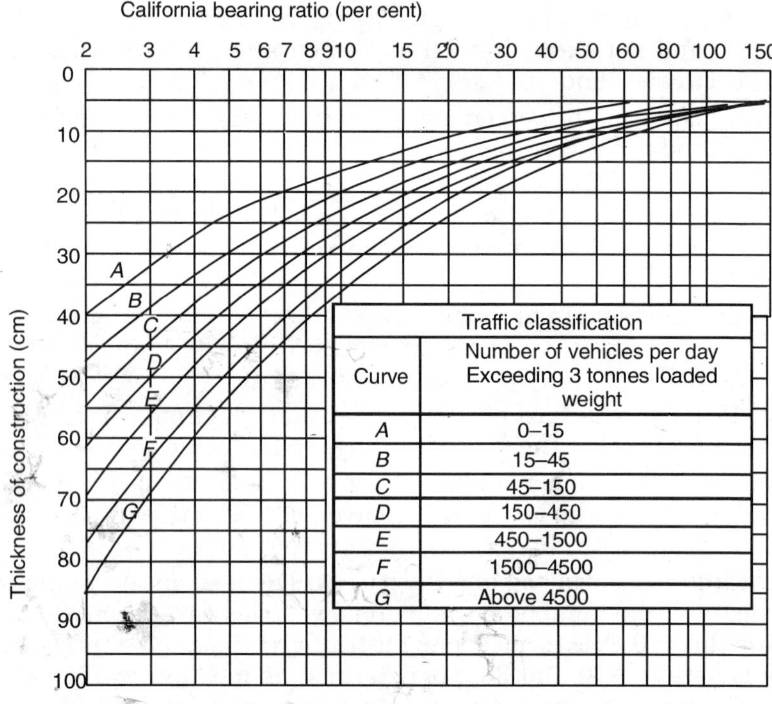

Figure 17.3

of test load required to force a cylindrical plunger of 3 m^2 or 19.355 cm^2 cross-sectional area into a soil mass at the rate of 0.05 cm or 0.127 cm per minute, to the load required for corresponding penetration of the plunger into a standard sample of crushed stone. The later load being known as *Standard Load.*

$$\text{C.B.R.} = \frac{\text{Test load}}{\text{Standard load}} \times 100$$

To design a pavement, the C.B.R. of the sub-grade material is first determined and the corresponding total thickness of the pavement required is read from the design curves. The total thickness so obtained is sub-divided into base, sub-base and surfacing by knowing the value of C.B.R. value for sub-base and base materials.

17.4 WHEEL LOAD

For the design of pavements, the wheel loads which the pavement is expected to withstand are decided. The pavement is designed from the point of view

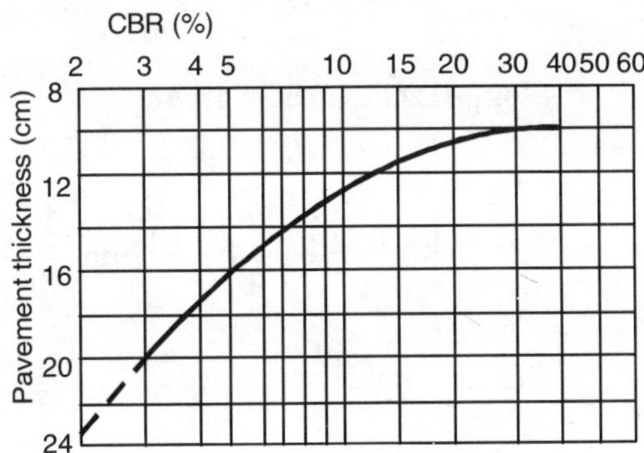

Figure 17.4

of maximum wheel load. The gross load of the vehicle has no importance and is not taken into consideration. Another important consideration is the effect of repetition of loads on the pavement which induce failure by causing comulative permanent deformations. To counter this effect an allowance is made by increasing the static wheel load by 10 per cent for medium traffic and 20% for heavy traffic.

In India, the bullock cart with steel-tyre wheels is a very common mode of transportation. A design curve proposed by N. Mohan Rao of Road Research Institute, New Delhi, gives pavement thickness for known C.B.R. value is shown in Fig. 17.4. Since the stresses caused due to bullock cart are very high particularly in the surface layers of the pavement. For bullock carts the top 10 cm thick layer should consist of very strong material.

In India, the static wheel load is taken as 2268 kg for commercial vehicles with contact surface of 12 cm and for bullock cart traffic the static wheel load is taken as 1270 kg and contact surface 4.5 cm.

17.5 DESIGN OF RIGID PAVEMENTS

Rigid pavements constructed with cement concrete will depend on their strength upon the slab action. The cement concrete being a rigid material, helps in spreading the load over a large area and that too uniformly over the sub-grade. It has generally been seen that the failure of rigid pavements tend to occur due to over stressing of the concrete itself than by the failure of sub-grade as is the case with flexible pavements. For the design of rigid pavements, Westergaard derived analytical formulae which were later modified by himself and Teller and Sutherland. The formulae are :

$$f_{(\text{centre})} = 0.275(1+\mu)\frac{P}{h^2}\ 4\log_{10}\ \frac{l}{b}$$

$$+\log_{10}\{12(1-\mu^2)\} - 54.54c_2\ \frac{l}{c_1}^{\ 2}$$

$$f_{(\text{edge})} = 0.529\ (1 + 054\ \mu)\ \frac{P}{h^2}\ \ 4\ \log_{10}\ \frac{1}{b}\ + \log_{10}\ \frac{b}{2.54}$$

$$f_{(\text{corner})} = \frac{3P}{h^2}\ 1 - \ \frac{a\sqrt{2}}{l}^{\ 1.2}$$

where

P = Wheel load in kg

h = Slab thickness in cm

l = Radius of relative stiffness which measures the stiffness of slab

$$l = \frac{Eh^3}{12\left(1-\mu^2\right)k_s}^{1/4}$$

μ = Poisson's ratio for concrete (0.1 – 0.35)

$f_{(\text{centre})}$ = Maximum tensile stress at the bottom of the slab due to loading in the centre (kg/cm^2)

$f_{(\text{edge})}$ = Maximum tensile stress at the bottom of the slab due to loading at the edge

$f_{(\text{corner})}$ = Maximum tensile stress at the top of the slab due to loading at the corner of the slab

REVIEW QUESTIONS

17.1. Differentiate between flexible pavements and rigid pavements as far as their design criteria is concerned.

17.2. What are the design criteria for the design of flexible pavements. Illucidate briefly.

17.3. Discuss those properties of cement concrete which are of vital importance in the design of concrete pavements.

17.4. Discuss the merits and demerits of the Group Index Method of flexible pavement design.

17.5. Design a suitable pavement for light traffic and calculate the Group Index for the following characteristics of a sub-grade soil :

Percentage passing through 75 micron sieve = 80, liquid limit = 30%, plasticity index = 15.

17.6. Elucidate any one method for the design for a flexible pavement of a highway.

18

Street Lighting

GENERAL

In the olden days, when there was not much of the fast moving vehicles, oil-lamps were used on the road-side. This practice is still in use in Indian villages. These oil lamps are only meant to show the path to the pedestrians and the bullock cart traffic. This was actually more a psychological than the actual benefit. With the advancement of fast moving traffic, hazards of travel for pedestrians and the automobiles have increased considerably due to dusk, darkness and impaired vision. In modern road development, efficient street lighting has become an important feature for the reduction of crime expedition of traffic.

18.1 FACTORS AFFECTING STREET LIGHTING

For efficient and proper street lighting the following factors should be kept in mind :

(i) The electric posts poles should be located at proper places.

(ii) The light from the source should illuminate the entire road surface uniformly.

(iii) The posts should be spaced uniformly.

(iv) The light should not produce glare.

(v) The height of lamps should be adjusted in such a way that it should produce such a shadow which may not darken the entire line.

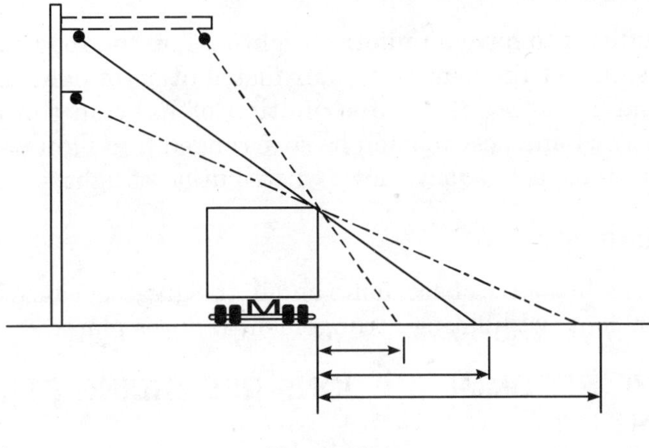

Figure 18.1

18.2 FACTORS EFFECTING VISIBILITY ON ROADS

Source of light Placing and distance of lamp posts distribution of light a lare.

Following factors determine the visibility on roads at night.

18.2.1 Source of Light

Source, power and type of light greatly effect the visibility on the road. Lighting on modern highways can be an ordinary filament lamp, sodium vapour lamp, mercury lamp or a fluorescent type lamp. Filament lamps are commonly used for ordinary roads for street lighting as they are cheap and easily available. Sodium vapour lamps produce distinctive yellow light are most suitable for hazardous localities. Mercury vapour lamps are being increasingly used on account of their long life and light efficiency. Fluorescent lamps are particularly used for overhead lighting.

18.2.2 Placing and Distance of Lamp Posts

Lamp posts can be located either on one side or staggered on both sides of the road or centrally overhung. Light posts staggered on both sides and centrally overhung are very efficient but costly and are used only on very important roads passing through busy and congested localities. On ordinary roads, lamp posts are placed on one side of the road only. The light posts should be spaced in such a way so as to produce uniform illumination on the road surface. It has been observed by experience that spacing between the light posts should be between 40 m and 60 m and the height of lamp should be between 3.5 m and 5.0 m.

18.2.3 Distribution of Light

It is very difficult to have a uniform brightness on the road surface during night hours, but at the same time fair distribution of light can always be achieved patchy and uneven distribution of light on the road is very dangerous. The lamp posts should be so arranged that light patches should just each other so as to avoid uneven distribution of light.

18.2.4 Glare

Glare is the main source of accidents and in the greatest cause of wastage of illumination contract lighting arrangement reduces glare.

18.3 SOME RECOMMENDATIONS REGARDING STREET LIGHTING

Street lighting plays an important role in maximising the use of highways with advantage. Properly designed street lights add to aesthetic appearance of the highway and minimise accidents. Uniform lightness of the road surface is an essential condition of street lighting. Professor William E. Barrow of University of Marrine suggests that if a particular street is wide enough, its brightness will depend on the location of the lamp posts with respect to the street as well as upon the amount of light delivered on the road surface. If the lamps are mounted on the road kerbs of wide lanes, the streak of light will be on the edges of the road and the central portion of the road with become dark. Better results can be obtained by hanging the lamps through long brackets so as to produce streaks of light as near the centre of the road, as possible. Better arrangement of street lighting will be to provide lamp posts on both sides in a staggered way. Overhanging lamps on the centre of pavement or electric poles arranged on the centre line of the road, will make the surface more attractive, produce uniform brightness and improve visibility. Road surfacing is also responsible for improving, visibility, brightness and uniform illumination. A light coloured surface of seal coat is very much desirable because it reflects light to the driver. A highway with uniform horizontal illumination throughout its length presents a large variation in brightness when viewed by drivers and pedestrians. For uniform brightness, the light must be distributed to take account of the reflection of light at small angles from the surface.

18.4 DESIGN OF HIGHWAY STREET LIGHTING

Lighting from one side of the road only usually is unsatisfactory except on bends and narrow roads. Due to economy single side lighting is usually adopted even for two lane roads. The maximum distance between two opposite rows of lights should not be more than 9.0 m. The roads upto 9.0 m width can be effectively lighted by lights mounted on the kerb-line in vertical line above the kerb.

When the width of the road is increased, brackets are used for overhanging the lights above the road. The maximum overhang of brackets over the kerb should be 1.8 m to avoid undue dark patches on the kerb and footways.

Following are the I.R.C. recommendations for horizontal clearance for lighting poles.

(i) For roads with raised kerbs (as in urban roads)	0.3 m minimum desirable 0.6 m from the edge of the raised kerb
(ii) For roads without raised kerbs (as in rural roads)	1.5 m minimum from the edge of the carriageway, subject to a minimum of 4.0 m from the centre line of the carriageway

The side mounted lamp posts can be arranged opposite to each other on both sides of the road or arranged staggered. The staggered arrangement is more efficient.

The usual mounting height of lamp post is 6.1 m to 7.6 m in 4.6 m (minimum) wide streets. The mounting height is between 7.6 m to 9.0 m in highways with 35 to 45 m centre to centre spacing. The ratio of spacing of poles to the mounting height is kept between 8 : 1 in congested areas and 10 : 1 in thickly populated areas. High mounting greatly reduces the blinding effect of direct glare.

The spacing of the light poles is reduced on the curves, bridges, level crossings etc. for better light requirements.

I.R.C. has specified the minimum vertical clearance required for electric power line upto 650 volts as 6.0 m. All the poles carrying overhead power and telecommunication lines except the urban areas, should be erected at least 10.0 m away from the nearest edge of the roadway. The distance of this line should be 0.5 m from the nearest line of avenue trees.

If the trees are interfering with the side lamps, they should be centrally suspended. On the S-curves the lamp posts should be changed from one side of the road to the other at some point in the middle of the bend.

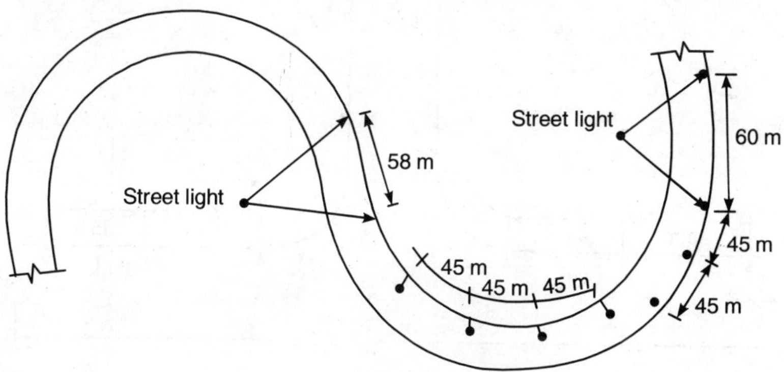

(a) Street light position on curves

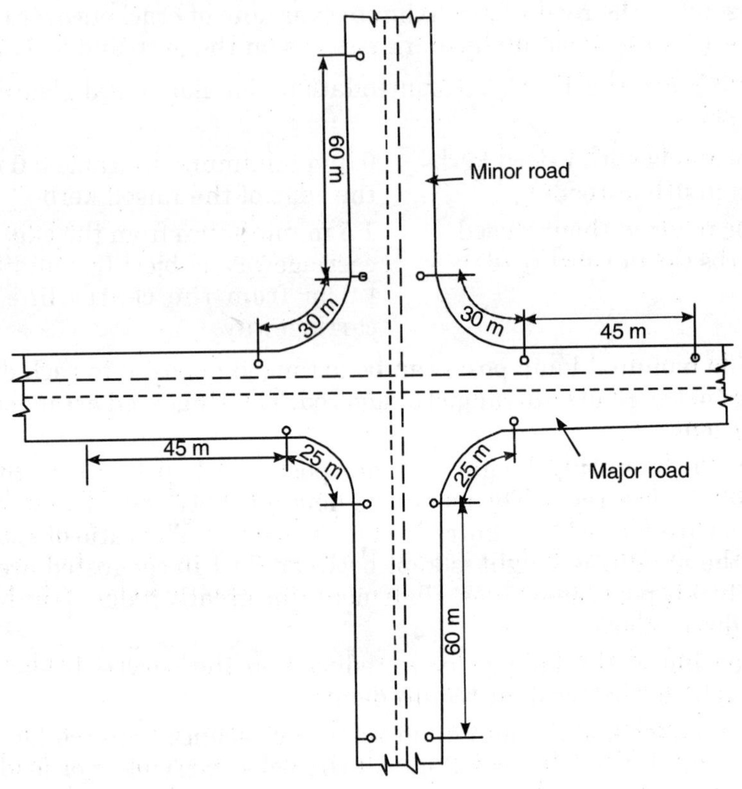

(b) Street light position on intersections

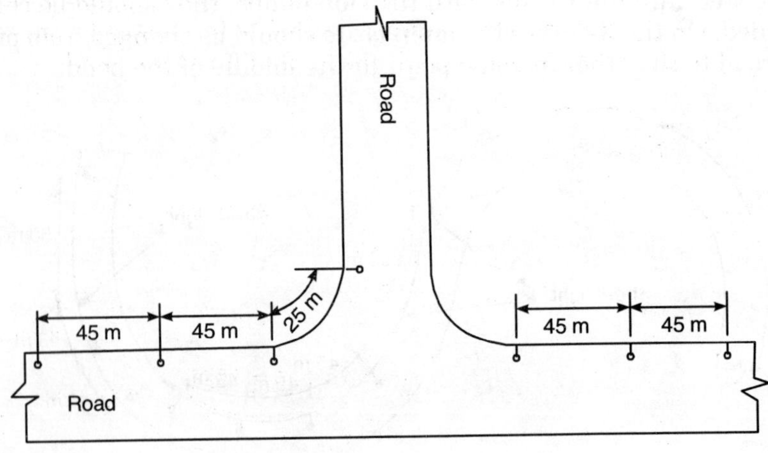

(c) Street light position on T-junctions

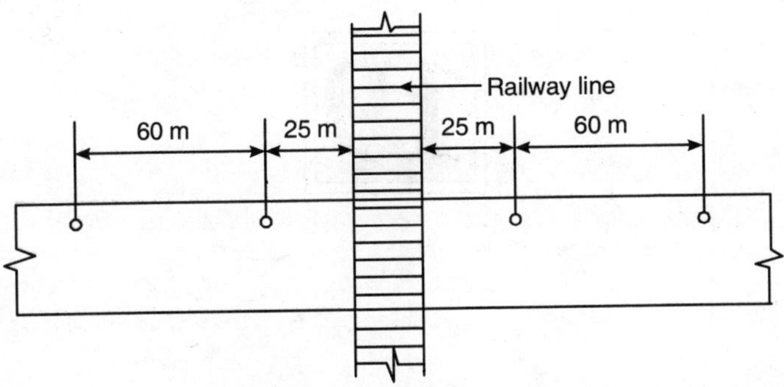

(d) Street light position on railway crossing

Figure 18.2

At all the road junctions and roundabouts, it is most important for safety to ensure that sources of light are so placed that a driver cannot only see from some distance away that he is approaching a junction but he can also see the route, he has to follow. At the small roundabouts upto 18.0 m diameter, a single light centrally mounted at a height of 9 to 10.6 m above the carriageway will be suitable. For bigger roundabouts, the light should be provided along its circumference.

At the mouth of the crossing, a bright patch must be provided. On the intersection of two roads, four light lamps at not more than 12.0 m along each road should be provided.

19

Airport Engineering

GENERAL

Roads act as feeders to aerodromes and airports. The airport engineering deals with the construction and maintenance of air strips, runways etc. In fact this is not all as far as airport engineering is concerned because it has now come into very much prominance and deals with planning and designing of runways, controls, signals and so many other things. But this chapter has been devoted mainly for the requirements, construction and maintenance of air-strips.

19.1 DEFINITIONS

19.1.1 Airport

It is the place where public facilities are provided for shelter, waiting lounge, restaurant etc. Airport also provides facilities for servicing and repair of aircrafts, refueling, receiving and discharging cargo etc.

19.1.2 Runways

It is a hard surfaced straight path, used for the movement of aeroplanes for taking off and landing within a landing area.

19.1.3 Landing Area

It is that portion of the airfield which is used primarily for taking off and landing of aircrafts. (The word aircraft is a general term used for aeroplanes, gliders, helicopter etc).

19.2 COMPONENTS OF AN AIRPORT

A large size international airport is in fact a city in itself. It has almost all the amenities of a modern life. It is normally spread over a vast area of open land. Following are the various components of a modern airport :

1. Runways
2. Lounge
3. Restaurants and shopping complex
4. Hangers
5. Fueling
6. Taxiways
7. Loading and unloading cargo yards
8. Aprons.

In this chapter only runways will be discussed as it is beyond the scope of this book to discuss about all the components of an airport.

19.3 LOCATION OF AN AIRPORT

For the location of an airport there are many a considerations such as political, geographical, aeronautical, military, economic and many other considerations. The following points should be particularly thought over :

1. It should be far away (minimum 3 kilometres) from the urban boundary or congested locality.

2. It should not have obstruction i.e., such objects which may obstruct the movement of flights.

3. It should not have restrictions for future expansions and developments.

4. It should be away from industrial hazardous, thereby limiting the operations of flights due to poor visibility etc.

5. The site of an airport should be away from the hillock, river and pond etc.

6. It should be well connected by a national highway or state highway.

19.4 TYPE OF AIRPORTS

In different countries, airports are classified in different ways. In India airports are classified as civil and military. The military airports are mainly meant for military aircrafts only. The civilian airports can be used for military aircrafts also. The civilian airports can be a minor, intermediate, major and international airports. For the development of airports and airways the Govt. of India under the Ministry of Civil Aviation, established two public sector enterprises namely Air India and Indian Airlines. The Air India International for foreign services and Indian Airlines for domestic flights. The airways and air services are fast growing world over and hence the airport engineering. The airways and airports are rendering creditable service to the mankind. There are various methods of classifying the airports. The International Air Transport Association (IATA) has fixed certain norms for the classification of airports.

19.5 ORIENTATION OF AIRPORTS

In the planning of airports, wind direction, its intensity, number of air strips, and metrological conditions, play an important role. The landing and take-off is done in the direction of prevailing winds :

Small aeroplanes are more sensitive than big commercial jets or boeings and airbuses, because of less wing load. The number of airways or air strips depends on the intensity of prevailing winds. While orienting runways, cross components of the prevailing winds should be calculated. Normally properly oriented runways may persuit aircrafts with cross component of wind as 20 km per hour (cross component of wind is defined as wind velocity multiplied by sine of angle between runways and direction of wind). At the time of orienting runways, direction and velocity may be obtained from metrological departments.

19.6 LENGTH OF RUNWAYS

For taking off and landing, the length of a runway may be divided into :

(i) Distance from the runway end to the starting point.

(ii) Distance between the starting point and the take off speed.

(iii) Additional distance for eventuality when the engine of the aircraft fails and the engine comes to a halt.

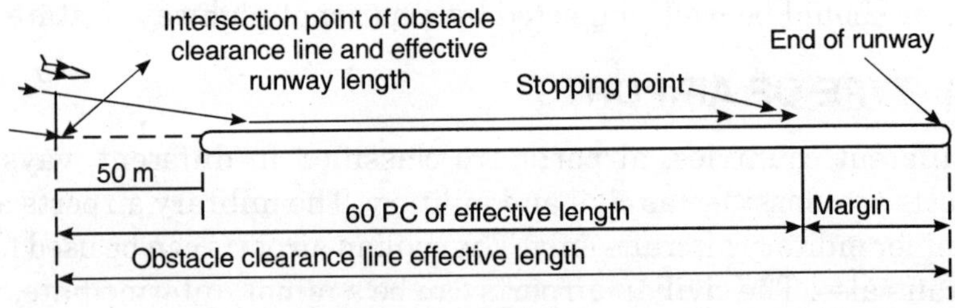

Figure 19.1 *Runway limitation for landing of planes*

The division of the length of a runway between different zones for landing and take-off are shown in Fig. 19.2. The take-off of the aircraft is more

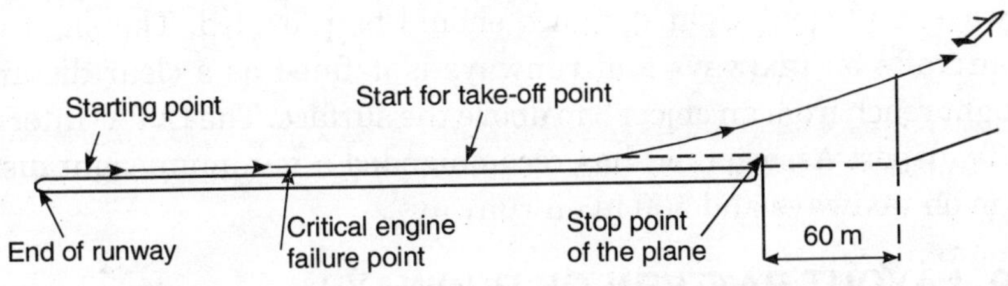

Figure 19.2 *Runway limitation for take-off of planes*

critical than the landing. Only 60% of the runway length is used for take-off and landing and the remaining 40% is used as a margin for unforeseen or eventualities.

19.7 TAXIWAYS

It is defined as a paved path provided for the purpose of allowing aircrafts to move to and from the runway and the apron. Movement of heavy aircrafts

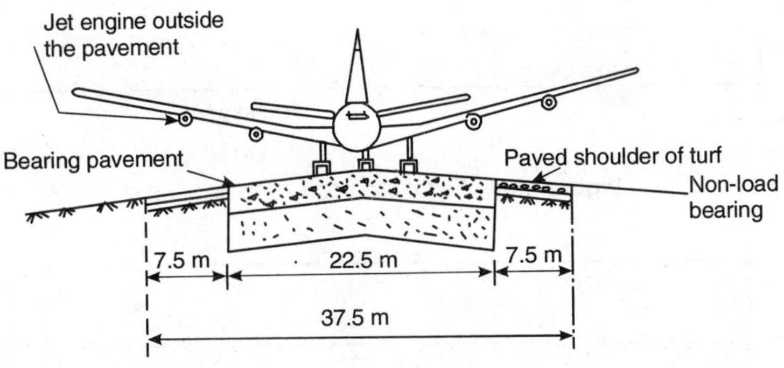

Figure 19.3 *Taxiway*

on the ground means heavy consumption of fuel. The length of taxiways should be as small as possible. No standard specification have been laid down by any organisation regarding the minimum length of taxiways. The speed of aircrafts is much less on the taxiways than on the runways. Due to this reason the width of taxiways is less than the runways and it varies from 12 m to 23 m depending upon the type of aircrafts using it. The centre to centre distance between two taxiways varies between 38 m to 100 m.

19.8 SIGHT DISTANCE

For smooth and uninterrupted movement of aircrafts on taxiways and runways, sufficient sight distance should be provided. The sight distance for aircrafts on taxiways and runways is defined as a clear distance on a straight reach from an object 3 m above the surface. The IATA (International Air Transport Association) has recommended a minimum sight distance of 300 m on taxiways and 500 in on runways.

19.9 LAYOUT PATTERN OF RUNWAYS

The layout pattern of runways may be classified as :

(i) One-way (ii) Two-way and (iii) Three-way.

The various type of runways are illustrated in Figs 19.4 to 19.12.

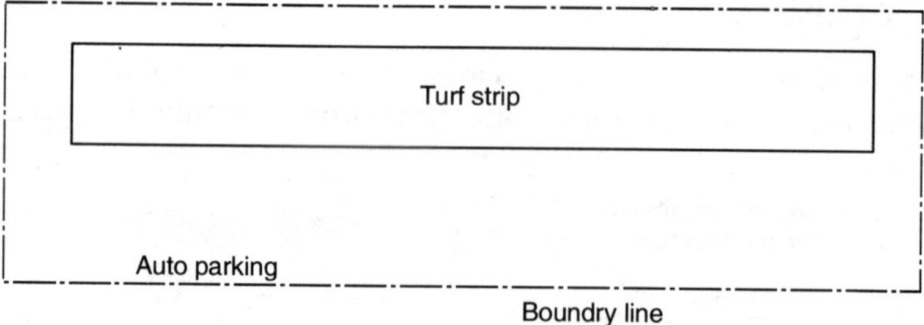

Figure 19.4 *Flight-strip one runway pattern*

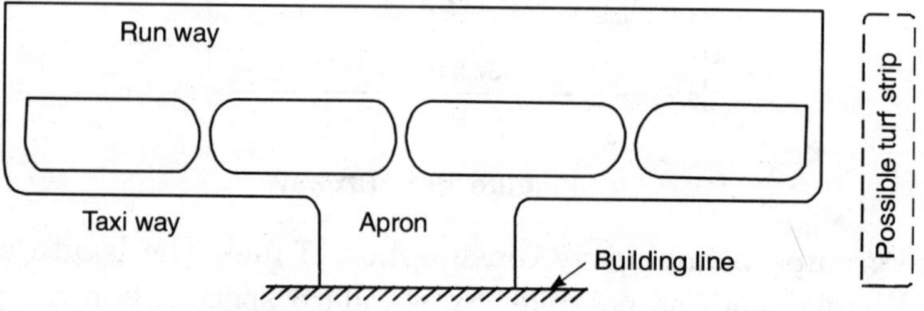

Figure 19.5 *Single runway pattern*

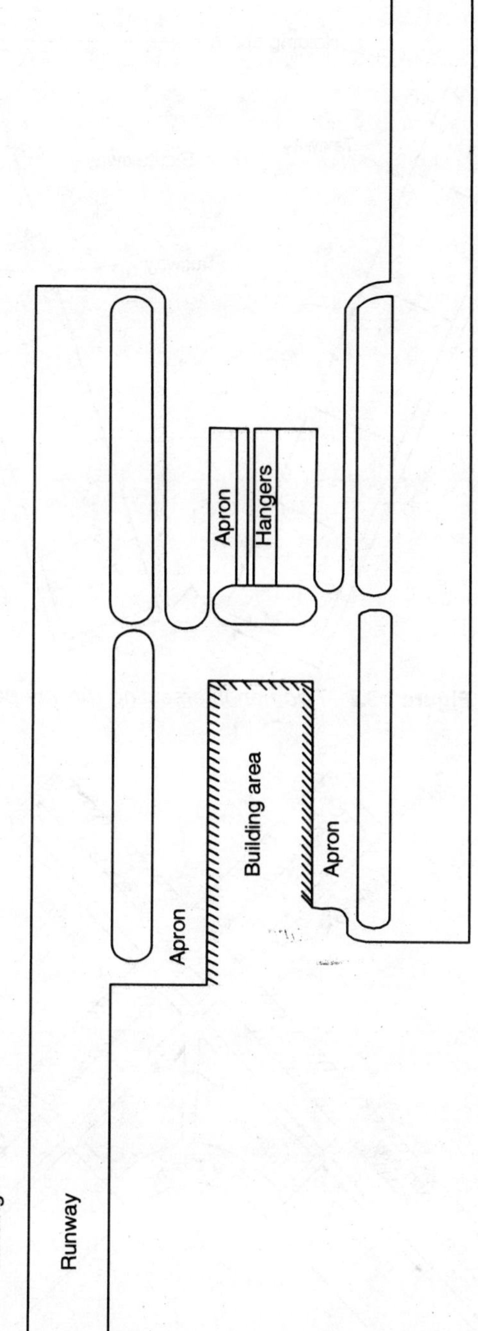

Figure 19.6 *Parallel runway*

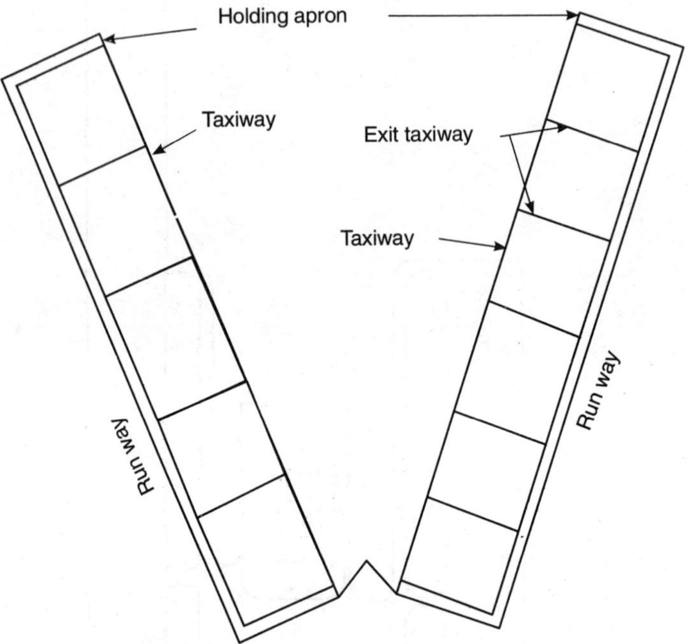

Figure 19.7 *Two non-intersecting runway pattern*

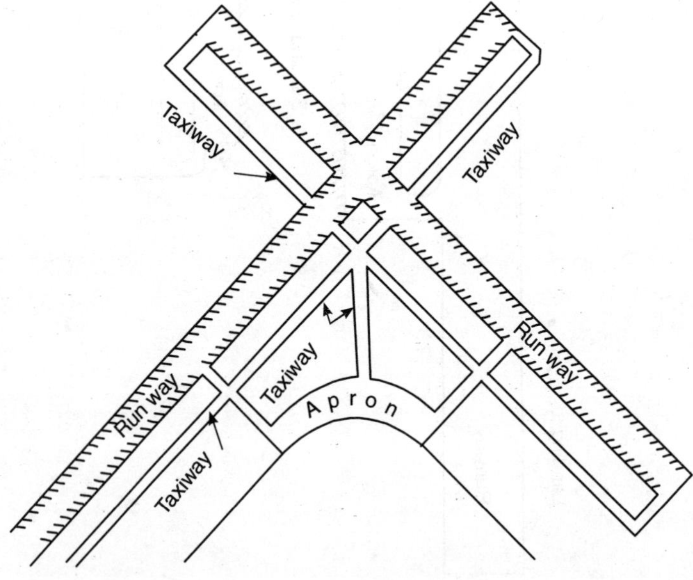

Figure 19.8 *Two non-intersecting runway pattern*

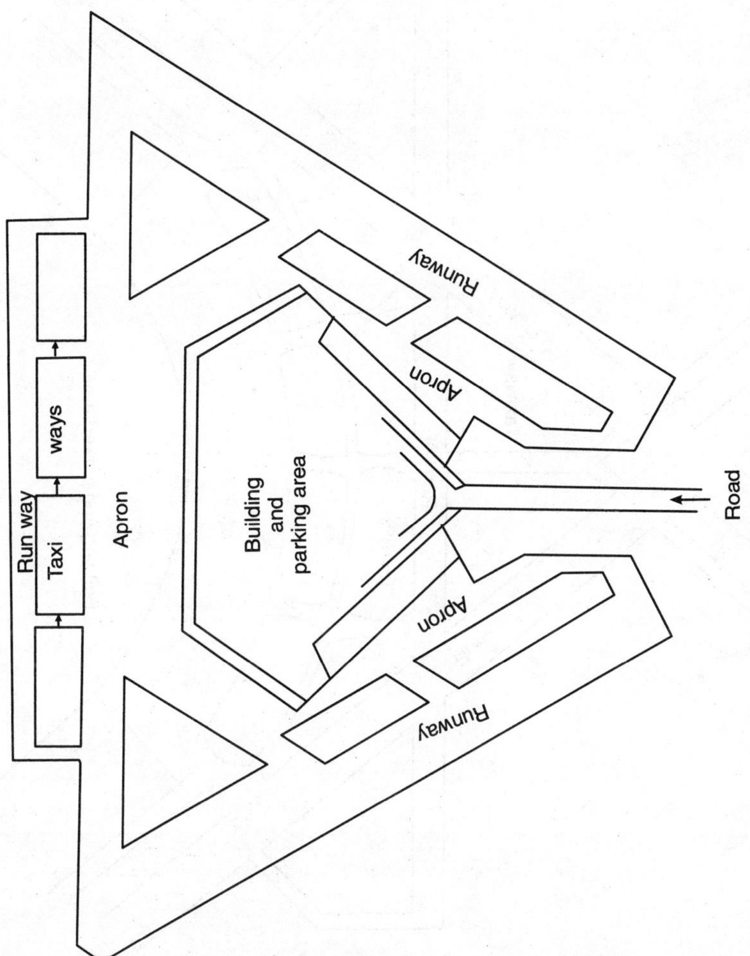

Figure 19.9 *Three non-intersecting runway pattern*

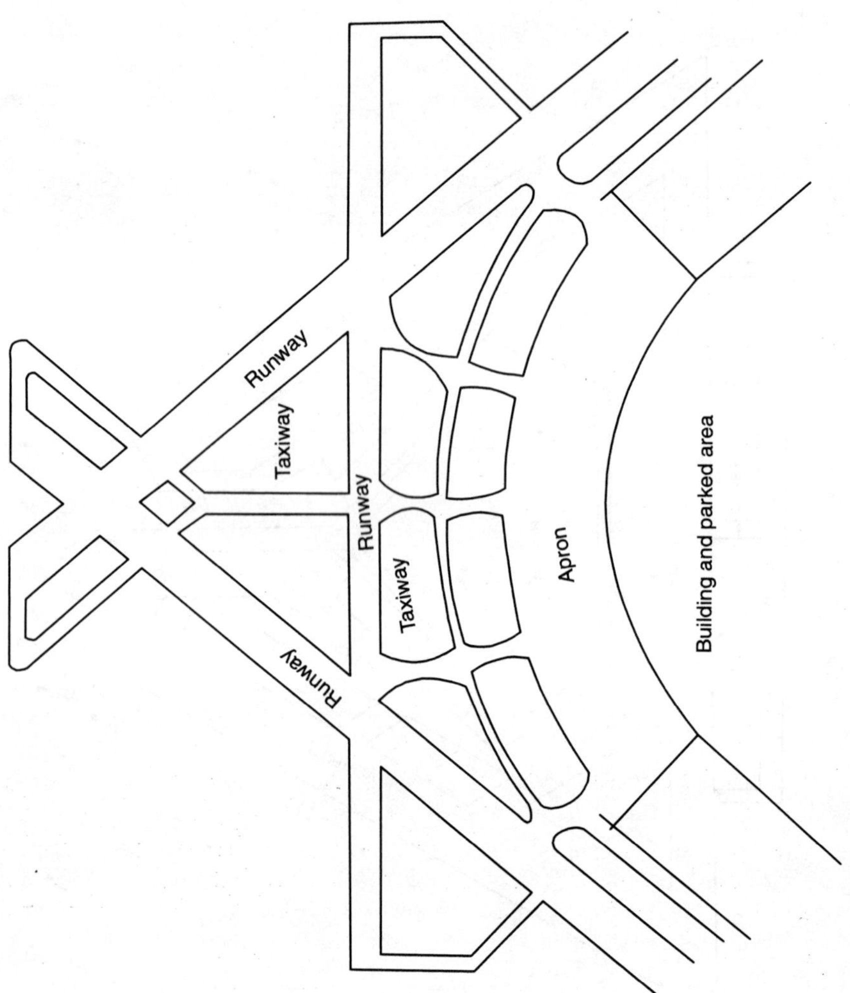

Figure 19.10 *Three intersecting runway pattern*

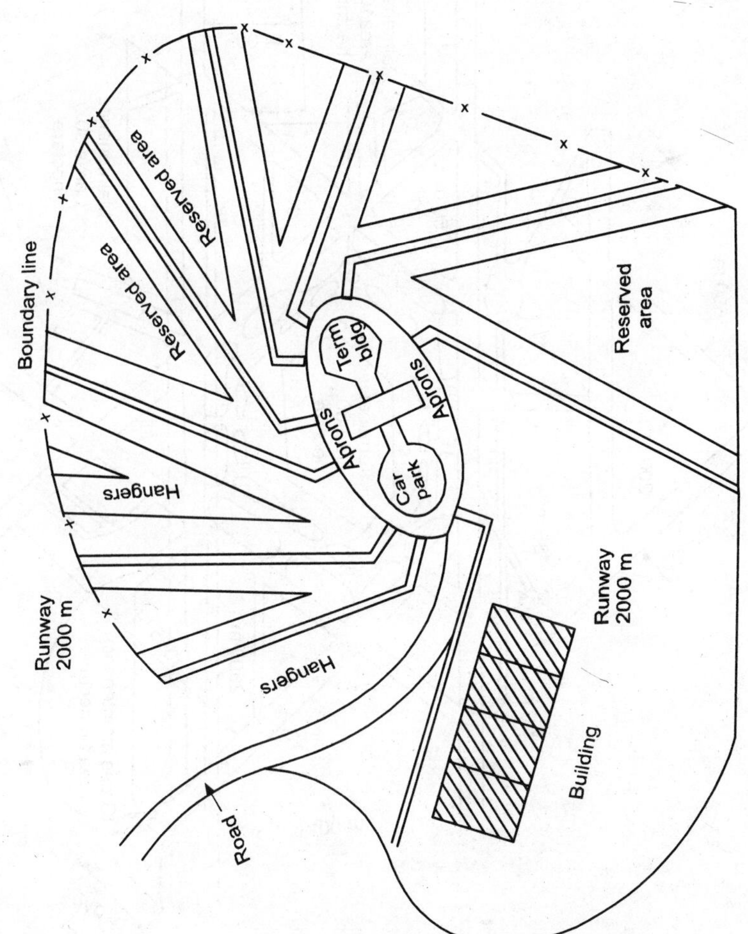

Figure 19.11 *Tangential runways pattern*

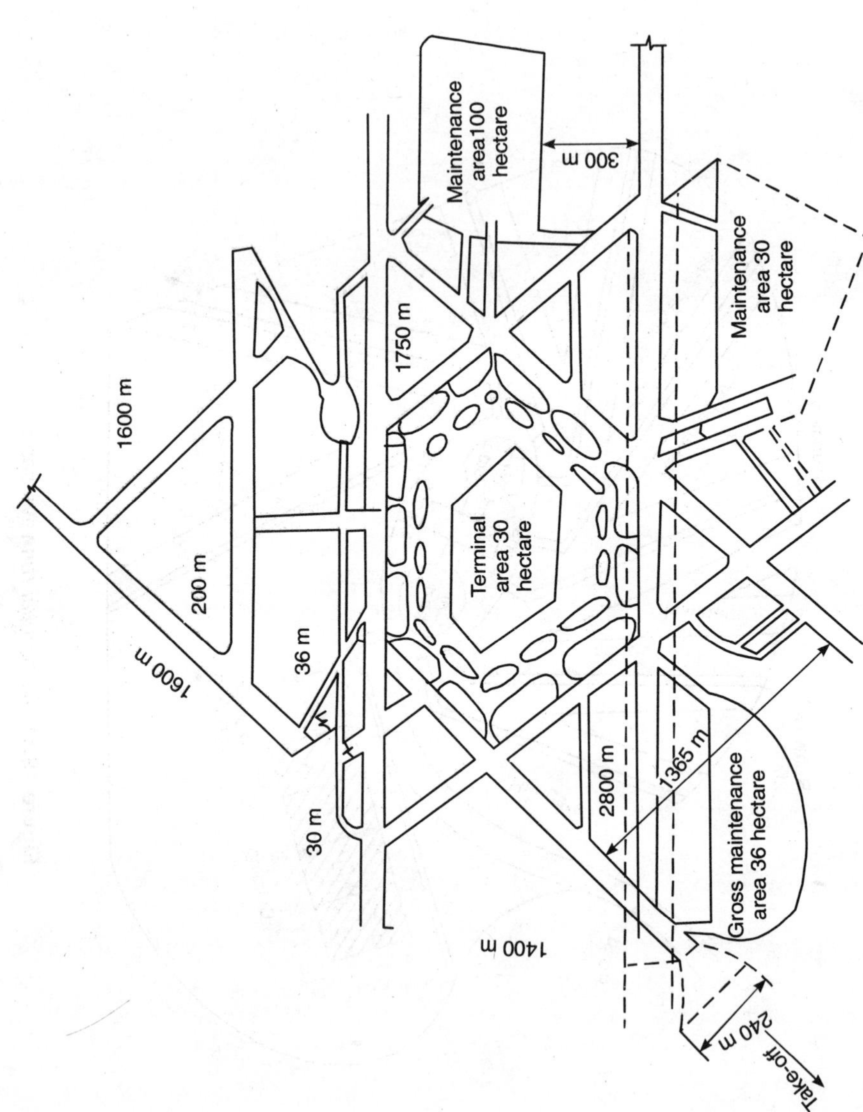

Figure 19.12 *Combination of various pattern of runways*

REVIEW QUESTIONS

19.1. Write short notes on :

Cross component of wind, Slight distance, Apron.

19.2. Elucidate how taxiways are planned in an airfield and what are their requirements.

19.3. How are runways oriented? What important factors mainly govern the orientation of runways?

19.4. Draw layout plans of any two type of runway patterns.

PART 2

Railway Engineering

Introduction

GENERAL

In ancient times Romans were first to try in the running of animal-drawn vehicles over two parallel lines of stones and bricks embedded in the ground. In the 15th century wooden planks and beams were used in place of stones in Europe and England, and such type of roads were known as 'Tramways'. Good speeds were obtained on such tramways by horse-drawn vehicles. But such wooden rails had short life because they wore out quickly and caused much difficulties. To increase the life of wooden rails, these were covered with iron plates and such covered paths were known as 'Plateways'. After some time a new idea, for preventing the lateral movement of wheels by using angle irons in place of plane iron sheets developed and the use of angle irons was started. It was the beginning of the development of the modern railways. Later on these angle irons were replaced by cast iron rails with raised flanges on outerside, because those raised flanges were more useful in preventing the lateral movement of the vehicles.

When it was observed that the animals can draw the vehicles on C.I. rails at good speed than on roads, the people of the 17th century started thinking about some mechanical device which could replace the animals. It is understood that in France Nicholas Cugnot was first to get success in 1771, in the construction of a steam locomotive. In Britain in 1786 William Murdock, a Scotman, also got success in the preparation of a steam locomotive model. In 1797 a Cornish Engineer Trevithick also did a lot of work on steam locomotive and got success in 1804 in the design and construction of a steam locomotive. In this way history tells us that a number of engineers tried to build a locomotive in the early days, but actually George Stephenson (1781–1848) was the first man to get complete success in this field.

George Stephenson completely designed, planned, constructed, and got success in running the first train of the world on 27th September, 1825 in England between Stockton and Darlington in the country of Durham.

After this the development of railways was started rapidly in all parts of the world.

20.1 DEVELOPMENT OF RAILWAY IN INDIA

Before the introduction of railways the communication means in India were extremely poor in nineteenth century. Actually it was only Lord Dalhousie, who insisted for the development of railways in India. The first proposal for the construction of railways was put up in 1844 by Mr. R.M. Stephenson to East India Company. The construction of first experimental line from Calcutta (now Kolkata) to Mirzapur was undertaken by East India Company in 1849. Similarly another contract to construct railway line from Bombay (now Mumbai) to Kalyan was given in 1850 to Indian Peninsula Railway.

The first train in India was run at 3.30 P.M. on 16th April 1853 between Bombay and Thana. It carried about 400 people in 14 coaches and was driven by three engines.

During 1850 to 1860 following eight different private companies were established in India for the development of railways :

1. The East Indian Railway
2. The Great Indian Peninsula Railway
3. The Madras Railway
4. The Bombay–Baroda and Central India Railway
5. The Eastern Bengal Railway
6. The Scindia Railway
7. The Calcutta and South Eastern Railway
8. The South Indian Railway

During 1871 a Selection Committee of the British Parliament was appointed to review the various schemes for the development of railways in India.

Due to disastrous famines from 1874 to 1879 for the supply of food to the famine affected areas, rapid development of the railway was demanded. Between 1881 to 1897 following new private companies were established :

1. Bengal Central Railway
2. The Rohilkhand and Kumaon Railway
3. The South Mahratta Railway
4. The Bengal North Western Railway

5. The Indian Midland Railway
6. The Bengal Nagpur Railway
7. The Assam Bengal Railway Company
8. The Burma Railway Company

In 1905 Railway Board was established under the Commerce and Industries Department to look after the railways. During First World War between 1914–1921 the railway fares were increased and some lines of strategic importance were constructed. The total length of lines laid in India by 1921 was 58,770 km.

In 1922 nationalisation of railway started. In 1937 Burma was separated from India, curtailing 3200 km line. During 1942 Second World War, 'War Transport Board' was formed and for the movement of military lines upto various strategic points were increased between 1939–1947.

During 1949–50 most of the railway companies were acquired by the Government. After the partition of India in 1947 the following zonal grouping of railways was done in India :

Railway Zones	*Headquarters*
1. Eastern Railway (E.R.)	Kolkata
2. South Eastern Railway (S.E.R.)	Kolkata
3. Northern Railway (N.R.)	Delhi
4. North-Eastern Railway (N.E.R.)	Gorakhpur
5. Southern Railway (S.R.)	Chennai
6. Central Railway (C.R.)	Mumbai
7. Western Railway (W.R.)	Mumbai
8. North East Frontier Railway (N.E.F.R.)	Maligaon (Gauhati)
9. South Central Railway (S.C.R.)	Secunderabad

The railway system of India is a biggest in Asia and second longest in the world under a single management, being next only to the Soviet Railways.

The total route of Indian railways was 65,217 km before partition but in 1947 after partition it came down to 54,149 km. At present total route is about 62,000 km. Indian Railways have about 11,800 engines, 37,000 coaches and 3,90,000 freight wagons. The Indian Railways carry nearly 6.8 million passengers and 6.3 million tonnes of freight every day. Indian Railways have about 7,100 railway stations scattered all over the country to serve the nation. It has given direct employment to about 1.45 million people. Indian Railways operate about 9,600 trains daily. The traffic density in certain parts of India is the highest in the world. The total distance covered by all the railway trains in India is more than 3.5 times the distance from the earth to the moon. After Independence there has been rapid development in the railway diesel traction, which was previously unknown. It now covers 10,000 km of track daily. Three production units have been set up : the Chittaranjan Locomotive Works which turned out the first locomotive in 1950; the Integral Coach Factory at Chennai which started production of passenger coaches in 1955; the Locomotive Works at Varanasi, which went into production in January 1964, and turned out its 100th diesel locomotive in January 1967. The wagon building industry in the private sector has capacity of manufacturing 35,000 wagons a year. The total investment which has been done on Indian Railways is about 3,320 crores of rupees. By the end of 1977 Indian Railways covered about 5,200 km of electrification lines, 28,500 km diesel traction and the remaining under steam locomotive traction.

For the development of railways the Indian Railway Board has been formed which looks after it.

20.2 DEFINITIONS

1. Adzing of sleepers. The sleepers are cut at the rail seat to provide a slope of 1 in 20 to the rails. The method of cutting the wooden sleepers or casting of concrete sleepers accordingly is known as adzing of sleepers.

2. Advance starter signal. The last stop signal of semaphore type for giving indication to the outgoing train from the railway station.

3. Articulated locomotives. These are locomotives which are formed by joints, for enabling them to take turn along the curves otherwise if manufactured as one rigid base they will damage the gauge and rails.

4. Ash pits. Long masonry pits constructed longitudinally inside and under the track to collect the ash discharge from steam locomotives.

5. Adhesion of wheels. It is the resistance offered by the friction between the metal surface of the rail and the wheels. This factor also limits the speed of the trains and the gradients of the track.

6. Audible signal or Fog signal. These are the signals which are kept on the rails, when the wheel passes over the track, there is explosion with a loud voice. In case of accidents or emergency these are used.

7. Ballast. It is the granular material which is used in packing under and around the sleeper for transferring the load to the formation.

8. Ballast-crib. The loose ballast between the two adjacent sleepers is known as 'ballast-crib'.

9. Bearing plates. To reduce the pressure intensity on the soft timber sleepers, steel plates are provided between the rails and the sleepers, which is known as bearing plates. With the bearing plate no adzing of sleepers is required, as the cant of 1 in 20 is directly provided through these plates.

10. Block instruments. For providing positive indications for the block stations to give or receive permission for the trains to approach, two instruments are provided at each block station, which are known as block instruments. These instruments release tocken which is given as an authority to enter the block.

11. Blocks. For providing the required gap for the flange of the wheels at the level crossings and other such places, steel blocks are provided in between the main rail and guard rails, which is known as 'blocks'.

12. Board gauge. The common widest gauge used in Indian Railway, which is 1676 mm between the inner faces of the top flanges of track.

13. Buckling of gauge. If during rise of temperature, the free longitudinal movement of the rails is checked, they will buckle sideway, which is known as buckling of gauge.

14. Bull-headed rail. The rail having similar head and bottom shapes is known as bull-headed rail. It is used with rail chairs.

15. Boxing. It is the process of filling the ballast around the sleepers.

16. Blowing joint. If the fish plate at the joint is loose and the ballast contains loose dust or sand particles, during the movement of the trains these loose particles are sucked up and are deposited on the surface. This is called blowing of joints.

17. Calling on signals. These are small signals having short arms fixed on the home signal posts under the main home signals. These signals allow the train to enter the station cautiously.

18. Centre bound sleepers. If the track is not properly maintained, the repeated applications of load causes the depression at the ends of the sleepers, which causes the sleeper to rest at the middle on the ballast. This defect is known as centre-bound sleepers.

19. Check rails. The rails provided on the inner sides of the main rails at the level crossing and other crossings for guiding the wheel flanges to pass are known as guard rails.

20. Cant. It is also known as super elevation, which is provided on the curves to counteract the effect of centrifugal force. In this method the level of the outer rail is raised above the inner rail. This raising of the level of the outer rail over that of inner rail is called cant or super-elevation.

21. Cant deficiency. The cant on the curves is provided on the basis of the average speed of the trains running on that track. This cant will fall short of the actual cant required for the higher speed trains. This shortage of the cant is called cant deficiency.

22. Chairs. For keeping the bull-headed rails in proper position, special devices are provided in between the sleepers and the rails, which are known as the chairs.

23. Colour light signals. On electrified trains, high intensity beam colour light signals are used instead of old semaphore and other arm signals operated mechanically.

24. Coning of wheels. The outer rims of the railway wheels are coned at a slope of 1 in 20, to prevent the rubbing of the wheel flanges with the sides of the top flanges of the rails, and also so that train may run in the middle. Provision of this slope of 1 in 20 to the wheel rims prevents the lateral movement of the axle with its wheels which is called coning of wheels.

25. Crossing stations. The single line railway stations, where a loop line is provided to allow a train to stay and the other to pass, are known as crossing stations.

26. Co-acting signals. On the curve lines, when the main signal is not visible due to obstructions in the line of sight, a duplicate signal is provided on the same main post at lower level which works with the same level with which main signal operates. This duplicate signal is known as co-acting signal.

27. Creep of rails. The longitudinal movement of the rails in a track due to various reasons is known as the creep of rails.

28. Crossing clearance. It is the clearance between the crossing rail and the wing rail.

29. Crossing number. It is defined as the ratio of spread (the distance between the point and spice rails at the leg of crossing) to the length of crossing measured from the theoretical nose of crossing.

30. Derailment. When the wheels of the trains or bogies get out of rails, it is known as derailment.

31. Derailing switch. By means of traps all the sidings and shunting are isolated from the running lines. The trap joint is known as derailing switch, as it will derail any vehicle which will try to pass over it when it is open.

32. Disc signals. These are small signals provided at the points and crossings and are used to indicate, whether points are set or not.

33. Detector. It is a safety device to ensure lowering of the correct signal for a set route.

34. Diversion temporary. The temporary shifting of a track alignment from its original position, in case of repair of the original track to the damaged caused by accident, wash out in floods, rebuilding of bridges etc.

35. Drop pits. The deep pits constructed in the locosheds for taking down wheels of the locomotive during repair.

36. Dynamometer car. This car is placed between the train and the locomotive. When the train moves, the equipment fixed in the car automatically plot the graph giving the condition of the track. These graphs are known as 'Hallade Charts' of the track profile.

37. End bound sleepers. Under a newly compacted track if the deflection of the sleepers takes place in the middle and only the ends take the load, it is called end bound sleepers.

38. Equilibrium cant. When the exact cant on the basis of the speed of the train is provided, it is called equilibrium cant.

39. Facing point. When a train moves in the facing direction, it is called facing point. In this case the wheels pass over the switches first and over the crossing later.

40. Facing direction. If the crossing is seen from toe of the switch, the direction is known as the facing direction.

41. Examination pits. These pits are constructed in the locomotive yards for examination of the engines from under neath side. Examination pits are rectangular tanks larger than the ash-pits and are lined from inside.

42. Fish plates. These are plates used for joining the rails at rail-joints. As these resemble in shape to the fish, therefore are named fish-plates.

43. Flange way clearance. The clear distance between the adjacent face of the stock rails and check or guard rails.

44. Flag stations. The railway stations without sidings, having no fixed signals to control the movement of the trains.

45. Flare. The gradual or tapered widening of the flangeway away from the gauge line.

46. Foul mark. The limiting point in the converging tracks, beyond which the train vehicle can stand safely without colliding with the train moving on the other track.

47. Flat-footed rail. The rails having wider or flatter base, for directly fixing on the sleeper, avoiding the necessity of chairs are also known as 'Vignole's Rail'.

48. Gauge. Gauge is the minimum distance between the running or gauge faces of the two rails.

49. Gang beat. The length of the track allotted for the proper maintenance to a particular gang is known as gang-beat.

50. Gang huts. The residential quarters or huts provided near the centre of the gong-beat for taking rest or living of the gang-workers.

51. Gathering lines. It is a track from which a number of parallel siding or branch takes off. The gathering lines are provided in the marshalling yards for sorting out the wagons for different destinations.

52. Goods stock. These are wagons which are used for the movement of heavy and bulky goods.

53. Goods yard. The yard in which goods wagons are shunted and sorted for loading and unloading.

54. Guard rails. These are additional rails provided over the bridge to prevent the damage and danger in case of derailment over the bridge.

55. Gradient. The slope provided on the track to reach at various elevations.

56. Grade compensation. On the tracks having gradients, the amount of gradient is reduced on curves for easy movement of the trains. This grade reduction on the curves is known as grade compensations.

57. Heel. The point where the tapered rail of switch is fixed to the main rail, is known as heel.

58. Heel divergence. It is the distance between the running faces of the stock rail. In other words, the gauge face of stock rail and gauge face of the tongue rail when measured at the heel of the switch.

59. Hauling capacity. It is the total load which a locomotive can haul.

60. High speed track. The track over which the trains more than 120 km/h are moving.

61. Hogged rails. The rails which get battered over the ends of the rails due to the action of the wheel flange. Such rails usually get bent down and deflect at the ends.

62. Hump yard. It is a place in the marshalling yard where a summit or raised portion is constructed on the track, from where the wagons are allowed to move under force of gravity in the various sorting line, during shunting operations.

63. Home signals. These are semaphore type singnals provided at the first diversion for loop lines or sidings. The incoming trains after passing the outer signals have to take permission for entering the station by the home signal.

64. Inter-locking. It is the mechanical or electrical technique for mutual locking between the levers for lowering the signals, setting of points and locking of gates of level crossings at a time. The later-locking also avoids all the possibilities of confusion and danger of pulling wrong signals and avoids accidents.

65. Junction station. The stations where railway lines from three or more stations meet. The lines may be of same of different gauges.

66. Keys. The tapered pieces of steel or timber used for fixing the rail with the sleeper or rail-chair.

67. Kinks. The joint points or rail joints which are not smooth due to lateral movement of the ends of the rail out of its original position, making the rough train movement. The kinds may be formed due to loose joints, defective gauge etc.

68. Lead or crossing lead. The distance from the heel of the switch to the theoretical nose of crossing measured straight.

69. Lead rails. The rail having the length from the heel of the tongue to the toe of the crossing.

70. Level crossing. The place where the road and the railway line cross each other at the same level.

71. Linking gang. The gang who is responsible for fixing the rails with sleepers and rails together with fish plates.

72. Loading gauge. The device fixed at the exit of goods yard, to check the loading dimensions of the wagons, whether they are within the fixed width and height or not.

73. Locomotive. The driving unit which hauls the trains.

74. Loop-line. When a branch line starting from the main line again joins the same main line. This line is provided at the crossing stations.

75. Material gang. The gangs which are responsible for unloading the materials from the wagons and spreading it on the formation or along the track for the construction of new track or repair of old track.

76. Marshalling yards. These are the yards in which wagons are sorted and new trains are formed according to the destinations.

77. Momentum gradient. When a falling gradient is followed by a rising gradient, the moment of the train moving on the falling gradient is utilized in rising the train on the rising gradient. Such gradients are called momentum gradients.

78. Negative cant. When a branch line takes off from a main line on curve on the opposite side, then the point from where both lines diverge, the cant is provided for the main line. Therefore at such situation the level of the outer line of branch line is lower than the inner line, such position is called negative cant.

79. Nosing action. While moving a train, the wheels of locomotives, wagons and coaches follow a zig-zag path moving from one rail to another within the wheel flanges. This striking of the wheel flanges on the sides of the rails at an angle in the forward direction of the motion is called nosing action.

80. Oscillating motion of locomotive. Vertical, transverse, longitudinal etc. motion of the locomotive in different planes is known as oscillation motion.

81. Outer-signal. The signal provided next to the warner signal on the station yard is known as outer signal. It is also of semaphore type.

82. Over-riding switches. The switches which separate rail sections for stock rails and tongue rails are called over-riding switches.

83. Packing. The process of ramming the ballast under-neath the sleeper is called 'packing'.

84. Packing gangs. The workers or gangs which do the packing of ballast are known as packing gangs.

85. Permanent track. The complete rail road consisting of rails, its fittings, sleepers and ballast laid over the prepared formation, is known as *permanent track.*

86. Plate-laying. The method of laying the permanent way or track over the prepared formation is known as plate-laying.

87. Point indicator. These are the indicators used to indicate the setting of points and switches, to the drivers and cabinmen.

88. Platform. It is the raised level surface from where passengers board in and get down from the trains.

89. Point indicator signals. All the less important points in the station yards which are not interlocked are provided by point indicators to indicate their positions. These signals are also called as disc signals.

90. Points and crossings. Points, crossings, cross-overs and turnouts, etc. are contrivances or arrangements, which allow the train to change from one route to another.

91. Power signalling. It is the electric colour light signals which are used on all the electrified routes of the railways. The operation of the points and crossings is also done by the electric power.

92. Pusher gradient. If one locomotive cannot haul the train on the high rising gradients, another locomotive is used to push the train to move on such higher gradients. Such gradients are known as pusher gradients, as extra locomotive is required to push the train.

93. Rack railways. In steep gradients, it is a three-rail system, in which both side rails are ordinary rails and the third middle one has tooths. The locomotive has tooth and pinion wheel, the teeth of which fits into the teeth of the central toothed rail and the train moves by the rotation of the rack and pinion wheel.

94. Railway station. This is the selected place where trains halt for exchange of goods and passengers. The changing of the running staff of the railway, filling the fuels etc. of locomotives are also done at railway stations.

95. Relaying of track. When after continuous use, the permanent way becomes very old, its rails, sleepers etc. are changed. The operation is known as relaying of track.

96. Railway zones. The Indian Railways have been divided into nine zones for efficient operation, maintenance and administrative control.

97. Repeater signals. When the line of sight of the signal is obstructed on curves due to bridge, buildings, etc. a duplicate signal of smaller size is provided at convenient place on the same post of main signal. Such signal is called repeater signal.

98. Roaring rails. At some places due to some defects, the top surface of the rails becomes corrugated. When vehicles pass over it, these rails produce roaring sound, therefore such rails are called roaring rails, or corrugated rails.

99. Rolling stock. Locomotives, goods wagons, passenger coaches, oil tankers etc. which move on the permanent way are known as rolling stock.

100. Ruling gradient. The maximum rising gradient, which is allowed on the railway track, is known as ruling gradient.

101. Router signals. At the location of the track, where branches off two or more different routes, a semaphore arm is used for showing the signals of all the routes. Such signals are called router-signals.

102. Sand hump. To prevent the vehicles from running off the track at the end of the siding, the track is embedded in the sand bed, with rising gradient at the dead end. Such device is called sand hump.

103. Saddle plate. This is the additional plate which is provided below the rail seat for strengthening the trough type steel sleepers.

104. Scissor packing. The method of packing of ballast under the sleeper diagonally is called scissor packing.

105. Selector. The mechanism used with detector by which two separate levers are combined into one and the pull is divided into two branches of which at a time only one is operated.

106. Semaphore signal. It is most common type of signal used in the railways.

107. Siding. It is a branch line which starts from a main line and terminates at the dead end with buffer stop or sand hump.

108. Sleepers. These are the members which are laid transversely under the rails, to support the rails and to transfer the load from rails to the ballast.

109. Sleeper density. It is the number of sleepers per rail length.

110. Sleeper crib. Sleeper crib is constructed to act as a pier for temporary diversion of the railway line during repair of culverts, small bridges or accidents.

111. Spikes. These are used for fixing the rails to the wooden sleepers. Various types of spikes are commonly used in the railways.

112. Starter signals. These are signals located near the platforms and give signal to the trains to start for the next destination.

113. Station yards. It is the place on the railway stations, where systems of tracks are laid for various purposes such as storing, sorting and despatching of vehicles.

114. Stock-rail. This is the main rail at the switch where the tongue rail fits against it.

115. Studs. These are the bent plates which are fixed between the stock rail and the tongue rail for preventing the lateral bending of the tongue rail. These are fixed to the web of stock rails by bolts.

116. Summit curves. These are the vertical curves in which gradient tangents meet above.

117. Switch. It is a complete set of stock-rail and tongue rail provided at the turn-over.

118. Switch back. To reach greater heights in the hill areas, to and fro movement of the trains is done by providing a number of switches at each turn. Such arrangement is called switch back.

119. Terminal station. This is the last station of the routes at which the track terminates.

120. Temporary track. The track constructed arid used in case of emergency and for the repair of other track is called temporary track.

121. Temperature compensators. These are the devices provided horizontally or vertically to nullify the temperature effect on the roads used for signalling and interlocking devices.

122. Toe. It is movable end of the tapered rail of the switch.

123. Throw of switch. The distance through which the toe of the tongue rail moves sideways, with heel of the tongue rail as the centre.

124. Through packing. The process of packing the ballast under and to the sides of the sleepers. This is done periodically.

125. Track alignment. This is the position and the direction of the centre line of the railway track on the ground.

126. Tongue rail. It is tapered rail used in the switch.

127. Track lifting. When due to constant running of the trains, the level of the track falls down, the whole track is lifted and the ballast packing is done underneath upto required limits.

128. Track circuiting. The length of the track which is connected by electric circuit and to the signal cabin, block telegraph equipments etc. and is required for indication of the light or bell, is known as track circuiting.

129. Transition curve. It is a parabolic curve introduced between the straight track and the circular curve or between two branches of a compound curve, for safety of train movement on curves.

130. Traverser. It is a device used to transfer the wagons or locomotives from one track to another parallel to each other.

131. Tube-railways. These are the underground railways, about 27 m or more in depth below the ground.

132. Tunnels. These are the passages through the mountains, closed at top by the mountain, for carrying railway train from one side to the other side.

133. Turn-table. It is a revolving device used for turning the direction of the locomotive.

134. Valley curves. These are the vertical curves in which the gradients meet below.

135. Vertical curves. The curves provided in the track in the vertical plane at change of gradients, are known as vertical curves.

136. Warner signal. It is a signal having fish-tailed semaphore. It is provided at the extreme end of the station which warns the driver to control the speed of the train.

137. Water column. It is the vertical pipe with swivel horizontal arm, fixed near the track and used for supplying water to the locomotives.

138. Way-side station. All railway stations on the way of the track, except the junctions and terminal stations are called way-side stations.

20.3 COMPARISON OF ROADS AND RAILWAYS

The rail-road differs from the highway in the following respects:

(i) The rail-road or track is subjected to very heavy concentrated loads of locomotives, coaches and wagons whereas only light loads move on roads. Therefore the track should be very strong so that it can take the loads which are coming on it.

(ii) Road is used by different types of vehicles such as motors, cycles, rickshaws, buses, trucks, pedestrians etc. but the track is reserved only for the movement of trains.

(iii) The train wheels can move only on the track in the restricted place, because lateral movement is checked by wheel flanges. The movement from one track to another track is done by special types of arrangements known as 'Points and Crossings'.

(iv) As the track is only meant for trains and there is no obstruction in the way, greater speeds are maintained in railways as compared to roads, on which it is impossible.

(v) In roads the vehicles move due to the tractive force of the engine between rubber tyre and road surface whereas trains move due to the tractive force between steel rail and steel wheel. The friction in case of rails is about $\frac{1}{5}$th of that between rubber and road, therefore it gives restrictions on gradients in railways.

(vi) The wheels of railway vehicles are rigidly fixed at fixed distance, therefore the rails are to be laid at the same distance and properly maintained because even slight change in distance will cause derailment and serious accidents. Therefore railway track requires constant maintenance as compared with roads which require only occasional repairs.

(vii) The railway is suitable for specific service only, whereas the road transport service can be as per customer's need.

(viii) Freight rates are not uniform and reflect cross subsidy in case of railway transport, whereas these are uniform in case of road transport.

(ix) Steeper gradients cannot be allowed in case of railways as in case of road transport.

(x) Packaging and inventory expenditures can be saved in case of road transport, whereas it is not possible in case of railway transport.

(xi) Railway transport can be delayed due to wagon shortage, whereas road transport is readily available at all times.

(xii) In case of railway transport there is no delay in movement due to octroi, toll tax posts, bridge taxes etc. as in case of road transport.

(xiii) Road transport has more employment potential than railway.

(xiv) For longer distances railway transport is cheaper than road transport.

20.4 GAUGE PROBLEM

The clear horizontal distance between inner faces of rails near their tops is known as the *Gauge* of the track. In England the first track was laid with 1.524 m (5 ft) gauge but it was the distance from outside to outside of rails and at that time the wheel flanges were on the outside of the rails. But later on outside flanges caused much difficulties, in changing the train from one track to another, therefore, it was decided to keep the flanges on inner side. But it was very difficult to change the gauge so only wheels were turned. It makes the gauge 1.435 m $4'8\frac{1}{2}''$ due to deduction of rail-head. Now this gauge of 1.435 m has been permanently adopted in Britain.

At the time of construction of railways in India, after long controversy 1.676 m (5' 6") gauge was adopted as standard gauge. The total cost of construction of railway directly depends on the gauge i.e., if gauge is wide it will be more costly. Therefore, wider gauge is only justified at such places where traffic is heavy. Therefore at the time of construction of railways the policy was adopted that gauge will depend on the intensity of traffic so that in future, running of railway should be economical and railway should earn some profit. Keeping in view the above policy three types of gauges were adopted in India. First one 1.676 m gauge, which is known as Broad Gauge (B.G.), was adopted for main cities and routes of maximum intensities. Second one 1.00 m gauge called Metre Gauge (M.G.) was used for undeveloped areas or interior areas where traffic is very small and future prospects are not very bright. In hilly areas and very thinly populated areas where it was much uneconomical to use metre gauge, Narrow Gauge (N.G.) of 0.762 m (2' 6") and 0.6096 m (2' 0") was provided. These narrow gauges generally serve the purpose of feeding at suitable places. In addition to the government owned tracks some tracks are laid by private concerns for their own use as in Steel plants, Aluminium factories, Sugar factories, Oil refineries, etc.

20.5 DIFFICULTIES DUE TO CHANGE OF GAUGE

One country should adopt only one gauge throughout its various parts. But since the policy of India and its economy lead to adopt various gauges in its different parts, it causes much inconvenience, which is described below :

1. At every change of gauge the passengers have to change the train, which causes much inconvenience to passengers.

2. The timings of trains at gauge-change points should coincided so that passengers of one gauge may go in another gauge train. If one train arrives late the other is required to be detained otherwise passengers will miss the train and will have to wait for another train which will unnecessarily waste their time.

3. At the junction the goods are to be unloaded from one train and are loaded in another, which required unnecessary extra labour and goods are likely to be damaged or dislocated. This requires extra godowns and sheds for keeping the goods, and careful attention at the time of loading, because if goods are wrongly loaded in other trains, it will cause much inconvenience to the parties concerned.

4. If the labour at junction goes on strike it will be impossible to send the goods to their destinations.

5. The owner will have to pay extra charges for this trans-shipment from one gauge to other gauge train causing price rise.

6. Rolling stocks and old engines of main lines cannot be used on branch lines.

7. Surplus wagons and engines of one gauge cannot be utilized on another gauge. During some festival in one part of the country if there is great demand for passenger trains or wagons in one gauge track then surplus rolling stock of other gauge cannot be utilized.

8. During war times change in gauge causes extreme difficulties to the army and checks their quick movement.

9. If the intensity of traffic becomes more and requires wider gauge, it will be impossible to change it, because change in gauge means changing of each and everything i.e., rails, locomotives, bridges, tunnels, platforms, etc.

20.6 ADVANTAGES OF RAILWAYS

Following are the social, political and economical advantages of railwavs :

20.6.1 Economical Advantages

(a) Due to railways industrial development in far off places is possible, increasing the land values and standard of living of the poor people.

(b) Due to easy movements of the products in all parts of the country, the price stabilisation could be possible, giving relief to the common man.

(c) During famines the essential goods, food and clothing can be speedily sent to the effected areas.

(d) Congestion of the cities has been relieved due to increase in mobility of the people, who can come from long distances daily to the places of work.

(e) Finished products of industries can be easily and cheaply distributed.

(f) Raw materials can be easily, economically and speedily brought to the industrial areas.

(g) Completion of big national projects such as Dams, Canals, Power Houses etc. was possible due to railways only, as labour and materials could be brought from long distances speedily.

20.6.2 Political Advantages

(a) Only due to effective countrywide network of railways, the Central Government administration has becomes easy and effective.

(b) Railway have created the national mentality among the people of different religions, areas, tastes, customs and traditions.

(c) During war the railway can easily and speedily move military from one place to another in time.

(d) Railway can also help in mass migration of people during emergency if required.

20.6.3 Social Advantages

(a) During travel as people of different castes and religions sit together the feeling of caste is removed.

(b) The inhabitants of far off places do not feel themselves isolated from the other parts of the country.

(c) People can easily reach their religious places even at very long distance easily.

(d) Railway is the most safe, comfortable and cheap type of transport than all other types of conveyances.

(e) Only the railway has made it possible for the people to visit all the parts of the country and make their social outlook broadened. People have started feeling pride of the greatness of their country.

20.7 CLASSIFICATION OF INDIAN RAILWAYS

Depending on the route, traffic carried and maximum permissible speed the Indian Railway Board has classified the railway lines in the following three main categories.

20.7.1 Trunk routes

Following routes have been classified in Indian Railway as trunk routes :

Broad Gauge
1. Delhi—Mughalsarai—Howrah
2. Delhi—Kota—Mumbai
3. Delhi—Jhansi—Nagpur—Chennai
4. Howrah—Nagpur—Mumbai
5. Mumbai—Guntakul—Chennai
6. Howrah—Vijyawada—Chennai

Metre Gauge
1. Lucknow—Gorakhpur—Gauhati
2. Delhi—Jaipur—Ahmedabad
3. Chennai—Madurai—Trivandrum

Railway Board has also fixed the following standards for the trunk routes :

S.No.	Description	B.G.	M.G.
1.	Max. permissible speed	120 km/h	80 km/h
2.	Rail section	55 kg/m or heavier	37 2 kg/m
3.	Sleeper density	$(n + 7)$	$(n + 7)$
4.	Ballast cushion	25 cm below sleeper	25 cm below sleeper
5.	Degree of curvature	$7\frac{1}{2}$	Suitable degree
6.	Design speed for new track	160 km/h	100 km/h

20.7.2 Main Lines

All railway routes other than trunk routes carrying 10 gross million tonnes per annum (G.M.T.) or more for Broad Gauge lines and 2.5 G.M.T. or more for Metre Gauge lines have been classified as main lines. In addition to the above all routes (except trunk routes) where maximum permissible speed allowed is 100 km/h for B.G. and 75 km/h. for M.G. are also classified under this category.

Railway Board has fixed the following standards for main-lines :

S.No.	Description	B.G.	M.G.
1.	Max. permissible speed	100 km/h	75 km/h
2.	Rail section	52 kg/m	37.2 kg/m
3.	Track relaying period	20 years	30 years
4.	Design speed for new track	120 km/h	75 km/h

20.7.3 Branch Lines

All the lines except the above railway lines are classified under branch lines. As a rule the old rolling stock of main and trunk routes is used in the branch lines. No Diesel Engines or Electric Engines are provided on these lines.

B.G. locomotives (WG/WP type) and Bob wagons are used at reasonable speeds on branch lines.

M.G. locomotive (YG/YP type) and wagons are used at reasonable speeds on branch lines.

Generally old passenger coaches and wagons are used on the branch lines.

20.8 LOADING GAUGE

It represents the maximum width and height upto which a railway vehicle may be built or loaded. The interior dimensions of bridges and tunnels are

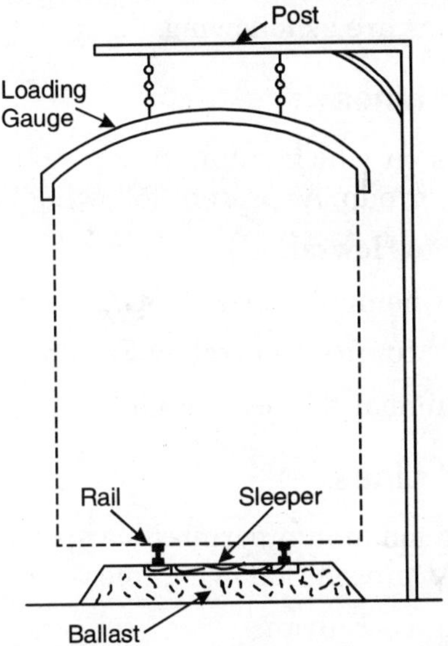

Figure 20.1 *Loading gauge*

fixed. If any vehicle is loaded more than the inner height of the tunnel it will strike against it, which will cause damage. Similarly, all the signalling posts, platforms, telegraph posts, water columns etc. are placed at a fixed distance from the track and if any vehicle is constructed or loaded in such a way that its body will strike against the post it may cause damage. Therefore, the maximum height of the loaded vehicle is fixed and it checked by an instrument known as *Loading Gauge,* which essentially consists of one vertical and one horizontal post. These loading gauges are fixed at the exit of goods yard and the height of every loaded goods wagon is checked by it. In India for B.G. track the maximum dimension above rail-level is fixed 4.72 m in height and 3.66 m in width. The corresponding dimensions for M.G. are 3.43 m and 2.59 m.

20.9 SPEED

The speed of the locomotive depends on the gauge. On wider gauge it can attain more speed as compared to narrow gauges. Thus the speed directly depends on the gauge. But gauge is not the only factor which governs the speed. Actually speed depends on other factors also such as power of locomotive, weight of the train, nature of formation under the track, strength of the track, gradients, degree of curves, maintenance of track, etc. The speeds in foreign countries are much as compared with the speeds of Indian Railways, because foreign countries are using more firm and strong tracks.

The classification of Indian Railways have also been done on the basis of the speed of the trains are as following.

20.9.1 Group 'A' Lines

All the trunk routes on which trains run at 160 km/h or more have been classified under this group. At present following lines are group 'A' lines :

(i) Delhi/N. Delhi to Howrah.

(ii) New Delhi to Chennai Central by G.T. route.

(iii) New Delhi to Mumbai Central by Frontier mail route.

(iv) Howrah to Mumbai VT via Nagpur.

20.9.2 Group 'B' Lines

Train routes having maximum permissible speed of 130 km/h come under this classification. At present following lines are group 'B' lines :

(i) Cennai—Raichur—Kalyan

(ii) Bhusaval—Itarasi—Allahabad

(iii) Wadi—Kazipet

(iv) Howrah—Bandel—Burdwan—Barharwa over Farakka Malda Town—Barsoi—New Jalpaiguri

(v) Mugalsarai—Patna—Kiul—Sitarampur

(vi) Kharagpur—Vijayawada

(vii) Kiul—Sahibganj—Barharwa

(viii) Ambala—Ludhiana—Pathankot

(ix) Jalarpet—Bangalore

(x) Arkonam—Erode—Coimbatorc

(xi) Baroda—Ahmedabad

(xii) Ambala Cantt—Lucknow—Pratapgrah—Mugalsarai

(xiii) Delhi—Ambala Cantt—Kalka

20.9.3 Group 'C Lines

All suburban routes of Mumbai and Kolkata come under this group.

20.9.4 Group 'D' Lines

All the B.G. and M.G. routes of railway where the maximum permissible speed at present is 100 km/h.

20.9.5 Group 'E' Lines

All the B.G. and M.G. lines having maximum permissible speed limit less than 100 km/h.

20.10 TRAIN RESISTANCE

The engine pulls the train due to tractive force which it develops. This force is equal to the resistance of the train. The resistance of the train is due to the following factors:

(a) *Resistance due to speed* occurs due to wind and oscillation and concussion resistance due to uneven track. The wind resistance depends on the direction it is acting on the train viz. side, head or tail. The side resistance depends on the length of the train and nature of vehicles. The head and tail resistances are approximately constant for all the trains. By constant maintenance the resistance due to oscillation can be reduced.

(b) *Resistance due to gradients.* This type of resistance can be easily calculated by mathematics.

$$\text{Resistance due to gradient} = \text{weight of train} \times \frac{1}{\text{rate of grade}}$$

If 1 in 150 gradient is provided, then this resistance will be equal to $\frac{1}{150}$ weight of train.

When a train is ascending on the slope, this resistance is to be overcome. If the train is descending, this force helps the locomotive in pulling the train.

(c) *Resistance due to friction and wave action* is developed between the rails and the wheels of the train. In the moving train, a wave action is set up in the rail causing creep and same power is utilised by it. This resistance is unavoided and its amount it generally 0.0025th weight of the train.

(d) *Resistance due to curves* develops only on the curve places of the tracks. On straight and level tracks it is not taken into account. But this resistance increases with the speed of train and degree of the curve. Its value is less on new rails and more on old rails. This resistance varies from 0.20 kg per metric tonne per degree curve to 1.0 kg per metric tonne per degree of curve. In India it is measured as follows :

For B.G. Line

Curve resistance = 4×10^{-4} × degree of curve × weight of train.

For M.G. Line

Curve resistance = 3×10^{-4} × degree of curve × weight of train.

For N.G. Line

Curve resistance = 2×10^{-4} × degree of curve × weight of train.

REVIEW QUESTIONS

20.1. Write a short note on the history of railways.

20.2. What do you know about the development of railways in India ?

20.3. What do you understand by the term 'Gauge'? What gauge is adopted in Indian railways ?

20.4. (a) What factors govern the selection of a suitable gauge ? Discuss.

(b) Will it be better to adopt one uniform gauge or different gauges ? Give your comments on it.

20.5. (a) What is loading gauge ? Describe in brief.

(b) Write short notes on the speed of locomotives. What points govern the speed, name them ?

20.6. What is meant by "Gauge of Track" ? What are the factors which have to be considered in determining it. Which of the two chief gauges used in India should be adopted for a new railway ?

20.7. Write a short note on political social and economic advantages of railways.

20.8. Write a short note on the classification of railways based on the importance of route, traffic carried and the maximum permissible speed.

20.9. Write in brief the development of railways in India.

20.10. Compare between the road and railway transportation.

Rails

GENERAL

The complete road of railway is known as *Permanent Way*, which consists of two parallel steel rails fixed on sleepers. The sleepers are further embedded in ballast which is spread on the formation of levelled ground. The rails on which train moves should be strong enough to bear the load and impacts which may come over them. These should be constructed in such a way that every portion can be easily repaired and replaced.

21.1 RAIL FUNCTION

Following are the main functions of the rails :

(i) For heavy concentrated moving loads of trains, rails provide hard, smooth and even surface with lowest friction between them and the steel wheels.

(ii) Rails transfer the load of the train from wheels to the sleepers, ballast and the formation between the safe limits.

(iii) Rails have to bear the vertical and lateral forces of the trains moving at high speeds.

(iv) Rails have to bear the thermal stresses due to change in temperature, as well as braking forces caused while stopping the trains.

(v) Rails are made of very high carbon steel, which has minimum wear even in the worst possible conditions of atmosphere and friction between the rail and the wheels.

21.2 COMPOSITION OF RAIL STEEL

Generally the rails are manufactured in the steel plants by the open-hearth process. The composition of High Carbon Steel rails is as under:

Item	*% for Ordinary Steel Rail*	*% for Rails on Points and Crossings*
Carbon (C)	0.55 to 0.68	0.05 to 0.60
Manganese (Mn)	0.65 to 0.90	0.95 to 1.5
Silicon (Si)	0.05 to 0.30	0.05 to 0.20
Sulphur (S)	0.05 or below	0.06 or below
Phosphorus (P)	0.05 or below	0.06 or below

21.3 RAIL EQUIPMENT

As the rails have to work as a continuous girder and have to bear the concentrated loads and impact of heavy fast running trains, their requirements are as follows:

(i) The section of the rail should be so designed that it should safely transfer the load of the train through their wheels to the sleepers under them.

(ii) The design of the rail-section should be such that it can safely withstand the heavy lateral forces caused by the fast moving trains.

(iii) Web should be so designed that it can safely bear the loads coming over it without its failure. It should also provide sufficient flexural rigidity in the horizontal plane.

(iv) The head of the rail should be properly designed having sufficient margin for the head wear for longer times.

(v) The foot of rails should have sufficient width making them stable against overturning as well as bringing the compressive stress on the wooden sleepers within safe limit.

(vi) The rail-section should be so designed, so as to keep the centre of gravity near the middle, so that tensile and compressive stress should be more or less equal.

(vii) The composition of the rails should be as given in Article 21.2 for minimum wear and long life.

(viii) The tensile strength of the rail material should not be less than 72 kg/mm^2.

(ix) The rails should conform to the latest Indian Railway Standards as laid down for the rails.

(x) The design of the rail section should be such that fish-plates can be efficiently fixed at the rail joints and should meet the requirements for maximum efficiency of the rail-joint.

21.4 RAIL SECTION

Originally *dumb-belt* or *double-headed sections* were designed in which both the heads were provided with the same cross-section. The main object in

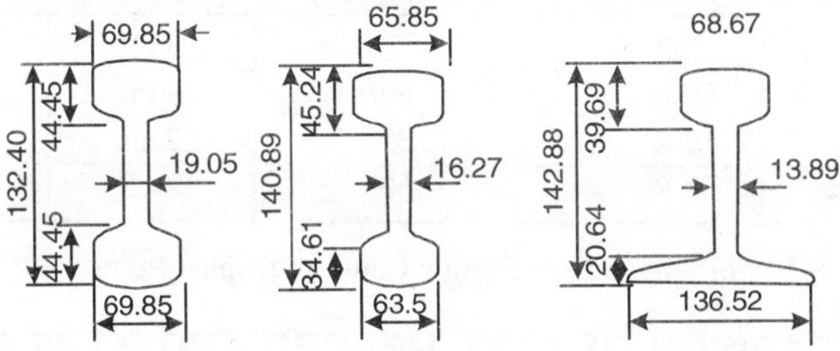

Figure 21.1 *Type of rails*

designing such a section was that when the one top section had worn out due to moving wheels, it could be inverted and re-used with lower section at the top. But when such rails were practically used, it was found that due to the impact of wheels the lower tread became dented and could not be used for required quantity of steel to keep the stresses within safe limit and was kept in bottom, and more metal was used in the top section which has to bear the rubbing action of the wheel. These rails are known as *Bull-headed rails* and are held in position by means of chairs fixed to sleepers. These rails are made of 9.144 m to 18.288 m in length and weight 29.77 to 49.92 kg/ metre.

At the same time when bull-headed (B.H.) rail was designed, Vignole invented other type of rail section known as flat-footed section. Vignole's main idea of developing such a rail was that this rail can be used directly on sleepers with small fastenings. In India both bull-headed and flat-footed rails are used which are foiled to B.S. Sections. When the F.F. Rails are directly laid on wooden sleepers, it is found that heavy train load sunk the rail in the sleepers, loosening the spikes. Therefore to avoid this sinking and for distributing the load on wider areas, steel bearing plates are used between sleepers and rails. The main advantage of the flat-footed rails is its lateral rigidity due to which it is mostly used in Indian Railways.

Generally the rail section is designed by assuming that it can bear 560 times its own weight per metre. Therefore for broad gauge main lines with maximum traffic loads, 44.7 to 56.8 kg/m rails are used. For metre gauge, 22.77 to 37.25 kg/m rail sections are used.

TABLE 21.1 *Weights of Standard Rail Sections*

	F.P.S Units		*M.K.S. Units**	
Gauge	*Type*	*Weight in lb/yd.*	*Type*	*Weight in kg/m*
Board Gauge	115 R	115	55 R	55
	90 R	90	45 R	45
	75 R	75	35 R	35
Metre Gauge	75 R	75	35 R	35
	60 R	60	30 R	30
	50 R	50	25 R	25
Narrow Gauge	50 R	50	25 R	25

* **Note** : M.K.S. units have been rounded upto multiple of five.

21.5 COMPARISON OF BULL-HEADED AND FLAT-FOOTED RAILS

1. Flat-footed rails are stronger in every direction than the Bull-headed rails for the same cross-sectional area and weight.

2. The fastenings used for fixing the flat-footed rails are cheaper, than that of B.H. rails, but it gets loose due to hammer blows and wave action caused by the moving trains. In F.F. rails bearing plates are required for soft wood sleepers.

3. Permanent way of B.H. rails requires more attention and maintenance than F.F. rails or track, because wooden keys are used in B.H. rails which require daily inspection.

4. The arrangements at points and crossings are cheaper and easy in case of F.F. rails than B.H. rails.

5. B.H. rails are fitted on chairs, hence provide more solid, smooth and better alignment than F.F. rails.

6. As chairs are used in B.H. rails soft wood sleepers can be used in B.H. rails as compared with F.F. rails.

7. F.F. rails are more rigid and can be used without any bearing plates. B.H. rails are less rigid than F.F. rails.

8. As there is no risk, daily inspection is not compulsory in case of F.F. rails, whereas in case of B.H. or D.H. rails daily inspection of wooden keys is necessary.

9. Replacement of B.F. rails is difficult than B.H. rails, as in latter case only after opening the fish plates and removing the keys the rails can be taken out easily. But in case of F.F. rails spikes are to be taken out and fish plates are to be removed, it disturb the position of sleepers.

10. Maintenance cost of F.F. rails is less than B.H. rails.

In addition to F.F. and B.H. rails various other types of rails are also designed and used in the world at various places, but those are not so important. Nearly 90% track in the world has been laid on F.F. rails.

21.6 SELECTION OF RAILS

The selection of a suitable rail section depends on the following factors :

(a) Gauge of the track.

(b) Axle load of the train and locomotive and the nature of the traffic.

(c) Speed of the train.

(d) Type of rail i.e., F.F., D.H. or B.H. type.

(e) Sleeper density.

(f) Maximum permissible rail wear on the top (generally 5% is allowed).

The load carrying capacity of the rail directly depends on its section ; bigger the section, higher shall be its carrying capacity.

In India the weight of the section of the rail is determined from the following formula

$$\frac{\text{Weight of the rail in tonnes}}{\text{Axle load of locomotive in tonees}} = \frac{1}{510}.$$

For example suppose we have to determine the suitable rail section for a locomotive axle load of 22.5 tonnes.

From the above formula

$$\frac{\text{Weight of the rail in tonnes}}{22.5} = \frac{1}{510}$$

or $$\text{Weight of the rail in tonnes} = \frac{22.5}{510}$$

$$\therefore \quad \text{Weight of the rail in kg} = \frac{22.5 \times 1000}{510} = 44.1\text{kg}$$

$$\text{Add 5\% rail wear} = 2.21 \text{ kg}$$

$$\therefore \quad \text{Total weight of rail section} = 46.31 \text{ kg}$$

But next section after 45 kg/m is 55 kg/m.

Hence adopt 55 kg/m rail section.

21.7 TRACK MODULUS

Track modulus is an index of measurement of resistance to deformation. It is defined as the load in pounds per inch or kg per cm of the rail length required to produce one inch or one cm depression of the rail table. The unit of the track modulus is lbs/inch/inch or kg/cm/cm and the higher this value, more stiffer is the rail.

Investigations at R.D.S.O. show that a track laid with CST-sleepers has a higher track modulus values than the track laid with wooden sleepers. The gauge of the track, section of the rail, the sleeper density and ballast cushion do not seem to have a significant effect on the track modulus values for all the practical considerations. The increase in sleeper density has some effect on the initial track modulus, but considering its effect on the absolute values of bending stress in the rail, it is only very marginal. A well maintained track with comparatively lesser voids or stacks has a higher initial track modulus as compared to a poorly maintained track.

The track modulus can be determined by the following two methods :

(i) By dividing the total load on one rail by the area of track depression curve.

(ii) By the formula :

$$Y_0 = \frac{P \times 10^4}{4\sqrt{64E.I.u^3}}$$

where

Y_0 = Depression of the track directly under the load in mm

P = Total load in tonnes

E = Young modulus of elasticity of rail metal in kg/cm^2

I = Movement of inertia of rail section in cm^4

u = Track modulus in kg/cm/cm

Following values of track modulus are used on Indian Railways :

	Track with wooden sleepers	*Track with CST-9 sleepers*
Initial Track Modulus u_1	50 kg/cm/cm	75 kg/cm/cm
Elastic Track Modulus u_0	250 kg/cm/cm	300 kg/cm/cm

21.8 RAIL JOINTS

It is impossible to construct track without rail joints, because rails are manufactured in such a length that it should be more economical in every respect, such as manufacturing, loading, unloading, and conveying etc. Rails are manufactured in suitable lengths and are jointed together after laying. The rails can be jointed together with the help of fish-plates and bolts or welding. The strength of rail at the rail-joints cannot be increased more

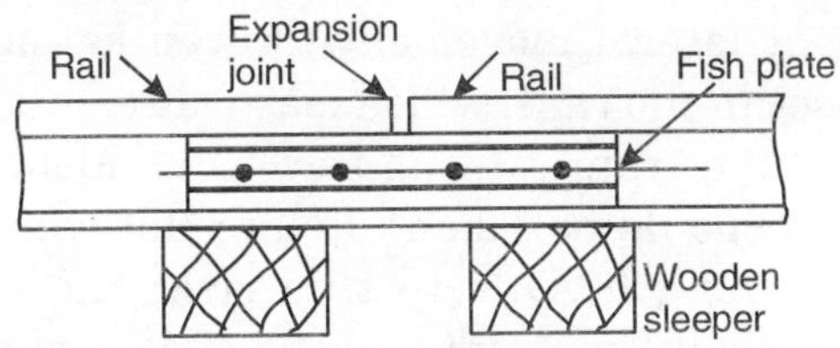

Figure 21.2 *Rail-Joint*

than 50% even if we use very strong fish-plates. Between two rails 1.5 to 3 mm expansion gap is provided for the free expansion of rails due to rise in temperature. Generally for giving more strength sleepers near rail-joints are placed closer together. *It should be noted that no sleeper is placed just below the rail-joint, otherwise it will cause more impact and discomfort to the passenger.*

A standard rail-joint should fulfil the following requirements :

(i) It should be as strong as the other portion of the track.

(ii) It should have just enough rail gap, between two rails for expansion.

(iii) Both rails at the joint should be laterally and vertically true in line when train is passing over it, because in this condition the wheel will not jump or change its direction of motion.

(iv) A rail-joint should have the same elasticity as the other portion of the track. If it does not have the same elasticity there will be impact at the rail-joint at the time of translation of wave.

21.9 WELDING OF RAILS

Nowadays it is common practice to weld the rails. The following are the main purposes of welding rails :

(i) To build up the worn-out parts of points and crossings and increase their life.

(ii) To build up the battered or worn heads of the rail ends, caused by the heavy blow of wheels at the rail joints.

(iii) Rebuilding of those portions of the rail-head, which are burnt due to slipping of wheels at the tune of applying brakes.

(iv) To increase the rail length by welding at joints, thus reducing the discomfort and weakness at the rail joints.

Experiments for lengthening of rails by welding show that it is very comfortable for passengers and maintenance cost is reduced by 20% to 40%. The total linear expansion in welded rail is far less than, we get by calculation. The reason for less expansion is that trace fittings hold the rails to the sleepers and prevent lateral movement. Prevention of lateral expansion causes heavy stresses in the rails, which may cause buckling of rails. As the buckling may derail the train causing serious accidents, therefore lateral strength of the train should be taken into account while welding the rails and increasing the length. Due to this risk, the maximum limit of continuously welding rails kept upto 5 rails of 12.8 m and thus making a total length of 64 m. Welding of rails give better results in track circuited and electrified tracks. Due to welding the construction cost of track is also reduced being less number of rail-joints.

It has been seen that after lengthening the rails by welding, the creep is reduced in the rails; due to less expansion, less number of blows are caused at rail-joints. If one long welded rail is provided over each span of long bridge there will be less impact on the bridge thus increasing its life.

After lengthening the rails by welding, its ballast section and sleeper density should be designed in such a way that it can prevent the buckling of rails. The buckling tendency directly depends on the number of rails welded together. Therefore sufficient care should be taken for it.

Generally the welding is done by the following methods :

(a) Oxy-acetylene welding.

(b) Chemical welding.

(c) Flash-butt welding.

(d) Electric arc welding.

21.10 LENGTH OF RAILS

As the weakest portion of the track is the rail-joint, hence their number should be as small as possible. To reduce the number of rail joints, the length of the rails should be increased. But the length of the rail is governed by the following factors :

(a) Transportation facilities available.

(b) Manufacturing facilities as well as economical cost.

(c) Loading, unloading, lifting and handling facilities available during transportation and track laying.

Keeping in view the above factors the following standard lengths of rails are used in Indian Railways :

For B.G. 12.8 m long rails.

For M.G. 11.89 m long rails.

The proposals to increase the length of rails to 25.6 m and 19.2 m are under consideration.

21.11 CONING OF WHEELS

The flanges of wheels are prevented from rubbing the inside face of the rail, by keeping the distance between the inside edges of flanges less than the gauge. If the rims of the wheels are kept flat, there will be lateral movement of the axle and the wheels will damage the inside edges of rails. To prevent this trend the wheels are given a slope of 1 in 20 which is called the *Zoning of wheels.* If the axle moves laterally towards one rail, the diameter of the wheel rim increases on that rail, whereas it decreases over the next rail. This brings back the axle to its normal position and prevents lateral movement.

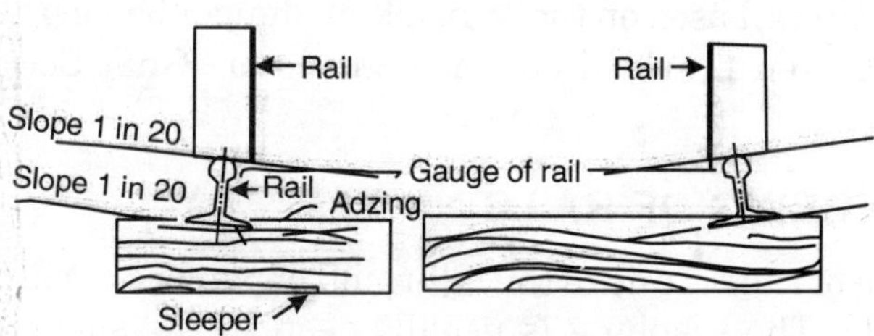

Figure 21.3 *Coning of wheels*

On curves, the outer wheel has to travel greater distance than the inner wheel, therefore, the axle moves sideways towards the outer rail and due to coning of wheels tread at outer rail increases while at inner rail it decreases. This helps the outer wheel to cover greater distance than the inner rail without any slip. Now, due to the coning of wheels, the pressure of wheel is always near the inner edge of the rail, therefore it wears out quickly. To reduce the wearing of rails in this way, the rails are also placed at an inward slope of 1 in 20. The slope of wheel cone is shown in Fig. 21.3.

21.12 RAIL END BATTER

It has been observerd that hammering action of moving trains wheels at the rail joints makes the rail ends battered in course of time. Due to the impact of the blows, the contact surfaces between rails also and sleepers get worn out, the ballast at joint sleepers gets shaken up, fish bolt become loose. These all factors further increase the rail end batter. The rail end batter is measured as the difference between the height of the rail at the end and at a point 30 cm away from the end. When the batter is upto 2 mm, it is classified as *Average*, and when it is between 2–3 mm, it is classified as *severe*. When the rail end batter is excessive, but the rail is alright, the ends of the rail can be cropped and the rail reused.

21.13 RAIL LUBRICATORS

Lubricators are provided on sharp curves where lateral wear is considerable. The main object of the rail lubricators is to lubricate or oil the running face of the outside rail for reducing the friction. It has been observed that rail wear is considerably reduced upto 50% by lubrication.

Various types of mechanical lubrication devices are used for lubrication purposes. These are attached to the rails. In these mechanical devices, the wheels of the moving trains normally cause the lubricant to flow on the side of the rail either by the action of wheels pressing the plunger up and down or by ramps on the outside of the rail being depressed by wheels. Sometimes the vibrations of the moving trains also cause the lubricants to flow. *P* and *M* type lubricators, based on the principle of plunger being pressed by moving wheels, have bsen provided on some curves on 'Ghat Section' of Central Railway.

21.14 SCABBING OF RAILS

The scabbing of rails is due to the falling off of patches or chunks of metal on the rail table. The scabbing is usually seen in the shape of an elliptical depression, whose surface shows progressive fracture with numerous cracks around it.

21.15 BENDING OF RAILS

At flat curves, where the degree of curve is less than 3, the rails are laid without bending. The rails are retained in the curved position by the sleepers.

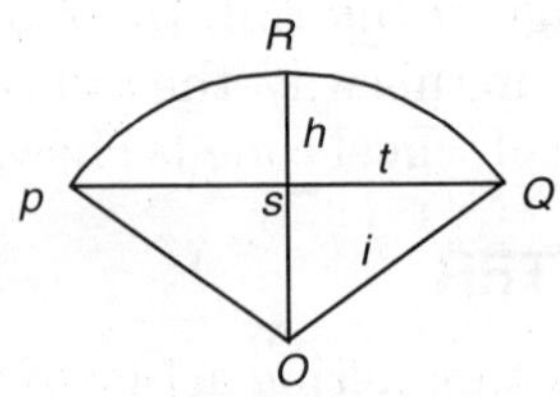

Figure 21.4

These sleepers are boxed carefully with ballast to prevent any lateral movement of the track due to the centrifugal force caused by the moving trains on the curves. But on curves of more than 3 degree, the rails are bent to the correct curvature before fixing them on the sleepers. If this bending is not done, elbows will be formed at joints and the alignment of the track will be disturbed.

The amount by which the rail is to be bent can be calculated mathematically by knowing the radius of the curve, and the distance between the starting and finishing point of the curve. Refer to Fig. 214, if

$PRQ = l$, length of rail on curve portion

$OQ = r$, radius of the curve

$PQ = 2t$, chord length and 5 versine $SR = h$

then

$$h(2r-h) = t^2$$

or

$$h = \frac{t}{(2r-h)}$$

or

$$h = \frac{t^2}{2r} \text{ (approximately)}$$

When the length of rail is small as compared to radius l may be taken as equal to $2t$

$$\therefore \quad h = \frac{(l/2)^2}{2r}$$

$$= \frac{l^2}{8r}$$

For example the value of h for a 12.8 m rail laid on a curve of 3°

$$h = \frac{12.8^2}{\frac{1146}{3}} \text{ m}$$

$$= 0.0536 \text{ m}$$

$$= 5.36 \text{ cm}$$

After calculating the required amount of bend in rails, the rails can be bent and laid on the curved portion of the track.

21.16 WIDENING OF GAUGE

If the rails are laid with correct gauge on sharp curves, after the movement of some trains, it will be found that the gauge is widened and the rail have been tilted outwards. The widening of gauge and tilting of rails is due to the rigidity of the wheel base. The distance L between two adjoining axles in a rigid frame is known as *wheel base distance* and its maximum value in Indian

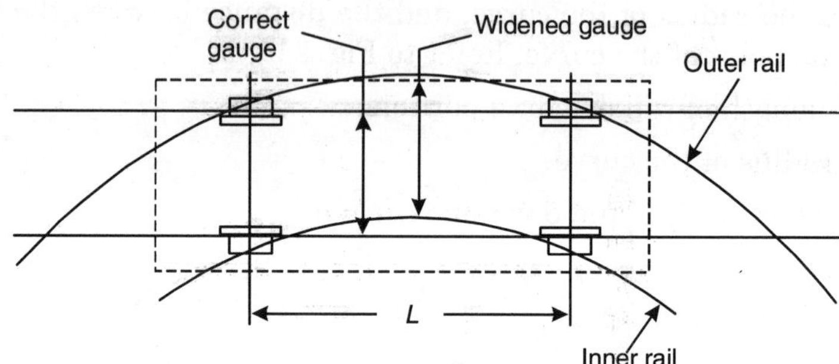

Figure 21.5 *Widening of gauge on curve.*

Railways on B.G. is 6.096 m. When a vehicle moves on curve, all of its wheel flanges occupy the position, shown in Fig. 21.5. Wheel flanges are butted against the rails and require wider gauge for free movement. At curves the axle comes in the position shown in Fig. 21.5 and requires wider gauge. To prevent the tilting of rails outside, the gauge is widened at curves to the correct amount after calculations. Generally the gauge is widened on curves more than $4\frac{1}{2}°$ curvature.

The gauge on curves shall be as follows :

Upto curves of 438 metre—Exact gauge.

Above curves of 438 metre—6 mm slack.

21.17 CORRUGATED RAILS OR ROARING RAILS

These are also called roaring rails, because in some rails their heads get corrugated and when the train passes over these rails, roaring sound is created. These corrugated rails have minute depressions at the surface of their heads, which are usually created at the braking points of trains or starting place of trains. The shape and size of these corrugations are not uniform nor their shapes and sizes are definite. Sometimes corrugations develop at a place in the rail and immediately after it the whole rail gets corrugated. Corrugations may occur on straight or curve track at any gradient.

Experiences show that corrugation in rails occur under the following common conditions:

1. *Factors due to physical and environmental condition of the track :*
   (a) Long tunnels
   (b) Steep gradients
   (c) Yielding formation
   (d) Electrified sections.
2. *Factors due to metallurgy and age of rails :*
   (a) Effect of oscillations at the time of rolling and straightening of rails.
   (b) High nitrogen contents of the rails.
3. *Factors due to atmospheric effect :*
   (a) Presence of sand
   (b) High moisture content in the air particularly in coastal areas.
4. *Factors due to train operation :*
   (a) At the starting points of train
   (b) At the braking points
   (c) High speed and high axle loads.

But the real cause for the corrugation is not exactly known. This may be due to excess of phosphorus in the rail steel or resiliency of the track or the defective shockers/springs of the trains.

The corrugation can be removed by grinding the rail head by fraction of a millimetre. On Indian Railway till date no method has been standardised to grind the rail surface. On German Railways following two types of equipments are normally used for rail grinding :

(i) Hand or motor driven trollies which moves on the rails at a slow speed and grind the individual rails one by one.

(ii) Rail grinding train which moves at a speed of 30 km/h and grinds both the rails simultaneously.

21.18 WHEELS BURNS

Wheel burns are caused by the driving wheel of locomotives slipping on the rail surface. While such slipping extra heat is generated and the surface of the rail may be torn resulting in a depression on the rail table. These burns are usually noticed where there are steep gradients or where there are heavy incidences of braking or near water columns.

21.19 CREEP

The longitudinal movement of the rails in a tracks is technically known as the *rail creep.* It varies from several centimetres to a negligible amount at different places in the track.

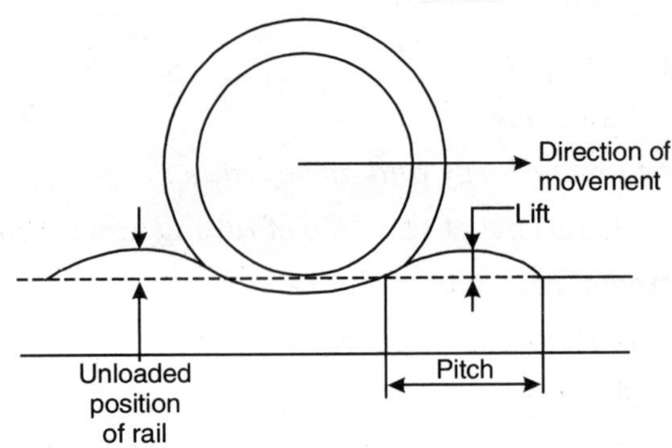

(a) Creep of rails

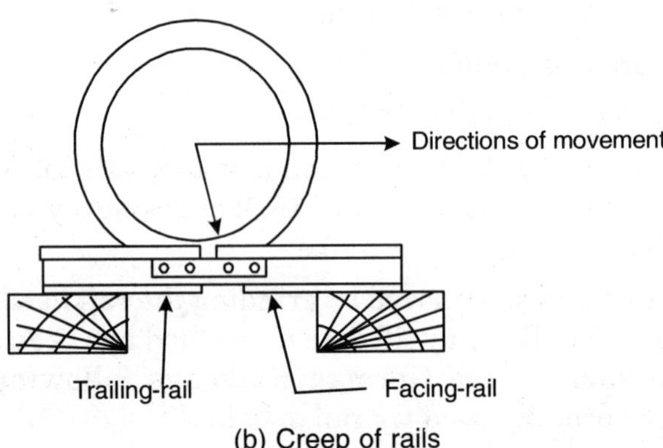

(b) Creep of rails

Figure 21.6

21.19.1 Causes of Creep

Following are the main causes of the creep :

(1) Train-wheels cause slight depression on the table of the rail due to their own weight. When the wheels move, the depression under it also moves and wave motion is set up in the track, which tends to move the rail in the direction of movement of train. This cause positive creep.

(2) When wheel passes over a rail joint, it gives a blow at the end of the rail, which causes positive creep.

(3) At the time of starting or accelerating a train, engine wheels give

backward thrust and tend to push back the rails, which causes negative creep.

(4) When brakes are applied to decrease the speed of the train, the wheels of brake-van slip on the rail and tend to push forward the rails, which causes positive creep.

(5) Steep gradients also cause creep in the downward direction.

(6) The expansion and construction of rails due to the change in temperature leave some joints open and some closed. This causes creep in the rails.

(7) On single line, if the intensity of traffic is greater in one direction than on another direction, it will cause creep in the direction of heavy traffic.

The above mentioned factors cause creep in the rails when track is in good condition. In addition to the above the following factors also cause creep :

(i) Insufficient and defective packing of ballast.

(ii) Sleepers not evenly and properly spaced.

(iii) Formation bed not properly compacted before laying track.

(iv) Improper gauge, at some places tight and at other places excessive slack.

(v) Rail joints not properly maintained.

(vi) Use of decaying and inferior quality of sleepers.

(vii) Proper rail sections not provided.

(viii) Sufficient expansion gap not provided.

21.19.2 Defects Caused Due to Creep

Creep affects the various parts of a track in the following ways :

1. The general surface and alignment of the rail track is destroyed.

2. Successive expansion space at rail joints start closing and at the starting point of creep, expansion space of joints start opening out.

3. The gauge of the track is tightened as the sleepers move out of square.

4. As the joints open out, heavy stresses are induced in the fish-plates and bolts, which may break the bolts and cause accidents.

5. The points and crossings on the main line either get pulled or pushed out of place, and it becomes difficult to keep them to proper gauge and alignment.

6. If due to some repair purpose any one rail is removed, the gap of this rail becomes too short or too long, which causes much difficulty at the time of refixing of the rail.

7. The ballast is forced out of place.

8. The bars connecting the points and crossing with the levers are bent and their smooth working is stopped.

21.19.3 Prevention of Creep

The following are the measures which are commonly adopted to prevent the creep:

1. The rails are pulled back to their original position by means of crow bars and hooks when the creep is set in the rail. Firstly, the starting point of creep and quantity of creep is noted and then rails are pulled back.

2. Sleepers are well packed with heavy angular ballast. Such type of sleepers are used which allow minimum creep. Steel trough sleepers allow least creep. If the number of sleepers per rail is increased, the creep will be prevented.

3. The creep can be prevented by providing anti-creepers. The patent anti-creep keys are the steel taper keys used with steel trough sleepers. Patent spring steel spikes are used in wooden sleepers to prevent creep. These spikes continue to grip the railfoot and do not allow creep. There are various types of anti-creepers used in railways and their number depends on the amount of creep.

4. The creep can be prevented by efficient maintenance of the permanent way.

21.19.4 Measurement of Creep

Creep can be measured by the following simple method:

After laying the track two small pieces of rails are driven on both sides of the track, in the formation perpendicular to the track. The top level of these posts should be equal to the top level of the sleepers. Two marks are made on the post and a thin string is stretched through them. On the bottom flange of rails, marks are made by chisel, where string touches it. After some time when the measurement of creep is required, again a string is stretched on previous marks. The distance between the string and the mark made on rail is equal to the creep.

21.20 KINKS IN RAILS

These are formed at the joints of rails, when the adjoining rails move slightly out of position. The following are the main causes for the formation of kinks:

(a) Due to defect in the gauge and thealignment.

(b) Due to loose packing of ballast in the track specially at the joints.

(c) When the ends of rails at the joints are not in cross level.

(d) Due to uneven wear of rail head at the joints.

Kinks effect and cause the uncomfortable running of trains. Following are the main effects of kinks:

(a) Unpleasant jerks occur to the trains passing over the kinks which cause uncomfortable journey as well as increase the maintenance cost.

(b) These jerks cause increase in the gauge of the track.

(c) On track curves series of kinks are formed at the joints of rails which may cause widening of gauge leading to the derailment trains.

To prevent the formation of the kinks in the track the alignment of the rails at the joints and curves must be correct, packing of ballast should be proper and the maintenance of the tracks should be properly done.

21.21 WEAR OF RAILS

The flow of rail metal due to abnormally heavy loads is called *wear of rails*. When the concentrated stress exceeds the elastic limit, the metal flows and causes the wearing of the rails. Elastic limit of the rail metal may be defined as the pull in tonnes on one square centimetre of metal, which will just stretch it. Broardly, wear of rails may be classified as:

(a) Wear on the head or top of the rail.

(b) Wear of the side of the head.

(c) Wear at the end of rails at joints.

Sometimes when metal flows, it takes the place on the inside of rail beyond the original rail section. The projection is known as *burrs* and care should be taken while checking the rail gauge, the distance between the burrs is not taken.

The following are the main causes of rail wear :

1. When the brakes are applied, the wheels slide on the rail and cause wear and abrasion of the rail head.

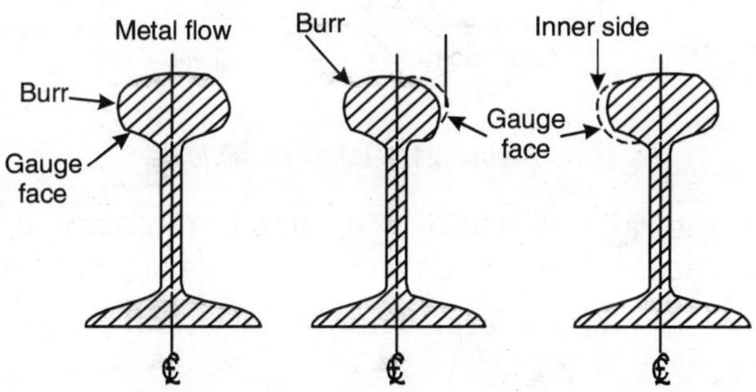

Figure 21.7 *Wear at rail sections*

2. Climatic conditions effect the rail wear. Rain reduces the wear by

lubricating the surface whereas sand and dust increase the near by blowing on the surface, due to their grinding action.

3. When train passes through the tunnel the gases from the engine are confined and attack the metal of rail, and cause wear.

4. In coastal areas, the sea-breeze acts on metal and causes wear of rails.

5. Due to loose packing of ballast, in the track there is heavy sinking of rails when the train wheels pass over them. This causes wear of rails.

6. When the wheel jumps over the gap of rail-joint, it gives a blow to the end of rail due to which the rail end is battered. This causes the top metal to flow towards the rail gap.

7. When train passes on curves which have no super-elevation it gives thrust on the inner side of outer rail and causes side-wear. If super-elevation is more than what is required, then there will be thrust on the inner rail and it will wear out. On the other hand, if super-elevation is less, more load will go to outer rail and it will be worn out due to the grinding action of wheel flange. In addition to the grinding action, the outer wheel flanges have a tendency to 'bite' into the inner side of outer rail-head, causing more wear.

8. On curves, if rails are laid at a tilt of 1 in 20 on the sloping sleeper, the tilt of wheel cone will not coincide with the slope of rail, therefore lesser

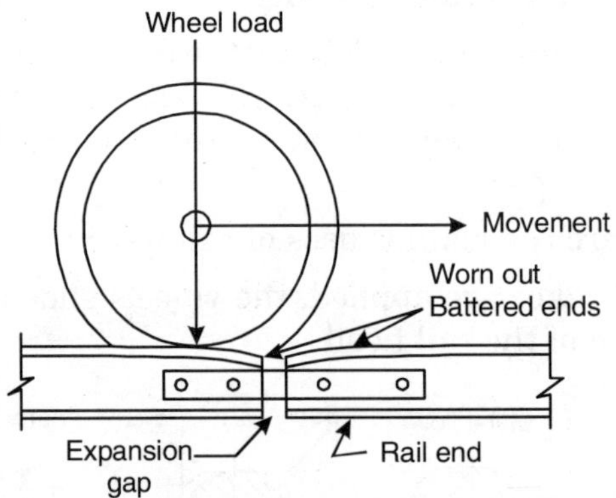

Figure 21.8 *Wear at rail joint.*

area of both, the rail and wheel, will come in contact and the intensity of pressure will exceed elastic limit and cause wearing of rails.

9. On curves, kinks are formed due to the lateral thrust; this also causes wear of rails.

10. Gradients in track also cause wear of rails.

11. The corroding action of acids of the refuse falling from the trains also causes wear of rails.

12. Due to sanding, which is done in the tunnels to reduce the dampness. This sand when comes in between the rails and the wheels do grinding action it will lead to the wear of rails.

13. Fluctuations in the gradients also cause rail wear.

14. Loose packing of ballasts or looseness between the rails and sleepers also increase rail wear.

15. On curves due to centrifugal force and super-elevation, the load on one rail is greater than that of other, which causes rail wear.

16. On curves, the sleepers are laid at a slope for the purpose of super-elevation, but when the rail is laid at a slope of 1 in 20, the resulting tilt is different from the slope of the wheel-cone, giving lesser area of contract between the wheel and the rail, causing rail wear.

17. Due to rigidity of the wheel base.

Measurement of rail wear. The measurement of rail wear is done by actual weighing of the rail or by measuring the cross-section of the worn rail. The weight of unworn rail is known and after weighing the worn-out rail the loss in weight can be easily determined, which gives the rail wear. In cross-sectional method the cross-sectional area of worn-out rail is calculated and the rail wear is calculated by the percentage loss in cross-sectional area. Weighing method is accurate, easier and better than cross-sectional area method, and so it is mostly used.

Reducing rail wear. The rail wear can be reduced by the following methods:

(i) Making long rails by welding at their joints.

(ii) Lubricating rails after some interval to avoid rust etc.

(iii) By using special alloys, the wear can be reduced, but it will be uneconomical.

(iv) By giving full attention to expansion gaps, fish-plates, etc. and maintaining the track in good order.

(v) Providing check-rail on curves.

(vi) By exchanging inner and outer rails on curves after some interval of time, the rail wear can be reduced.

Limit of rail wear. The capacity of rail to wear load of vehicles directly depends on its own weight, because rail sections are designed on the assumption that it can take 560 times its own load. If the longitudinal stress in rail sections do not exceed the permissible limits, the rails can be allowed to wear upto 10% of its weight. But generally, full wear takes place only in the head of the rail, therefore percentage wear in the section may be allowed upto 25%.

On curves when the wheels start fitting snugly on the worn rails, even if the wear is less than the permissible limits, the rails should be changed or renewed by welding the worn out portion. On curves in no case the wheel flange should be allowed to touch the fish-plates after wearing out of rails, otherwise it may cut the fish-bolts.

In India permissible limit of rail wear is 5% by weight i.e., when it loses its weight by 5% it should be replaced and used on branch lines. But rails which have lost their weight by 10% should not be used anywhere under any circumstances.

21.22 HOGGED RAILS

Hogging in rails is caused due to battering action of wheels over the ends of the rails, which causes the rails so bent down. Sometimes poor packing of ballast under the joints of rails or loose fish-plates also cause the hogging. Hogged rails cause rough and uncomfortable riding of trains, which also increases the maintenance cost of the track as well as trains. Following various methods are adopted to rectify the defects of hogged rails.

(a) *Dehogging.* In this method the defective ends of the rails are straightened by means of jim crow or dehogging machine. But this is not effective measure.

(b) *Changing.* In this method the defective rails are completely changed.

(c) *Welding.* In this method the defect is removed by welding.

(d) *Cropping.* In this method the defective portion of the rails are cut, fresh holes are drilled for fish-plates, the rails are pulled back and properly joined. The job of cutting of rails and drilling holes may be done in the workshops or at the site.

21.23 RAIL FAILURE

Sometimes without any notice the failure or rails occurs. After long research and investigation, it had been noticed that under such cases the following are the main causes of rail-failure :

(a) *Manufacture defects.* If the rails are not properly manufactured, it may cause head crushing, transverse fissure, splitting of head or horizontal fissures in the rail head. Sometimes manufacturing defect causes vertical or inclined shearing of the rail also.

The head of the rail may also be crushed due to the following reasons :

(i) Slipping of wheels at the time of applying brakes.

(ii) Due to skidding of wheels flat spots are formed on the wheels, these may cause rail head crushing.

(iii) Loose packing under the rail-joint or loose fish-bolts.

(b) *Fatigue caused by shearing stresses.* When the train moves on the rail it causes constant reversal of stresses. For one portion of wheel there is compressive stress at top and tensible at bottom, but the other position of the wheel has tension at top and compression at bottom. The moving train causes so rapid reversal of stresses, that if proper material is not used, it will cause failure of rails.

(c) *Splitting of head.* In this type of defect cracks are developed in the middle of the rail head or splits occur from the side to the end of the head. When the surface of the crack after opening appears smooth and dark, such type of defect is known as 'pipe rail' and this is formed during manufacturing or shrinkage of metal.

(d) *Web-splitting.* In this failure cracks are developed in the web of rail.

(e) Horizontal cracks between the rail head and the web occur due to improper packing or use of worn fish-plate.

(f) *Defects due to fault of the rolling stock and abnormal traffic effects.* Such rail failure occur due to flat spots in tyres, engine burns, skiding of wheels, severe braking etc.

(g) *Excessive corrosion of rails.* Excessive corrosion in the rail takes place generally due to weather conditions, sub-soil containing corrosive salts like chlorides etc. When the rails are constantly exposed to moisture near water columns, ash pits and tunnels. It has been noted the corrosion normally leads to development of cracks in regions of high stresses concentration.

(h) *Improper maintenance of track.* Rail failure also occur due to poor maintenance of joints, improper packing of joint sleepers and loose fittings etc. As a result of ineffective or careless maintenance of track or due to delayed renewal of track, the rail failure may occur.

(i) *Derailment.* These failures occur due to derailment of the train.

(j) *Defects in welding of joints.* These are failures occurring due to defects in welding the joint. This defect arises either because of improper composition of thermit weld metal or defective welding of the joint.

21.24 RAIL FLAW DETECTION

A defect in rail which will ultimately lead to the fracture or breakage of rail is known as *Flaw.* It is most necessary to detect these flaws and take timely action to remove them for the safety of the trains. The detection of rail flaws can be done either by visual examination of rail ends or by rail flaw detection equipments.

21.24.1 Visual Examination of Rail Ends

Before doing this examination, the joint is opened after removing the fish plates. Now the rail ends are cleaned by kerosene oil. Now the visual examination is done by means of magnifying glass for any hair cracks. For identifying the flaw more clearly, sometimes the chalk is rubbed over it. If the light is insufficient, more light is thrown by means of mirror from sun.

21.24.2 Ultrasonic Rail Flaw Detector

On the Indian Railways nowadays ultrasonic rail flaw detectors are commonly used. This test is also known as non-destructive method of testing the rails. Following four types of flaw detectors are used in Indian Railways.

(i) Soni-rail flaw detector

(ii) Mistubishi-rail flaw detector

(iii) Audi-gauge-rail flaw detector

(iv) Krant Kramer-rail flaw detector.

Equipments from Sr. No (i) to (iii) work on the simple principle of frequency modulation, whereas the last one works on the advanced principle of pulse echo. It has been noted that the Krant Kramer-rail flaw detector is best suited to the Indian conditions.

21.25 CLASSIFICATION OF DEFECTS

The defects noticed by rail flaw detectors are classified as follows :

(a) *REM Defects.* The defects requiring immediate removal are marked with red 'REM'. REM has been taken from the word REMOVE.

(b) *OBS Defects.* The defects which are not so serious in nature and are kept under observation are known as OBS defects. These defects are marked with yellow paint. OBS has been taken from the word OBSERVATION.

Nowadays in India Soni-rail Detector in electronic device is used for testing of rail defects.

REVIEW QUESTIONS

21.1. (a) Write short notes on Rail-sections.

(b) Draw neat sketches of Bull-headed and Flat-footed rails. Discuss the relative merits and demerits of both types of rails.

21.2. What are the requirements of a standard rail-joint ?

21.3. What are the different types of rail-wear ? How can they be reduced ?

21.4. What is the 'Creep' of rails ? What are its causes and how does it affect the various parts of a track ? What measures are commonly adopted to prevent creep ?

21.5. What do you know about the welding of rails ? What are the main purposes of it?

21.6. Write short note on 'Coning of wheels'. What purpose does it serve in wheels ?

21.7. Write short notes on the following :

(a) Bending of rails (b) Widening of gauge

(c) Wear of rails (d) Corrugated rails

(e) Kinks in rails (f) Hogged rails

21.8. What are the methods of rectifying creep ? What are the results of creep ?

21.9. Give sketches of the creep during starting and during stopping of train.

21.10. Write short note on the causes of failure of rails.

Sleepers

GENERAL

Rails require some support on which they can be laid and fixed. The support should keep the rails apart at required distance. This is done by means of sleepers. The chief function of sleepers is to support the rails, keep the two rails at correct gauge and distribute the load coming on the rails to the ballast. The sleepers should counteract the tendency of forces which tend to disturb the track and should act as elastic medium between the rails and the ballast in order to absorb all shocks and vibrations of the trains.

22.1 FUNCTIONS OF SLEEPERS

The following are the main functions of the sleepers :

1. Sleepers hold the rails at proper gauge, exact on straight track and loose at curves as per standard.
2. Sleepers transfer the load of the trains from rails to the ballast or girders of the bridge.
3. Sleepers provide stability of the track.
4. Sleepers act as an elastic medium between the rails and the ballast.
5. Sleepers hold the rails in proper level or transverse tilt on curves.
6. Sleepers also hold the each rail at 1 in 20 tilt (i.e., the slope of the wheel cone).

22.2 REQUIREMENTS OF GOOD SLEEPERS

The following are the requirements of good sleepers:

1. They should maintain correct gauge.

2. The rails can be easily fixed and taken out from the sleepers without moving them.

3. The sleepers should provide sufficient bearing area for the rail.

4. They should have sufficient weight for stability.

5. They should be sufficiently strong to act as a beam under loads.

6. They should provide sufficient effective bearing area on the ballast.

7. They should not be pushed out easily of their position in any direction even under maximum forces of the moving trains.

8. Their design should be such that packing and tamping may not damage them.

9. They should be economical in initial as well as maintenance cost.

10. The fittings of the sleepers should be such that rails can be easily adjusted during maintenance operations.

11. If track-circuiting is required, it should be possible to insulate them from the rails.

12. They should be able to bear the stresses which will come over them.

13. They should not be too heavy nor light in weight.

14. The design and spacing of the sleepers should be such that ballast packing can be easily and effectively done in less time.

22.3 TYPES OF SLEEPERS

Different types of sleepers are used in Indian Railways depending on their availability, suitability, economy and design. These different types of sleepers can be classified as follows :

1. Wooden sleepers
2. Steel sleepers
3. Cast iron sleepers
4. R.C.C. sleepers

22.3.1 Wooden Sleepers

These sleepers satisfy all the requirements and are only-suitable for track circuiting. In India as the climate changes from place to place, it is difficult to choose best wood for sleepering. The life of wooden sleepers depends on the quality of timber used and its ability to wear decay, resistance to white ants and resistance to atmospheric actions. Teak is the best wood for sleepering purposes, but due to its heavy cost it is used only in girder bridge for sleepering. Sal wood is harder and strong than teak as well as cheap in cost, therefore, it is mostly used for sleepering purposes. Deodar and Chir wood sleepers are also used in the zones where these timbers are easily and cheaply available.

Following solutions are used for the preservation of timber sleepers :

(a) Chloride of zinc ($ZnCl_2$)—method is known as Burnettising.

(b) Creosote solution at 49°C and 24 kg/cm^2 pressure.

(c) Salt solution of Bichloride of mercury-salt solution ($HgCl_2$) method is known as Kyanizing.

The number of sleepers used for rail vary from $(n + 2)$ to $(n + 6)$, where 'n' is the length of rail in metres. Minimum packing space of 30–35 cm is provided between two sleepers. Timber is thoroughly seasoned before using it as rail sleepers.

The standard sizes of timber sleepers are as follows, which are used in different gauges:

2.74 m × 25 cm × 13 cm on B.G.

1.83 m × 20 cm × 11 cm on M.G.

1.52 m × 15 cm × 10 cm on N.G.

Rails are directly laid over hard-wood sleepers and fixed with spikes. Adzing is done in the sleepers to give a slope of 1 in 20. But in the case of soft-wood sleepers steel-bearing plates are used to distribute the rail-load over wider area of the sleepers. Figure 23.1 shows the cross-section of track with wooden sleeper.

Composite Sleeper Index (C.S.I.)—It is the index to determine the suitability of a timber for use as sleeper. The mechanical strength of the timber is measured in this index. The C.S.I. is determined by the formula

$$\text{C.S.I.} = \frac{S + 10H}{20}$$

where

S = Strength index of the timber at 12% moisture content.

H = Hardness index of the timber at 12% moisture content.

Following minimum values of C.S.I. have been allowed on Indian Railways :

Place where timber sleeper is to be used	*Allowable minimum C.S.I.*
Track sleepers	783
Crossing sleepers	1.352
Bridge sleepers	1.455

22.3.2 Steel Sleepers

These consist of steel troughs made out of about 6 mm thick steel sheets, with its both ends bent down to check the running out of ballast. Generally two types of steel trough sleepers are used on Indian Railways. In the first type lugs or jaws are pressed out of metal and keys are used for holding the rails. In the second type holes are made in the sleepers and clips and bolts are used for fixing the rail. At the time of pressing, cant of 1 in 20 is also provided towards the centre as shown in Fig. 22.1, Clip, bolt and four key type sleepers can also prevent creep and do not allow change in gauge.

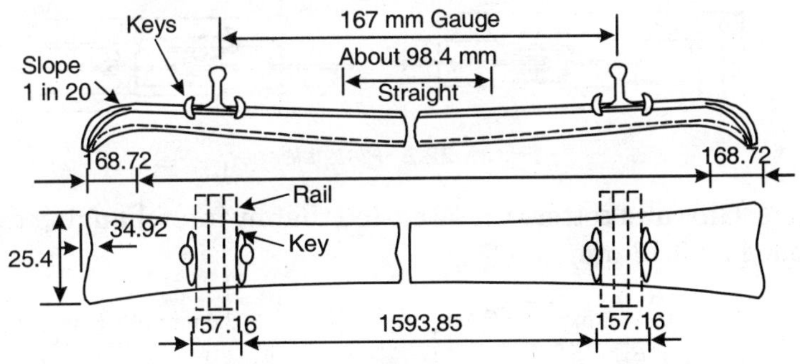

Figure 22.1 *Steel sleeper*

For fixing the rails to the sleepers, rails are first inserted in lugs and wedges or keys are fixed on both sides of the rail. The gauge can be accurately adjusted with the help of four keys fixed in lugs. The method of fixing the rail by clips and bolts is similar to this.

The main disadvantage of steel sleepers is that they get rusted very quickly and lugs sometimes get cracked or even broken and cause much difficulty.

22.3.3 Cast Iron Sleepers

Cast iron sleepers are extensively used in India. They are generally of the following types :

1. Pot sleepers
2. Plate sleepers
3. Box sleepers
4. C.S.T.-9 sleepers
5. Duplex sleepers

Pot sleepers are in the form of two bowls placed under each rail and connected together by a tie-bar. The total effective area of both the pot sleepers is kept 0.46 m^2 which is equal to the effective bearing area of a wooden sleeper. Two holes are provided in each pot sleeper for inspection and packing of ballast. The rail seat is given a slope of 1 in 20. Both pots are connected together by a tie bar with necessary fittings such as keys, gibs

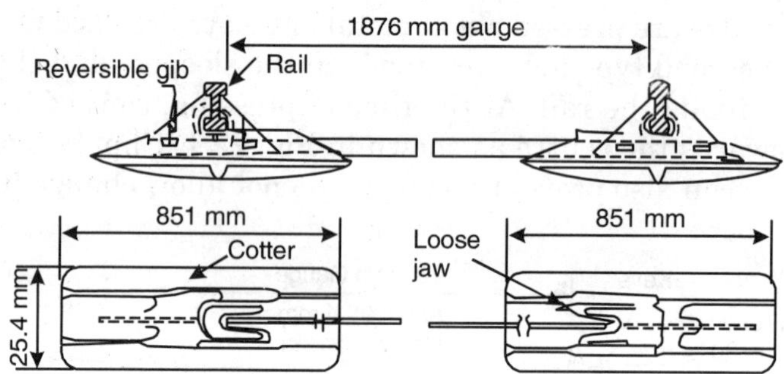

Figure 22.2 *Plate sleeper*

and cotters. Gibs and cotters are so casted that by interchanging them gauge is slackened by 3.18 mm.

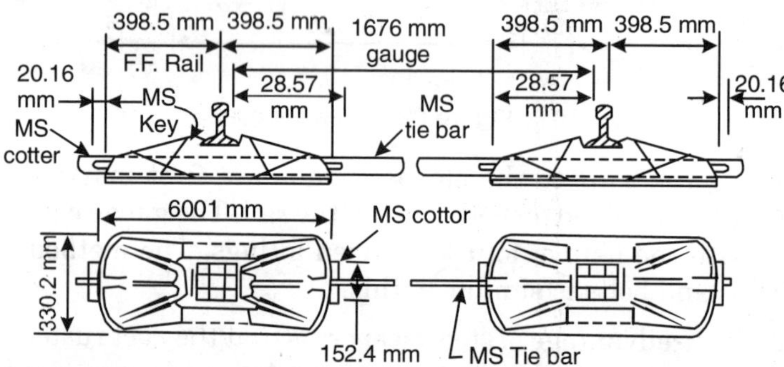

Figure 22.3 *C.S.T.-9 sleepers*

Plate sleepers consist of a plate of 851 mm × 254 mm in dimensions, with 254 mm side parallel to the rails. Both sleepers provide an effective bearing area of 0.46 m^2 under each rail. The plate is provided with projecting rib in the bottom to provide a grip in the ballast to check the lateral movement of the sleeper. At the top of the plate stiffners are provided to increase the strength. For fixing the rails suitable arrangement is done as shown in Fig. 22.2. The sleeper plates are connected across the track by means of a tie-rod.

C.S.T.-9 sleeper is more satisfactory than other types of C.I. sleepers, therefore, it is mostly used. It is actually a combination of plate, pot and box sleeper. It essentially consists of a triangular inverted pot on either side of the rail seat as shown in Fig. 22.3. Suitable rail seat or rail chair is provided at the top to hold rails at 1 in 20 cant. Two pieces of sleepers are connected together by means of a tie-rod.

The main disadvantage of cast iron sleeper is that they are liable to crack and break. But broken pots have a salvage value and can be melted further for casting new ones. The weight of C.I. sleeper is 113.4 kg.

22.3.4 Reinforced Concrete Sleepers

There are two types of R.C.C. sleepers. Fixst type is like a wooden sleeper i.e., on piece sleeper. In the second type two R.C.C. slabs are jointed together by means of a tie-bar generally of a Tee-section. There are various methods for fixing rails of R.C.C. sleepers. In the beginning spikes were driven in wooden plugs embedded in concrete, but they were not found satisfactory. At some places coach screws are fixed in wooden plugs. In the third method bolts and clips are inserted in the sleeper. To avoid disintegrating of concrete sleeper under heavy traffic, metal bearing plates are provided at the rail-seat with a shock absorbing pad.

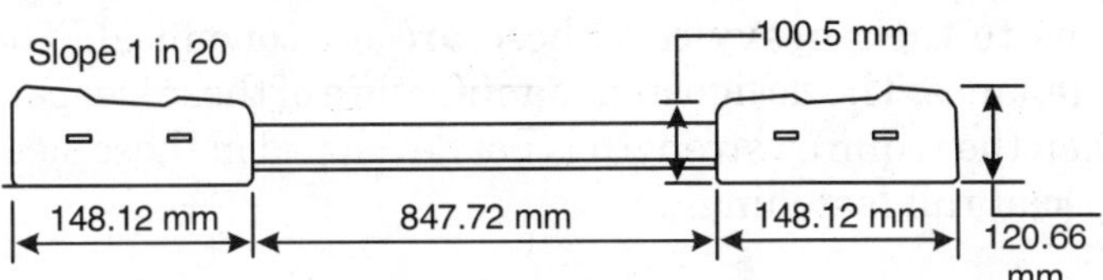

Figure 22.4 *R.C.C. sleeper*

The concrete sleepers are not affected by natural decay or insects etc., hence they have long life of 40–60 years under normal conditions. These are not affected by the chemical action of the cinder, ballast and sub-soil salt. Track-circuiting is possible in case of concrete sleepers. These sleepers provide greater stability to the track due to their heavy weight. These sleepers also help in minimising joint-maintenance and good resistance against temperature. Due to high elastic modulus these sleepers can withstand the higher stresses of fast moving trains.

The concrete sleepers have also some disadvantages. Their heavy weight (2.5 to 3 times than wooden sleepers) possess great difficulties in transporting, laying and handling. Pads and plugs are required for fixing the spikes. During packing the bottom edges are damaged. If the concrete sleeper is cracked or broken the scrap value is almost nil.

22.4 PRESTRESSED CONCRETE SLEEPERS

These sleepers are costly initially, but are very cheep in the long run. There are two types prestressed concrete sleepers :

22.4.1 Pre-tensioned

In this method high-tension steel wires are passed through so many moulds placed in one straight line. These wires are stretched by hydraulic jacks to give necessary tension in the wires. After it, the concrete is poured in the moulds and when it is cured hardened, wires are cut and sleepers are ready for use.

22.4.2 Post-tensioned

In this method thin steel tubes are embedded in the concrete sleepers, through which high tension steel wires are passed and stretched after hardening the concrete. After stretching the wires, the tubes are filled with cement grout under pressure.

Most of the disadvantages of concrete sleepers are eliminated in case of pre-stressed concrete sleepers. But even then these have so many disadvantages. In case of train derailment these are totally damaged and require replacement. Special ballast bed is required for the laying of these sleepers. Due to their heavy cost these are uneconomical. These require costly maintenance. The design and manufacture of these sleepers is difficult and even then the required strength is not developed in these sleepers. These requires special rail fastenings.

22.5 TYPES OF CONCRETE SLEEPERS

Following types of concrete sleepers are being manufactured for B.G. system on the Indian Railways.

22.5.1 Mono Block Prestressed Concrete Sleepers with Pandrol Clips

This sleeper is similar to German B-58 type of sleeper, having an overall length of 2750 mm and weight of 260 kg approximately. It has trapezoidal cross-section having 154 mm top width, 250 mm bottom width and 196 mm height at the rail seat. A cant of 1 in 20 is provided for a distance of 175 mm on the top surface of the sleeper on either side of the centre line of rail. The prestressing of the sleeper is done by 20 number high tensile steel bars of 3 × 3 mm dia. and 12 number 6 mm dia. mild steel links. The crushing strength of the concrete at 28 days should not be less than 525 kg/cm^2.

The rails rest on a grooved rubber pad with grooves parallel to the axis of the rail. For 52 kg rail pandrol clips 401 are used for fastenings of rails. These sleepers are known as PCS-2 sleepers. PCS-1 sleepers were earlier version of monoblock prestressed concrete sleepers, which are not manufactured now.

PCS-3 sleepers are the latest type of prestressed monoblock concrete sleepers. The sizes of these sleepers are 2750 × 290 × 270 mm. The width and height at the centre are 230 mm and 130 mm respectively. These are reinforced with 22 nos. of 5 mm dia high tension steel wires, with 6 mm dia m.s. links 12 nos. The high tension wires are tensioned with an initial tension of 2040 kg/wire.

22.5.2 R.C.C. Two Block Sleepers for use with I.R.N-202 Clips for B.G. 90 Rails

This sleeper is similar to French R.S. type of sleeper having its overall length 2478 mm. The sleeper mainly consists of two R.C.C. blocks each weighing approximately 90 kg. The two blocks of R.C.C. are joined together by means of an angular tie-bar 7575010 mm size. Size of each block is 725 × 295 × 251 mm. Total weight of one assembly of sleeper is about 215 kg I.R.N. 202 fastenings, anchored bolts are used for joining rails with sleepers. The bolts are fixed of the angles to tie-bars and the nuts are tightened to press the clip against the rail seat. A rubber pad is provided to absorb the shocks and dampen the vibrations at each rail seat. Nowadays RCS-2 sleepers are used, RCS-1 was the earlier version of two block concrete sleeper designed with conventional loose jaws and key type fastenings.

22.6 MAINTENANCE OF CONCRETE SLEEPERS

The following points should be kept in view while doing maintenance of concrete sleepers:

(i) Only 30 sleepers spaces should be opened out at a time between two packed stretches of track of 30 sleepers length each.

(ii) The spacing of the sleepers should be uniform.

(iii) The maintenance of concrete sleepers should normally be done with heavy 'On-Track' tampers. If mechanical maintenance is not possible, measured shovel packing should be adopted. The size of chips for M.S.P. should be 15.25 mm. Manual maintenance of concrete sleepers should not be done.

(iv) The sleepers should be compacted well and uniform for good riding surface. Central 800 mm of the sleeper should not be hard packed to avoid centre binding of monoblock concrete sleepers.

(v) As far as possible the laying and maintenance should be done by mechanical equipments.

(vi) The inspection and maintenance of elastic fastenings, rubber pads and insulators should be done periodically.

(vii) When casual renewal of concrete sleepers is to be done and 3 consecutive sleepers are to be renewed at a time, the speed restriction of 15 km/h should be imposed.

22.7 SLEEPER DENSITY

Sleeper density is the number of sleepers used per rail on the track. This density depends on the following factors :

(a) Lateral thrust of the locomotives.

(b) Axle load of train.

22.8 COMPARISON OF VARIOUS TYPES OF SLEEPERS

Various types of sleepers are mention in Table 22.1.

TABLE 22.1

S.No.	*Comparison Point*	*Wooden Sleepers*	*Steel Sleepers*	*Cast iron Sleepers*	*Concrete Sleepers*
1.	Cost-(each sleeper)	Rs. 42	Rs. 55	Rs. 50	Rs. 40 to Rs. 70 depending on design
2.	Life	10–15 years for untreated 15–20 years for treated sleepers	40–50 yrs.	35–40 yrs.	50–60 yrs.
3.	Weight (each sleeper for B.G.)	55 kg	77.5 kg	113.5 g	130–160 kg
4.	Track fitting requirements	Not more	Not more	More	Not more
5.	Handling difficulties	No, can be rough handled	No, can be easily handled	Yes, can break or crack in rough handling	Yes, may break or crack in rough handling
6.	Maintenance cost	Higher than all other types	Minimum	Moderate	Moderate
7.	Overall economy	Economical in initial cost	Costlier in initial cost	Costlier in initial cost	Under trial in Indian Railways

Contd.

Contd. Table 22.1

S.N.	*Comparison Point*	*Wooden Sleepers*	*Steel Sleepers*	*Cast iron Sleepers*	*Concrete Sleepers*
8.	Laying and Relaying	Very easy	Easy	Difficult	Difficult
9.	Track Elasticity	Good	Moderate	Not so good	Moderate
10.	Regional Suitability	Suitable in all areas except wet and white ant or insect areas	Suitable only with stone ballast, not suitable for yards	Suitable only with stone ballast, not suitable for station yards	Suitable for all areas
11.	Track circuiting	Best	Insulation pads are necessary	Insulation pads are necessary	Good
12.	Maintenance of gauge	Not very good	Moderate	Moderate	Moderate
13.	Gauge adjustment	Diifficult	Easy	Easy	Easy
14.	Creep	Large nos. of anchors required to prevent creep	Very small	Very small	Very small
15.	Scrap value	Less	High	Highest	Nil

(c) Methods of providing rail joints viz., staggered or both opposite each other. In case of staggered joints extra sleepers are required as on either side of joint sleepers are provided.

(d) Speed of train.

(e) Type and section of the rail.

(f) Type of sleeper and type of ballast used. In India the number of sleepers per rail varies from N + 3 to N + 6 for main tracks where N is the length of the rail in metres. For example, for 128 m long rail, 18 sleepers are required for [N + 5 = (12.8 + 5)]. This 18 number is the sleeper density.

In India, generally the sleeper density of 18 sleepers per rail length is provided, whereas in USA 23 to 25 sleepers per rail are provided. But nowadays in India the sleeper density on main lines is increased due to fast running trains such as Rajdhani Express and Satavadi Express etc.

It is not necessary to provide all the sleepers at equal distances. At rail-joints the spacing of sleepers is more than on the other portions of the track to avoid loosening of ballast due to impact.

If the joints on the curves are staggered, one additional sleeper is provided for each joint for making suspended joints.

REVIEW QUESTIONS

22.1. Name and explain with sketches the different types of sleepers used on Indian Railways, including the method of fixing rails to them.

22.2. (a) What are the requirements of good rail sleepers ?

(b) What are the advantages of steel sleepers over wooden sleepers ?

(c) Why are wooden sleepers used on track circuited lengths ?

22.3. Write short notes on cast iron sleepers with the help of neat sketches.

22.4. Write a short note on sleepers-density.

22.5. Compare various types of sleepers.

22.6. What are the functions of sleepers in the track.

22.7. Write short note on mono-block prestressed concrete sleepers.

22.8. Describe in brief R.C.C. two block sleeper.

22.9. What points should be kept in view while doing maintenance of concrete sleepers.

Ballast

GENERAL

Layer of broken stone or any other suitable material which is spread on the top of railway formation and around the sleepers is known as *Ballast.* Some portion of the ballast is tightly rammed under the sleeper to transmit the load of train from the sleeper, which is known as *packing.* Another portion of the ballast known as *boxing* is loosely filled on slopes and thrown around the sleepers to prevent the lateral and longitudinal movement of sleepers.

23.1 FUNCTIONS OF BALLAST

Following are the chief functions of providing ballast in the railway track.

1. To uniformly distribute the load from sleepers over a large area of formation.

2. To hold the sleepers in position by acting as their foundation and preventing their lateral and longitudinal movement.

3. To provide elasticity and resilience to the track, the ballast acts as an elastic mat between formation and sleepers due to which there is resilient running.

4. To provide an easy method for track adjustments such as alignment and gradient, without any disturbance to formation.

5. To drain the rain water from the track. There is a large number of voids in the ballast, due to which water cannot remain in the track. If the

water is not drained out quickly the formation will become soft, causing settlement of track and decay of wooden sleepers.

6. To prevent the growth of weeds inside the track.

23.2 CHARACTERISTICS OF GOOD BALLAST

To fulfil the above objects of providing ballast, it should have the following characteristics:

1. It should not be brittle and have sufficient strength to resist crushing under heavy loads of moving trains.
2. It should have angular and rough surface so that it may give lateral and longitudinal stability to the sleepers and prevent their movements against creep and centrifugal forces.
3. It should be sufficiently durable to resist the abrasion and weathering action.
4. It should be cheaply available in the required quantity along the track.
5. When laid on formation, it should have sufficient voids to drain-off the rain-water.
6. It should have good workability, so that it can be easily laid on formation.
7. Due to the impact of train, if a portion of ballast becomes powder, then it should be easily removable.
8. It should not have any chemical effect on the rails and the sleepers.
9. It can be easily packed in position with hand tools.
10. The size of the ballast for wooden sleepers should be 5 cm to 3.8 cm, for metal sleepers. At turnouts and cross-overs ballast of 2.5 cm is desirable.

23.3 BALLAST MATERIALS

Materials which are generally used on Indian Railways are broken stone, gravel, sand, ashes, moorum, kankar, brick-bats and in some cases earth. The best material for ballast is non-porous, hard and angular stones and therefore the stone ballast is used on all important tracks. In American and English railways high-grade slag is also used as ballast material.

23.3.1 Broken Stone

It is the most expensive but best type of ballast. Due to its high interlocking action, it holds the track to the correct alignment and gradient. It possesses all the characteristics of a good ballast. It is not subject to abrasion, therefore, it provides very good drainage to rain water. The stones which are non-porous, hard and do not flake on breaking should be used. Granite, quartzite and trap are the best varieties of stones for using as ballast. Graded stones from 5 cm to 2 cm provide maximum stability to the track.

23.3.2 Gravel

The best material after stone for ballast is gravel. It consists of smooth rounded fragments obtained from river beds. Gravel which is obtained from pits usually contains earth which has to be removed by washing before used. Gravel obtained from river beds with rounded edges varies in size from very small to very large pieces. Only the required size gravel is used and smaller pieces are discarded after screening. Larger pieces of ballast are broken into required size. The pieces which have round edges are broken to increase their interlocking action. Due to round edges and corners gravel ballast gets easily centre-bound. Gravel ballast has very good drainage property and holds the track to correct alignment and gradient. At certain places where formation is unstable, it is easy to use gravel ballast than stone ballast.

23.3.3 Sand

It is a very cheap material which can be used for ballast. It has good drainage properties when it is free from soil particles. If has no stability and gets distributed by the vibration caused by moving trains, therefore, mostly it is covered by a layer of stone ballast. Sand particles sometimes get into the moving parts of the rolling stock and cause heavy mechanical wear thereby increasing the maintenance cost. The sand ballast also causes wear of rail-seats and keys. Due to all the above difficulties, sand ballast is not used on main and branch lines. It is used on unimportant lines, sidings and marshalling yards. As it has good drainage properties, it is used on light traffic sections for keeping always yards dry in rainy season.

23.3.4 Coal Ashes

These are waste products obtained from steam locomotives. Ashes or cinder have very good drainage qualities due to more void spaces. This is the cheapest material for railways, but it cannot withstand heavy traffic because under heavy loads it crumbles to powder. Due to this reason it is used only in siding and unimportant lines. On account of its good drainage properties, coal ashes are used in railway yards, because it keeps the yard dry and non-slippery during rainy season. Coal ashes are available in large quantities, therefore, these can be used in cases of emergency like settlement, embankment, wash aways etc.

23.3.5 Moorum

It is easily available in many parts of India. It is obtained by the decomposition of laterite and has red or yellowish brown colour. Every type of moorum crumbles to powder under heavy moving trains therefore it is used on unimportant lines and sidings. Moorum is a good material for using as blanket for new embankments or old banks in black cotton soil, because it acts as a water seal and prevents the entry of water into the embankments.

23.3.6 Kankar

It is a natural material in the *form of nodules* from which lime is prepared. It is also soft aggregate like ashes, brick ballast and moorum, therefore it also crumbles to powder under traffic and is used in restricted places.

23.3.7 Brick Ballast

At some places where stone or good ballast material is not available over-burnt bricks are broken into suitable size and are used. But it becomes powder under heavy traffic and makes the track dusty; therefore rails laid over such ballast get corrugated. Like other soft aggregate, brick ballast has good drainage properties.

23.3.8 Earth Ballast

When new tracks are laid on new formation then in the beginning sleepers are packed with earth for a few months. After some time when formation is consolidated and surface becomes hard, good type of ballast is laid. The main function of using earth ballast in the beginning is to prevent the loss of good and expensive ballast by sinking into soft formation.

23.3.9 Slag

It is a waste product obtained from the blast furnaces of heavy metal plants. Slags of good varieties have good drainage properties, are strong and hold the track to correct gradient and alignment. High grade slags have all the characteristics of a good ballast, therefore it is widely used in foreign countries.

23.4 SPECIFICATION OF STONE BALLAST

Following are the specifications as followed on Northern Railway for stone ballast:

(i) *Quality*. The stone ballast should be durable, hard, resilient to impact and free from adherent coatings. It should not contain more than 10% by weight of quarry dust, rubbish or decomposed matter which can pass through 5 mm sieve. The faces of ballast should have resulted from crushing operation and only one smooth surface may be allowed.

(ii) *Size*. Following are the standard sizes of ballast for various sleepers.

Sue of ballast	*Type of sleepers*
50 mm	Wooden and C.I. pot sleepers
40 mm	CST-9 and trough sleepers
25 mm	Points and crossing

(iii) *Grading*. The ballast should be well graded to give the following sieve analysis.

Nominal size of the ballast	*Sieve size with square mesh*	*% by wt. that passes through the sieve*
50 mm	65 mm	100
	50 mm	90 to 100
40 mm	50 mm	100
	40 mm	90 to 100
25 mm	40 mm	100
	25 mm	90 to 100

(iv) *Sampling*. The sample of the ballast is collected at the rate of 1 m^3 for every 2000 m^3 of ballast. Square sieves of the nominal size given above are used. The percentage passing through the sieve is determined by volume.

(v) *Over-sized ballast*. When more than 10% ballast is retained on the nominal size sieve, it is considered to be over-sized. Over-sized ballast are not accepted without prior approval of the engineer.

(vi) *Stacking*. The ballast should be stacked along the quarry siding on a level bed in a convenient number of zones marked on the ground. The height of the stack should not be less than 120 cm. The payment for the ballast is based on the stack measurement made by A.E. and after deducting a shrinkage allowance of 8%.

23.5 DEPTH OF BALLAST SECTION

Depth of the ballast section in straight tracks may be calculated by the following formula:

$$D = \frac{S-b}{2}$$

where

D = Depth of ballast section

S = Sleeper spacing

b = Width of sleeper

Depth of the ballast-section will come between 20–25 cm from the above formula.

23.6 SECTION OF BALLAST

Figure 23.1 shows the standard ballast section for B.C., M.G. and N.G. tracks. On straight tracks depth of ballast remains uniform throughout as shown in Fig. 23.1, but on curves additional ballast is required to make up the

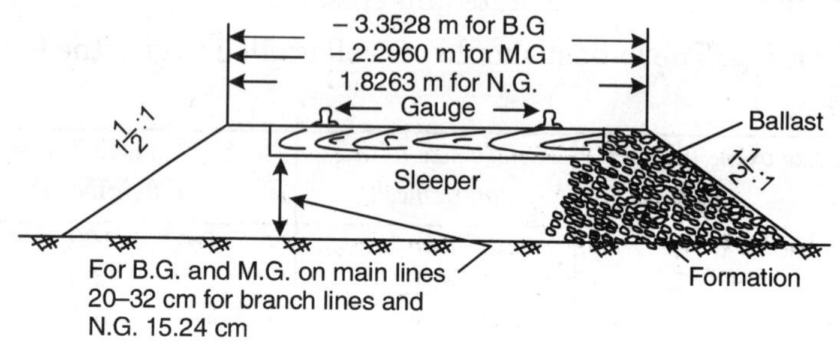

Figure 23.1 *Typical cross-section of track showing section of ballast on straight lines*

super-elevation as shown in Fig. 23.2. It is the practice in India to provide the recommended minimum depth of ballast under the inner rail-seat of the curve as shown in Fig 23.2. The depth of ballast under the outer rail is increased so as to give the required super-elevation. To counteract the increased lateral thrust on curves, extra shoulder of 15 cm is provided on the outside of curves.

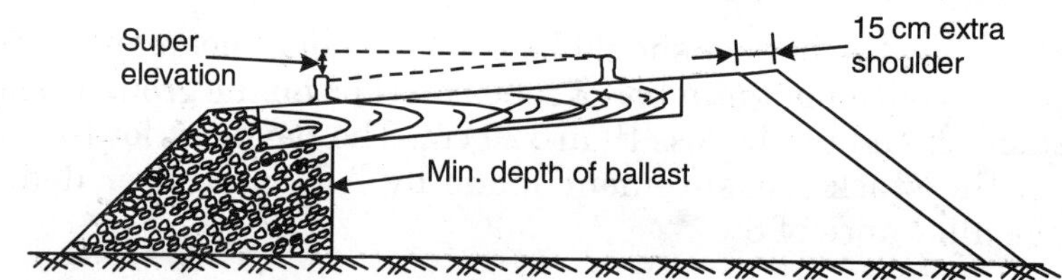

Figure 23.2 *Typical cross-section of track showing section of ballast on curves*

Table 23.1 gives the details of ballast sections for various types of curves.

TABLE 23.1 *Details of ballast sections*

S.No.	*Gauge*	*Width of Ballast Section cm*	*Depth of Ballast below sleeper cm*	*Quantity of Ballast per metre length of track cu.m*
1	B.G.	335.3	20–32	1.11
2	M.G.	229.6	15–25	0.767
3	N.G.	183.0	15–17	0.530

Lateral stability of the track depends partly on the quantity of ballast at the ends of the sleepers. The load bearing capacity and uniformity of distribution of load on formation mostly depends on the depth of ballast provided below the sleeper. Wooden sleepers require less quantity of ballast than metal sleepers.

23.7 BALLAST RENEWAL

Due to continuous movement of the trains on the tracks, the ballast gets crashed and their sizes continuously reduce. The crushed particles of the ballast accumulate in the voids of ballast and continuously reduce them, causing obstruction in the drainage of rain water. The ballast depth also reduces due to penetration of ballast in the formation and crushing under heavy loads.

For the efficient functions, the loss of ballast quantity is renewed from time to time. While doing renewal, the crushed ballast particles are removed by screening the ballast through the screens. The additional quantity of ballast is added to make up the desired depth and section of the ballast.

REVIEW QUESTIONS

23.1. What is the object of providing 'Ballast'on the track?

23.2. What are the different varieties of ballasts ordinarily used in a railway track? Discuss the merits and demerits of each of them with special reference to their suitability for various types of sleepers.

23.3. Draw a cross-section of the track showing ballast section for various tracks.

23.4. Write short note on the ballast renewal.

23.5. Compare the different ballast materials with respect to merits, demerits and suitability of each material as ballast.

23.6. Write short notes on:

(i) Size of ballast. (ii) Quantity of ballast. (iii) Depth of ballast.

23.7. What do you understand by the periodic renewal of ballast? How it is done ?

24

Rail Fastenings

GENERAL

Rail fastenings and fixtures are used to connect the rails and sleepers together in their proper positions. These fixtures also set the points and crossings properly for changing the trains from one track to another track. Rail fastenings joint the rails with each other and also fix them with the sleepers at the desired gauge. The rails may be fixed with the sleepers directly or through bearing plates or chairs as the case may be.

Following are the common fastenings and fixtures used in the track fittings :

1. Fishplates
2. Fish-Bolts
3. Rail-Chairs
4. Bearing-Plates
5. Spikes
6. Anchors or Anticreepers
7. Blocks
8. Keys

24.1 FISH-PLATES

Rails are manufactured in certain fixed lengths keeping in view the economy and ease in transportation. These rails are then joined together to form one long continuous rail. At every joint it is most essential that the tables of both the rails should have the same horizontal and vertical planes in order to preserve the continuity of the rail. For joining the rails fish-plates are used, therefore they should fulfil the above requirement i.e., hold the adjoining ends of rails in correct horizontal and vertical planes. As under the variations of temperature the rails contract or expand, the fish-plates should also allow free longitudinal movements of rails due to change in temperature. Therefore small clearance is left between the ends of the rails for expansion and this gap is known as expansion-gap. From the above statement we see that the fish-plates should be designed for:

1. Bearing the vertical and lateral stresses which come at joints without any distortion.
2. Fish-plates must support the underside of the rail and top of the foot.
3. Allowing free contraction and expansion of rails for this purpose they should not touch the web.
4. Taking all types of wears.
5. Easy renewal and replacement of rails in case of wear and damage.

Two types of fish-plates are shown in Fig. 24.1 which are commonly used on Indian Railways for joining F.F. and B.H. rails. Each fish-plate is 45.72 cm long with four nos. of 31.75 mm diameter holes at 114.3 mm centre to centre. Originally fish-plates were manufactured of wrought iron, but nowadays it has been replaced by steel. Two fish-plates are fixed at each joint with four bolts. These fish-plates are so designed that they fit the underside of the rail head and the top of the rail foot in case of F.F. rails. The holes which are drilled in the rails are slightly wider so that rails can contract or expand under variations of temperature. Generally these fish-plates are oiled occasionally for the free movement of rails.

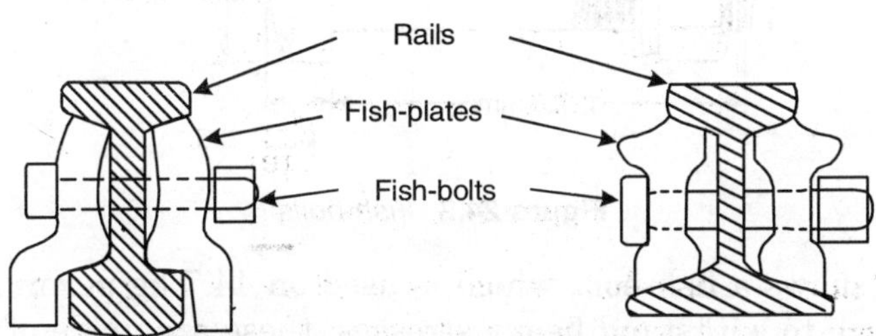

Figure 24.1 *Fish-plates*

The wearing of fish-plates starts when sand enters between the contact surface of rail and fish-plates, but oiling reduces this tendency. Second type of wearing starts when the rail head is worn out and the wheel flange starts touching the fish-plate.

The wearing of fish-plates due to impact of wheels, expansion and contraction of rails is unavoidable, therefore the section of the fish-plate should be such, which can be adjusted by means of further tightening the fish-bolts. The slope of fish-plates is kept flatter (1 in 4) than the contact plane of the rail which is 1 in 2.75.

Fish-plates generally fail due to abrasion on their top, reversal of stresses due to their large length and cracks along their sections.

Table 24.1 gives the details of standard fish-plates used on Indian Railways far different rail sections.

TABLE 24.1

Rail Section	*Total length of fish-plate in mm*	*Dia. of hole for fish-bolt in mm*	*Approx. wt. per pair in kg*
52 kg	610	27	28.71
90 R, T 1 (M)	610	27	26.11
90 R, T 2 (M)	610	27	25.93
90 R, T 059 (M)	460	27	19.54
75 R	420	27	13.58
60 R	410	24	9.975
50 R	410	20	83.07

24.2 FISH BOLTS

At every rail-joint fish-plates are connected to the rails with the help of fish-bolts. With each pair of the fish-plates four or six fish-bolts are used. But on Indian Railways four number of fish bolts are used.

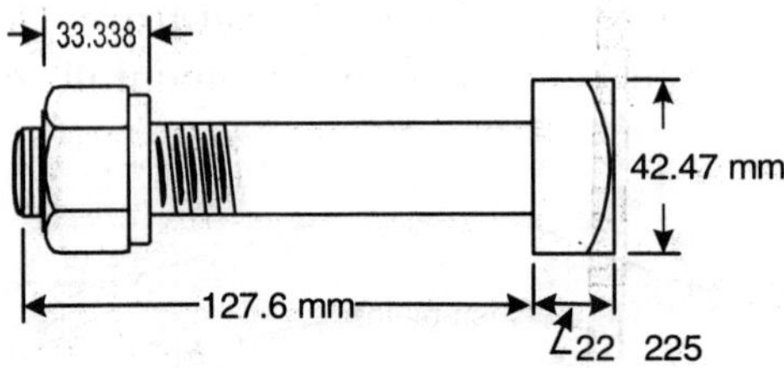

Figure 24.2 *Fish-bolts*

Figure 24.2 shows a fish-bolt, which is used on 44.7 kg per m rail. As the fish-bolts have to withstand heavy stresses, these are made of medium or high carbon steel. The length of fish-bolt depends on the type of fish-plate. These bolts generally get loose by the vibrations caused by the moving loads and require tightening from time to time; therefore, for maintenance its incharge should check it. Over-tightening of fish-bolts is prohibited because it will jam the fish-plates with rails and will prevent the free expansion of rails. In practice the nut is loosen by a quarter turn after thorough tightening. Nuts of these bolts are made of sufficient length so that they can provide good grip on the bolt. At some places lock-nuts and spring washers have been tried with good success.

Table 24.2 gives the details of standard fish-bolts used on Indian Railways for different rail sections.

TABLE 24.2

Rail	*Thickness of Nut/Dia of Bolt in mm*	*Width of Bolt head in mm*	*Length of bolt in mm*	*Approx. weight with ordinary head in kg/piece*
52 kg 90 R & 75 R	25	41	130	0.94
60 R	22	36	105	0.654
50 R	18	32	90	0.420

24.3 RAIL CHAIRS

Flat footed rails can be directly fixed on the sleepers, but Bull-head or Double-head rails cannot be fixed directly and require some special types of devices, which may keep them in the required position under heavy loads of trains. Such type of devices are known as chairs, which distribute the load from the rail, on the sleeper uniformly. The chairs are generally made of Cast Iron. The weight of each chair is about 22 kg.

Figure 24.3 shows two types of rail-chairs which are commonly used. The rail is placed between the jaws in such a way that the inner jaw should

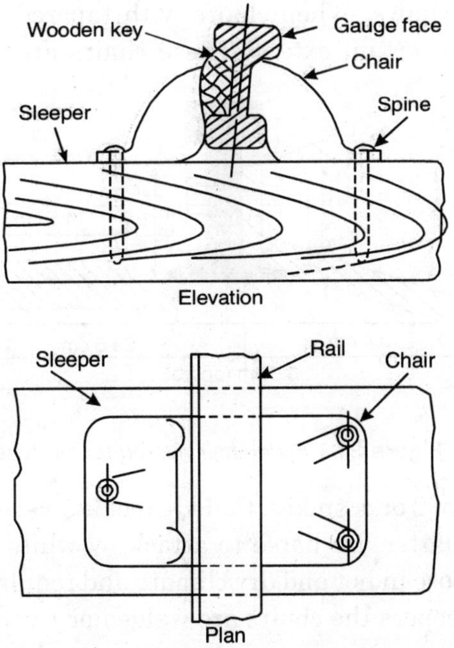

(a) Rail-chairs with wooden key

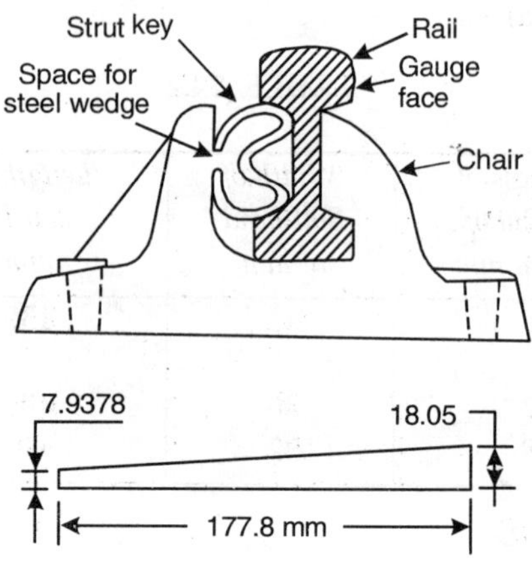

(b) Rail-chair with spring key

Figure 24.3

remain in contact with the web of the rail as shown in the figures. Jaws are either straight or tapered on the inside face. In the case of inside tapered face the chairs are classified as right hand or left hand chairs. In the space which is left between outer jaw and the web of rail slightly tapered steel or wooden key is driven. Fig. 24.3 (a) and 24.3 (b) show both types of keys fixed in position in the chairs. When chairs with tapered jaws are used, they prevent creep upto certain extent. These chairs are fixed to the wooden

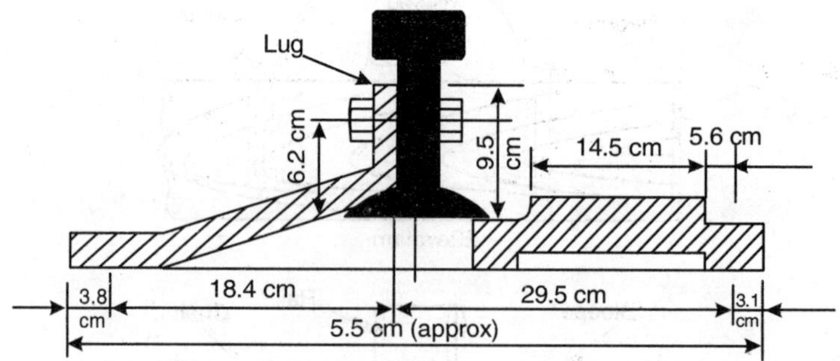

Figure 24.4 *Slide-chair riveted up lug type*

sleepers by means of 2 or 4 spikes. Coiled-steel keys are better than wooden keys, because the latter are liable to attack by white ants, and to thefts. Wooden keys get loose in hot and dry climate and require more attention. In the case of steel sleepers the chairs are welded or riveted to the sleepers.

Slide-chair. For changing the trains from one track to another track on turn out points special shape slide-chairs are used. The stock rail is bolted to the projecting arm known as lug of the chair and the tongue rails slides laterally. The slide chairs are 12–15 cm wide and of length increasing towards

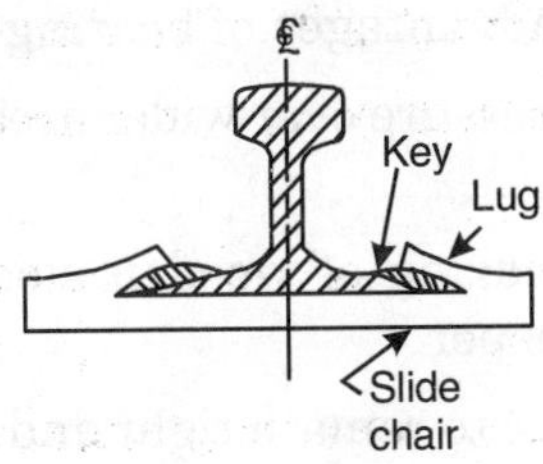

Figure 24.5 *Slide-chair with key-pressed up lug type*

the heel. Two types of slide-chairs are used. First is pressed up lug type and the second riveted lug type. Figures 24.4 and 24.5 show these types of chairs.

24.4 BEARING PLATES

Flat-footed rails can be directly fixed on sleepers of hardwood, but, very heavy loads of trains cause the rail to sink in the sleeper and thus loosen the spike. In the case of soft wood sleepers the timber fibres are crushed under loads. As the flat-footed rail does not provide sufficient bearing area on the timber-sleepers, the intensity of pressure is more than the bearing capacity of the timber fibres, which causes crushing on the timber. To overcome this difficulty bearing plateis are used under flat footed rails, which distribute the load over a wider area and bring the intensity of pressure within safe limits. In the case of B.H. and D.H. rails bearing-plates are not required, because these rails are fixed in chairs which distribute the pressure over a wider area.

Bearing-plates of various patterns are used on the Indian Railways, but the most common one is shown in Fig. 24.6. Bearing plate is manufactured

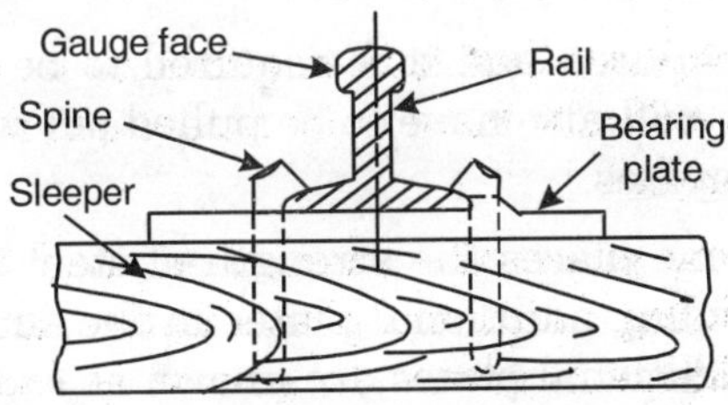

Figure 24.6 *Bearing plates*

in such a way that it gives the required 1 in 20 slope to the rail directly and no adzing is required to be done in the wooden sleepers. These bearing plates are fixed to sleepers by spaces. The bearing plates may be of cast iron or steel. In the case of C.I. bearing-plates, small projection is casted on the underside which penetrates in the timber sleeper under heavy loads and resists the lateral movement of the bearing plates.

Following are the main advantages of bearing-plates :

1. They distribute the pressure over wider area and prevent the crushing of sleepers.

2. They eliminate the adzing of wooden sleepers which require more labour and weakens the sleeper.

3. They enable the spikes to remain tight and require less maintenance.

4. They prevent the widening of gauge on curves.

5. They increase the life of timber sleepers by preventing the rubbing action of the rail.

Table 24.3 gives the length and weight of Bearing plates used on Indian Railways.

Bearing plate with	*Rail section*	*Gauge mm*	*Length in mm*	*Width in mm*	*Approx. wt. of each bearing plate in kg*
One-key	52 kg and 90 R	B.G.	285	205	10.76
Two-key	52 kg and 90 R	B.G.	285	205	11.26

In addition to the above advantages the bearing plates also have some disadvantages, which are as follows:

1. If the bearing plates become loose due to settlement of ballast and sleeper, moisture is likely to enter in between the bearing plate and the sleeper, which will cause sleepers to wear.

2. If any spike is injured and it is required to be redriven at another place, all other spikes will also have to be pulled out which will reduce the holding power of the spikes.

Saddle-plate. At some places the strength of steel trough type sleepers are increased by providing additional plates in the shape of the saddles at the rail-plates. The additional plates are known as *saddle-plates.*

24.5 BLOCKS

On curves, level crossings and bridges another rail is provided on the inner side of the track for providing flange-gap for the wheels. To hold this check rail at the required distance, small blocks of steel are inserted in between the two rails. Depending on the requirements, these blocks may touch either the webs or the fishing faces or both.

24.6 SPIKES

For fixing the rails to wooden sleepers various types of spikes are used. Three different types of spikes are shown in Fig. 24.7 which are most commonly used.

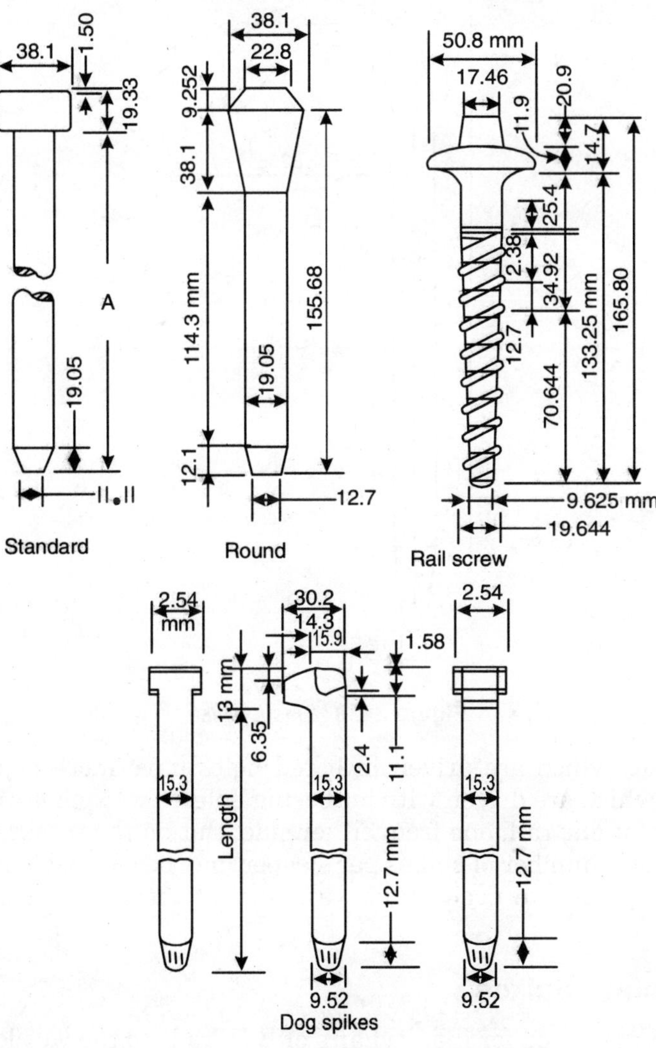

Figure 24. 7 *Common rati spikes*

24.6.1 Dog-spikes

This is the cheaper type of spikes which hold the rails at correct gauge, and can be easily fixed removed. But mostly this type of spikes become loose under the wave action caused by the moving trains and require more attention during maintenance. These are square spikes, at the top of which *lugs* are provided for extraction. The lower end of the spike may be pointed, blunt or chisel shaped, but mostly the blunt is in common use. The holding power of this spike depends on the friction between the spike and the wooden

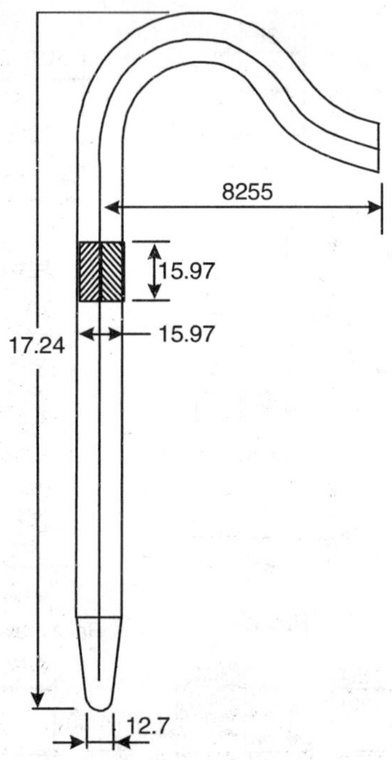

Figure 24. 8 *Elastic spike*

fibres. Spikes which are driven in bored holes have more holding power than those which are driven without boring holes. Two spikes are driven in one sleeper for one rail, one from either side. But on the tracks with heavy traffic the total number of spikes per sleeper may be 6 or 8. Generally more number of spikes are used on curves, so that they may counteract the centrifugal force.

24.6.2 Round Spikes

These spikes are used for fixing chairs of B.H. rails to the wooden sleepers. These spikes are also used for fixing the slide chairs of points and crossings. The length of these spikes depends on the type of chairs and the gauge of the track. These types of spikes are not used for fixing F.F. rails to wooden sleepers. These are blunt at the bottom and their use is very much limited.

Table 24.4 gives the length and weight of Dog spikes used on Indian Railways.

TABLE 24.4

S.No.	Description of Dog spike	*B.G.* Length mm	*B.G.* Weight per 100 spikes kg	*M.G.* Length mm	*M.G.* Weight per100 spikes kg	*N.G.* Length mm	*N.G.* Weight per 100 spikes kg
1.	For use with × ring timber (special)	160±3	38.51	135±2	33.49	120±3	30.48
2.	For use with bearing plates	135±3	33.49	120±3	30.48	110±3	28.47
3.	For use without bearing plates	120±3	30.48	110±3	28.47	110±3	28.47

24.6.3 Screw Spikes

These spikes are just like coach-screws. The holding power of these screws is more than Dog-spikes. The length of these spikes depends on the gauge. This screw spike is driven in the sleeper first by boring a hole and then by revolving it with the help of a spanner. It takes more time in driving and extraction from the sleeper as compared with Dog-spikes. Also gauge adjustment is very difficult in the case of screw-spikes. But these spikes have more lateral rigidity than Dog-spikes and also do not spoil the sleeper. But being more costly and taking more time in driving, these are not popular on Indian Railways.

24.7 KEYS

These are small pieces of timber or steel which are used to fix the rails to the chairs. Figure 24.3 (a), (b), 24.5 and 24.9 show these keys.

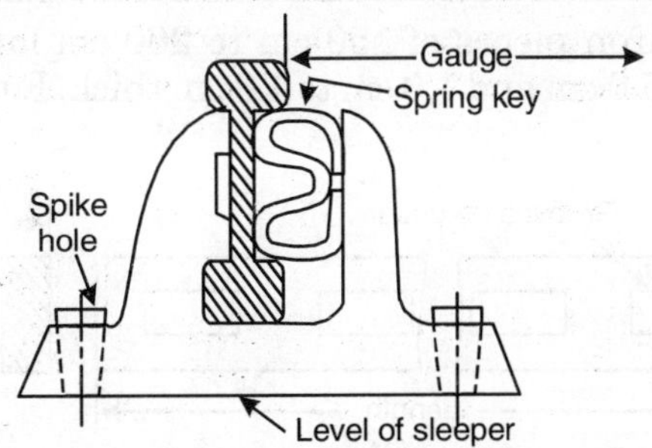

Figure 24.9 *Spring key*

Morgaa keys are about 18 cm long and tapered 1 in 32. These keys are most suitable for the C.I. chairs, plate sleepers and steel sleepers. Two keys are used for joining the rail with the sleeper chair. Morgan keys are light in weight and versatile in nature.

24.8 COTTERS AND TIE-BARS

C.I. sleepers are kept at the required distance by means of tie-bars which are connected by means of cotters. The cotter is a wedge-shaped plate with a split in the horizontal or vertical plane for expansion against prevention of coming out from the slit of the tie-bar.

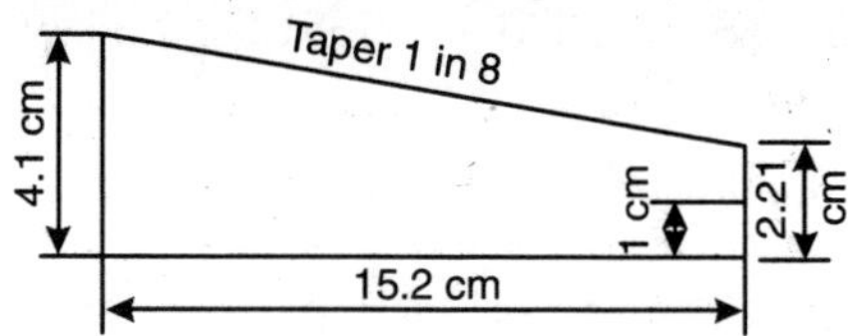

Figure 24.10 *Cotter (side-split)*

Table 24.5 gives the approximate weights of jaws, Two-way keys and Tie-bar various gauges.

TABLE 24. 5

Rail section	*Gauge*	*Approx. wt of 100 jaws in kg*	*Approx. wt. of 100 two-way M.S. key in kg*	*Approx. wt of 100 M.S. cottor in kg*	*Approx. wt of one Tie-bars in kg*
52 kg	B.G.	28.8	48.5	37	12.62
90 R	B.G.	28.8	48.5	37	12.62
75 R	B.G.	28.1	48.5	37	12.62
75 R	M.G.	23.1	48.5	37	12.62
60 R	M.G.	21.5	43.0	37	6.45
50 R	M.G.	21.5	43.3	37	6.45

Tie-bars are flat iron pieces of 150 cm to 260 cm long. The width of the flat-iron is 4.5 cm to 5.0 cm and 1.0 cm to 1.3 cm thick. Four holes are provided

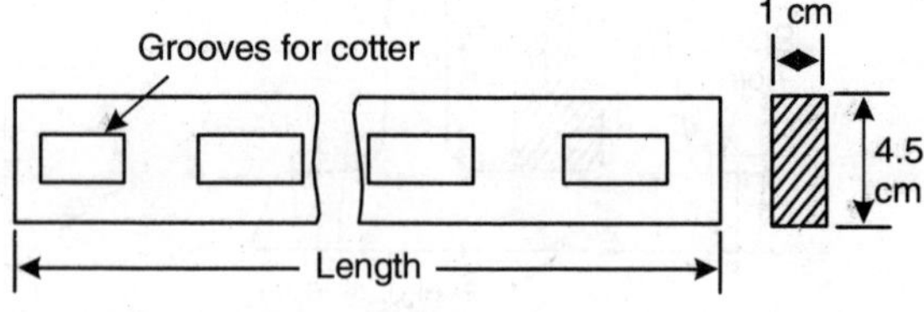

Figure 24.11 *Tie-bar*

in the tie-bars for keeping the sleepers at both the ends at the proper distances by means of cotters.

24.9 ANCHORS OF ANTI-CREEPERS

We have seen in Chapter 21 that if the longitudinal movement of the rails is not checked, it will cause *'Creep'*. At the time of discussing about the preventive measures for creep, it had been mentioned that it can be checked

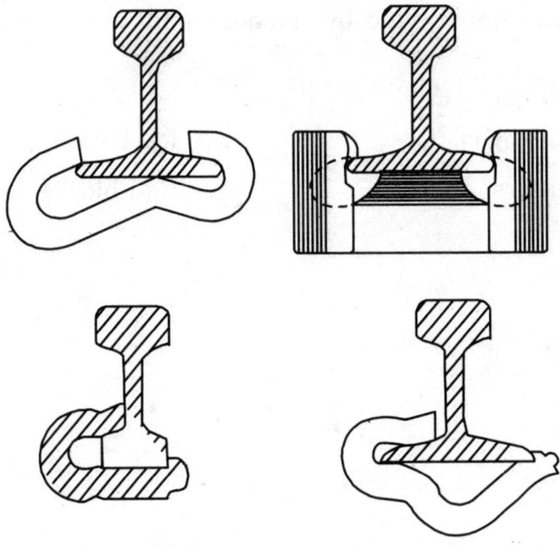

Fig.ure 24.12 *Rail anchors.*

by using *anchors* or anti-creepers. Actually anchors are fastenings which are fixed to the sleeper and foot of rails. Steel trough sleepers which have a clip and bolt for fixing rails does not require any extra anchors, because their fittings are sufficiently anti-creeper. Anchors are fixed at some required intervals in the rails. This distance or interval depends on the traffic and place in rails, because on curves and at points and crossings more number of anti-creepers are required. Generally the number of anchors vary from a minimum of two to a maximum of twice the number of sleepers in the panel. Even after fixing anchors in the track, after some time the pulling-back of the track is required. Sometimes the direction of creep is reverse and at such places the anchors are fixed on both directions in the track. Some of the anchors which are used in Indian Railways are shown in Fig. 24.12.

REVIEW QUESTIONS

24.1. Why bearing plates are provided below rails ? What points are kept in mind while designing bearing plates ?

24.2. At what place, fish-plates are used in railways ? What are their main functions ? Draw a neat sketch of fish-plates fixed in position in rails.

24.3. Write short notes on :

(i) Fish-bolt (ii) Rail-chairs

(iii) Bearing-plates.

24.4. What are the advantages and disadvantages of bearing-plates?

24.5. With the help of neat sketch write down short notes on spikes.

24.6. What do you understand by 'anchors' and what are their functions in railways?

24.7. Write short notes on :

(i) Cotters and Tie-bar (ii) Keys

(iii) Slide-chairs (iv) Rail-blocks.

25

Geometrical Design of Track

GENERAL

The trains may derail or accident may occur due to the defects in the track, defects in the vehicles and the faulty operation of the trains including their control from various places. Civil Engineers are mainly responsible for the track defects. Therefore it is most essential that they must keep strict watch and keep the track free from all types of defects.

Various track defects may be classified as follows:

(A) Defects on Curved Tracks

1. Improper radius of the curve
2. Improper super-elevation
3. Improper speed
4. Improper unequal distribution of load on the inner and outer rails

(B) Defects at Crossings and Turnouts:

1. Due to inadequate fitting toe-lifting
2. Loose crossing bolts or wing-rails
3. More gap at points
4. Use of excessive wear switch
5. Excessive tight gauge
6. More clearance at the nose of crossing

(C) Defects on Straight Track:

1. Improper alignment
2. Incorrect gauge
3. Defective-joints
4. Defective cross-levels

In this chapter we will study about the geometrical design of track which is related to Civil Engineering will be dealt. Track should be designed in such a way that trains with heavy loads can run over it at maximum possible speed. To achieve this object the design of the following is required :

1. Cross-section of the railway track
2. Gradients and grade compensations
3. Horizontal and vertical curves
4. Speed of train
5. Widening of gauge on curves

Super-elevation or cant.

25.1 CROSS-SECTION

The railway tracks are laid to regular and suitable gradients at a level consonant with the requirements of local conditions. As the track passes in every part of the country, it meets with natural undulations of the ground, which have to be evened out by constructing embankment or cutting at high places. The top of embankments and the floor of the cutting is known as the *'Formation'*, over which track is laid. Therefore, before laying of the track, first a formation is prepared. The width of the formation depends upon the gauge of track, number of track and the space between tracks. The formation should be well compacted so that it can bear the loads coming over it safely. Side slopes in embankments and cuttings should suit the soil conditions and should not cause settlement of the formation level. On both sides of the embankment and cutting sufficient drains should be provided to carry away the rain-water. If the ground level has continuous slope along the cross-section, drains may be provided only on one side of the embankment and at the top of cutting as shown in Figs 25.1, 25.2, 25.3 and 25.4.

Indian Railway Board have recommended the following dimensions for Broad gauge:

Minimum distance, centre to centre of track ...4.725 metres

Minimum formation width in embankment for single-line ...6.100 m

Minimum formation width in cutting (excluding side drains) for single-line ...5.49 m

Minimum formation width in embankment for double line ...10.82 m

Minimum formation width in cutting (excluding side drains) for double line ...10.058 m

Four typical cross-sections of railway tracks with dimensions have been shown in Figs. 25.1, 25.2, 25.3 and 25.4 which are self-explanatory. Side slopes in these figures have been taken as 2:1 for embankments and $1\frac{1}{2}:1$ for cuttings. These slopes should be adopted on the type of soil and the height of cutting or embankment. If possible the barrow pits be placed on the upper side of the cross-slope of the ground and the soil banks on the lower side.

For calculating permanent land-widths all the horizontal components of the slopes of banks and cuttings, side-drains, formation width etc. are added together. In the figures, the permanent width in each case has been clearly shown. Outside the permanent land, sufficient temporary land should also be acquired for taking soil for embankments, or depositing from cuttings. In the case of single-line track, sufficient land should also be acquired for future extension, when the single-line will be converted into double-line track. All the outside boundaries of railway track should be clearly marked by boundary stones to avoid disputes at any stage.

25.2 GRADIENTS

The rate of rise or fall of the track is known as the *gradient.* It is expressed as the ratio of vertical to horizontal. Thus is any track rises 1 m in 200 m horizontal length, its gradient is expressed as 1 in 200 or 0.5 per cent. Similarly if another track falls by 1 m in 100 m, its gradient will be expressed as 1 in 100 or 1 per cent.

In every country there are natural undulations and different places are at different heights or elevations. If the track is laid truly horizontal, it will be impossible to reach other stations situated at higher or lower altitudes. Therefore in order to reach different stations, tracks are laid to some gradients. Gradients are also provided in roads, but in Railways these are classified as follows :

1. Ruling gradient or maximum gradient.
2. Momentum gradient.
3. Pusher gradient.
4. Station yard gradient.

25.2.1 Ruling Gradient

It is defined as the maximum gradient to which the track may be laid in a particular section. It mainly depends on the load of the train and the additional power of the locomotive, because on gradients extra power is required to pull the train, the value of which is $W \cdot \sin\theta = W \cdot \tan\theta = W \times$ Gradients. In this way it is seen that the extra power required to pull the

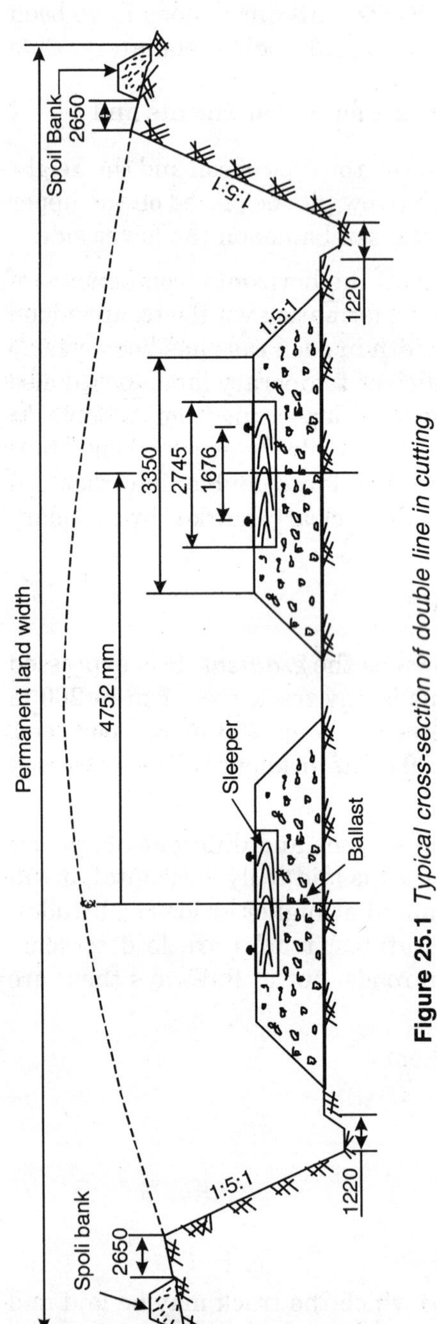

Figure 25.1 *Typical cross-section of double line in cutting*

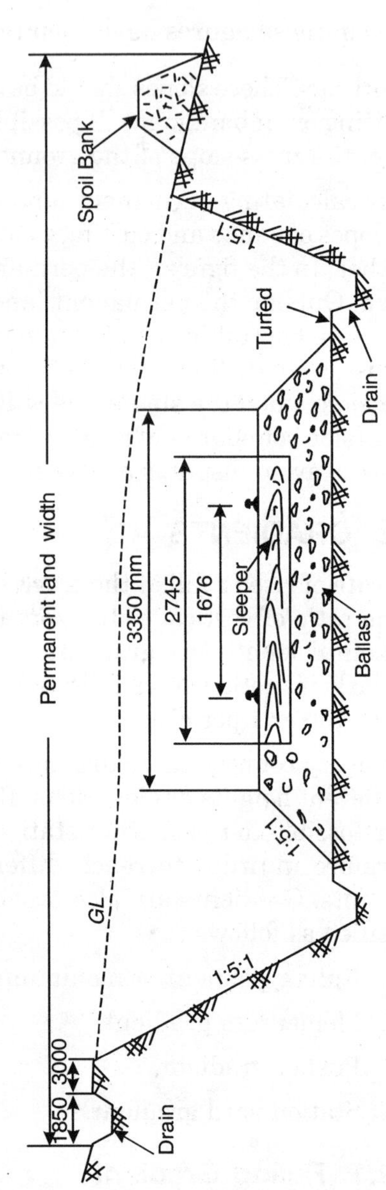

Figure 25.2 *Typical cross-section of single line in cutting*

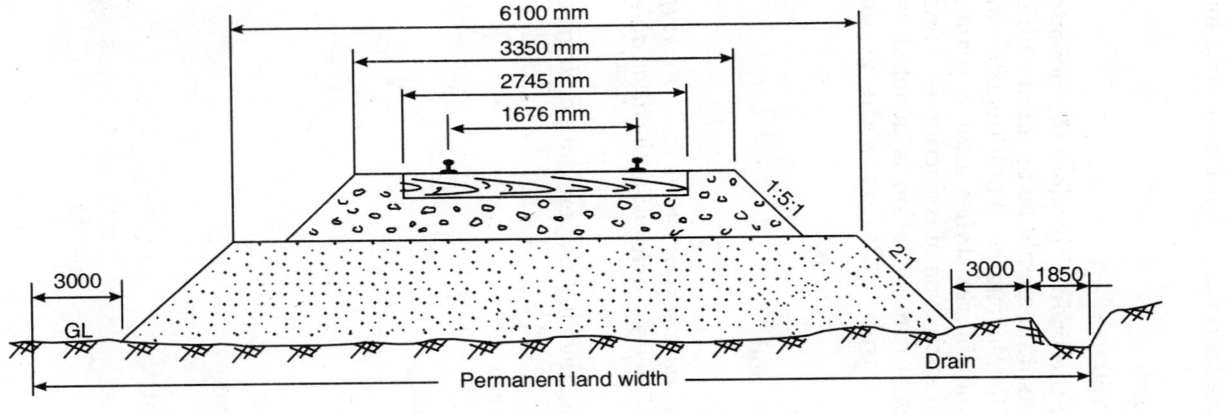

Figure 25.3 *Typical cross-section of single line in embankment*

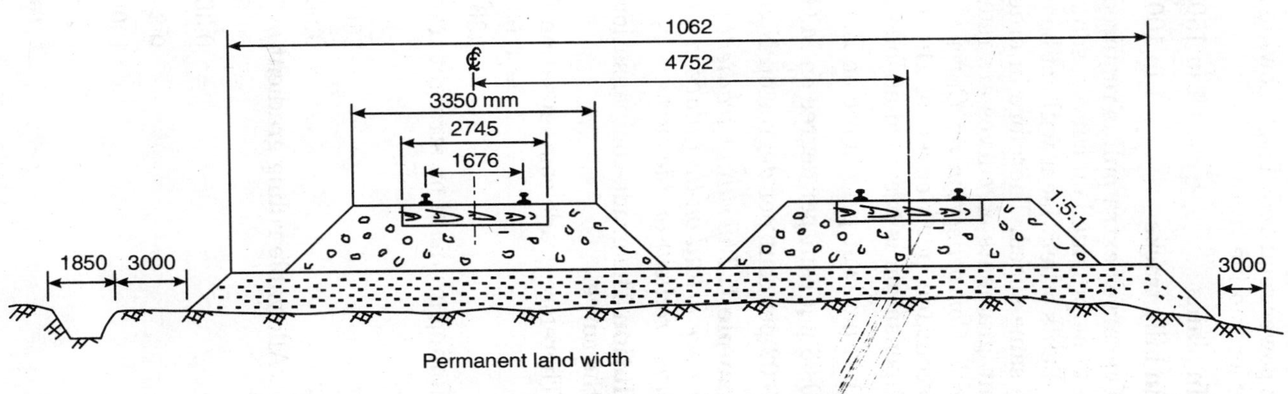

Figure 25.4 *Typical cross-section of double line in embankment*

train directly depends on the gradient. Steeper the gradient more will be the power required. The following ruling gradients are provided in India for one locomotive train :

In plains	1 in 150 to 1 in 200
In hilly tracks	1 in 100 to 1 in 150

On curves extra pull is required to pull the train. Therefore if gradients are to be provided on curves, some compensation should be given in ruling gradients otherwise it will not be possible for the engine to pull the train at the same speed. Therefore in order to avoid the resistance after a certain limit, gradients on curves are reduced and such type of reduction is called *Grade Compensation on Curves.* This grade compensation is denoted by percentage per degree of the curve. On Indian Railways the grade compensation is provided as follows :

0.04 per cent per degree on B.G. curves

0.03 per cent per degree on M.G. curves, and

0.02 per cent per degree on N.G. curves

Example 1. *The ruling gradient is 1 in 200 on a section of the B.G. track. If the track is laid on that place in a curve of 5 degree, determine the allowable ruling gradient on the curve.*

Solution The grade-compensation on B.G. track is 0.04 per cent per degree of the curve.

Therefore grade compensation for 5° curve

$$= 0.04 \times 5$$

$$= 0.20 \text{ per cent.}$$

Ruling gradient on straight track

$$= \frac{1}{200} \times 100$$

$$= 0.50 \text{ per cent.}$$

∴ Allowable ruling gradient on the curve

$$= 0.05 - 0.20$$

$$= 0.30 \text{ per cent}$$

$$= \frac{0.3}{100}$$

i.e. 1 in 333.3

or **1 in 334**

25.2.2 Momentum Gradient

On falling gradients train gets momentum and this momentum can be utilized for rising gradients. Therefore in hilly tracks, laid in valleys, the momentum gained by the train on falling gradient is utilized for rising gradients and at some places this enables the train to overcome such gradients which are steeper than ruling gradients. In plains also every falling gradient as far as possible is followed by a rising gradient. Such types of gradients, where the advantage of momentum is taken, is known as *Momentum Gradients.* That is why on such gradients no signals are provided, which may stop the train and bring down the momentum to zero.

25.2.3 Pusher or Helper Gradients

In mountains to avoid heavy cutting or for reducing the length of the track, at certain places, steeper gradients than the ruling gradients are provided. If a ruling gradient is provided in the longer length of the track, the load carrying capacity of the engine shall be reduced, because for the larger portion of the journey the full use of the locomotive capacity would not be done. In such cases it is better to concentrate the gradient instead of limiting the load of the train, which will be more economical. At such places one engine cannot pull the train and extra engine is required to push the train. Such gradients where extra engine is required to push the train are known as *Pusher Gradients.* Before deciding the pusher gradient in a particular length of the track, the capacity or the engine and the maximum load should be taken into account. In such cases optimum conditions should be attained so that no capacity is wasted. At Bhor Ghat between Poona and Bombay a gradient of 1 in 37 has been provided for a continuous length of about 4 km. In Darjeeling Railways gradient upto 1 in 20 is provided at some places.

25.2.4 Station-yard Gradients

On the station-yards gradients are avoided as far as possible due to the following reasons :

1. In station-yards, when bogies are standing on gradients, if heavy wind blows, it might start the movement of the bogies, and cause accidents.

2. During the starting time of the trains, the locomotive will require extra force to pull the train on gradients.

Keeping in view the above difficulties gradients are completely avoided on platforms and maximum limit is fixed as 1 in 400 in station-yards. But Indian Railways have recommended that a gradient of 1 in 1000 should be provided for easy draining of station-yards.

25.3 HORIZONTAL CURVES

It is completely impossible to lay the track throughout the country in one straight line, because in its way it comes across so many obstructions. Tracks are so laid that it may pass through important places. Therefore to change the direction of track curves are introduced to take it to the required places. Similarly to avoid kinks vertical curves are introduced in the track at each point where a rising gradient meets the failing one.

A *curve* is represented by the length of its radius or by the degree of the curvature subtended by a chord of 30.48 m. The radius R of any curve can be directly calculated if its degree of curvature D is known by the following formula:

$$R = \frac{1746.5}{D}$$

Thus from the above relation 1° curve has a radius of 1746.5 m, 3° curve has a radius of 582.17 m, and so on.

Following types of horizontal curves are used in the alignment of the railway track:

(i) Simple curve (ii) Compound curve

(iii) Reverse curve (iv) Transition curve

25.3.1 Simple Curve

This is the simple type of curve, which is generally represented by the degree of angle subtended at its centre by 30.48 m. The degree of the curve can be calculated by the formula given above.

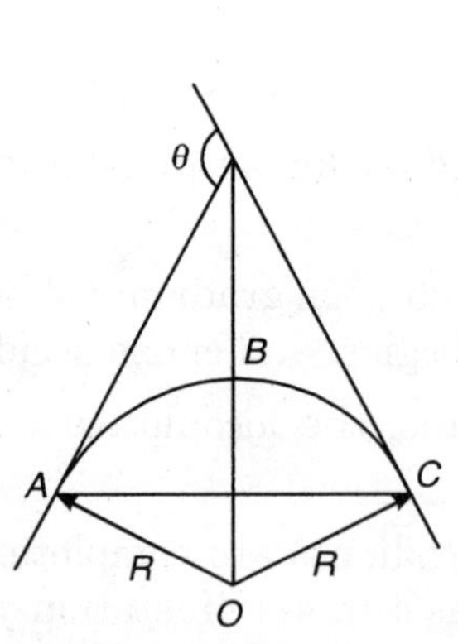

Figure 25.5 *Simple curve*

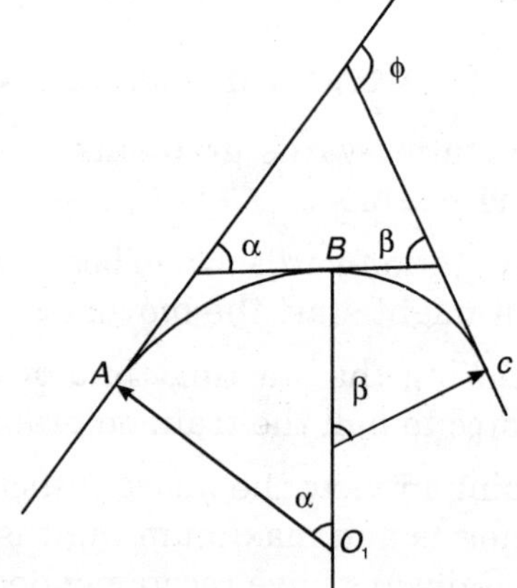

Figure 25.6 *Compound curve*

25.3.2 Compound Curve

This is the curve which comprises two or more simple curves. Figure 25.6 shows a compound curve which includes two simple curves of radius R_1 and

R_2. This type of curves are used to avoid excessive cutting or filling at the sites. Such curves are not very common on Indian Railways.

25.3.3 Reverse Curve

This type of curves are provided mostly on diversion on the railways, when the track is to be diverted temporarily for repairing the main track or in case of derailment or accidents. This curve mainly consists of two simple curves joined together in reverse direction as shown in Fig. 25.7. Both the simple curves have common tangent *DE*. Sometimes both the simple

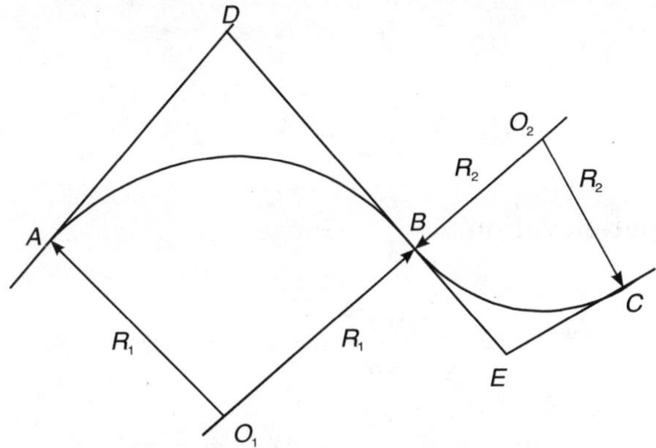

Figure 25.7 *Reverse curve*

curves are joined by means of straight length of the track. In such cases the curve is known as *deviation curve.* The deviation curve is mostly used in case of major repair works of culverts, worn out track, accidents etc. The trains are allowed to pass through the deviation track till the repair of the main track

Transition curves have been discussed in Article 25.7 of this chapter.

25.4 SUPER-ELEVATION

When any vehicle moves on curve, it is subjected to centrifugal force :

$$F, \text{ which is equal to } \frac{WV^2}{GR}$$

where

W	wt. of the vehicle
V	speed of the vehicle
R	radius of curve
G	acceleration due to gravity.

Therefore if centrifugal force F is not counteracted, it will exert a great horizontal force on the outer rail. The outer rail will have to bear more than half the load of the train and the wheel flange will start rubbing against its head. This horizontal force and uneven load on the rails from the top will cause derailment. The centrifugal force can be counteracted by raising the outer rail, which is known as *Super-elevation* or *Canting*.

Figure 25.5 shows how super-elevation is determined. Centrifugal force F and weight of the vehicle W are acting at the C.G. of the vehicle. From the figure :

$$\frac{BC}{AB} = \frac{F}{W}$$

or $$BC = \frac{F}{W} \times AB$$

or $$\text{Super-elevation} = \frac{F}{W} \times \text{Gauge}$$

But $$F = \frac{WV^2}{GR}$$

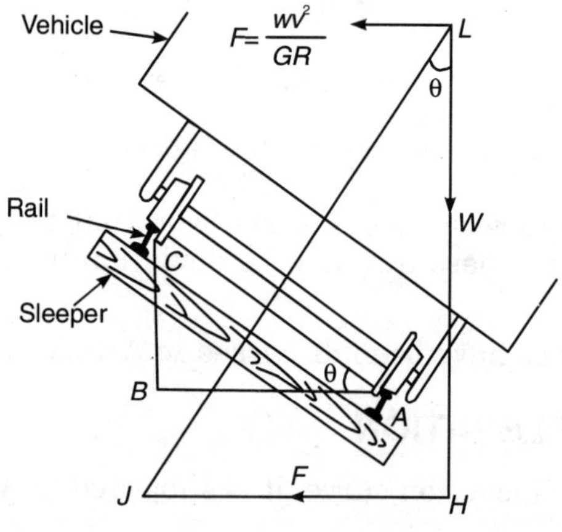

Figure 25.8 *Providing super-elevation*

$$\therefore \text{Super-elevation} = \frac{WV^2}{G.RW} \times \text{Gauge}$$

$$= \frac{V^2}{GR} \times \text{Gauge}$$

If the velocity is expressed as V km per hour,

$$\text{Super-elevanon} = \frac{1.676}{9.81}\left(\frac{10}{36}\right)^2 \frac{V^2}{R}$$

$$= \frac{V^2}{75.25R} \text{ metres}$$

$$= 1.316 \frac{V^2}{R} \text{ cm}$$

where K is the radius of curve in metres.

Therefore we can easily calculate the amount of super-elevation by the above formulae. If we provide super-elevation for the speed of mail trains, it will not suit the goods train and *vice versa* because the speed of mail trains is very high as compared with the goods trains. Therefore the *cant* is calculated by the average speed of trains moving on that track.

If the super-elevation is provided on the actual speed of the train, it is called as *Equilibrium Cant'*. But on tracks cant is reduced to meet the demand of slow moving trains and this reduction is known as *Cant deficiency*. Therefore cant deficiency is the actual amount by which the *equilibrium* cant falls short. On Indian Railways following *cant deficiencies* are allowed:

On B.G. Track—7.6 cm

On M.G. Track—5.1 cm

On N.G. Track—3.7 cm

The maximum limit of super-elevation has been fixed on Indian Railways as follows :

On B.G. Track—16.51 cm

On M.G. Track—10.16 cm

On N.G. Track— 7.62 cm

Example 2. *A curve of 5° is situated on a section of B.G. If the maximum permissible speed on the section is 60 km/hour, determine the amount of equilibrium cant. If for meeting the demand of slow moving trains 7.6 cm cant deficiency is to be provided, determine the super-elevation to be provided.*

Solution The radius of curve can be determined by

$$R = \frac{1746.5}{D}$$

where D is degree of curve

$$R = \frac{1746.5}{5} = 349.3 \text{ metres}$$

Equilibrium cant $= 1.36 \dfrac{V^2}{R}$ cm

$= 1.39 \dfrac{60^2}{349.3}$

$=$ **14.2 cm.**

Super-elevation = Equilibrium-cant, minus cant-deficiency

$= 142.76$

$=$ **6.6 cm.**

25.5 TRANSITION CURVES

It has been concluded that super-elevation is to be provided on curves to prevent the wear and tear of rails and smooth riding. The centrifugal force starts acting on the vehicle at the same time when it enters the curve, therefore, full amount of super-elevation is required from the beginning point of the curve to counteract the effect of centrifugal force. But it is impossible to provide full amount of *cant* at one time or at one point. The *cant* can be provided gradually in certain length of the track by two methods which are as follows :

1. To start the super-elevation on outer rail on straight line before the tangent points and gradually increasing it to the required quantity at tangent point.

2. To provide transition curves on both sides of the curve.

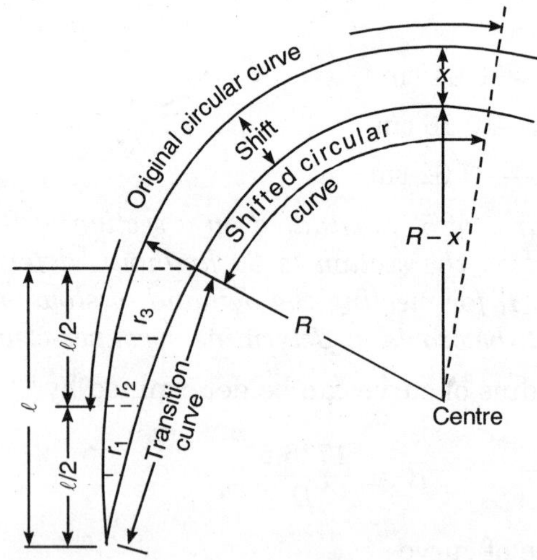

Figure 25.9 *Method of providing transition curves on circular curve*

In the first method, as the super-elevation has been started on outer rail on the straight track before the curve, the inner rail has to bear more load and consequently worns out quickly. Again it does not give smooth running of the trains. Due to this difficulty nowadays transition curves are provided on every curve.

Actually transition curve is a spiral, the curvature of which starts gradually from zero on the straight length and attains full degree of curvature at the beginning of circular curve. Therefore when the transition curves are provided, the circular curve shifts from its original position as shown in Fig. 25.9.

25.5.1 Length of Transition Curve

It is the length of the transition curve along the centre line of the track from its meeting point with the straight line to that with the curve. In Fig. 25.9, L is the length of the transition curve.

Indian Railways have specified the following formulae for determining the lengths of the transition curves :

(a) $L = 7.20e$

(b) $L = 0.073\ D.V_{max}$

(c) $L = 0.073\ e.V_{max}$

where

L length of the transition curve in metres

e actual super elevation in cm

D cant deficiency for maximum speed in cm

V_{max} Maxima speed of the train on the section

Railway code has also recommended the following formula for the length of the transition curve:

1. $$L = 4.4\sqrt{R}$$

where

R Radius of the simple curve in metres.

2. $$L = \frac{3.28v^3}{R}$$

where

v speed of the train in m/sec.

3. At the rate of change of super-elevation of 1 in 360 i.e., providing a gradient of 1 cm in 3.6 metres from super-elevation to straight line.

Now the super-elevation is gradually increased on the whole length of transition curves. This method uniformly distributes the loads on both the rails and at every place the super-elevation remains in required quantity, and thus eliminates the lurching and jolting. Mostly following three types of transition curves are provided :

1. *Spiral curve.* This is the ideal curve which satisfies all the requirements of the transition curve. The radius of this curve is inversely proportional to the length traversed

i.e. $$\text{Radius of curve} = \frac{1}{\text{Lenght of the curve traversed}}$$

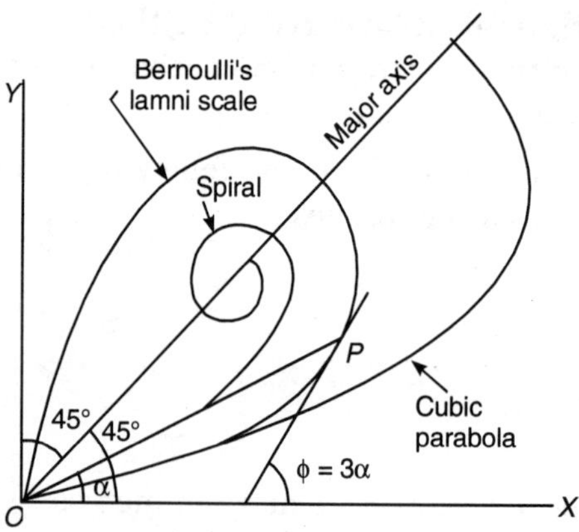

Figure 25.10 *Various types of transition curve*

The rate of change of acceleration in this curve is uniform throughout its length.

2. *Bernoulli's lemniscade curve.* In this type of curve the radius of the curve decreases as the length increases and therefore the radial acceleration goes on falling but the fall is not uniform beyond 30° deflection angle.

Mathematically, the curve can be expressed as

$$L = k\sqrt{\sin^2 h}$$

where

L length of the curve

k a constant

h polar angle in radians.

3. *Cubic parabola.* In this type of transition curve the rate of decreases of radius is low from 4° to 9° but beyond 9° there is rapid increase in the radius of the curve. This curve is easy in laying, hence commonly used on Indian Railways. Mathematically, this can be expressed as

$$Y = \frac{X^2}{6RL}$$

where

X and y are the co-ordinates of any point P on the curve

B Radius of any point P on the curve

L Length of the curve from the starting point to any point

25.6 NEGATIVE-CANT

At certain places branch line meets the main line on curves. The outer rail of main line curve meets the inner rail of branch line curve. But as the outer-rail of main line will be at higher elevation than the inner rail, the inner rail of branch line will have to be kept higher than outer-rail. Actually on curves the outer-tail remains at higher elevation that the inner rail. But here in the case of a branch line the outer-rail becomes at lower level than the inner rail which is reverse of the requirement. Hence the super-elevation

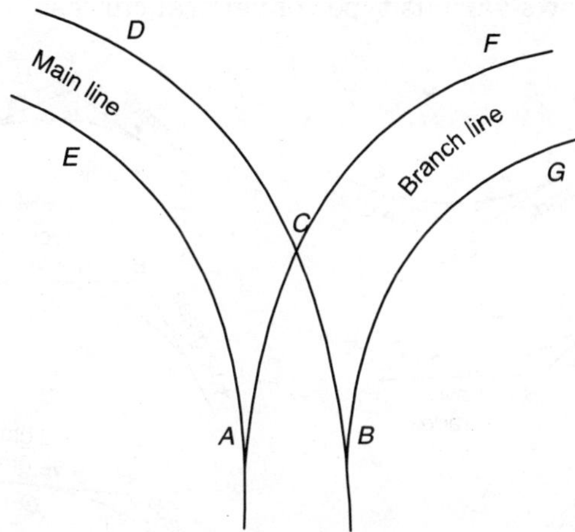

Figure 25.11 *Negative cant*

provided on branch line is negative. Such type of super-elevation is known as *Negative Super-elevation* or *Negative Cant.* Therefore on such branch lines there are restrictions on the speed of trains. In Fig. 25.11 point A is lower than point B, because super-elevation for the main line has been provided and due to which line BD is at higher elevation than line AE. For the branch line AC, line should be at higher level than BG, but as point A is lower than B, the line BG becomes at higher level than line AF.

25.7 VERTICAL CURVES

At every change of gradient vertical curves are provided for smooth running of trains by avoiding kinks. If these curves are not provided there will be very bad lurching in the train which will cause discomfort to the passengers and sometimes may cause accidents.

There are two types of vertical curves (a) Summit curves, (b) Sag or valley curves. When a train climbs a certain upgrade at a uniform speed and crosses the summit of the curve, due to the weight of the train, an acceleration starts acting on the train immediately after passing the summit and the speed of the train starts increasing. Similarly when a train passes over a sag the engine side of the train is ascending the up-grade while the back side vehicles tend to compress the couplings and buffers. At the point of passing the sag or summit the vehicles are subjected to the jerks, which can be removed only by providing suitable curves.

Generally parabolic curves are used in vertical curves in the tracks. A parabolic curve is set out with its apex at a level half way between the points of intersection of the tangents to the two intersecting grades.

Figure 25.12 shows various types of vertical curves.

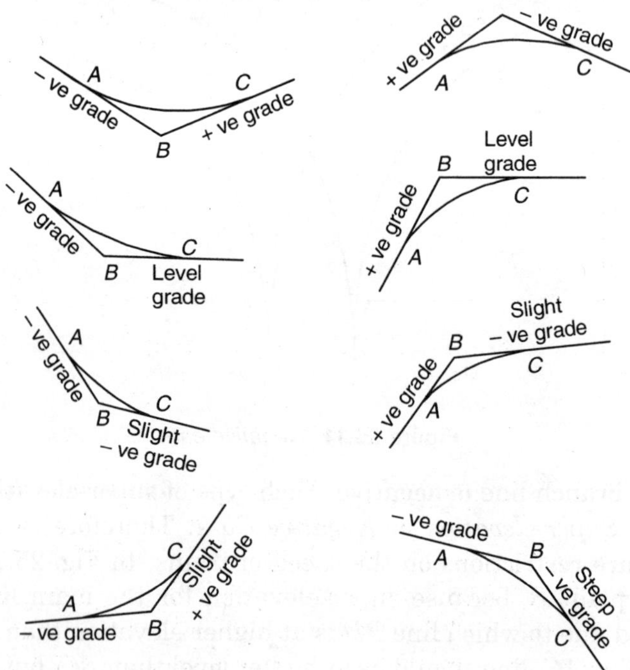

Figure 25.12 *Various types of valley and summit curves*

The length of curves will depend on the algebraic difference in the grade and on its position that whether it is in sag or summit. All vertical curves are generally in the form of parabolas. Mostly, sags are avoided in cuttings and apex on embankments.

For first class track Indian Railways has standardised the rate of change of 0.10 per cent on summits and 0.05 per cent on sag curves. But at few places these rates are exceeded.

25.8 SPEED OF TRAINS

The maximum permissible speed of trains for good order track on transitioned curves should be calculated by the following formula :

$$V = 4.4\sqrt{R-70} \text{ for B.G. and M.G. tracks}$$

and

$$V = 3.6\sqrt{R-60} \text{ for N.G.}$$

where

V Maximum speed in kilometres per hour

R Radius of curve in metres.

25.9 CANT-BOARD

Figure 25.13 shows a cant-board which is used for providing super-elevation on tracks. Cant-boards are graduated in 6 mm or 5 mm and these are made

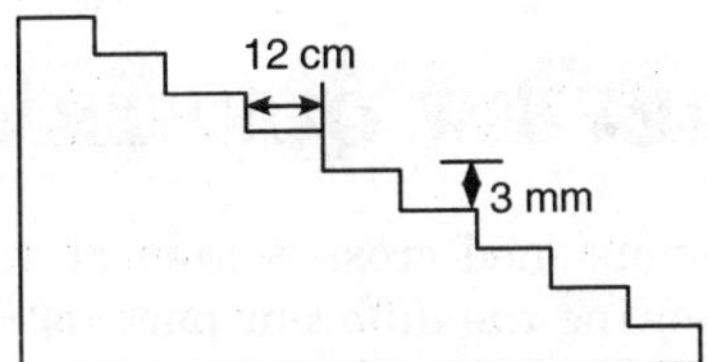

Figure 25.13 *Cant-board*

of unserviceable tie bars of railways. At the time of using it, cant-board is kept on inner rail and one long bar with spirit level on its top is kept with its one end on outer rail and other end on cant-board. In each case the spirit-level should be in the centre, because only in this position correct super-elevation can be provided. For the method of providing super-elevation, please see Fig. 25.11.

25.10 WIDENING OF GAUGE ON CURVES

The wheel base of the train vehicles is rigid. While passing on the curves, with the outer wheel of the front axle strikes against the outer rail, the outer wheel of the inner axle bears a gap with the outer rail. While laying the track on the curves, sufficient care should be taken otherwise the rails may tilt outward. The increase or widening of the gauge on the curves should be just sufficient, otherwise if more widening is done, it may lead to derailment.

The increase in the gauge at the curves is calculated by the formula :

(i) $$b = \frac{13(B+l)^2}{R}$$

where

b extra width of the gauge

B width of rigid wheel base in metres (6 m for B.G. and 4.88 m for M.G.)

R radius of the curve in metres

l top of the flange in metres

(i) $$b = 0.02\sqrt{h^2 + D.h}$$

where

b $0.02\ \sqrt{h^2 + D.h}$

h depth of wheel flange below rails in cm

D diameter of wheel in cm

REVIEW QUESTIONS

25.1. Draw a neat dimensioned cross-section of a double-line B.G. track in embankment, showing the different parts of the structures of the track including the drains. Also mark the width of the permanent land.

25.2. Write short notes on:

(a) Ruling gradient

(b) Gradients in station yards

(c) Speed of trains on curves

25.3. What do you understand by the 'Momentum' and 'Pusher' gradients ? Explain, at what places these are provided ?

25.4. (a) Why vertical curves are introduced at the junction of two gradients ?

(b) Write short notes on transition curves.

25.5. (a) What is the object of providing super-elevation on the curves of railway track ?

(b) What do you understand by 'Equilibrium super-elevation' and 'Cant deficiency' ?

25.6. (a) What are transition curves and what is the necessity of providing them in tracks ?

(b) Write a short note on 'Negative Cant'.

25.7. A curve of 7 degree is situated on a section of B.G. Find out the super-elevation to be provided for average speed of 45 km/hour. (*Ans.* 11 cm)

25.8. If on the curve of question No. 7 an equilibrium-cant of 4.5 cm has been provided, determine the permissible speed in km/hour.

(*Ans.* 28.9 km/hour)

25.9. A 5 degree curve branches off from a 3 degree main curve in the opposite direction in the layout of a B.G. yard. If the speed restriction on main line is 30 km/h, determine the permissible speed on branch line, assuming the cant-deficiency as 4 cm. (*Ans.* 30.6 km/h)

25.10. If at a place on Narrow Gauge, a super-elevation of 3.8 cm is provided on 15° curve, determine the permissible speed of train. Assume the super-elevation at equilibrium cant. (*Ans.* 27.37 km/hour)

25.11. On a main B.G. line curve at 4°, of a branch of 7° is taken out in opposite direction the speed limit on main line is 5.7 km/hour. Determine the maximum permissible limit on branch line. (*Ans.* 20 km/hour)

25.12. Write short note on the widening of the gauge.

26

Plate Laying

GENERAL

Plate laying is the operation of laying out and connecting up sleepers and rails on the prepared formation of a railway track. In other words, construction of new tracks including carrying of materials to the site, and assembling operation, is known as *Plate laying.*

Rail-head is the point upto which track has been laid at any time. The point from where the track laying is started is known as *Base.*

At the time of laying new track, ballast is not laid in the beginning, because in the beginning formation sets or settles down and if the ballast has been used in the beginning it will sink in the formation. Therefore, in new tracks, sleepers are directly laid on the formation, and after few months when it become consolidated or compacted, the track is lifted and ballast is spread on formation and packed around the sleepers.

26.1 METHODS OF PLATE-LAYING

Following are the methods which are used for plate-laying :

1. Side Method
2. Telescopic Method
3. American Method

26.1.1 Side Method

This method is used for doubling the railway-lines. When one track is already existing and one more track is required to be constructed, this method is used. In this method all the required materials are taken from the central depot in material-trains on the existing track and are spread on the formation of new track. When all the required materials have been spread on the formation, the assembling is started from one end by manual labour.

When no line exists for distributing the track materials, temporary tram-line is laid by the proposed track. Therefore sometimes the side method is also known as the *Tram-line Method.* Sometimes instead of tram-line a service road is constructed for transporting the materials.

Side-method is also used in flat countries for plate-laying, when there is no existing track, one service road is constructed parallel to the track and all the required materials are transported by bullock-carts or trucks and are spread on the formation.

26.1.2 Telescopic Method

This method of plate laying is extensively used in India. In this method first of all a large central depot is constructed near the junction of the existing railway and the proposed railway line. Plate-laying is done totally by manual labour in India; therefore gangs are formed and different duties are allotted to each gang. The following are three categories of gangs.

(i) *Material-gangs.* The duty of these gangs is to unload the materials from the train and then carry it to the rail-head and supply it to the linking-in-gangs. Rails are very heavy in weight, therefore, these require special rail-carriers to carry them easily. Anderson rail carriers may be used for conveying rails from rail head to the actual site. Material-gangs distribute the sleepers, rails, fish-plates, fish-bolts, etc. at approximately required places.

(ii) *Linking-in-gang.* When the materials have been distributed on the formation, the next work is to place them in position and link. This work is done by linking-in-gangs. These gangs mark the centre line and place the sleepers at required distances, over which rails are placed on chairs or on

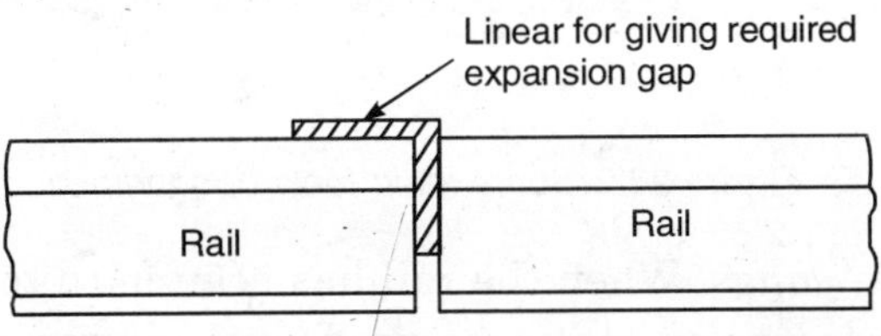

Figure 26.1 *Method of providing expansion gap*

bearing plates and jointed together by fish-plates and bolts after leaving suitable expansion gaps. For giving uniform expansion gap at each joint *liner* or *shim* is placed between the ends of two rails as shown in Fig. 26.1. After joining rails, these are fixed to the sleepers with the help of spikes. At the time of fixing rails to the sleepers, the gauge is checked by *Rail gauge* which is a gangmen's tool shown in Fig. 26.2.

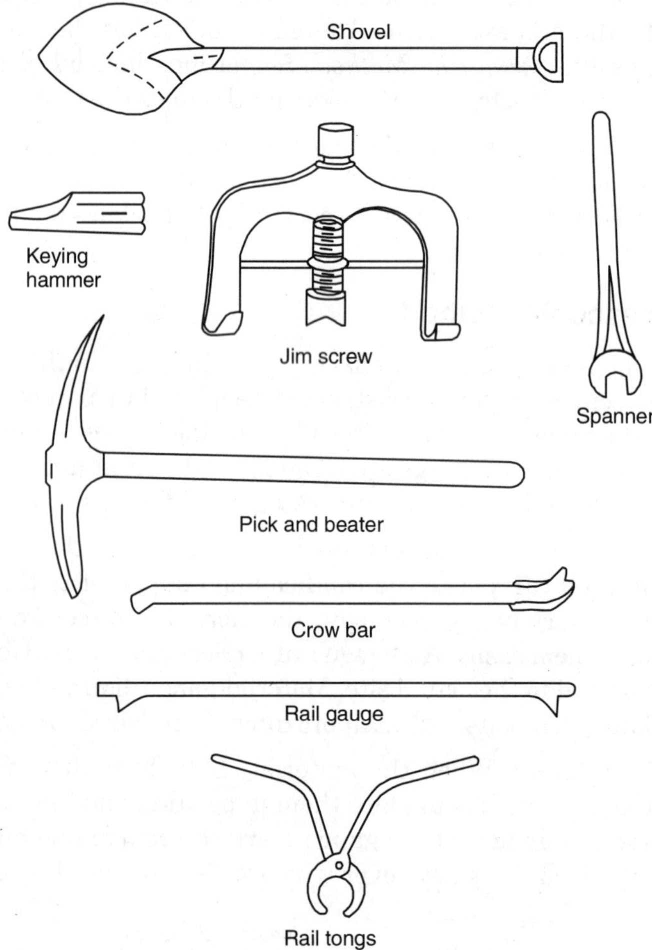

Figure 26.2 *Some main tools of gangmen*

(iii) *Packing-in-gangs.* When the rail has been fixed to the sleepers, the next operation is to bring it to the required level and gradients by packings. This operation is done by packing-in-gangs. These gangs pack the earth below and around the sleepers and bring it to the required level.

The transport of materials from the central depot to the rail head is carried out in wagons and trollies. In the beginning when the rail-head is within 2 km from the central depot, the material can be carried in trollies. But when the rail-head reaches at more distance, wagons and engines are employed for the cartage of materials from the depot. Generally two wagons and one engine is used for this purpose, so that one wagon is loaded at central depot and the other wagon should be unloaded at the rail-head. When the track has been laid upto considerable length from where engine cannot make one complete trip per day, another subsidiary depot is established near the rail-head at convenient place and all the materials are stored in that depot.

If the following precautions are taken much time can be saved, and good progress can be obtained during plate-laying :

1. All the preliminary preparatory work such as cleaning and oiling fish-bolts, adzing and boring holes in wooden sleepers etc. should be done at the central depot.

2. At the time of unloading rails from the material train, much care should be taken, so that rails may not be bent or twisted. In no case the rails from the train should be dropped on hard rock or boulders because this will cause bending of rails, and it will require a lot of labour and time in straightening them.

3. No rail bent or hogged should be sent to the rail-head from the central depot, without completely straightening it.

4. No cracked sleepers in case of C.I. should be sent to the rail-head. Therefore at the time of loading material-train at central depot, everything should be checked.

26.1.3 American Method

This method involves all mechanical work and is not used in India, because it would be uneconomical when labour at cheap rates is available here. In this method rails are fixed to the sleepers and the whole unit is lifted up by heavy cranes and placed in position at rail-head. The only work remains to do at the site, is to join the rails by fish-plates and bolts.

26.2 LAYING OF SWITCHES AND CROSSINGS

The usual practice is to construct the switches and crossings and also to fix them to the sleepers in the workshops. Now these completed fixed units are taken to the site on trollies and placed in position. For smooth sliding the moving portions of rails are greased. Sometimes first sets of switches and crossings are prepared in workshops and then these are taken in parts to the site, where they are assembled.

Figure 26.3 *Ballast fork or wire claw*

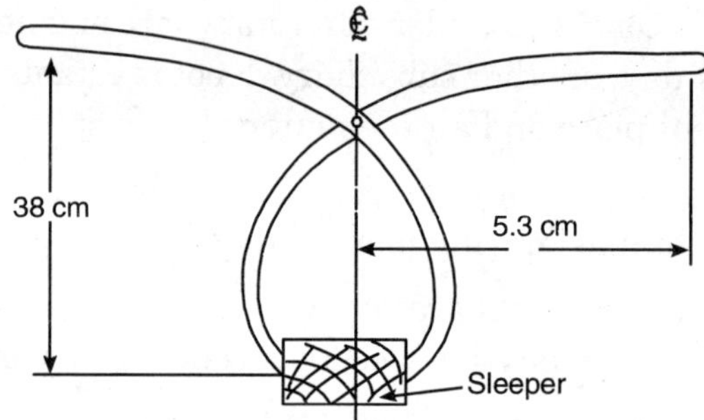

Figure 26.4 *Sleeper tongs*

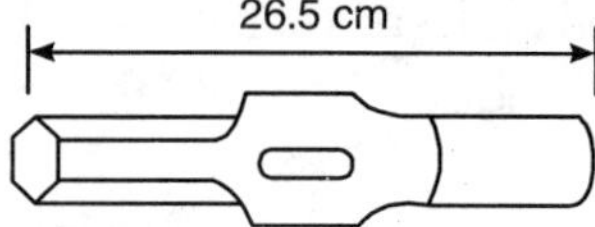

Figure 26.5 *Spiking hammer*

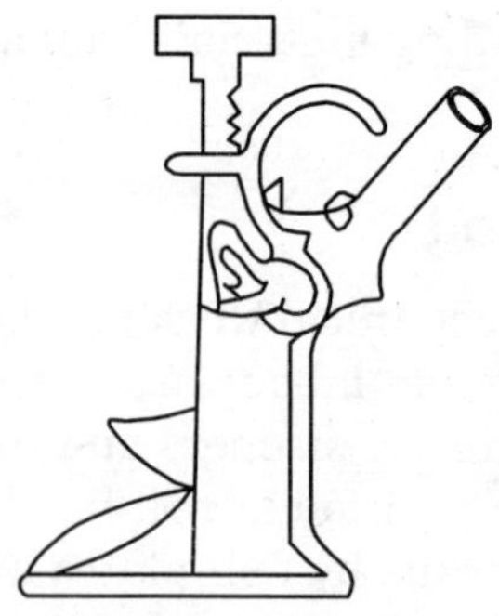

Figure 26.6 *Simplex lifting jack*

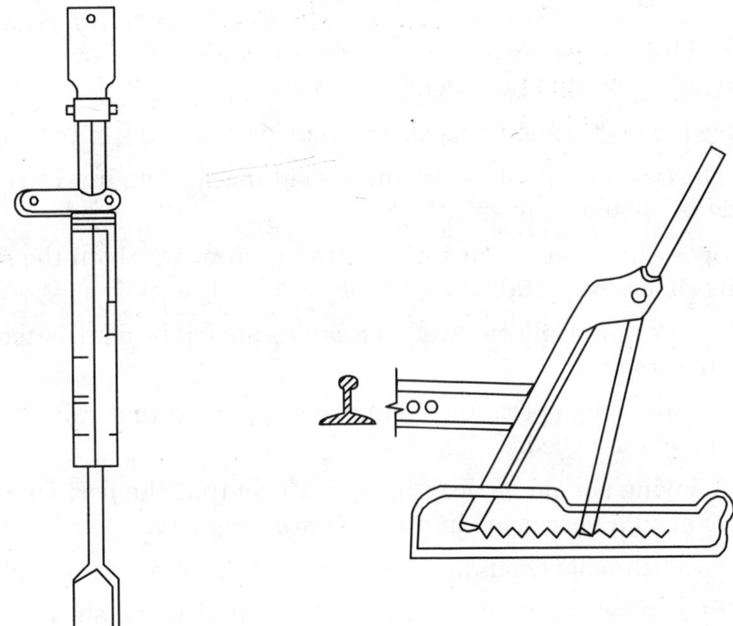

Figure 26.7 *Electric tie tampler*

Figure 26.8 *Tack liner*

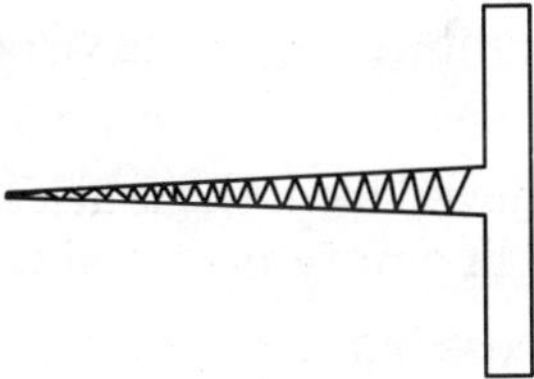

Figure 26.9 *Auger pin*

26.3 RELAYING OF TRACK

When the materials of the track are worn out or lighter track is to be replaced by heavy one, it requires relaying. There are many methods of relaying track, but the section of the method depends on local conditions, density of traffic and plant available at site.

If there are more than one track at the site, the best way is to divert the traffic by means of turn-outs from the track, to be relayed. After it, the track can be relayed easily.

On the other hand if the back is single-line, the work of relaying is more difficult than new plate-laying, because short lengths of the track are to be relayed during the periods available between the train timings, depending on the frequency of train service. If the traffic is very heavy in day time, the work will have to be done during nights. As such before the actual starting of relaying work, it is thoroughly planned.

Following points should be observed during relaying of the track :

1. The caution signals at 400 m beyond the section, in which relaying is to be done, should be installed.

2. Speed limit boards at the required positions on both the sides of the sections should be installed.

3. All the materials required in relaying should be distributed along one side of the track.

4. The relaying operation should be so planned that rail traffic is least affected and disturbed.

5. Relaying should be done in such a way that the portion of the track removed should be relayed in the same working day.

6. Two fish-bolts should be first removed from each joint and oiled.

7. Two spikes and sleeper fittings from alternative sleepers should be first removed.

8. Trains with restricted speeds can be allowed to pass over the track whose fittings from alternative sleepers have been removed.

9. Correct gauge on straight as well as curve portion should be maintained in relaying.

10. The old materials removed from the track should be properly arranged and sent back to the stores in empty material trains.

11. The *crib ballast* and cess ballast must be removed, screened and then relayed.

12. For easyness in relaying the position of the sleepers is generally marked on the rails.

13. All the critical activities connected with the relaying work must be clearly marked and checked.

14. Adequate labour force should be arranged before starting the relaying work. For relaying one kilometre of B.G. track about 650 labourers are required and for M.G. about 550 labourers are required.

Generally the work of relaying is divided in three phases :

(i) Preliminary work.

(ii) Actual relaying work.

(iii) Consolidation of relayed track and removal of released old materials.

26.3.1 Preliminary Work

In this phase all the required materials to be changed are spread along the track at suitable intervals. The train timings are noted and the available time for relaying track is calculated. Warning signals are placed on both the ends of the section, of which the track is to be relayed. At the time of distributing materials along the track it is advisable to distribute it on one side only. Sleepers should be placed at right angle to the track at the time of distributing, because it is easier and quicker to pick them at the time of relaying.

26.3.2 Actual Relaying Work

Various operations of relaying work should be given to each gang, which should be responsible for it, because it will avoid confusion and work will be done in less time. At the time or removing old-track only those fittings are removed which are essential, e.g. at each joint two fish-bolts from one rail should be removed; and spikes from one side of each rail may not be removed. After removing necessary fittings the old rails are turned over and placed on the shoulder of the cess. The released sleepers are thrown aside on formation slope. The old ballast is levelled and the new sleepers are placed in position. One line of rail is laid and fixed by two fish-bolts and fish-plates. The spacing of the sleepers is marked on the rails and position of sleepers is corrected. After it, second line of rail is laid at proper gauge and the rails are fixed to alternate sleepers. Now the track is brought to the correct alignment by raising or lowering sleepers. After it the remaining fitting is done and the gauge is finally checked. The packing around the sleepers is done by the ballast. Now the track is checked by moving gang-trolley over it.

26.3.3 Consolidation of Relayed Track and Removal of Released Old Material

In the third phase the track is thoroughly packed for second and third time subsequently and if extra ballast is required it is to be taken from the depot and used. The trains which brought ballast, while returning would pick up the released materials and carry it to the depots.

Restriction of speed on relayed-track. Speed restrictions are done on relayed sections of the track till the track is thoroughly consolidated.

26.4 MATERIAL REQUIREMENTS

It is the duty of the Engineer-in-charge to work out the exact quantities of all the materials required for the construction of new tracks or relaying of old track. The excess of the material will cause uneconomy in transportation and shortage will delay the work and will make laying expensive.

For one kilometre B.G. track the materials are calculated as follows :

(A) *Rails*—using 12.8 m long rails @ 44.7 kg/m.

No. of rails $= \dfrac{1000}{12.8} \times 2 = 156.6$

say 157 Nos.

Weight of rails $= \dfrac{157 \times 12.8 \times 44.7}{1000}$

= **90 t.**

(B) *Sleepers*—using (N + 5) sleepers density or 18 sleepers/rail of 12.8 m.

No. of sleepers $= \dfrac{157}{2} \times 18$

= **1413 Nos.**

(C) *Fish-Plates*—No. of Fish-plates/km

No. of Fish-plates = 2 × rail numbers per km

= 2 ×157

= **314 Nos.**

(D) *Fish-Bolts*

No. of fish-bolts per km

= 4 × Nos. of rails per km

= 4 ×157

= **628 Nos.**

(E) *Bearing-Plates*

No. of Bearing plates per km

= 2 × No. of sleepers per km

= 2 × 1413

= **2826 Nos.**

(F) *Dog-Spikes*—(For timber sleepers)

No. of Dog-Spikes per km

= 4 × No. of sleepers

= 4 × 1413

= **5652 Nos.**

(G) Number of anti-creepers will depend, whether they are provided in alternative sleepers or after fourth or fifth as the case may be.

If they are provided after fourth sleeper their number will be

$$= \frac{1}{4} \text{ No. of Bearing Plates}$$

$$= \frac{1}{4} \times 2826$$

$$= 706.5$$

$$= \mathbf{707 \text{ say.}}$$

26.5 CHECKING OF RAIL LEVEL

The level of both the rails on straight lengths should be the same. This is checked by a straight edge and spirit-level. Straight edge is placed on the

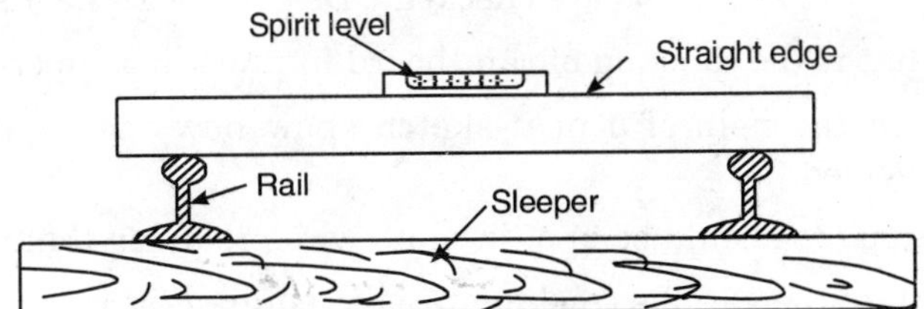

Figure 26.10 *Method of checking rail level*

rails and the spirit-level is kept in the middle of it as shown in Fig. 26.10. When both the rails are at the same level the air-bubble in the spirit level will be in centre.

26.6 PROVIDING SUPER-ELEVATION

Super-elevation on curves is provided by means of cant-board, straight-edge and spirit-level. The cant-board is placed on the inner-rail and the straight-edge is placed on outer-rail and on cant-board at required super-elevation height. The spirit-level is placed on the straight-edge as shown in Fig. 26.11.

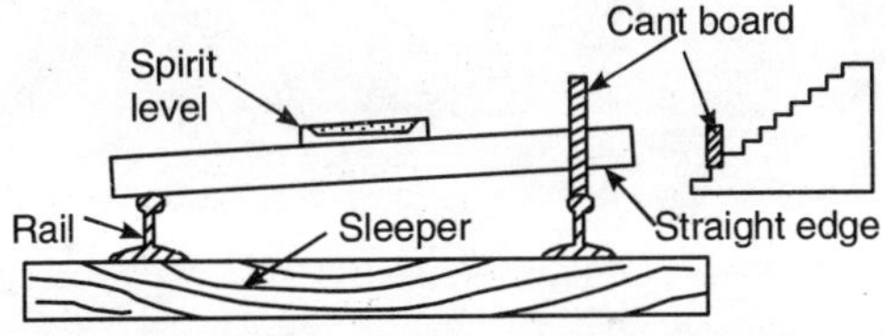

Figure 26.11 *Method of providing super-elevation*

When the outer-rail level will be above the inner-rail level by the amount equal to super-elevation provided by the cant-board, the bubble of spirit-level will be in the centre.

REVIEW QUESTIONS

26.1. What do you understand by the terms 'plate-laying' and 'rail-head' ? Describe the principal operations of plate-laying.

26.2. What are the various methods of plate-laying ? Describe telescopic method in detail.

26.3. Write short notes on the duties of the following gangs :

(i) Material-gang

(ii) Linking-in-gang

(iii) Packing-in-gang

26.4. Write short notes on relaying of track.

26.5. (a) How the rail levels are checked ? Describe with neat sketch.

(b) What is the function of cant-board in providing super-elevation ?

With the help of a neat-sketch show how the super-elevation is provided.

26.6. What points should be kept in mind while relaying the track.

26.7. Calculate the various track materials required for laying one kilometre track.

27

Maintenance of Track

GENERAL

When heavy trains move over the track at high speeds, it is diversely affected by vibrations, because it is an elastic structure. The modulus of elasticity of steel is very high as compared with ballast and sleepers. Therefore, under constant vibrations of moving trains, sleeper-fittings may become loose and the elasticity of the ballast may also be reduced. The track is also affected by the weathering action of the wind, sun, rain, and frost etc., which cause wear and tear of it. For the efficient running of trains with safety, strict in time should be the policy in removing it immediately so the track should be in well-condition. If any defect is noticed in the track, it should immediately be removed with a "stitch in time" policy, because it is a matter of safety of all the people moving in the train. Even if one defect is overlooked, it may cause serious accident. Therefore, it is most important to keep the track in well efficient running condition by proper maintenance.

Following are the advantages of good track maintenance:

1. The life of track and rolling stock is increased
2. The operational cost of the trains is reduced.
3. The train journeys become more comfortable and more safety to the delicate goods carried in the trains.
4. The security and safety against derailment and accident becomes more.
5. High speeds of the trains can be maintained.

27.1 MAINTENANCE

Maintenance of the track is divided into two parts :

1. Daily maintenance.
2. Periodical maintenance.

Railways have employed permanent staff for the daily maintenance of the track. Different gangs work under the direct supervision of Permanent Way Inspector (P.W.I) or Assistant Permanent Way Inspector (A.P.W.I). Different duties are allotted to each man in the respective gang, which each person does in a routine way daily. One set of the required tools is supplied to each gang with which they maintain the track in proper condition. All the periodical maintenance work is done after an interval of 2–3 years. This may be done by permanent-staff or on contract basis through outside agency.

Maintenance of track may be divided into following five classes:

1. Maintenance of surface levels of track.
2. Maintenance of alignment.
3. Maintenance of drainage.
4. Maintenance of track materials.
5. Maintenance of bridges and approaches.

27.2 MAINTENANCE OF SURFACE

The top level of the rails over which wheels move, is called the surface of the track. During the laying of track, the surface level of both the rails is kept at one level on straight portions and at curves, the outer-rails are kept at some fixed higher level known as *super-elevation*. When the trains move on the track, vibrations disturb the top level of the track. Generally maintenance of the lop level of the track is most important and takes up more than 50% time of the maintenance gang.

The surface level of the rail is disturbed by the displacement of ballast from the sleepers. When the ballast moves out below the sleepers, they sink and cause sagging. If at one time slight sagging occurs, it increases very rapidly and causes uncomfortable running of trains. The surface level of the track can be brought to its original level by packing the ballast below the sleepers.

Generally the packing is done by the following method :

Required quantity of ballast is collected by means of forks or shovels near the sleepers. Then sleepers are lifted by crow-bars or jacks up to required level and the ballast is packed below the sleepers. Generally the packing is done by the diagonal method. Below the each sleeper, the ballast is packed thoroughly by beaters. In the case of pot sleepers the ballast is packed through holes.

The width of the ballast under the rail seats is 45.7 cm, 35.6 cm and 25.4 cm for B.C., M.G. and N.G., respectively. The depth of ballast, which is added, is generally from 5 cm to 7.5 cm.

While doing maintenance of the surface of the track, the gauge is also checked and corrected. Following maximum tolerances are allowed including rail wears:

(A) For B.G. Track— For straight and curves upto 4°, for 3 mm tight to 6 mm slack gauge is allowed.

(B) For M.G. Track— For straight and curves upto 6°, from 3 mm tight to 6 mm slack gauge is allowed.

(C) For N.G. Track— For straight length from 3 mm tight to 6 mm slack gauge is allowed.

For curves from 4° to 20°, 6 mm to 19 mm slack track.

For curves sharper than 20°, 19 mm slack track.

The following points should be kept in mind, while packing the ballast:

1. At the time of lifting sleepers, it should be raised from both sides at the same time, because if lifted only from one side, the track may slip laterally towards one side.

2. After packing, the level of the track should be checked by sighting boards which are like boning rods. Three sighting-boards are used at a time.Two are placed at about 17 m and the third is placed in the midway. The levels at the place of outer sighting boards are first corrected and the level between them is checked by moving the third sighting board on the rail.

3. On curves first the level of the inner rail is corrected and then required super-elevation is provided to the outer rail by thorough packing.

4. The gauge of the rail should be checked at some intervals.

5. The packing below rail-joints should be done frequently, because generally they become loose at the points.

6. When more packing is done in the center of the sleeper than that at ends, sometimes this causes topling of sleepers. At such places the sleepers are said to be *centre bound.* This should be avoided by breaking the consolidated ballast from the centre.

7. Sufficient care should be taken in packing below the points any crossings. In no case packing should be done in such a way that sleepers may rock or spring.

8. All the dust particles should be removed from the ballast because these may be sucked up by the rolling-stock and cause wear.

9. If the formation consists of black cotton soil, a layer of moorum, clinkers or coal ashes should be spread over it before laying ballast, because on absorbing moisture black cotton soil expands and the blanket of the above materials prevent the raising of the track.

10. After the packing work the sleepers should be thoroughly boxed on all sides by ballast.

11. The level of both the ends of the rails at the rail-joint should be checked by means of spirit level after properly packing the ballast and lightening the fish-plates.

12. After doing packing under the sleepers, the ballast is lightly packed and levelled in the cribs i.e., the spaces between the sleepers is brought to the required section. The method of laying the ballast between the sleepers is known as 'boxing'.

If the rail joints and the surface are not properly maintained, following defects may occur in it:

1. Blowing joints.
2. Pumping joints.
3. Buckling of track.
4. Centre bound track.
5. Corrugated rails.
6. Hogged rails.

27.3 MAINTENANCE OF ALIGNMENT

When the track has been brought to the correct level the next operation is to correct the alignment. The track does not remain in the centre and shifts sideways actually due to the following reasons :

1. When the centrifugal force acts on curves, it pushes the track aside.

2. When wheels jump over expansion gap at each joint, it causes hammer action, which pushes the track forward. Sometimes this force is concentric and sometimes eccentric. This shifts the track from its original position.

3. If the ballast, below one rail becomes loose, it will cause tilt of the track and when trains move over it, it will be shifted aside due to non-uniform pressure on the rails.

4. If the formation consists of black-cotton soil, on absorbing moisture from the atmosphere, it will expand and lift the track. This lifting is known as *Heaved Track.*

5. When the quantity of ballast used is insufficient.

6. When creep is acting in one direction.

7. Heavy temperature stress.

To maintain the track or to bring it to the original alignment, the track is pulled back horizontally by means of crow-bars or special pulling devices. *Slewing* is the technical term which is used to denote the pulling back of the track. After bringing the track back to its original alignment the sleepers are thoroughly packed and boxed.

Following methods are used for the realignment of the track :

(a) *By Crow-bars.* The less distributed tracks can be realigned by pushing the track at the original place by means of crow-bars. For pushing the track at a time more than 8 persons should be employed. After realignment the ballast is properly packed.

(b) *By Track-liners.* When the alignment of the track has been disturbed more, the realignment can be economically and efficiently done by means of track-lines.

After doing the realignment, its correctness can be checked by means of naked eye, theodolite or by string line method.

27.4 MAINTENANCE OF DRAINAGE

Quick draining of water from the track is very essential. Because if the water does not drain out quickly, it will be absorbed by the formation, which will loose its bearing power and will settle down. If the embankment becomes more wet, there are fair chances of its sliding. Due to the above difficulties the drainage-system of railways should always be maintained in good working condition. The capacities of the drains provided on both the sides should be sufficient to carry away the rain water as early as possible. In no case weeds or grasses should be allowed to grow on the track or formation and inside the drains. On the top of cuttings, extra drains should be constructed to take the water from the upside of hills. Sufficient waterway should be provided below culverts and bridges. All the banks should be prevented from erosion by doing pitching works. In those areas where rainfall is heavy, underground drain-off pipes should be laid, which may take out the water absorbed by the formation. Sufficient width of cess or shoulder should be provided at the toe of the embankment. Dirt should not be allowed to accumulate on the ballast because it interferes with the draining of water. The formation should have ridge in the centre and slopes on both sides, for easy and early removal of water from the top.

The portion between the toe of ballast and the edge of formation known as *cess* should be cleaned and maintained so that water may not collect over it. The cess should be given an outward slope so that water run away.

In the case of double-tracks, generally a natural channel is formed between both the shoulders of ballast. The water in between this channel should be drained off by providing cross-drains at suitable intervals.

If the formation has excessive moisture or water and there are possibilities of water-logging, underground drains should be provided to drain-off the water.

27.5 MAINTENANCE OF TRACK MATERIALS

Maintenance of track materials includes renewal of rails, oiling fish-plates and fish-bolts, renewal of decayed wooden or cracked C.I. sleepers etc. and it is done in the following ways :

1. All the worn out main rails and check rails are removed and new ones are relayed.

2. The fish-plates and fish-bolts are occasionally oiled so that rails may have free longitudinal movement during temperature changings.

3. All the sliding chairs at points and crossings are oiled for the free movement of sliding rails and bars.

4. In the yards as rails are much affected by impurities of coal-ashes, they get rusted, and so, should be cleaned.

5. All the decayed sleepers should be replaced. Similarly all the cracked sleepers of iron should be cleaned.

6. All the worn out fish-plates should also be changed

7. All the loose spikes should be repaired by providing wooden plugs in the widened holes.

8. All the worn out points and crossings should be renewed by welding the worn out places.

9. If a rail is found broken or cracked, then a sleeper should be immediately placed below the fracture and tightened with spikes, until the rail is replaced.

10. If a kink is found in a rail, it should be immediately removed by using jim-crow.

11. If the excessive wear is found on the inner or check rails on curves, then inner rail should be immediately replaced, otherwise it may affect the outer rail.

12. To protect the rails from corrosion from ashes, sea-breeze etc. in station yards, waste oil should be applied to the foot and web of the rails.

27.6 MAINTENANCE OF BRIDGES AND APPROACHES

This has been discussed in Part 3—Bridge Engineering Ch. 48 of this book.

27.7 MAINTENANCE OF POINTS AND CROSSINGS

As the switches are the weakest points of the track and most of the derailment occurs at the points and crossings only. Therefore it is most essential to maintain the switches most efficiently. The inspection of points and crossings should also be done under moving trains, because only in this condition their actual defects shall be noticed. The points and crossings may go out of adjustments due to creep of track and wear of fittings.

The maintenance of points and crossings should be done as follows :

1. If creep has been noticed, it should be removed. Anti-creepers should be checked.

2. The condition of wear at points and crossings on the top and side of the stock should be thoroughly examined. All the badly worn, damaged and buried rails and tongue rails should be completely changed.

3. Less worn tongue rails or crossings may be rectified by welding.

4. For the given angle of crossing the leads and radius of the turn-out should be checked. If there is any variation it should be corrected.

5. The displacement of sleepers, if any, should be checked and rectified.

6. All the bolts should be inspected and tightened daily.

7. Periodically the ballast should be screened and repacked.

8. The clearance between check-rail, wing-rail and tongue-rail should be checked as per Indian Railway Standards and if any deviation is found, it should be rectified.

9. Drainage around the points and crossings should be in efficient condition.

10. All fouling marks should be regularly pointed.

11. All interlocking connections should be cleaned and oiled regularly.

27.8 MAINTENANCE OF LEVEL CROSSINGS

At the level crossing the road crosses the track at the same level. Wheel flange grooves are left between the rail and the guard rails. These guard rails are kept slightly higher than the main rails to protect their surface from damage from the road traffic. All the rails are spiked to the sleepers, which are in turn embedded in the road.

Level crossings should be maintained as follows :

1. The rails and its fittings should be inspected by opening out the level crossing.

2. Water bound macadam road should be constructed at the level crossing for easy removal and replacement during maintenance.

3. If the road traffic is heavy the water-bound macadam road should be pointed with bituminous pavement.

4. Drainage at the level crossing should be inspected, and no water should be allowed to stagnate at or around the level crossing, which may weaken the road or the formation.

5. The approaches of the level crossings should be inspected and maintained in efficient condition, for smooth running of the road traffic.

27.9 SIGNALLING DURING MAINTENANCE WORK

When the maintenance work in a particular section of the track is in progress, it is most essential to warn the train drivers about the conditions ahead. This warning is achieved by providing various types of signals and boards on both the sides of the repairing or maintenance section.

Following methods are generally adopted for signalling during track maintenance works:

1. If the maintenance work is of less than three days, then only flags as temporary signals are used for warning the drivers.

2. If the heavy maintenance work is to be done whose durations is more than three days, the semaphore type signals are provided on both the sides of the section to be repaired. If required additional detonating or audible or fog signals may also be provided. When the repair work is over these signals are removed.

3. If the repairing work is done in specified hours only, then they should be installed and removed as required, so that trains are less affected.

4. Provisions should also be made for signal lightings with standard indicators for night.

5. With the progress of the repair work the location, spacing and change of signals should be done as per requirements.

6. Temporary signal incharges should be provided with lamps with coloured glasses, kerosene oil, match-boxes, flags, detonators etc. as per the requirements.

7. Standard hand signals may also be used if required. Various types of signals which should be used have been discussed in chapter 31 of this book.

27.10 SPEED RESTRICTION ON TRAINS DURING MAINTENANCE WORK

As the restriction in speed reduces the efficiency of the railways, it should be avoided as far as possible. But during major maintenance works it cannot be avoided. Following points should be kept in mind while putting speed restriction on the trains :

1. No speed restriction should be done on the trains during daily maintenance work.

2. During the periodical maintenance work the speed indicators as given in chapter 31 of this book should be installed at proper places.

3. On weak bridges, culverts and at reverse curves the speed restrictions are done permanently as it is unavoidable from safety point of view.

4. All possible efforts should be made to avoid the speed restrictions but it should not be done at the cost of safety.

5. The lifting of speed restrictions should be done gradually with progress of the work and the perfection of the track. It should be removed completely only when the track has reached the standard perfection.

27.11 TRACK MAINTENANCE ORGANISATION

Indian Railways have complete track maintenance organisation. Head of all the construction and maintenance works is the chief engineer. The deputy chief engineer is the head of the maintenance works under the chief engineer. Flow chart 27.1 illustrates the Track Maintenance Organisation.

FLOW CHART 27.1 Track maintenance organisation

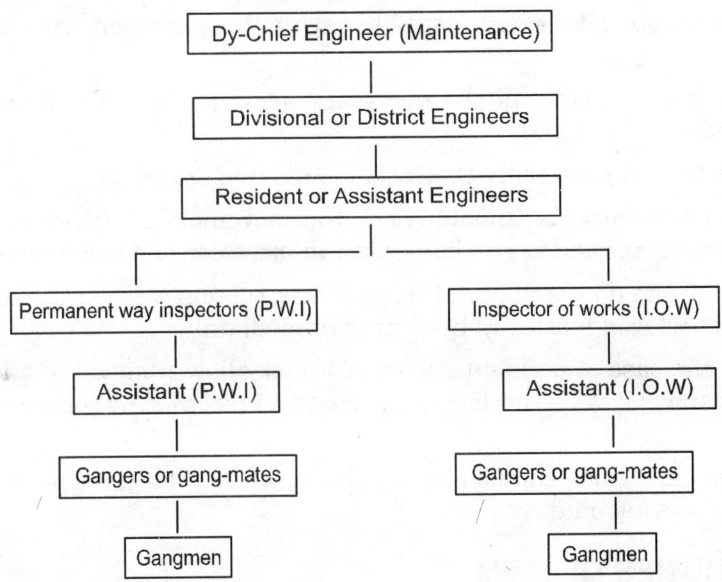

Each Divisional Engineer is the incharge of about 300 to 600 km of the railway track including all the stations on the track. The length of the track under Divisional Engineer mainly depends on the density, type of traffic, importance of the track and lines etc. Number of P.W.I, and I.O.W.'s work under each Assistant Engineer. Each P.W.I, is the incharge of about 80–120 km of the track. The I.O.W.'s are mainly incharge of the buildings, culverts etc. Three to four A.P.W.I.'s work under each P.W.I. Each A.P.W.I. supervises the work of 10–15 gangs directly under him. A ganger or gang-mate is the incharge of each gang. With each gang one keyman is attached who checks the keys of the track.

The length of the track under a gang depends on the traffic density, type and speed of traffic, types of sleepers number of crossings, turnouts, age and condition of the track, type of soil in the formation, length of sidings etc. The railway track is divided into sections of 5–6 km and is maintained by a gang of 8–10 persons.

The actual maintenance and repair work of the track is done by the gangmen, while the other staff do the work of supervision, arrangement of materials, preparation of the records, stores etc.

27.12 DUTIES OF GANG-MATE OR GANGER

The following are the duties of a gang-mate, who is the head of his gang :

1. He is responsible for the good running condition of his section.

2. He should allot specific duty to each of his gangmen and check their work.

3. He has to make all the necessary arrangement of tools and other equipments for his gang.

4. He should personally check the points and crossings.

5. In emergency, he should check the movement of trains by showing temporary signals and to give immediate information of it to his senior officer.

6. If anyone does any unauthorised construction work in his section, he should report and matter to his officer immediately.

7. In the case of accidents, he should not allow removal of any broken thing of railways, till it has been inspected and recorded by some responsible person.

8. The gang-mate should also watch the safety of his gang persons, when they are working on duty.

27.13 DUTIES OF P.W.I

The following are the duties of a P.W.I.:

1. He is fully incharge of his section and, also fully responsible for the upkeep of his section in good running condition.

2. He should redistribute the works to the gangs and check them.

3. He should keep a clear account of his store.

4. He should keep the full record of the renewal of track-materials.

5. While going on inspection he should check each and everything very carefully and instruct the workers accordingly.

6. He should develop team-work spirit in his gangs, because it will be very difficult for him to perform his duties without it.

7. He should personally check the gauge, the super-elevation, the depth of the ballast, the creep and the super-elevation on curves.

8. He should always have very good behaviour with his staff and should try to win their hearts, by keeping them happy.

9. In the case of accidents, he should make necessary arrangements to start the movement of trains as early as possible by clearing the lines immediately. He should also set up an enquiry to know the cause of the accident.

REVIEW QUESTIONS

27.1. Why the maintenance of track is required ?

27.2. How the maintenance of rail surface can be done ? What points should be keep in mind while doing so ?

27.3. What do you understand by alignment maintenance ? How is it done ?

27.4. Write short notes on maintenance of drainage.

27.5. What are the duties of a P.W.I. ?

27.6. What do you understand by the packing of ballast ? What are the important points to be observed during packing under the sleepers ?

27.7. Why are alignment of the railway track gets disturbed ? Describe the method of realigning the track on straight and curved sections.

27.8. How the maintenance of track fittings is done ?

27.9. What precautions should be taken during the maintenance of points and crossings?

27.10. Write short note on the Track Maintenance Organisation.

27.11. Write short notes on :

(a) Boxing (b) Through packing (c) Maintenance of gauge.

27.12. Write short note on the signalling during track maintenance.

28

Station and Yards

GENERAL

A *Station* is defined as a place where railway trains come to a halt in order to take up and get down passengers and goods. At stations traffic is booked and an authority is given to travel in the trains.

Following are the main purposes of railway stations :

1. To take up passengers and goods.
2. To control the movement of trains.
3. On the single line track, to enable the trains to cross each other.
4. To enable the express and mail trains to overtake the passenger and goods trains.
5. For filling water, coal, diesel in the locomotives.
6. For changing the engines and running staff.
7. For adding or detaching bogies in the trains as per the requirements.
8. For sorting out the wagons and making goods trains as per the requirements.
9. For repairing engines and changing its direction etc.
10. To hold in the passengers and trains in case of emergencies, Melas, accidents or other necessary works.

28.1 SELECTION OF SITE

The following are the main points which are considered while selecting the site for a railway station :

1. It should be close to the town or village.

2. Sufficient vast area should be available for future development of the station.

3. Site should have fairly level ground.

4. It will be preferred if roads already exist at the site connecting towns and villages.

5. Site should not be situated on the curve of a railway line.

6. It should have sufficient quantity of water in the nearby water-sources.

7. It should not be in low-lying area, otherwise water will be filled in the yards.

8. It should suit to the civil and military authorities.

9. Stations should not be located on slopes, because at such places locomotive has to apply more tractative effort to start the train.

10. There should be sufficient arrangements for the drainage of water from stations and their yards.

28.2 EQUIPMENTS IN STATIONS

The equipments of a railway station may be conveniently divided into following five categories:

1. Equipments for the public.
2. Equipments for traffic department.
3. Equipments for locomotive department.
4. Equipments common to all.
5. Further development of the railways.

28.3 EQUIPMENTS FOR PUBLIC

The passengers or the public generally require :

1. Booking office, where tickets can be got and goods can be booked.
2. Platforms for entraining and detraining.
3. Waiting-halls or sheds.
4. Water-supply taps for drinking water.
5. Bathrooms.

6. Retiring rooms.
7. Refreshment stalls.
8. Public telephone.
9. Name-boards of the station.
10. Lighting arrangement during nights.
11. Microphones with announcer for announcing the timings of arrival and departure of trains.
12. Enquiry-office.
13. Refrigerators for cold water in hot weather.
14. Big board showing railway time-table.
15. Boards showing reservation charts.
16. Guides to help the illiterate passengers.
17. Goods sheds.
18. Police office to help the passengers.
19. Clocks for correct time.
20. Guide map of the city.

28.4 EQUIPMENTS FOR TRAFFIC DEPARTMENT

Traffic department requires following equipments for the easy movement of the traffic:

1. Machines for holding and dating tickets.
2. Weighing machines.
3. Machinery to control and record the movement of trains.
4. Sufficient number of sidings to hold loose wagons and for easy crossing of trains.
5. Seating arrangement for the staff of traffic department.
6. Sufficient number of godowns for storing goods.
7. In case of big stations underground passages leading to various platforms.

28.5 EQUIPMENTS FOR LOCOMOTIVE DEPARTMENT

These include the following :

1. Water-columns for supplying water to engines.
2. Arrangements for supplying fuel to locomotives.
3. Ash-pits for removing coal ashes.

4. Pits for the inspection of locomotives.
5. Hydraulic jack for lifting locomotives.
6. Turn-table for turning the direction of engines.
7. Arrangements for the residential, resting, or waiting accommodation of staff of locomotive department.

28.6 EQUIPMENTS COMMON TO ALL

These include the following :

1. Roads connecting railway station with the surrounding areas.
2. Clock to show time.
3. Coolies for carrying luggage.
4. Escalators.
5. Big waiting halls.
6. Foot-bridges leading to various platforms.

28.7 TYPES OF STATIONS

Railway stations have been classified as follows :

1. Wayside-railway stations.
2. Junction stations.
3. Terminal stations.

28.8 WAYSIDE STATIONS

On these stations trains can move only in two directions, on upside or downside. These are situated on running lines at some suitable places. Wayside stations are of the following types :

28.8.1 Halts

This is a simple type of stopping place, having no building or staff. Halts have usually only one platform with a name board at either side. At Halt some trains stop to enable passengers to entrain and detrain. During night the trains stop to see the simple light signs at the stations. The tickets to the passengers are issued by Travelling Booking Clerk or by Travelling Ticket Examiners or sometimes by a contractor at the station itself. For the facility of issue of tickets a separate compartment is provided in such trains where contract is not given to the contractors for sale of tickets. Generally the contractor sells the tickets on commission basis. Halts are usually provided on light traffic sections and long block sections. Halts provide facility to passengers of neighbouring villages for removing their inconvenience in going to adjoining stations for catching trains. One of the main disadvantage of halts is that a large number of passengers entrain without tickets as there are no ticket collectors at the stations.

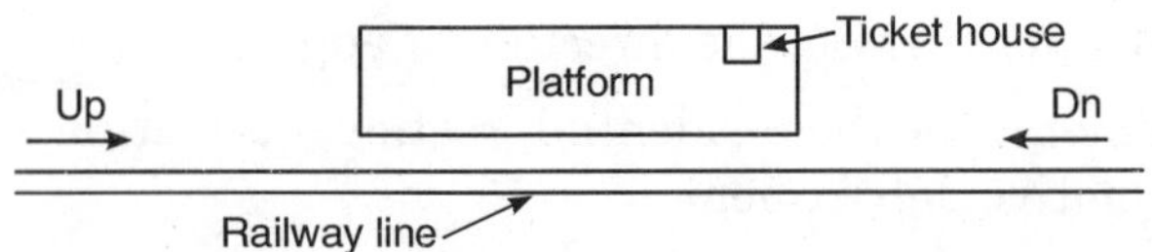

Figure 28.1 *Halt*

28.8.2 Flat Stations

These are next after Halts, having buildings and staff. At some Flag stations telegraph is provided, but no line clear work is done, because there are no crossing facilities. Flag stations may or may not have sidings. Some Flag stations have sidings in the form of a loop.

Figure 28.2 shows a typical layout of a Flag stations. It has one-way sidings which can be used only by trains moving in one direction. If any wagon has to be detached from trains moving in other direction, then this wagon has to be carried over first to the next block station and then brought back again.

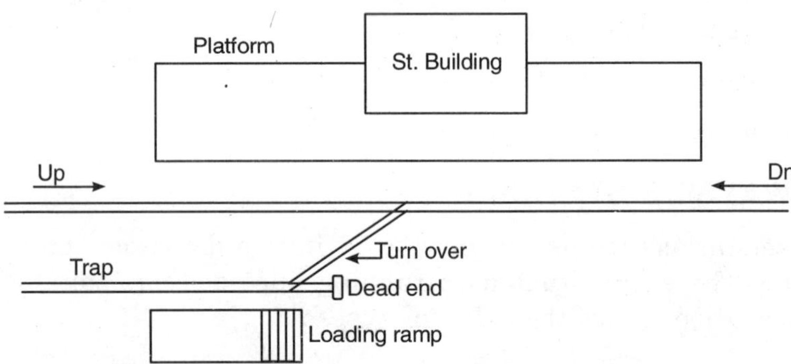

Figure 28.2 *Flag station with dead end siding and loading ramp*

The movement of trains at Flag stations is controlled by showing flags, that is why these are called *'Flag-Stations'*.

28.8.3 Crossing Stations

After Flag Stations, the next are simple crossing stations with two lines. These stations are provided with one loop, so that while one train is standing

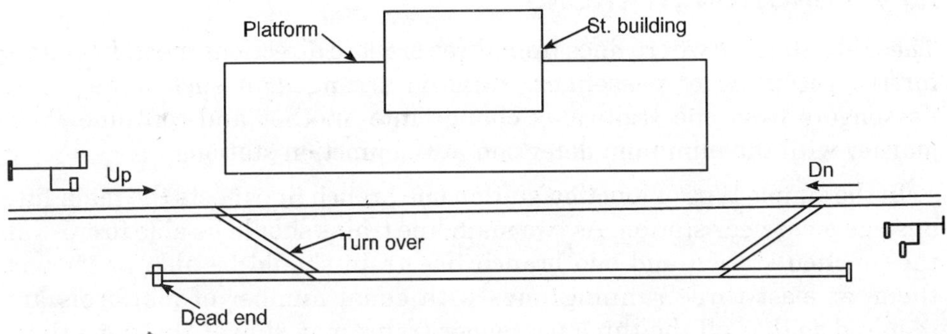

Figure 28.3 *Crossing station with dead ends on the loop*

on loop, the other train which has not to stop at this section, can cross it. The loop may or may not have dead end sidings. Figure 28.3 shows the layout of a crossing station with dead end sidings on both sides. Whereas

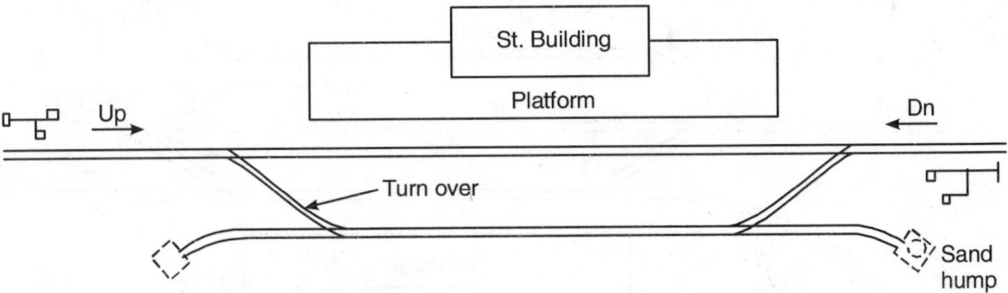

Figure 28.4 *Crossing station with sand humps on the loop*

Fig. 28.4 shows a crossing station with sand humps on the loop. Figure 28.5 gives the typical layout of a three-line station with staggered platforms. The layout can permit simultaneous reception with directional working.

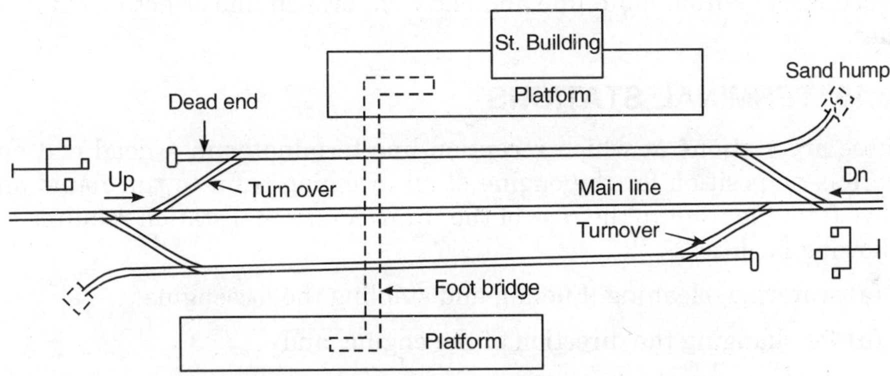

Figure 28.5 *Three line station with staggered platforms*

28.9 JUNCTION STATIONS

These are stations where lines from three or more directions meet. Facilities for the reception of passenger trains is arranged in such a way that passengers from one train may change into another and continue their journey with the minimum detention at the junction stations.

In the simple type of junction station one branch line meets the main line at some point near station. As two main line trains should be able to cross at the junction station and one branch line train should be able to connect them, at least three running lines with equal number of platforms are required so that all the three passenger trains may stop at the same time. In addition to this, the station should have good siding, engine shed, turn-table etc. according to requirements.

Figure 28.6 above shows an improved lay-out for a junction station. In this lay-out two trains can simultaneously be received from both the

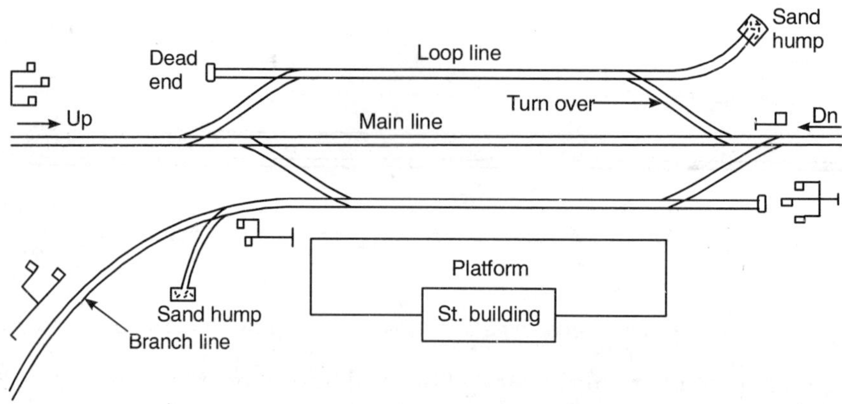

Figure 28.6 *Layout of junction station*

directions, one from main line and one from branch line or both from main lines.

28.10 TERMINAL STATIONS

These are stations at which reception line terminates in a dead end and there is no position for the engine of an incoming train to run round and move from the front to the rear of the train. A terminal station requires the following facilities:

(a) watering, cleaning, Fueling and stabling the locoengine,

(b) for changing the direction of the engine, and

(c) dealing with goods traffic.

On very important terminal stations all types of facilities are provided. But on unimportant branch line terminal stations only required facilities, depending on the traffic, are provided.

In Fig. 28.7 a terminal station with several platforms has been shown. The access to each platform is from the concourse as no over-bridge for this

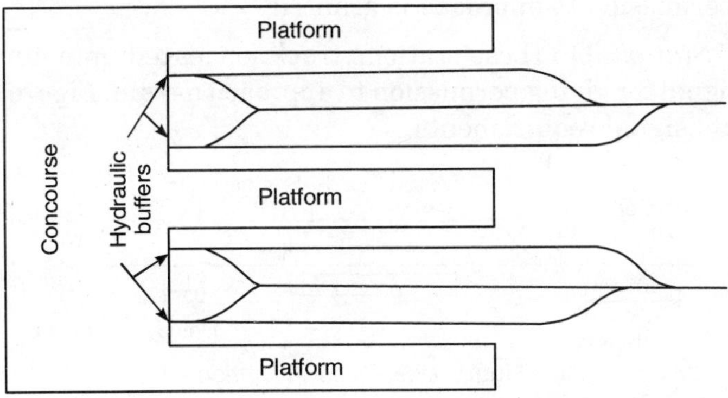

Figure 28.7 *Terminal station*

purpose has been provided. In this type of a typical lay-out the engine of an incoming train can be released via central spare line between the platforms and can run-round from either platform.

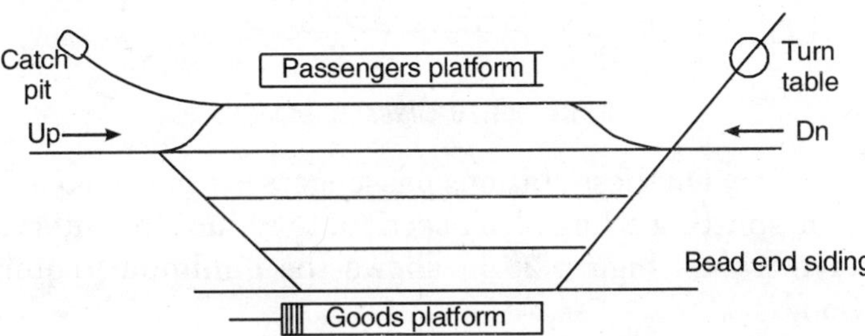

Figure 28.8 *Goods traffic at way-side station*

Figure 28.8 shows a terminal section with goods, shed, turn-table and ship siding.

28.11 CLASSIFICATION OF STATIONS

Indian Railway have classified the stations depending on their operational importance as follows :

(i) Block-stations—class A, class B and class C stations.

(ii) Non-block stations—class D stations and flag stations.

(iii) Special class stations.

28.11.1 Block Stations

The stations made at the end of the block sections are block stations. Authority to proceed in the shape of the token is given at these stations. Similarly permission to approach is granted.

Class 'A' Station. On these stations track is cleared upto 400 m beyond the home signal for giving permission to approach a train. Figure 28.9 shows its minimum signal requirements.

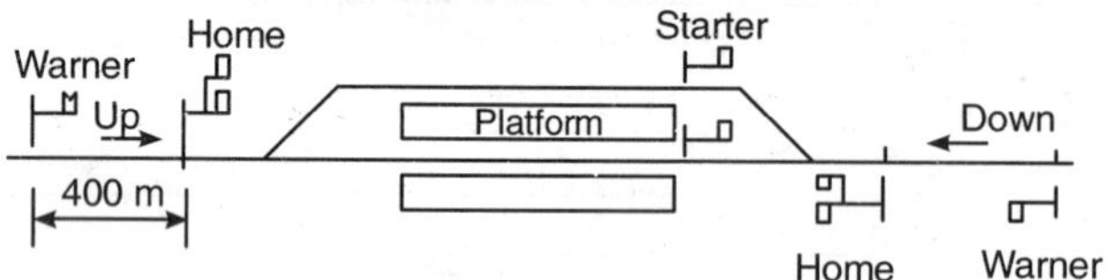

Figure 28.9 *Class 'A' station*

Class 'B' Station. In such stations the other signal is provided at about 580 m from the home signal. Figure 28.10 shows the minimum requirements of such a station.

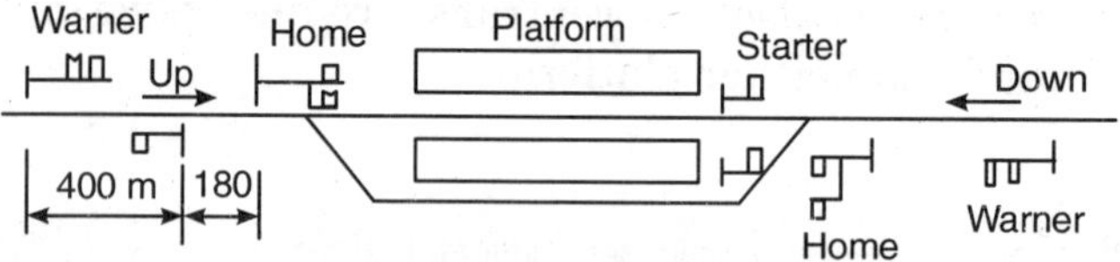

Figure 28.10 *Class 'B' station*

Class 'C' Station. On these stations passengers are not booked. It is simply a block hut for splitting a ling block-section, to reduce the interval between the successive trains. Figure 28.11 shows the minimum requirements of such a station.

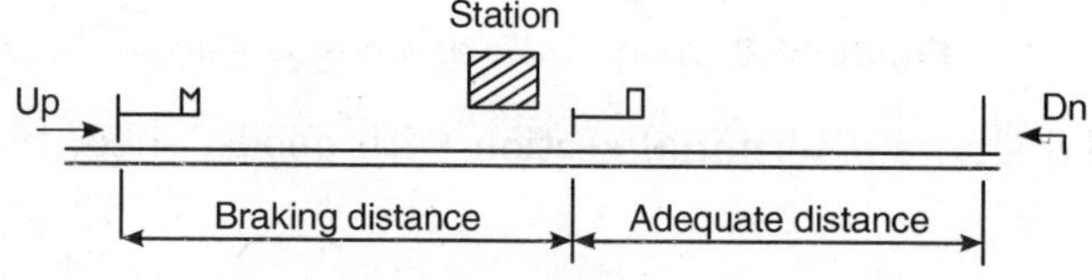

Figure 28.11 *Class 'C' Station*

28.11.2 Non-block Station

These are also known as 'D' class stations or 'Flat' stations. These are situated between two consecutive block stations. These may not be telegraphically connected to the adjacent stations. No equipments or staff is provided for controlling the movement of the trains. Trains are stopped by showing flags only.

28.11.3 Special Class Station

Those stations which do not come under the above class A, B, C, D and non-block stations are called special class station.

28.12 STATION YARDS

These are systems of tracks laid within limits for various purposes, over which movement of trains is controlled by prescribed rules, regulations and signals.

Railway yards have been classified as follows:

1. Passenger yards.
2. Goods yards.
3. Marshalling yards.
4. Locomotive yards.

28.13 PASSENGER YARDS

This yard provides all the facilities for the convenience of the passengers. The chief requirements of a passenger yard are as follows:

1. Booking office, cloak-room and luggage booking room.
2. Space for parking all sorts of vehicles outside the station.
3. Enquiry office.
4. Signals for the reception of the trains from different directions at the same time, so that all the trains, which are required to be connected with one another may be received and connected in a short time.
5. Signals for the departure of trains in various directions at a time.
6. Facilities for passing a through-train at the full speed without interference.
7. Sufficient number of shunting necks for withdrawal and placement of rakes without causing interference to other trains.
8. Sufficient number of platforms.
9. Sufficient number of sidings with platforms, so that extra carriage after detaching from trains, can be kept till they are attached to other trains.
10. Facilities for charging batteries of trains.

11. Washing lines, sick-lines and stabling-lines in required numbers.

28.14 GOODS YARD

The chief requirements of a goods yard are as follows :

1. Approaches with roads for the movement of trucks and other vehicles to each platform with ramps for loading and unloading of the goods.
2. Gathering line taking off the main line of loop line with a number of parallel dead-end-sidings with buffer stops.
3. Platforms with sufficient height so that loading and un-loading from wagons can be done easily. The heights of the platforms for B.G., M.G. and N.G. are 1 m, 0.7 m and 0.6 m respectively above rail level.
4. Booking office.
5. Sufficient number of godowns for storing goods.
6. Cranes for loading and unloading very heavy goods.
7. Cart-weighing maching for weighing full carts.
8. Weight-bridge to weigh wagons.
9. Loading-gauge to check the height and width of loaded wagons.
10. Vacuum testing machine.

28.15 MARSHALLING YARDS

One author has described marshalling yards as "hearts that pump the flow of commerce along the track". These are places where trains are born. The unit of loading on railways is usually wagon, but the unit of transport is a train. Therefore, before despatch, all the wagons are associated in the form of a train. All the wagons should be arranged station-wise so that there should be no difficult in detaching them at stations. This is the function of marshalling yards i.e., to short out and combine isolated wagons into train loads.

Marshalling yards are very costly in construction and maintenance. The following are the drawbacks of marshalling yards :

(a) Delay in the transit of wagons.

(b) Damage of wagons while shunting.

(c) As the shunting work is done continuously in the yard approaches to yards are often places of traffic congestion.

(d) Much time of engines is wasted in terminating and originating trains.

From the above drawbacks, we observe that marshalling yards are largely unproductive, and the work done is only indirect. Therefore marshalling yards are necessary without which it is impossible to run goods trains. The efficiency of marshalling yards depends on the time taken for sorting out wagons and forming it into a train.

Fot the efficient function of the marshalling yards, the following factors should be taken into consideration :

1. The shunting operation should not interfere with the time table of the regular trains. Though goods train should be dealt properly.

2. These yards should be provided with all the facilities to place and withdraw the wagons from various sides.

3. The design of the marshalling yard should be such that shunting and forming of trains can be easily done in the minimum possible time.

4. For more efficiency of the marshalling yards, the sorting work of the terminating trains and the load of the internal yard should be maximum in a given time period.

5. The design of the yard should be done in such a way that it can be increased or developed in future depending on the increase in the goods traffic.

6. Tranship platforms for exchange of goods should be separately provided on one or more sidings of the marshalling yards. Such sidings are generally reserved for the defective wagons, whose goods can be exchanged to the other wagons through these platforms.

7. To increase the efficiency of the marshalling yard facilities of water column, ash-pit, coal stock, weight bridge etc. should also be provided in the yards.

8. If the traffic is more it will be better to have separate marshalling yards for up and down trains.

9. The marshalling yards should be constructed at all the important railway stations.

10. Goods yard should be near the marshalling yard.

The following are the main works, which are done in marshalling yards :

1. After change of engine and train examination with occasional shunting, through-trains are allowed to pass.

2. Terminating trains are broken up and sorted out.

3. To form through trains for farthest possible points.

4. To place in and draw wagons from various points in local area i.e., goods shed, loco-shed, sick line and factories or industrial establishments.

5. To form, work and van trains for the different sections which are served by the yard.

6. When the main lines are not clear, to hold the trains and wagons.
7. To move train engines and pilots between the yard and the loco-shed.
8. To stable surplus and condemned stock.

The layout of a marshalling yard should be such that it can fulfil all the above requirements with the minimum detention of wagons with minimum consumption of shunting engine-hours.

28.16 TYPES

There are three types of marshalling yards :

(a) Gravity (b) Hump (c) Flat

28.16.1 Gravity-yard

In a gravity type yard the wagons move into various dead end sidings by the force of gravity only. From the beginning of gathering lines at the time of sorting, wagons are released one by one so as to go in different sidings.

28.16.2 Hump-yard

In a Hump-yard the wagons are pushed up to a summit by a shunting engine from where they roll-down to opposite slope, under the force of gravity, into fan-shaped sorting sidings. Nowadays this type of yard is provided mostly everywhere, because the shunting operation can be done more quickly than gravity or flat yards. Figure 28.12 shows a Hump-yard.

The pushing is provided with a rising gradient of 1 in 150 to 1 in 175 for about 175 metres. 2–3 metre portion at the top is kept level in the middle

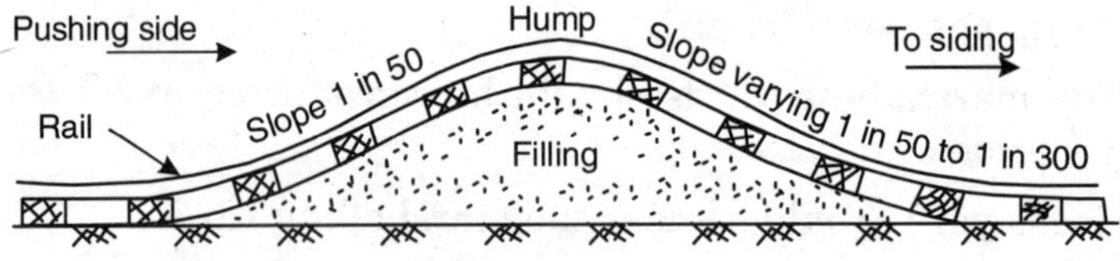

Figure 28.12 *Hump yard*

of the hump. In the third portion the vehicle is allowed to fall in gradients of 1 in 50 to 1 in 300 and then level in the yard as shown in the figure.

By means of shunting engine the group of wagons or individual wagons are pushed from the pushing side, upto hump level and then allowed to fall under force of gravity on the slope. Usually the men run along the wagons and apply the brakes at the exact place to stop the vehicle. On Indian Railways skid method is also used to stop the moving vehicle in marshalling yards. In this method rolling wagon drags the skid and the friction thus developed stop the vehicle.

28.16.3 Flat-yard

In flat-yards all the sorting work is done totally by shunting-engines. Therefore, it requires more number of hours of a shunting engine. At the places where space is limited and other types of sorting yards cannot be provided, flat-yards are provided. These are not commonly used.

28.17 LOCATION

Figure 28.13 shows a typical arrangement of the up and down goods reception and departure lines in relation to the sorting yard and the marshalling-cum-departure lines. The trains from up and down lines are taken into separate

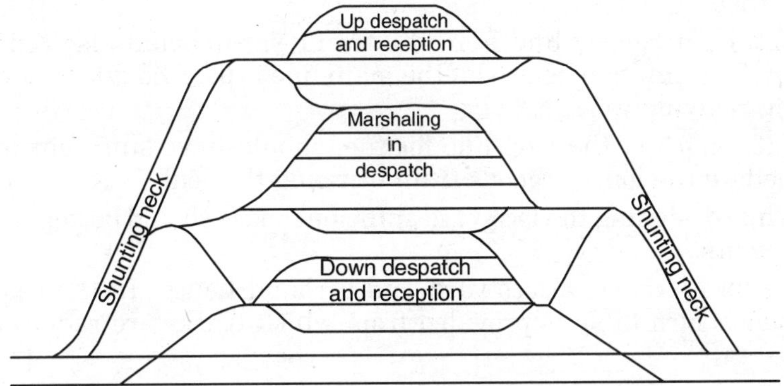

Figure 28.13 *Typical arrangement of reception and sorting lines*

sorting yards. Trains from up and down can enter or leave marshalling yard simultaneously without any interference to each other.

28.18 LOCOMOTIVE YARDS

The following are the chief requirements of a locomotive yard :

1. It should be located near the passenger-yard and goods-yard, so that both incoming and outgoing engines may take less time.
2. Escape line to admit train engines immediately after a train has arrived.
3. Water column.
4. Ash pits, inspection pits.
5. Hydraulic jack for lifting the engines while repairing.
6. Turn-table.
7. Engine-shed.
8. Space and equipments for loading fuel in engines.
9. Sick-sidings.

10. Over-head tank with loco-well.
11. Space for future expansion of the yard.

Following points should be kept in mind while designing the Loco-yard :

1. The arrangement of the tracks should be in proper sequence, so that no interference is caused to any engine during the servicing.
2. The turn-table should not act as an obstruction, the line from the traffic yard to the turn-table should be kept vacant.
3. The siding leading to the coal stock and ash-pits should be easily approachable from the traffic yard.
4. The sick-siding should also be easily approachable from the loco-yard.
5. The location of the loco-well and over head tank should be done near the loco-shed.
6. Extra emergency line from the traffic-yard should also connect the loco-yard. This line can be used if the main line is blocked due to derailment or during relaying work.
7. The length of the loop line for fuel should have sufficient length to accommodate the longest goods train bringing the coal.
8. While designing the loco-yard sufficient space should be kept for future developments.

Figure 28.14 shows a loco-yard and a round house. In the centre of a round house turn-table is provided from which tracks are radiating in all

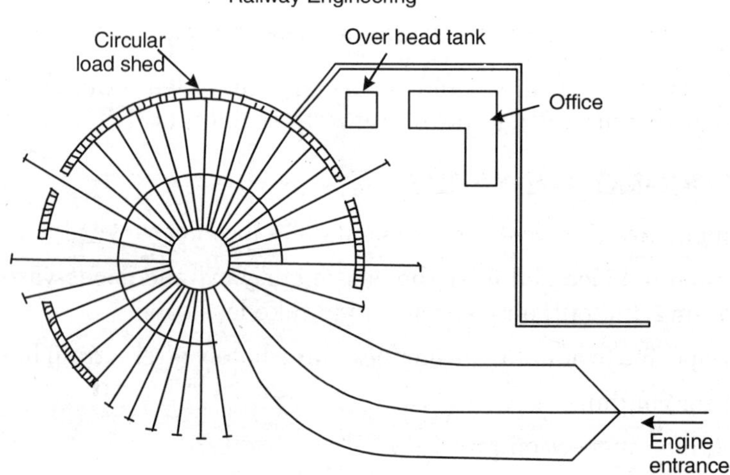

Figure 28.14 *Loco-yard*

directions. Over-head tanks and railway staff office have also been provided on one side. But, this technique is not in use because the of the new technology to change the track has been adapted throughout the rail net.

REVIEW QUESTIONS

28.1. What is a railway station ? What factors affect the selection of site for a railway station ?

28.2. What equipments are required in railway stations ? Describe them in detail.

28.3. What are the main classifications of stations in India ? Write short notes on each type.

28.4. What do you understand by the term 'station yard' ? What are the different types of yards ? Describe any one in detail.

28.5. Write short notes on :

(i) Halt
(ii) Flag stations
(iii) Junction stations
(iv) Terminal stations.

28.6. (a) What are the chief requirements of a passenger-yard ?

(b) What are the different types of marshalling yards ?

28.7. Draw a typical layout of a locomotive-yard and write its chief requirements.

28.8. Write short note on the operational classification of railways.

28.9. Write short note on the functions of the railway stations.

28.10. What points should be kept in mind while designing a loco-yard.

28.11. Write short note on the hump-yard.

29

Station Equipments

GENERAL

For the efficient running of trains, safety of the traffic and locomotives, different types of station equipments are required for the following purpose :

1. Repairing, cleaning and inspection of locomotives
2. Fueling, watering and reversing the direction of engines
3. Inspection and repair of coaches and wagons
4. Weighing, loading and unloading of goods
5. Checking the entrance of shunting vehicle into main running lines
6. Diverting the train from one track to another
7. Starting and stopping the trains
8. Conveying to engine drivers the speed restrictions on different sections
9. Safe movement of traffic on level-crossings
10. Facilities to passengers at stations

We will study about equipments in this chapter and some subsequent chapters.

29.1 PLATFORM

The sills of coach doors of trains are considerably higher than the rail level. Therefore for entering and getting down from trains suitable higher places should be provided. Such places which are provided for the movement of passengers using the railway are known as 'platforms'. The length of a platform should be more than the longest train which is moving on that section. The minimum length of a platform should be 183 metres, but for B.G. railway it should not be less than 300 metres. Platforms should have

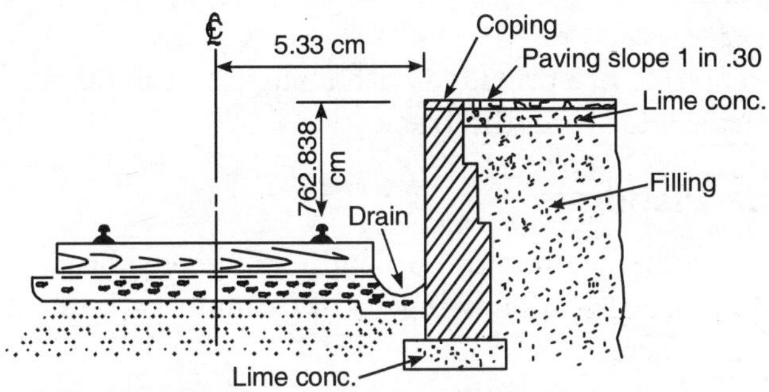

Figure 29.1 *Platform*

sufficient width for the easy movement of passengers, and keeping the goods etc. Under no circumstances the width of a platform should be less than 4 metres. Figure 29.1 shows the cross-section of a B.G. platform with all dimensions. The platform should be provided 1.676 metres away from the centre line of the track for B.G., 1.346 metres for M.G. and 1.219 metres for N.G. The height of the platform above the rail surface should be as follows:

For B.G. — 76.2 cm to 83.8 cm

For M.G. — 30.5 cm to 40.6 cm

For N.G. — 22.9 cm to 40.6 cm

The platforms should be provided with sheds for at least 60 metres of their length and 3.66 m in width to protect the passengers from sun and rains. Platforms should be paved throughout on their tops and at both ends ramps should be provided. Platforms should be well lighted during nights to enable the passengers to use them. Names of the stations should be written on big boards placed at right angle to the track on both the ends of the platform.

The minimum width of the passenger platform is 3.66 in front of the station building, but it varies as per the importance of the station.

The longest platform in the world exists in India on Sonepur (N.E.R.) station whose length is 736 metres. The other important stations having longer platforms are as follows:

1. New Lucknow (N.R.) — 686 metres
2. Jhansi (C.R.) — 617 metres

3. Kharagpur (E.R.) — 716 metres
4. Bezwada (S.R.) — 674 metres
5. Trichinapally (S.R.) — 471 metres
6. Ranaghat (R.R.) — 464 metres
7. Dakor (W.R.) — 448 metres

Passenger platforms are provided with a slope of 1 in 30 across its width for drainage purpose towards the track.

29.1.1 Goods Platforms

The platforms used for loading and unloading of goods for goods trains are known as goods platforms. Following points should be kept in view while planning these platforms :

1. For easy handling of the goods their height should be near the wagon floors.

2. Weighing arrangements should be provided at the goods platform for weighing the goods at the time of booking it.

3. Goods shed should be provided on the platform to protect the goods from rain, sun etc.

4. Direct access facilities from the goods platform to the goods siding and to the marshalling yard should be provided.

5. The minimum width of 3.1 m is kept in case of goods platform.

6. The permissible platform heights from the rail level are 1.07 m, 0.69 m and 0.61 m for B.G., M.G. and N.G. respectively.

7. The minimum distance to any building from the centre line of the track should be kept as 4.72 m, 3.18 m and 3.05 m for B.G., M.G. and N.G. respectively.

29.2 WATER COLUMN

These are provided at every 30–50 km on main lines to supply water to locomotives. At main junctions more than one number of water columns are provided. These are situated near the ends of platforms so that engines of running trains can take water when halting on stations. Coal-ash pits are also provided near the water columns so that engines can remove its ashes simultaneously while taking water.

Figure 29.2 shows a water column. It essentially consists of vertical pipe with one horizontal or swan-neck shaped pipe at its top. The horizontal pipe can be moved around the vertical pipe in a horizontal plane. One valve is fitted in it which can be operated by wheels with chains as shown in

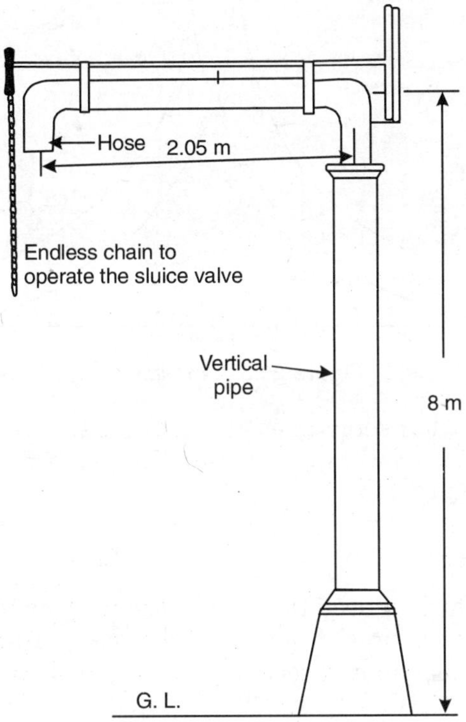

Figure 29.2 *Water column*

Fig. 29.2. This valve is used for regulating the flow of water through the water column.

Generally a light funnel is provided to direct the water to fall in locomotive-tank without splashing. Necessary arrangements for draining the surplus water are also made near water-column site.

29.3 TRIANGLES

These are used for changing the directions of engines. Turn tables are very costly and cannot be provided everywhere for turning the direction of engines. At small stations where small number of engines are to be turned, a triangle can be used. This consists of three short lengths of tracks laid to form a triangle as shown in Fig. 29.3. These tracks are connected to each other by three pairs of points and crossings. Generally, two tracks are laid in curves

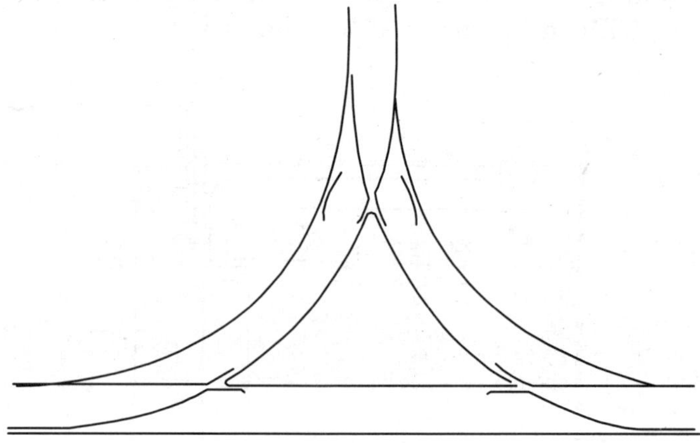

Figure 29.3 *A triangle*

and third as straight, but sometimes all the three tracks are laid in curves. If one engine moves completely round the triangle, its direction is automatically changed.

29.4 TRAVERSER

This is the device which is used to transfer the wagons, passenger coaches or locomotives (one at a time) from parallel tracks without any shunting. Figure 29.4 shows the plan and sectional elevation of traverser.

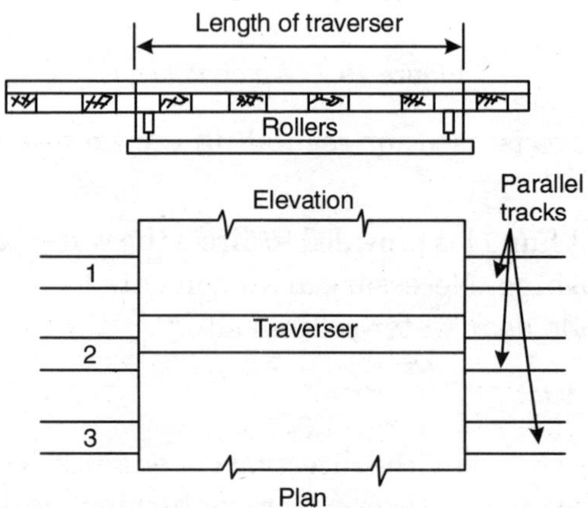

Figure 29.4 *Traverser*

Traverse mainly consists of a platform mounted on small wheels or rollers which moves at right angle to the parallel tracks. When any vehicle is required to be changed from one track to another, it is placed on the traverser and the traverser is moved sideways and the vehicle is transferred to the desired track.

Traversers are used only on loop or sidings. These are never provided on the main lines or branch lines.

29.5 ASH-PITS ASH-PANS, EXAMINATION PITS AND DROP-PITS

After burning coal in locomotives a large quantity of coal-ashes is produced, which is to be dumped at some place. Ash-pits are used to collect the ashes from the locomotives. As the ashes are to be collected from the locomotives of the running trains, ash-pits should lie on the main lines. Generally these are provided in main lines after the end of platforms near water-column, so that a locomotive can dump its ashes while taking water.

Figure 29.5 shows the longitudinal section of an ash-pit. These are generally 1.067 metre deep and larger than the length of the longest locomotive moving on that section. Ash-pit consists of two masonry walls over which longitudinal beams of timber, concrete or steel are fixed to which rails are fixed. No cross-connections are done between the longitudinal beams supporting the rails, because ashes have to be dumped in space between the

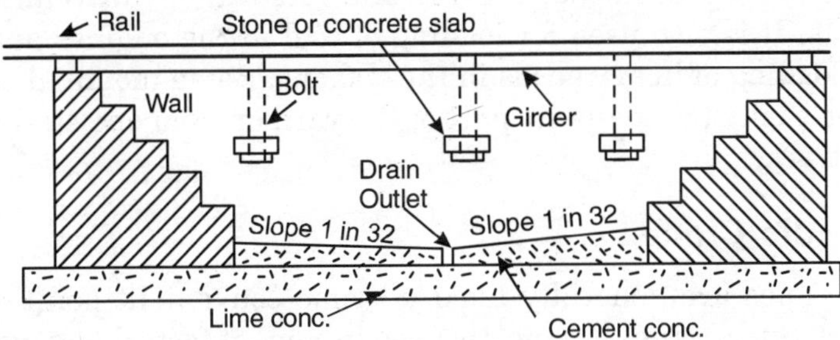

Figure 29.5 *An ash-pit*

rails. As no cross-connections are done between the rails, the gauge is likely to be disturbed, therefore, it should be checked occasionally and tie-bars should be provided on both ends of the ash-pits. The beams should be fixed to the holding bolts embedded in the longitudinal walls to prevent change in gauge. The timber-beams should be covered with iron or steel sheets to protect them from burning by hot ashes. After extinguishing the coal ashes by water, they should be removed from the ash-pit. The floor of the ash-pit is provided with a slope from both sides towards centre, where drain hole is provided to remove the water.

29.5.1 Ash-pans

These are used for the same purpose as ash-pits. These are constructed with precast U-shaped R.C.C. sections placed side by side to form a shallow pit. The rails are directly laid on the side walls of such pans.

29.5.2 Examination-pits

These are similar in construction to ash-pits and are provided in loco-yards to inspect the engine mechanism from underneath. The length of these pits is usually longer than ash-pits.

29.5.3 Drop-pits

These pits are constructed at right angles to the track within which hydraulic jacks are fixed to lift the locomotive.

29.6 CRANES

On big railway stations cranes are used for loading and unloading of heavy articles in the wagons. Following types of cranes are most commonly used :

29.6.1 Mobile Crane

This is very useful type of crane mounted on wheels and moves on the railway track. It is used in lifting, shifting and removal of rail vehicles in case of accidents. It is also used for loading of coal in the engines and for loading and unloading of heavy goods in the wagons. Being mounted on the wheels it can be easily taken in any part of the yard or from one station to another on the track.

29.6.2 Jib Crane

These cranes are generally fixed at some convenient point on the goods platform. Such cranes are mostly used in cement factories, sugar industries, steel industries and other big industries where railway siding goes in the factory for transporting the goods of the factories.

29.6.3 Gantry Crane

This type of crane is also commonly used in loco-sheds, large factories for hoisting the engine components during repairing or assembling purposes. This consists of two parallel beams supported on vertical posts. The hoisting machinery travels on the parallel beams. Wagons or trucks are brought under this crane for loading and unloading purposes. These cranes are also suitable for loading and unloading from one wagon to another on parallel tracks, thus save shunting time.

29.6.4 Goliath Crane

This crane is also similar to the gantry crane with the difference that it consists of two parts and a beam, with the lifting machinery installed on the beam. The frame of the crane travels on the rails laid on the ground.

29.7 TURN-TABLES

This is a device used for changing the direction of engines. The given Fig. 29.6 shows the most common type of turn-table used in India. The turn-table is installed in a circular conical pit with a pump for pumping the water of a drain. It is supported on three points, central pivot or disc bearing and two wheels one at each sides end. The level of the rails on the turn-table is kept the same as that of the tracks radiating from the edge of the pit. For changing the direction of an engine, the turn-table is revolved and brought in the direction of the track on which engine is standing, and locked to avoid relative movements. Now the engine is moved on the turn-table and it is rotated manually or by power till the engine is brought in the desired direction and on the desired track. After it the turn-table is again locked and the engine is moved on the track.

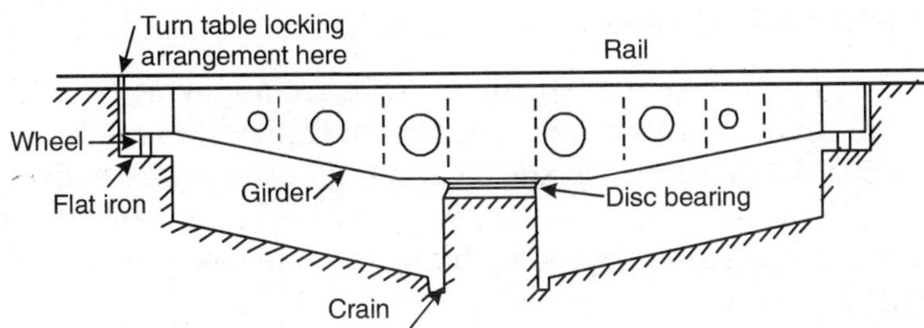

Figure 29.6 *Turn-table*

For changing the direction of very heavy locomotives, non-continuous types of turn-tables are used. Flangeless wheels roll on circular rail fixed along the circumference as in continuous turn-tables, but there is a joint in the centre and disc or roller type bearing is provided in place of pivot on which two halves of girders roll. The standard diameter of a turn-table for B.G. is 19.8 metres, but it may vary from 26 to 30.5 metres for longer engines.

29.8 END-LOADING RAMP

For loading the road vehicles such as trucks, cars, military vehicles, earth moving machines into the railway open wagons, loading ramps are constructed in the goods yards.

These are constructed at the end of a siding and are built to the height of the wagon floor as shown in Fig. 29.7. The top of the ramp is given a sight slope towards the road-side. Buffer-stops are provided in front of this small platform to protect the wagons from striking against the ramp. The gap between the wagon floor and the platform is bridged by hinged steel plates.

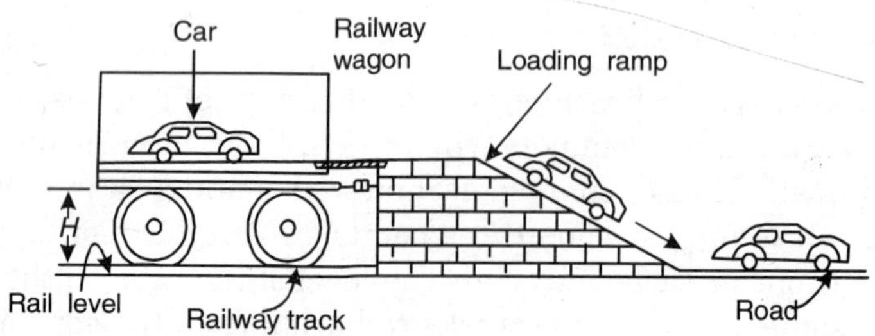

Figure 29.7 End-loading ramp

Sometimes old railway sleepers are used to fill this gap while loading the road vehicles.

The height of the end-loading ramp above the rail level is kept as 1.3 and 0.86 m for B.G. and M.G respectively.

29.9 BUFFER STOPS

These are provided to prevent the vehicles from moving beyond the end of rails at terminals or sidings. Simple type of Buffer-stop consists of a timber-beam of 30 × 13 cm section at the level of buffers on vehicles, fixed to the two vertical rail posts bolted to the track rails on other ends. These beams and posts are further strengthened by fixing them to pieces of bend rails, which are again bolted to the track rails.

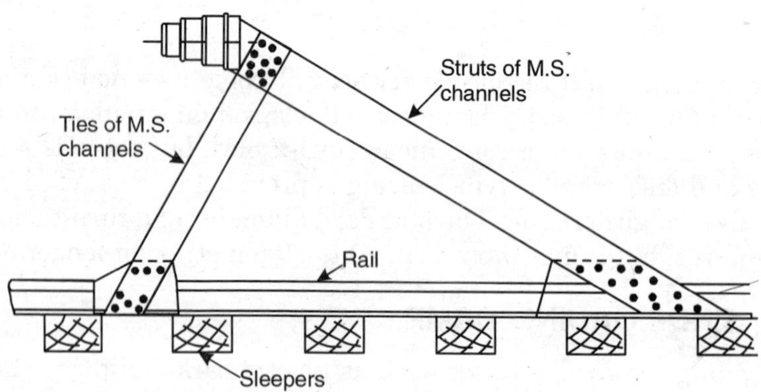

Figure 29.8 *Buffer stop*

Figure 29.8 shows the longitudinal section of a buffer stop. It is constructed with steel channels fixed to the rails as shown in Fig. 29.8. Spring buffer-stops are fixed at the top.

29.10 WEIGH BRIDGES

For weighing the load of the loaded wagons, weigh bridges are provided in the goods yard. These are level platforms with rails, which are connected with the track. The level platforms are supported on the beams which are located in the pit under the track. The loaded wagon is brought on the platform and its weight is taken through the graduated scale attached to the levers, by sliding the weights on the scale.

This weigh bridge is similar to that used for taking the weight of the trucks installed on the roads or gates of the big steel, sugar or other industries.

Care should be taken to provide the weigh-bridge only in the separate siding, meant only for weighing the wagons, otherwise it may be damaged or may cause obstruction the movement of the traffic. Wagons should be allowed to move very slowly while coming on and going from this platform.

29.11 CATCH AND SLIP SIDINGS

These are derailing switches used to stop the movement of vehicles which are out of control, by derailing them. On hilly railway flatter slopes are provided only in railway yards, because at other portions it is not economical. On hills if any standing vehicle or loose bogie gets movement due to heavy wind, it will start moving downwards along the slope at a very high speed. As no person is sitting in it, if its movement is not stopped, it may strike against another train and cause accident.

Now, suppose if one vehicle starts unauthorised movement from one station by the force of wind towards other station downwards, it should be checked. Generally, for this at every railway hill-station two derailing

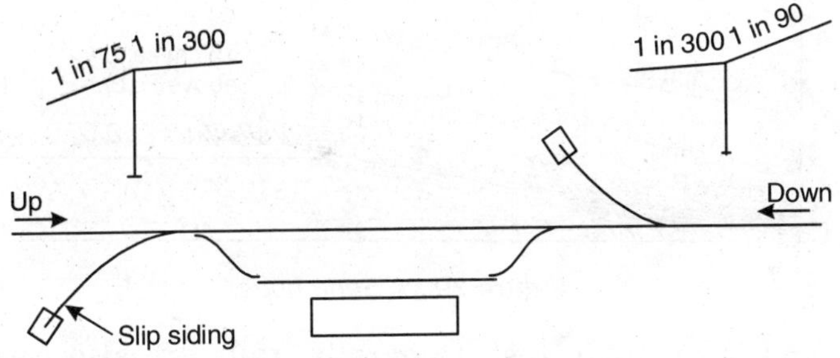

Figure 29.9 *Catch and Slip sidings*

switches are fixed, one on the upper-side and the other on lower-side. When information of the movement of a vehicle is received by the station downwards it will open both the switches to derail the vehicle. The upper-side derailing switch is known as *Catch Siding* because it tries to catch the vehicle. If by chance the vehicle has been slipped by the catch siding it will be derailed by the lower-side derailing-switch. As the lower-side derailing switch can catch the vehicle only if it has been slipped by the catch-siding, which is known as *Slip Siding.*

29.12 SCOTCH BLOCKS

These are also used for preventing the movement of vehicles beyond the dead end slidings. These consist of wood or steel block placed on the rail and locked at suitable position. It becomes an obstruction in the way and prevents the movement of vehicles. At the time of using siding the lock is opened and

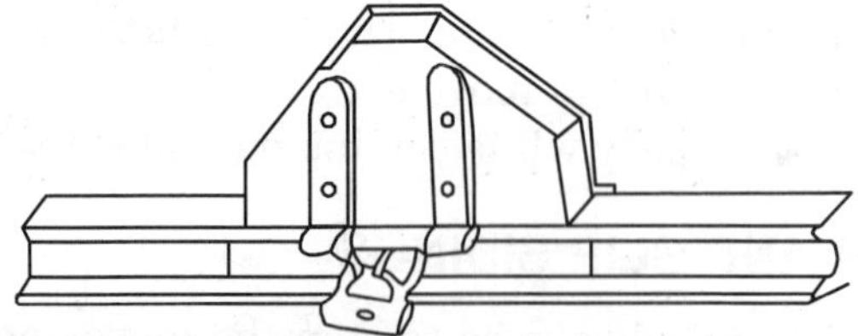

Figure 29.10 *Scotch block*

the scotch-block falls on one side of the rail. This arrangement was made in olden days but nowadays it is not used.

29.13 SAND HUMP

This is also a device to check the movement of a vehicle. The rails in dead end sidings after some fixed distance are embedded in sand. When any vehicle tries to go beyond the limit on dead end sidings, its wheels are embedded in

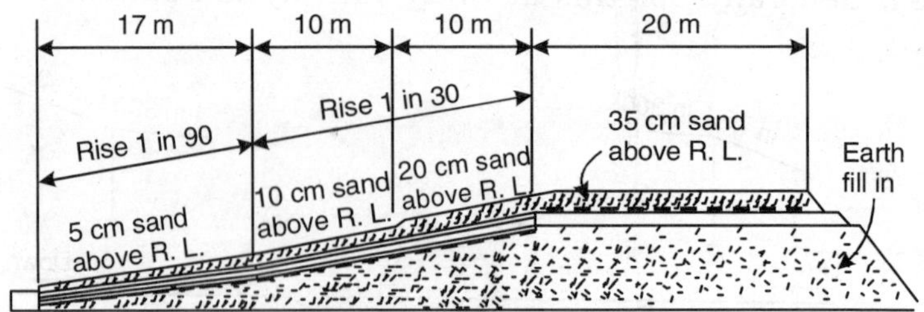

Figure 29.11 *Sand hump*

the sand, and it comes to rest. Generally, rails are also raised at some gradients inside the sand layer, which also helps in obtaining the function of the sand hump.

29.14 FOULING MARKS

When more than one tracks are running parallel to each other, the minimum distance maintained between them is fixed. But at such places where tracks cross each other or converge, the distance centre to centre constantly remains reducing and ultimately it becomes zero at the crossing. Now, suppose if a

Figure 29.12 *Fouling mark*

vehicle is standing at a place where the centre to centre distance is less than the minimum, it will strike the other vehicle passing on the next track. Therefore, at such places it becomes necessary to mark the place from where the distance becomes less. These marks are known as *Fouling Marks* and usually consist of stone or concrete blocks. To distinguish them in nights they are painted in black and white. The total number of wagons which can be accommodated in siding are generally marked on these fouling marks.

29.15 COW-CATCHER

It is the arrangement provided in front of railway engine to push a side the animals from the track. This is a part of the rolling stock. It is in the shape of a grill fitted at an inclination of 60°–70° with the horizontal, about 5–10 cm above the rail level.

REVIEW QUESTIONS

29.1. Write short notes on :
(i) Water-column (ii) Turn-table (iii) Triangle.

29.2. (a) Describe with the help of a suitable sketch the construction and working of a buffer-stop.
(b) Briefly describe the function of 'catch and slip siding'.

29.3. What are fouling marks in railways? Why they are provided? Give a neat sketch of a fouling mark.

29.4. Draw a neat cross-section of a platform and write short notes on it.

29.5. (a) Why ash-pits are provided? With the help of a neat sketch describe an ash-pit.
(b) Write short notes on 'Scotch Block' and 'Sand Hump'.

29.6. Write short note on weigh-bridge used in railways.

29.7. What are the functions of the end-loading ramp. Describe it with the help of neat sketch.

29.8. Write short note on the traverser.

29.9. Why cranes are required? Describe in brief the various types of cranes.

29.10. What factors should be kept in view while designing and planning the platforms?

Points and Crossings

GENERAL

These are the special arrangements which divert the trains from one track to another track. As the wheels of railway vehicles have flanges on their inner side which prevent the lateral movement of vehicles and allow them to move only in one defined way on the rails, special devices are required to divert from one track to another track. These special devices are known as *'points and crossings'*.

30.1 TECHNICAL TERMS

Following are the main important technical terms related with the points and crossings:

Check Rails. These are provided on the opposite side of the crossing for checking the tendency of the outer wheel to climb over the crossing.

Crossing Clearance. This is the clear distance between the wing rail and the crossing rail. Practically it is slightly more than what it comes through calculation.

Crossing Number. It is defined as the ratio of spread (i.e., the distance between the point and spice rails at the leg of crossing, generally 30 cm) to the length of the crossing measured from the theoretical nose of crossing.

Facing Direction. When the crossing is seen while standing at the toe of switch, the direction is known as facing-direction.

Facing Point. A point is called a facing point when a train runs in facing direction only. The wheels of the trains pass over the switch first before passing over the crossing.

Flangeway Clearance. The distance between the adjacent faces of the stock rail and the check or guard rail is known as flangeway clearance.

Flangeway Depth. The vertical distance between the top of the surface of the rail to the top surface of the blocks provided for keeping the flangeway.

Flare. It is the tapered widening of the flangeway which is formed by bending and splaying the end of the check rail or wing rail away from the gauge line.

Heel. Tapered rails where they are fixed to the main rails are known as heel.

Heel Divergence. It is the distance between the running faces of the stock rail and gauge faces of the tongue rail when measured at the heel of the switch.

Lead or Crossing Lead. The distance from the Heel of the Switch (H.S.) to the Theoretical Nose of Crossing (T.N.C.) is known as the lead. It should be measured in straight line.

Lead Rails. In the turnouts, lead rails are the length of rails from the heel of the tongue to the toe of the crossing.

Left Hand Turnout. It is the turnout which turns the train towards left hand direction.

Nosing Action. The striking of the wheel flanges at the nose of the crossing is known as nosing action.

Over-riding Switches. These are one of the types of the switches in which separate rail sections for stock rail and tongue rail are adopted. The light tongue rail rides over the foot of the heavy stock rail.

Stock Rail. The straight track rails along which tongue rail fits is known as stock rail.

Stretcher Bar. This is the bar connecting the toe of both the tongue rails and allows both the tongue rails to traverse the same distance at a time.

Studs. The bent plates fixed between the stock rails and tongue rails to prevent the lateral bending of the tongue rails. These are fitted with the web of stock rails.

Switch Angle. This is the angle between the running faces of stock rail and tongue rail.

Throw of Switch. It is the sideway distance moved by the toe of the tongue rail with the heel of tongue rail as the centre of rotation.

Toe. It is moveable end of the tapered tongue rail.

Tongue Rail. It is the tapered rail having toe at one end and heel at the other end, along which it can rotate.

Throat of Crossing. This is the place where the converging rails are at the nearest position.

Nose of Crossing. The point of intersection of the point rails and the splice rail.

Trailing Point. These are the points where train passes over the crossing first and then over the switches.

Slide Chairs. The plates used to support the tongue rails throughout their length and allow them to slide laterally while changing the points.

30.2 SWITCH OR POINT

The combination of a tongue rail and stock rail along with other necessary fittings are known as switches or points. Stock rail is a fixed rail against which the tongue rail fits and changes the train from one track to another. Switch mainly consists of a pair of stock rails, a pair of tongue rails, stretcher bars, heel blocks, slide chairs, switch tie plates and stops or studs.

Following three types of switches are used in railways. But only the first two are used on Indian Railways:

1. Stub switch.
2. Split switch.
3. Wharton's safety switch.

30.2.1 Stub Switches

Figure 30.1 shows this type of switch, in which main lines are securely fixed to the sleepers upto point *A* and *A*′. Between the point *AB* and *A*′*B*′, the

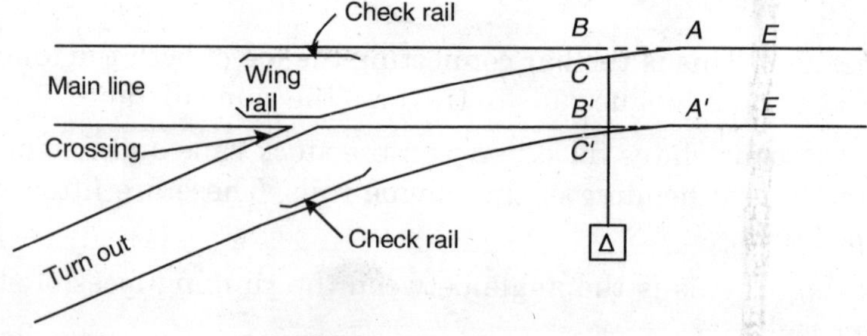

Figure 30.1 *Stub Switch*

main lines are placed on sliding chairs which can be operated by the lever *L*.

When the L is pulled laterally the rails stretch and bend in an elastic arc as shown in Fig. 30.1. The radius of the stretched rails is supposed to be equal to that of turnout, but actually it is not the same.

The stretched rails occupy the position shown in the figure and points BB' come at CC', and thus connecting the main lines with turnout. Now, when the rails are in this position, if any train comes it will move on the turn-out from the main line. At the point D where turnout rail crosses the main line, one crossing is introduced.

The lateral strength of the turnout mainly depends on the strength of the stretcher bar operated by the level. Therefore it cannot withstand high centrifugal force caused by heavy moving trains at high speeds. It is not used nowadays at any place. Of course at temporary works it can be used for diverting the trains during maintenance or construction of new railway track.

30.2.2 Spilt Switches

This is the standard switch which is mainly used on all modern railways. In the most common form of it, one turnout rail and one main line rail are made movable.

The given Figs. 30.2 (a) and (b) clearly show the working of a split switch. The switch rails AB and CD can swing about the point A and C and their *toes* are machine-tapered for some length, so that they can snugly fit against

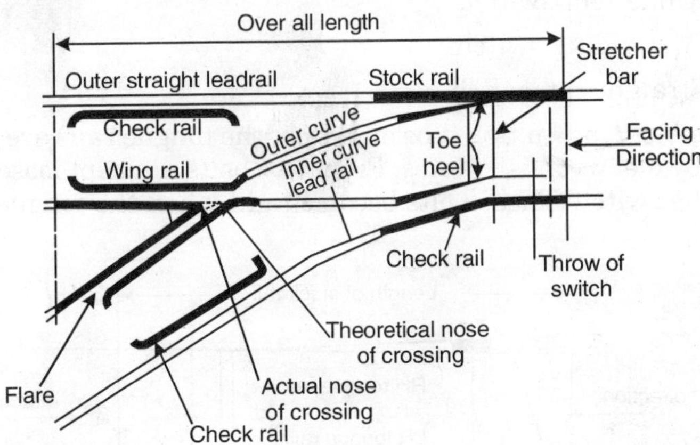

Figure 30.2 (*a*) *Split switch* (*points set for main line*)

the outer stock rails. The end of the rail about which switch rotates or swings is known as *heel* and the tapered end as *toe*.

The figures which are given here are self-explanatory. As both the outer rails are fixed and there is no movement, this switch gives more lateral rigidity to the turnout. Therefore, these switches can be used by the trains moving at high speeds.

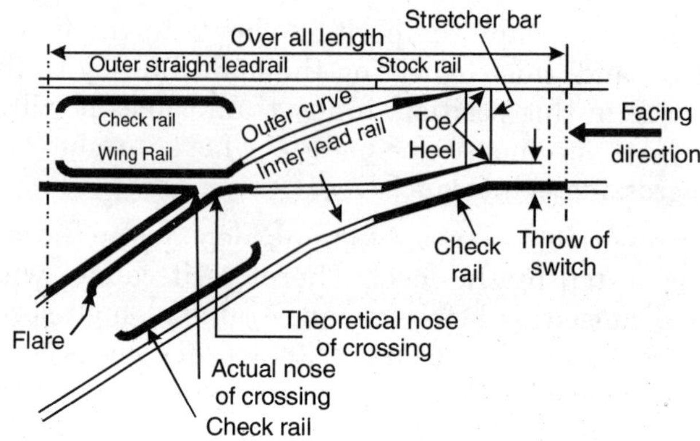

Figure 30.2 (*b*) *Split switch* (*points set for turnout*)

Split switches can be further classified as:

(a) On the basis of fixation of the heel:

(i) Loose heel type.

(ii) Fixed heel type.

(b) On the basis of cut of the switch :

(iii) Under-cut switch.

(iv) Over-riding switch.

(v) Straight cut switch.

(*i*) *Loose heel type.* In this type of switch, the tongue rails are joined to the load rails by means of fish plates. Front two bolts are kept loose to allow the throw of the switch. These bolts become tight when the tongue open.

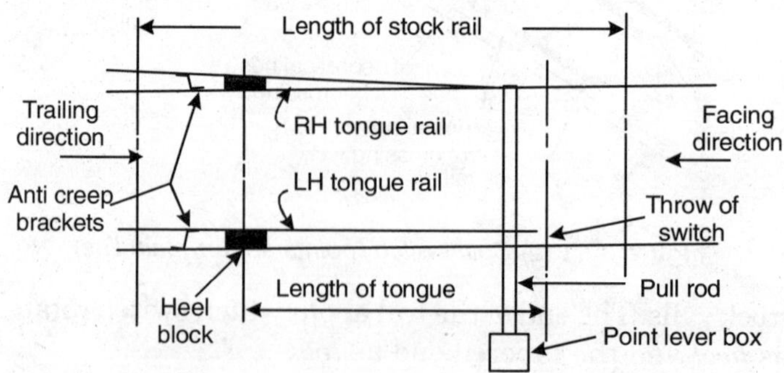

Figure 30.3 *Loose heel type switch*

(*ii*) *Fixed heel type.* This type of switches are in common use as these are the improvement over the loose heel type. All the bolts remain tight when the tongue is closed.

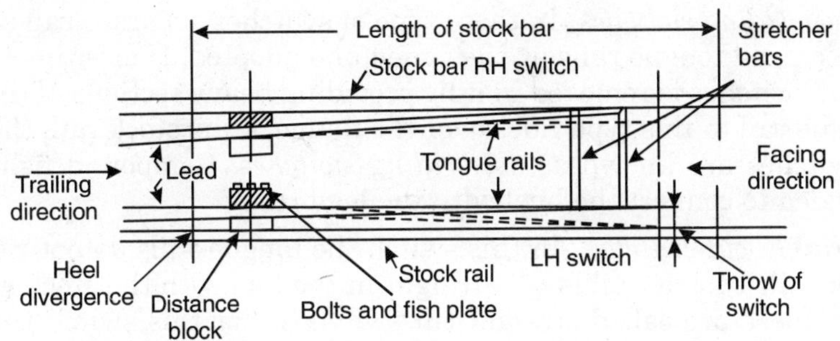

Figure 30.4 (*a*) *Fixed heel type switch*

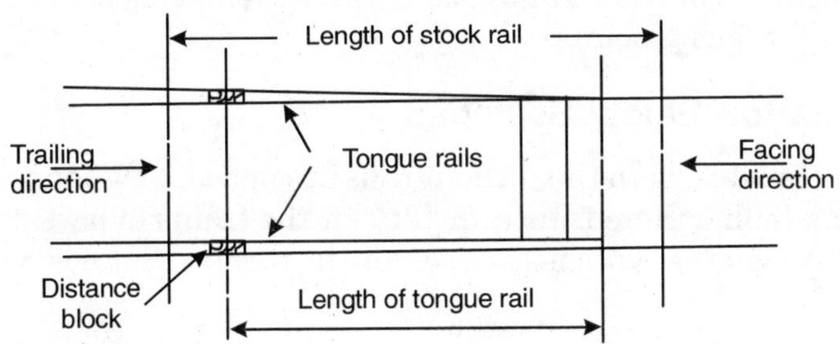

Figure 30.4 (*b*) *Line diagram of fixed heel type switch*

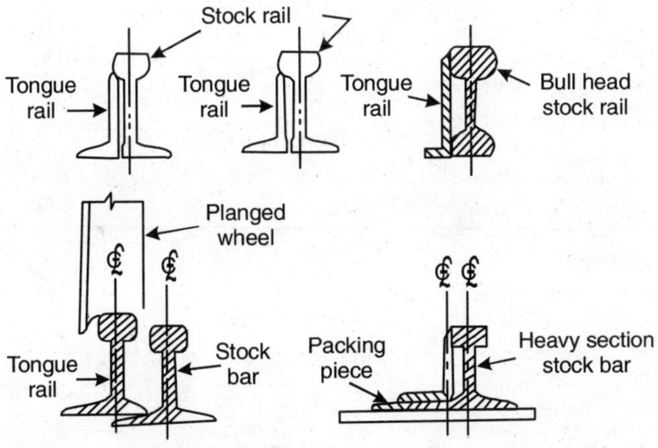

Figure 30.5 *Various types of split switch rails*

(*iii*) *Under-cut switch.* In this type the height of the top of the stock and tongue rail is the same. A portion of the foot of the stock rail is cut to receive the tongue rail. The main disadvantages of these types of switches are that due to cutting of the foot of main stock rails, they become weak.

(*iv*) *Over-riding switches.* In these type of switches separate rail sections of the stock and tongue rails are designed and adopted. Hence the defect of under-cut switches is removed in it by providing heavy sections of the rails. The tongue rail in this type rides over the flange of the stock rail, therefore these switches are known as *over-riding switches.* Compound fish-plates are provided to connect the heel with the lead rail.

(*v*) *Straight-cut switches.* In this switch the tongue rails do not ride over the stock rail. Tongue rail is cut straight in the line with the stock rail and therefore these are called straight-cut switches. For this switch the stock rail is jaggled near the switch toe during manufacturing to fit the toe flush in the stock rail. This type of switches are more suitable for B.H. rails. As the track gauge increases over the jaggled length of the stock rail, it is common practice to provide trailing-cut switches for facing points and under-cut switches for facing points.

30.2.3 Wharton Safety Switches

In this type of switch, both the turnout rails are movable. When the switches are set for main line there is no difficulty for the trains to pass as shown in Fig. 30.6 (a).When the switches are set for the turnout there is no clearance

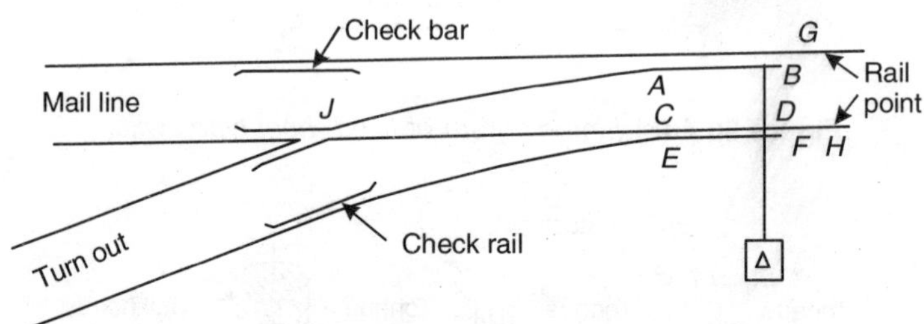

(a) Wharton Switch set for main line

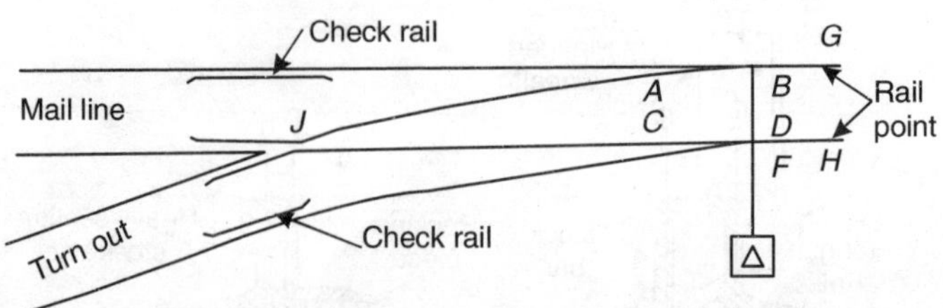

(b) Wharton Safety Switch (points set for turnout)

Figure 30.6

at the *toe* of inner rail of turnout. Therefore, in order to reach the turnout main line wheel flange has to cross the main line. If the portions of the rails *AB* and *EF* are ramped gradually and their level is raised upto the track level, the wheel will rise along the ramp provided and will cross the line. These switches require very careful attention and maintenance. These are not used in India.

30.3 CROSSINGS

It is a device which allows the crossing of rails with each other without causing any obstruction to the flange of wheel. The given Fig. 30.7 clearly shows 1 in $4\frac{1}{2}$ acute crossing with names of all its various parts.

A crossing essentially consists of the following :

(i) V-portion built up by one pair of rails by machining them.

(ii) One pair of wing-rails which check the lateral movement of the wheels.

(iii) Some special types of chairs.

On Indian Railways the common number of crossing are 1 in 6, 1 in $8\frac{1}{2}$ and 1 in 12. But in few cases 1 in 16 is also provided. At present most common are 1 in $8\frac{1}{2}$ and 1 in 12. As more the angle of crossing, lesser will be permissible speed and *vice versa.*

The acute angle and crossings are designated by either the angle which the gauge faces or makes with each other or generally by the crossing number

The number of crossing (N) is defined as the ratio of

$$\frac{\text{The speed at the leg of crossing}}{\text{The length of crossing from T.N.C}}$$

The crossing number are calculated by the *Right angle and Cole's Method.*

From Fig. 30.8,

$$\tan\alpha = \frac{1}{N}$$

$$N = \frac{1}{\tan\alpha} = \cot\alpha$$

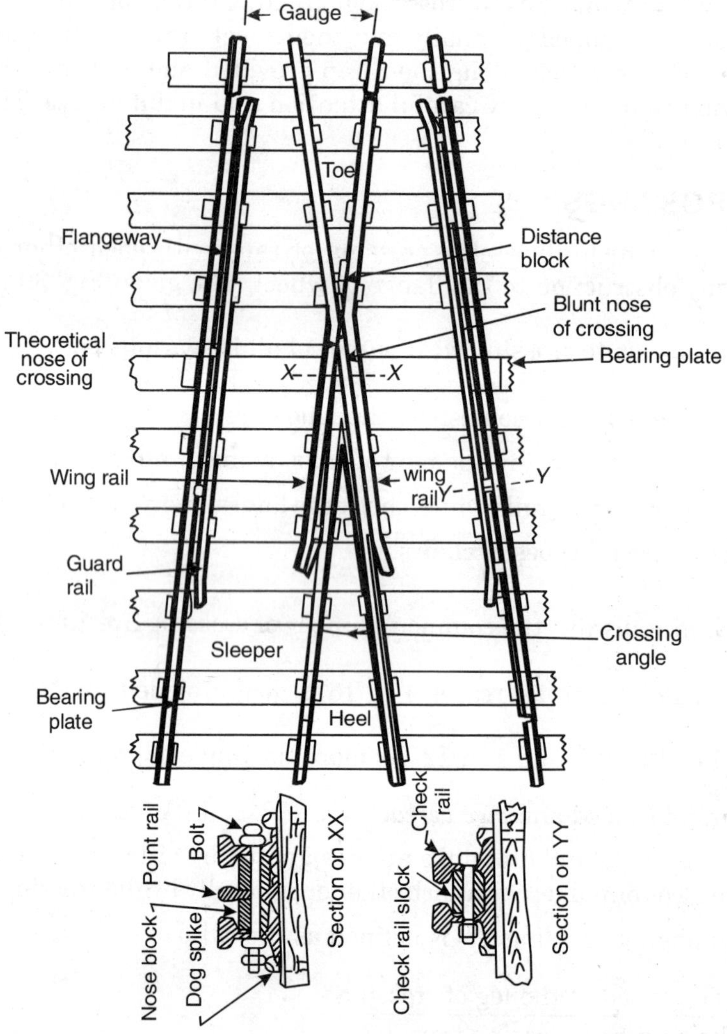

Figure 30.7 *Acute crossing of 1 in* $4\frac{1}{2}$

This is the standard method adopted on Indian Railways.

Following types of crossings are used.

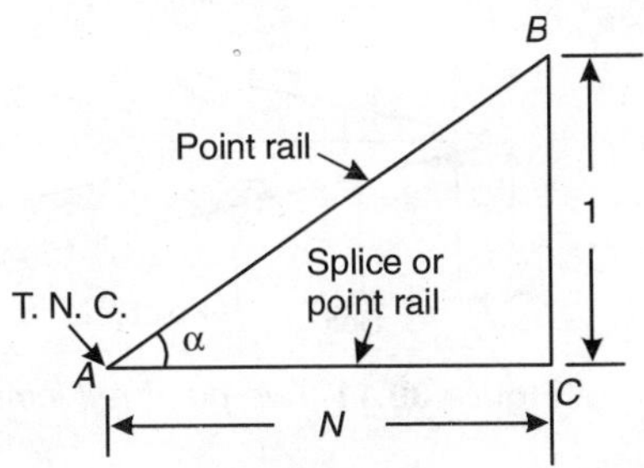

Figure 30.8

30.3.1 Acute Angle Crossing

This type of crossing is mostly used on Indian Railways. When a left-hand rail of one track crosses a right-hand rail of another track or *vice versa,* this crossing is obtained. If the angle of intersection of the approaching rails is acute angle, the crossing is known as acute angle crossing. This crossing mainly consists of point and splice rails, check rails and wing rails.

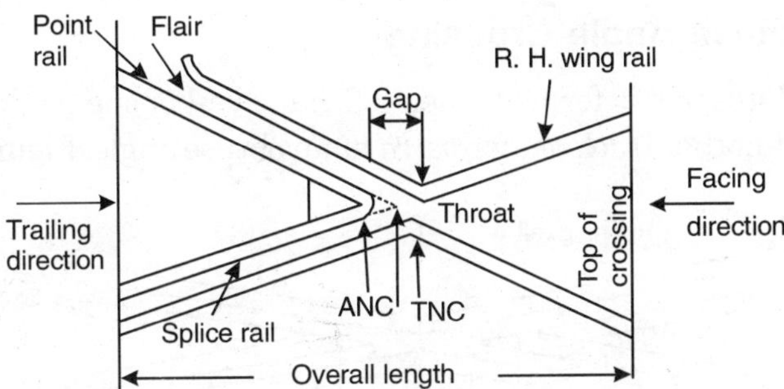

Figure 30.9 *Acute angle crossing*

Point and Splice rails. Figure 30.9 shows the acute angle crossing. The acute angle is formed either by a point rail and a splice rail or by two point rails.

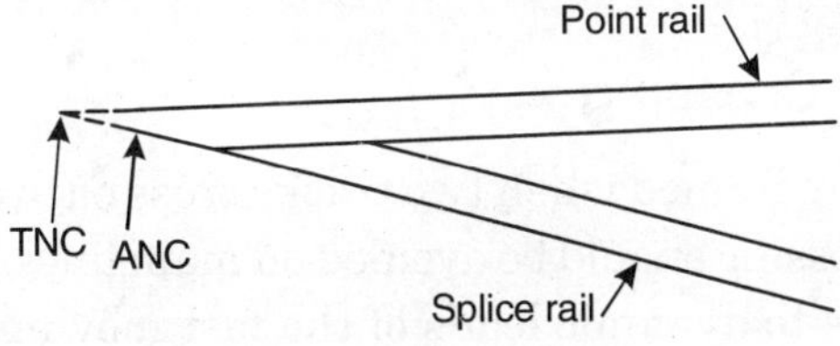

Figure 30.10 *Point and splice joint*

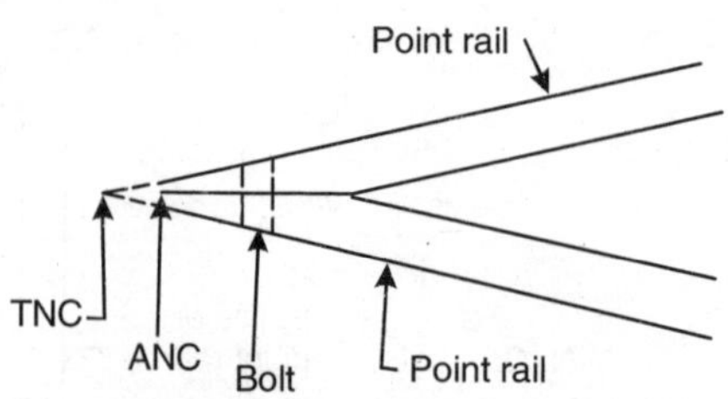

Figure 30.11 *Two point rail joint*

Pair of wing rails. These rails are bent at the ends. One end of the wing rail is connected to the lead rails, and the other end is flared. This flaring is necessary for entry and exit of the wheel flanges through the gap.

Pair of Check rails. These are parallel laid rails to the running rails. These check rails are also flared at their ends for guiding the wheel flanges while passing through the crossing. These are provided on the opposite sides of the crossing angle.

30.3.2 Obtuse Angle Crossing

This type of crossing is formed when left-hand rail of one track crosses right-hand rail of another track or *vice-versa* at an obtuse angle. Figure 30.12 shows

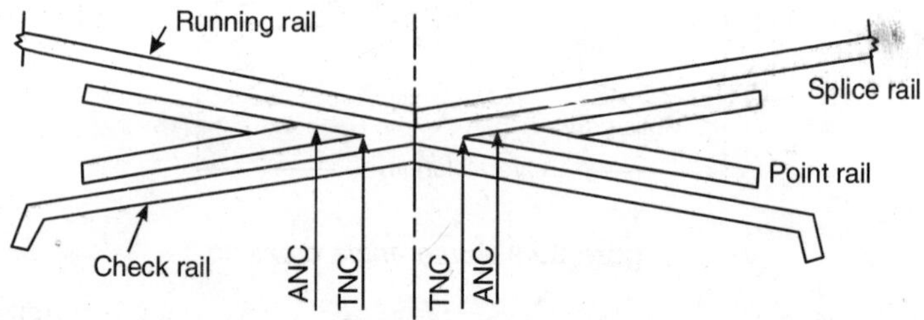

Figure 30.12 *Obtuse angle crossing*

this type of crossing. In this crossing the long wing rails do not carry the wheel as in case of acute angle crossing, but they act as check rails.

30.3.3 Square Crossing

Square crossing are formed when two tracks cross each other at right angles. But this type of crossing should be avoided on main lines, as they are subjected to heavy wear due to dynamic loads of the fast moving trains.

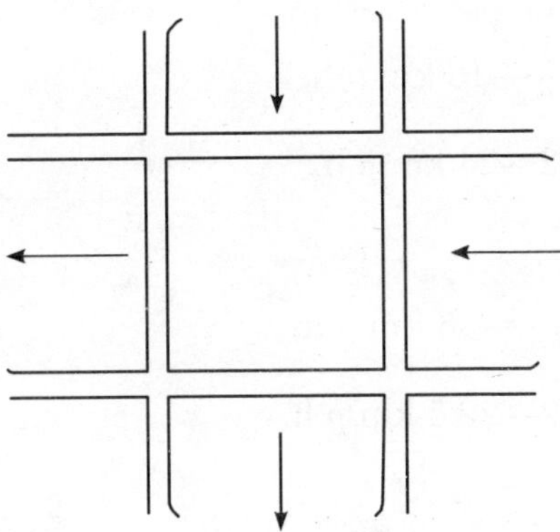

Figure 30.13 *Square Crossing*

30.4 REQUIREMENTS OF GOOD CROSSINGS

Following are the requirements of goods crossings :

1. Sole plate and turned bolts should be used for connecting the point and splice rails.

2. The foot-flanges of the wing rails should be riveted to the sole plate.

3. Wing rails opposite the nose of crossing should be protected.

4. The nose of the crossing should be of special steel, so that it wear the impact of the wheels.

5. The overall assembly of the crossing should be rigid and should withstand the severe vibrations and impact caused by the heavy fast moving trains. The components should not become loose under the worst possible conditions.

6. The length of the crossing should be as long as practicable, because short crossing has a tendency to rock at one end and lifting of the unloaded end.

7. From practical point of view the nose should have thickness, which may vary from 6 mm to 15 mm.

30.5 SPEED ON CROSSINGS

The following maximum permissible speeds have been recommended for the Indian Railways:

On B.G. Tracks

For crossing 1 in $8\frac{1}{2}$—16 km p.h.

For crossing 1 in 12—24 km p.h.

On M.G. Tracks

For crossing 1 in $8\frac{1}{2}$—16 km p.h.

For crossing 1 in 12—22.5 km p.h.

On N.G. Tracks

For crossing 1 in $8\frac{1}{2}$—16 km p.h.

For crossing 1 in 12—16 km p.h.

30.6 TURNOUT

Turn-out is a track assembly used for diverting the trains from one-track to another. The track from which train diverts is known as *main line* and the branch line is known as *turnout.* If it diverts the train towards right, it is known as *right-hand turnout* and if it diverts towards left, it is called *left-hand turnout.*

Every complete set of turnout essentially consists of the following :

1. One pair of switches with stretcher bars.
2. One complete crossing.
3. Tongue rails set.

30.7 COMBINATION OF POINTS AND CROSSINGS

Points and crossings are combined together to divert track in desired direction. The following are the main combinations:

1. Cross-over
2. Diamond crossing
3. Scissors crossing
4. Single slip
5. Double slip
6. Three-throws
7. Gauntlet track
8. Ladder track or gathering lines

30.8 CROSS-OVERS

When two tracks run at some distance parallel or at any angle, the train from one track can be diverted to other track by *cross-over.*

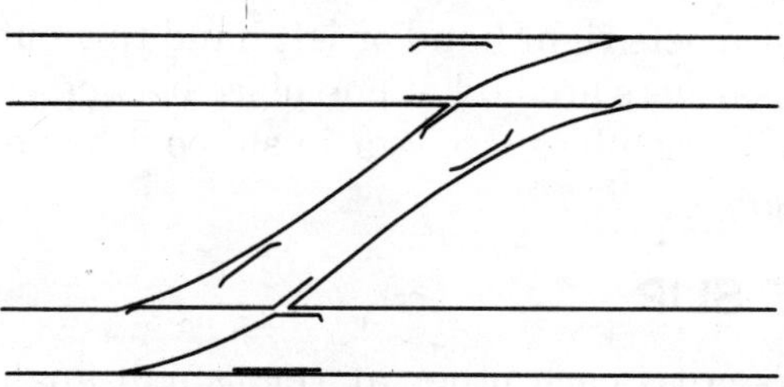

Figure 30.14 *Cross-over*

A cross-over requires two sets of switches and two crossings as shown in Fig. 30.14. In other words, it requires two turn-overs joined together by a straight length of the track.

30.9 DIAMOND CROSSING

When two tracks of same of different gauges cross each other at any angle, diamond crossing is provided. It essentially consists of four crossings. The length of the gap at crossing between two noses of a diamond crossing increases as the crossing angle decreases.

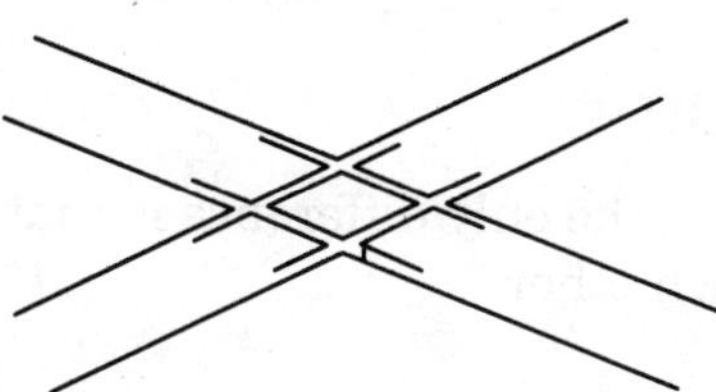

Figure 30.15 *Diamond crossing*

30.10 SCISSORS CROSSING

In cross-overs the trains moving on both the directions on one track cannot be diverted to other track. It can be diverted only when it is moving in such

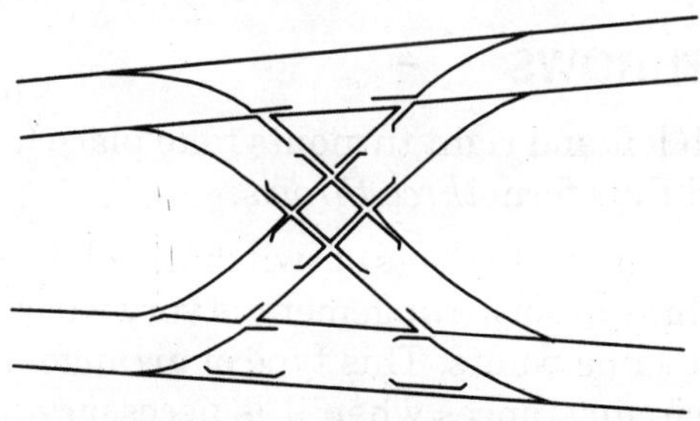

Figure 30.16 *Scissors cross-over*

a direction that it meets right hand or left hand turnouts. To overcome this difficulty two cross-overs are laid at one place between two tracks as shown in Fig. 30.15. As it resembles a scissor in shape it has been given the name as *Scissors crossing.*

30.11 SINGLE SLIP

When two tracks cross each other at very small angle, the traffic of one track can be diverted into another by providing slip as shown in Fig. 30.17. Actually at the time of crossing, the train slips into another track, therefore, it is called slip. When the train from only one track can slip into another track, it is called *Single slip.*

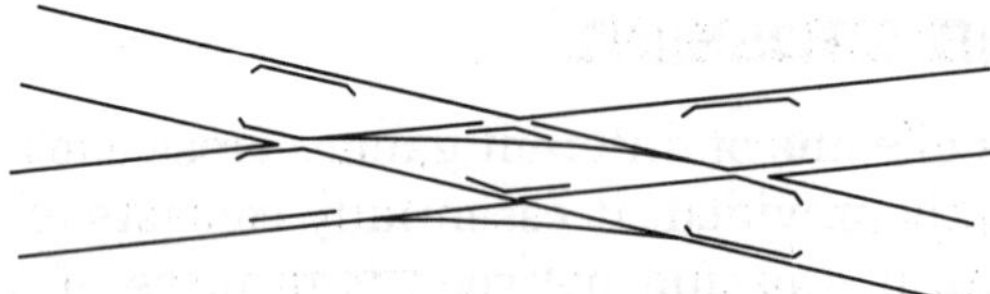

Figure 30.17 *Single slip*

30.12 DOUBLE SLIP

It is just like single slip, the only difference is that the trains from both the tracks can slip into each other.

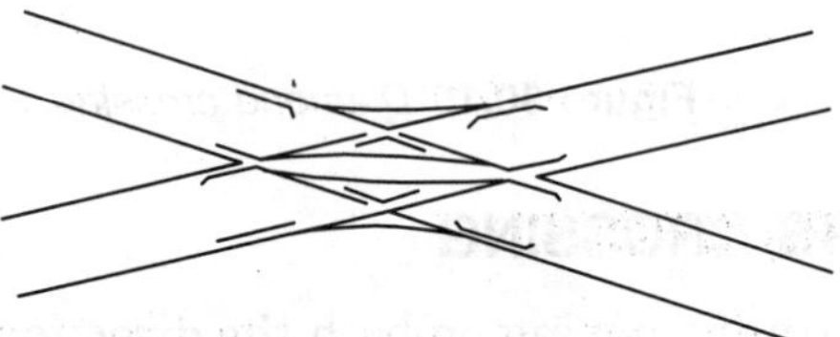

Figure 30.18 *Double slip*

30.13 THREE -THROWS

At some places both left and right turnouts take place from one main line at the same point and thus form *three-throws.*

One three-throw requires two sets of switches and three crossing as shown in Fig. 30.19. But this type of arrangement is very weak, because two sets of switches overlap at same points. This type of arrangement is generally not used. Under such circumstances when it is necessary to take both left and

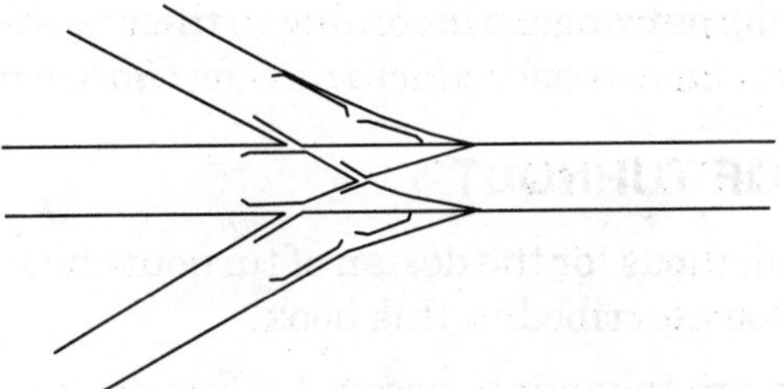

Figure 30.19 *Three-throws*

right turn-outs, these are staggered a little so that switches may not overlap and form the weak structure,

30.14 GAUNTLET TRACK

When two tracks of the same of different gauges are running parallel to each other, for economy purpose only single-line bridges are constructed and at the bridge only one track is laid. Therefore, this is a temporary deviation of a track. If the gauges of both the tracks are same it will require two sets of turnouts. When the gauges of both the tracks are different it will

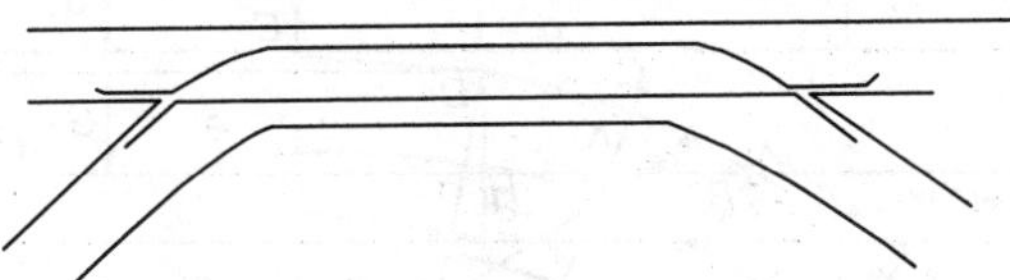

Figure 30.20 *Gauntlet Track*

require only four crossings, if the shorter gauge track is passed inside the wider with the same centre line. On the other hand if only three lines are laid over the bridge, it will require two crossings as shown in Fig. 30.20.

30.15 LADDER TRACK OF GATHERING LINES

In marshalling and station yards number of siding branches are taken out from one inclined or diagonal track. This type of track layout is very

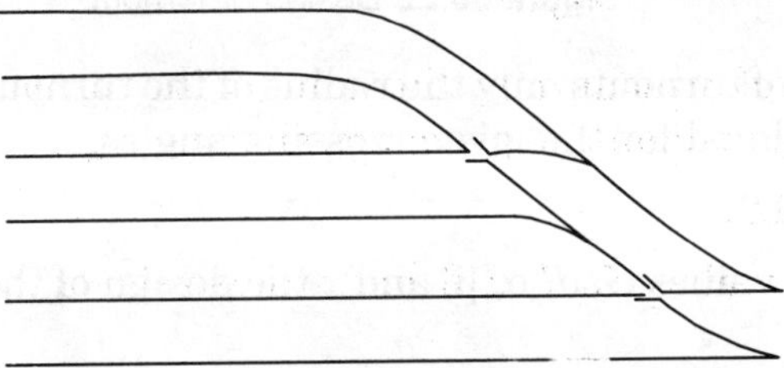

Figure 30.21 *Ladder track or gat...ng lines*

convenient for sorting out wagons according to their destinations. Such types of arrangements are known as *Ladder track* or *Gathering lines.*

30.16 DESIGN OF TURNOUT

There are various methods for the design of turnouts but the most commonly used method shall be described in this book.

Following points are taken into account while designing the turnouts:

1. The straight length at the crossing is provided.
2. One end of the curve is tangential to the tongue rail and springs from the heel of the switch and the other end springs from the toe of the crossing and is tangential to the straight length of the crossing.
3. There are no kinks at the toe of crossing and at the heel of the switch.
4. Only one kink exists at the toe of the switch.
5. Built up tongue rails and crossings of heavy rail sections are used.

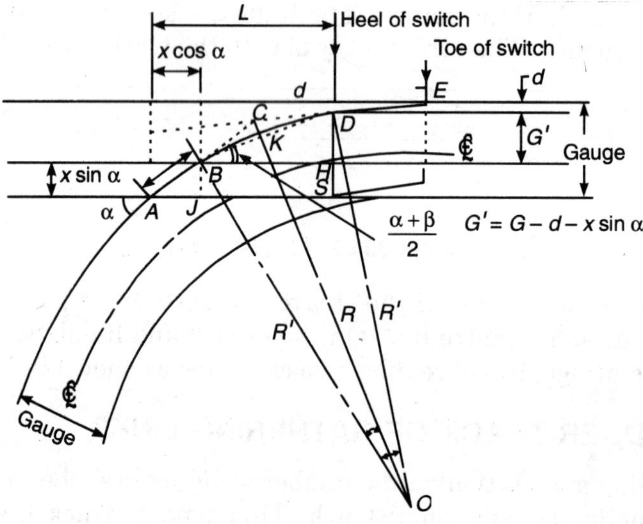

Figure 30.22 *Design of turnout*

In the design of turnouts only the radius of the turnout R and the crossing lead L are calculated for the given crossing angles.

Ref. to Fig. 30.22.

For the given values G, d, α, β, and x the design of the turnout is done as follows:

Calculation of radius R

$$\sin HBD = \frac{HD}{DB}$$

$$DB = HD \text{ cosec } HBD = HD \text{ cosec } . \frac{\alpha+\beta}{2}$$

$$\therefore \quad DK = BK = \frac{1}{2} DB = \frac{1}{2} HD \text{ cosec } . \frac{\alpha+\beta}{2} \qquad (1)$$

From triangle *OKB*, where *O* is the centre of the curve

$$\sin BOK = \frac{BK}{OB} = \frac{BK}{R'}$$

$$\therefore \quad R' = BK \text{ cosec } BOK$$

$$= BK \text{ cosec } \frac{\alpha-\beta}{2}$$

$$\therefore \quad R' = \frac{1}{2} DH \text{ cosec } \frac{\alpha+\beta}{2} \cdot \text{cosec } \frac{\alpha-\beta}{2}$$

[Putting value of *BK* from Eq. (1)]

$$= \frac{DH}{2\sin\frac{\alpha+\beta}{2} \cdot \sin\frac{\alpha-\beta}{2}}$$

$$= \frac{DH}{\cos\beta - \cos\alpha}$$

or

$$R' = \frac{G-d-x\sin\alpha}{\cos\beta - \cos\alpha}$$

[Putting $DH = (G - d - x \sin \alpha)$] (2)

$$\therefore \quad R = R' - \frac{G}{2} \qquad (3)$$

Calculation of crossing lead (L)

$$L = AJ + JS = AJ + BH$$

$$= x \cos\alpha + DH \cot \frac{\alpha+\beta}{2}$$

$$= x \cos \alpha + G' \cot \frac{\alpha + \beta}{2}$$

$$\therefore \qquad L = x \cos \alpha + (G - d - x \sin \alpha) \cot \frac{\alpha + \beta}{2} \qquad (4)$$

From Eq. (4) heel divergence

$$d = x \sin \alpha - G - \frac{L - x \cos \alpha}{\cot \frac{\alpha + \beta}{2}} \qquad (5)$$

Example. *Design the, turnout of a BG track having following data :*

Crossing angle = *6°42′40″*

Angle of switch = *1 °36′30″*

Length of switch rail = *4.75 m*

Heel divergence d = *11.43 cm*

Length of straight arm x = *0.86 m.*

Solution The following data have been given :

$$\alpha = 6°\ 42'40''$$

$$\beta = 1°\ 36'30''$$

$$G = 1.676 \text{ m}$$

$$d = 11.43 \text{ cm} = 0.1143 \text{ m}$$

$$x = 0.85 \text{ m}$$

$$\text{Value of radius } R' = \frac{G - d - x \sin \alpha}{\cos \beta - \cos \alpha} \text{ From Eq. (2)}$$

$$= \frac{1.676 - 0.1143 - 0.86 \times 0.1143}{0.9999 - 0.9932}$$

$$= \mathbf{218.16 \ \ m}$$

$$\therefore \qquad R = R^{\circ} - \frac{G}{2} = (218.16 - 0.838) \text{ m}$$

$$= \mathbf{217.32 \ m}$$

Length of crossing lead (L)

$$L = x \cos \alpha + (G - d - x \sin \alpha) \,.\, \cot \frac{\alpha + \beta}{2}$$

From Eq. (4)

$$= 0.85. \cos 6° \, 42'40'' + (1.676–0.1143 – 0.85 \sin 6° \, 42'40'') \times \cot 4°9'35''$$

$$= 0.85 \times 0.9932 + 1.4617 \times 13.76$$

$$= \mathbf{20.957 \ m.}$$

30.16 METHOD OF LAYING SLEEPERS BELOW POINTS AND CROSSINGS

On Indian Railways two types of methods are commonly used in laying the sleepers below points and crossings which are shown in Figs. 30.23 and 30.24.

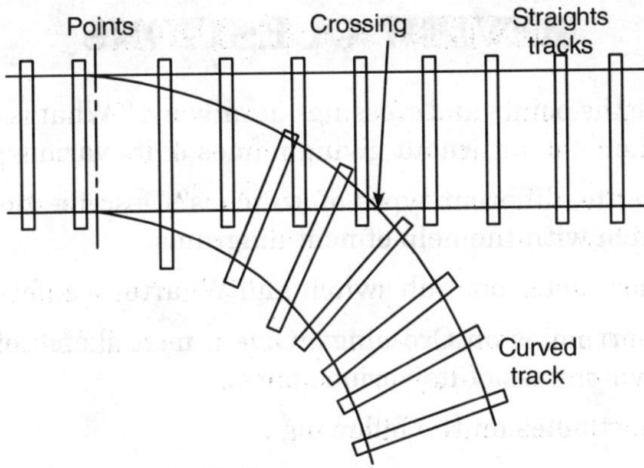

Figure 30.23 *Method of laying interlaced sleepers*

30.16.1 Through Sleepers

This method is the best method as it maintains both the tracks (straight and curved) at the same level. But there may be difficulties in the procurement of longer sleepers as well as in transporting.

30.16.2 Interlaced Sleepers

This method of providing sleepers under points and crossings is adopted only when longer sleepers are not available. But as both the tracks are laid on different sleepers, the curved track often deforms and causes difficulties in the maintenance. These sleepers also possess great difficulties in the proper packing of the ballast.

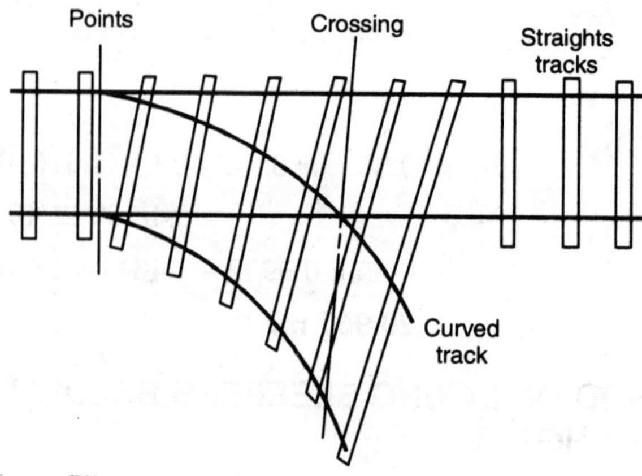

Figure 30.24 *Method of laying through sleepers*

REVIEW QUESTIONS

30.1. What are the points and crossings in railways? What is a turnout? Draw a neat sketch of a turnout giving names of its various parts.

30.2. What are the different types of switches? Describe the working of the split switch with the help of neat diagrams.

30.3. Write short notes on stub switch and Wharton's safety switches.

30.4. Write short notes on Crossings. Give a neat sketch of a crossing and write down names of its various parts.

30.5. Write short notes on the following :

(1) Cross-overs (2) Ladder track

(3) Three-throw (4) Slips

(5) Diamond crossing.

30.6. What are the function of points and crossings in the railway track?

30.7. Explain the working principle of the turnout.

30.8. What do you understand by the crossing? What are the main requirements of the good crossing?

Describe, various types of crossings in common use on Indian Railways.

30.9. Determine the crossing angles of 1 in $8\frac{1}{2}$ 1 in 12 and 1 in 16 number of crossings respectively.

[*Ans.* 6°45′35", 4°45′49" and 3° 34' 35"]

30.10. Determine the angle of switch and the theoretical length of the switch when the following data are given :

Thickness at the toe of tongue rail = 6.30 mm

Heel divergence = 13.3 cm

Actual length of tongue rail = 640 m. [*Ans.* $\alpha = 1° 8'$. 7.72 m]

30.11 Write short notes on :

(i) Speed on crossings

(ii) Requirements of good crossing.

31

Signalling and Interlocking

GENERAL

Signalling is the device by which movement of trains is controlled. This is most urgent from the safety point of view of the trains. In the beginning policemen were employed for clearing the track and regulating the movement of trains, but after it, signals were used for this purpose. Through the signalling the trains are operated efficiently, utilization of the tracks and trains is the maximum, the safety of the passengers, staff and the trains is maintained. Signalling includes the signals, points, block instruments and other necessary equipments.

31.1 OBJECT OF SIGNALLING

Following are the main objects of signalling :

1. To maintain a safe distance between trains running in the same direction on one line or crossing or approaching each other's path. Thus, if the forward train has been stopped due to any reason, the train which is following it should also be slopped to avoid accident.

2. In the case of repair works of track, to run the trains at restricted speeds.

3. At the junctions where tracks are converging, to protect the trains against collision and derailing.

4. At the diverging places, to give indication in which direction trains have to move.

5. In marshalling yards, to provide safety to shunting operation.

6. To provide facilities for the maximum use of the track and the rolling stock.

7. To facilitate the safe and efficient shunting operation in the yards.

31.2 CLASSIFICATION OF SIGNALS

Various types of signals have been classified as follows :

(A) *Depending on the operational characteristics*

1. Fog or Audible signals or detonating signals.
2. Visual indication hand signals.
3. Visual indication fixed signals.

(B) *Depending on the functional characteristics*

1. Stop or semaphore type signals.
2. Warner signals.
3. Shunting signals (Disc or ground signals).
4. Coloured light signals.

(C) *Depending on the location characteristics*

1. Outer receptional signals.
2. Home receptional signals.
3. Starter signals.
4. Advance starter signals.

(D) *Special Signals*

1. Routing signals.
2. Calling-on signals.
3. Point indicators.
4. Repeater or co-acting signals.
5. Modified lower quadrant semaphore signals.
6. Miscellaneous signals.

31.2.1 Detonating Signals

During the foggy or cloudy weather when hand or fixed signals are not visible, detonating signals are used. These are in the form of detonators which are fixed on the top of the rails, when the engine passes over it, these are exploded with loud sound, which draw the attention of the driver to stop the train. These signal are also used in case of emergency when the track is blocked due to derailment of other trains or other reasons. Generally the detonators are placed 400–500 metres ahead. 3–4 detonator signals are fixed at intervals of about 10–15 metres. Such type of signals are kept with the level crossing guards and all other signal controlling cabins.

31.2.2 Hand Signals

These signals are given either by coloured flags (red and green) or by bare arms (when flags are not available), during the day time. During nights kerosene lamps fitted with movable green, red and yellow coloured glasses are used. These signals are given by railway guards, small station masters, cabin-man, gang-man, key-man, points-man or any other authorised man.

Green signal indicates — proceed

Red signal indicates — dead stop

Yellow signal indicates — proceed with caution.

31.3 SIGNALLING SYSTEMS

Following are the main signalling systems :

1. Absolute block system.
2. Space interval system.
3. Time interval system.
4. Pilot guard system.

31.3.1 Absolute Block System

This system of signalling is mostly used in India. In this system, the railway line is divided into block sections. Telegraph plays a very important part in this system. Under this system only one train can move at a time in one block section. Trains are protected throughout their journey. Mostly the distance between two successive stations forms one block section.

For understanding this system, take three successive stations *P*, *Q* and *S*. The distance between *PQ* and *QS* forms one block. Now suppose any train is reaching station *P*, the information of which has been already received by the man at *P*. Now first signalman at *P* gives 'BE READY' signal to station *Q* and also indicates the nature of the approaching train such as goods, mail, passenger etc. with its number. Now the signalman at *Q*, if the previous train has passed and the section between *P* and *Q* is clear, repeats the same signal to the station S. The signalman at S, if has not already received any intimation from next station about the approach of train towards him from other side, answers telegraphically 'YES NO OBJECTION' to the station Q. Then the man at Q gives a 'LINE CLEAR' signal to P. Now only after receiving line clear signal from Q, signaller at P lowers his signal and allows the train to enter block section PQ.

Actually the lowering of signal is possible only after putting the token in the lock instrument. At each block station two block instruments are installed. One instrument is interconnected by electric communication to a station behind and the second with the station ahead. Now when the signaller at Q gives line-clear signal to P, he has to operate a lever of his block instrument which is in connection with block instrument of P. When the man at Q operates his lever and gives signal, it is received by the man at P through his block-instrument. When train reaches station P, the guard gives a token to the signaller, which he puts in the block-instrument and then can remove one other token from his block-instrument and hand it over to the guard as an authority to move in the block section PQ. Signaller at Q then confirms that train enters block PQ and informs S and similarly S warns the next station. Between this time train reaches station Q and the same processes continues.

In this way we see that every train guard has one token with him which he hands over to the signaller on reaching the station. The signaller puts it in his block-instrument. Now his block-instrument will release such another taken, only if it has already received line clear signal from the next station. Signaller hands over the released token to the guard and he moves. In this way the train can move safely in block sections.

31.3.2 Space Interval System

This system is used on single-line working. In this system between two trains there is always one clear section. To understand it consider successive stations P, Q, R and S, PQ, QR and RS groups form one section. Between these sections one section should always be clear. If the train is between RS the next train following will be between PQ stations and QR section will be clear or it will have no train. Under any circumstances two trains cannot occupy successive sections.

31.3.3 Time Interval System

This system may be used in emergency when the block system is not in working order. In this system only when telegraphic message is received from the next station that line between those stations is clear the train is allowed to leave for the next station. If two trains are moving in the same direction, the follower train will be allowed to leave for next station when the forward train has left that station.

31.3.4 Pilot Guard System

This system was used in early days of railways but is used nowadays even in the case of failure of telephone and telegraph system. When the message from one station cannot be conveyed to the next station by telephone or telegraph this system is adopted. In this system one Pilot guard goes with one train to the next station and comes back with another train and goes again with another train in the same direction. In this way the train is allowed to leave the station only when Pilot guard comes with another train which confirms the line clear.

31.4 REQUIREMENTS OF SIGNALLING

1. While designing the signalling system, it should be simple in design, operation but must be reliable and economical.

2. All the signalling systems should be such that if any defect has come in its any component, it should not work nor give incorrect signals in any case. Alternative techniques to check the failure or faults of signals should also be provided.

3. Same signals should indicate the same indication every where on all the routes under all climatic conditions.

4. Signals should be such that they must be visible from long distances during day as well as nights.

5. The number of signals should be as minimum as possible.

6. The indications of the signals should be simple and easy in implementation.

31.5 SEMAPHORE SIGNALS

This signal consists of a vertical post on which a movable arm is pivoted at the top. This can be kept horizontal or inclined making an angle of 45° with the horizontal. The outer end of the arm is about 2.45 cm broader than that at the post. Figure 31.1 clearly shows a semaphore signal with all its components. The movement of the arm is controlled by levers from the cabins. Spectacles of red and green colour are fixed in the arm as shown in Fig. 31.1. In the back of the spectacles one lamp is fixed. The moving arm of the signal is pivoted on the horizontal pin which is known as spindle near the top of the post. The length of the arm is from 1.2 m to 1.7 m and 23 cm to 25 cm wide at the inner edge. At the outer edge the arm is 25 cm to 35 cm wide. The height of the centre of the arm is about 7.47 m above the ground level. The semaphore signals are fixed on the left hand side of track with spectacles towards driver. On Indian railways one universal rule has been fixed that horizontal arm indicates "DANGER-STOP" and the inclined arm indicates "CLEAR-PROCEED". In the day time the position of arm indicates the signal.

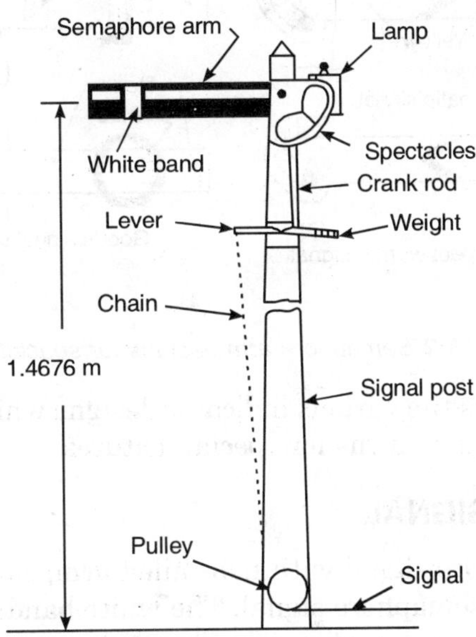

Figure 31.1 *A semaphore signal with all its parts*

In the night time the light of the lamp after passing through the spectacle gives signal. When the arm is in horizontal position, red spectacle remains before the lamp and when the arm is inclined, green spectacle remains before the lamp. Thus red light indicates danger and green light indicates line-clear. Also in the night the back side of the signal gives white light when the front light is red and no light when the front light is green. The left hand end of the movable arm is painted with thick vertical band surrounded by thin vertical red lines on both sides, so that it can be seen easily.

For moving or operating the signal arm lever are provided. At one end of the lever weight is attached and the crank rod is connected to the lever through a cam. The other end of the lever is connected with the wires which are controlled from the cabin passing over the rows of pulleys.

As the height of the moving arm is more, a ladder is provided along the vertical post to fix the lamps in the nights and for the repairing purposes.

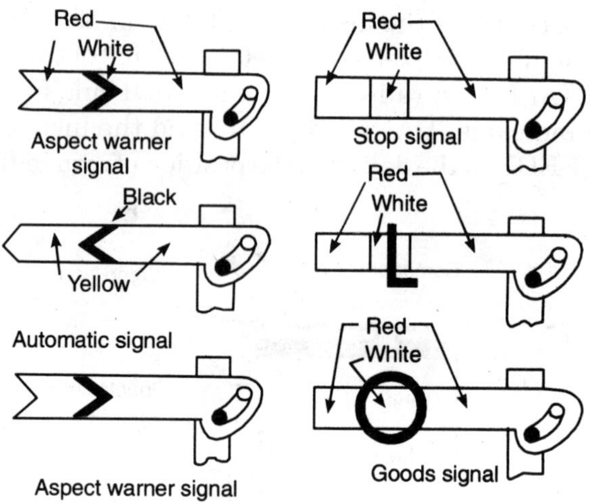

Figure 31.2 *Semaphore arm designs for special features*

Figure 31.2 shows the various indicator designs which are generally put on the semaphore signal arms for special features.

31.6 WARNER SIGNAL

This is like semaphore signal with fish-tailed arm, so that it can be easily distinguished from semaphore signal. The white band at the end is also V-shaped similar to the end as shown in Fig. 31.3.

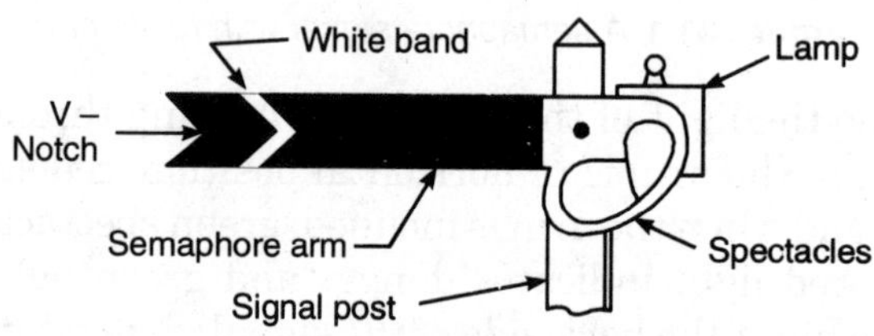

Figure 31.3 *Warner signal*

These signals are placed ahead than the semaphore signals to warn the driver before entering a railway station, therefore, these are called warner-signals. When warner signal remains horizontal, it indicates that the signal ahead is in stop position. Thus it gives warning about the stop signal, whether it is in danger-position, or line-clear position. If the warner signal is in horizontal position, the driver can take his train very cautiously upto the semaphore signal. Generally warner signals are placed 540 metres away from the first stop signal, where the line is approximately flat. At some places in India, warner signals are provided with yellow lights instead of red lights to distinguish them from semaphore signals during nights.

Sometimes warner and semaphore signals are placed on the same pole, the semaphore arm being at top and warner arm about 2 m below it. When both the signal arms are horizontal, it indicates *'stop, line not clear'*. When semaphore is lowered and warner is horizontal, it indicates *'proceed with caution,* i.e., section upto station is clear but next section is not clear. When both the arms are lowered down, they indicate 'proceed on with confidence'

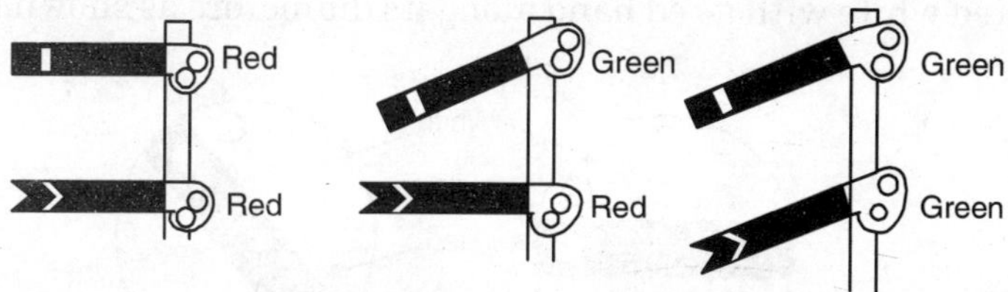

Figure 31.4 *Warner and semaphore signals*

i.e., this section and next section both are clear. In this way these signals tell the driver that he has to stop at this station or not.

31.7 REPEATER OR CO-ACTING SIGNALS

When the mail or express trains pass through small stations without stopping, the trains have to pass through five signals in sequence, WARNER, OUTER, HOME, STARTER and ADVANCE STARTER. If the track is throughout straight, these are visible sufficient before passing them to the driver. But if the vision is obstructed by an over-bridge, sharp curves etc., it

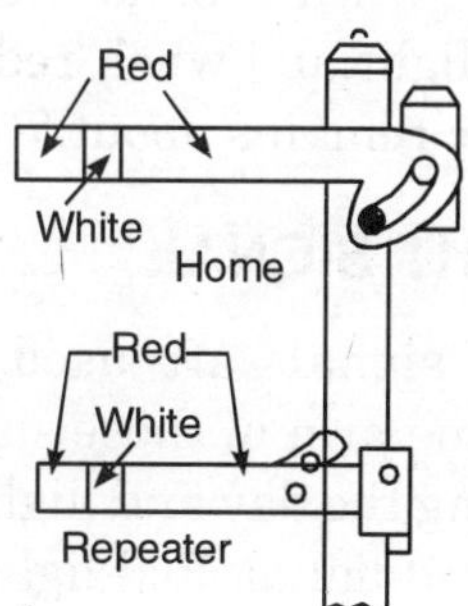

Figure 31.5 *Repeater or co-acting signal*

will not be possible for the driver to see signals on the other side of the stations. In such cases duplicate signals having arm of smaller size are provided as duplicate signals on the previous visible signal. As these signals are duplicate of the signal ahead and repeat the indication of the signal ahead these, are called *repeater or co-acting signals.* These signals are linked with the main signal and when the lever of the main signal is pulled for lowering them, these also get lowered alongwith the main signals.

31.8 DISC SIGNALS

These are also known as *shunting signals* or *miniature semaphore signals* and are used during shunting operations. These signals consist of circular disc painted white with a red band along its diameters as shown in Fig. 31.6.

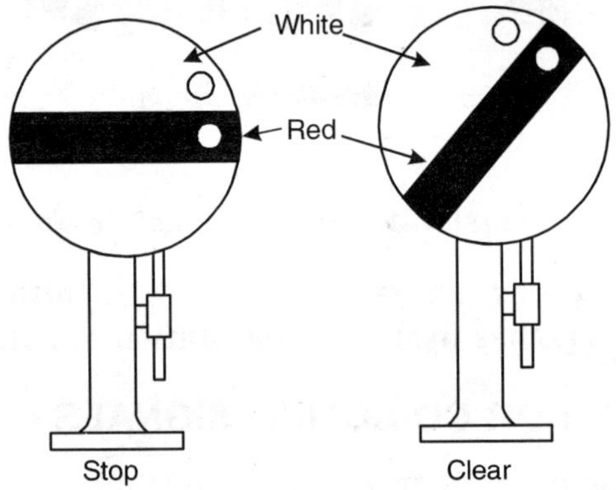

Figure 31.6 *Disc signal for shunting operations*

When red band is horizontal it indicates 'stop' and when inclined it indicates 'proceed'. Similar to semaphore signals these are provided with lamp and coloured, glasses. In the night when band is horizontal, lamp is in the back of red glass and gives red light and when red band is inclined it gives free light. The centre of the disc remains about 70 cm from the ground level.

31.9 COLOURED LIGHT SIGNAL

Nowadays these types of signals are used for automatic signalling on railways. There is no moving arm in these signals and they give indication by electric lights both during the day and night. Generally these signals are provided with three coloured lights. Red light indicates 'Stop', yellow light indicates 'Proceed with caution' and green light indicates 'Proceed with confidence'. These signals are provided with lenses and hoods to emit a beam of light which can be visible from long distance even during day time.

Semaphore type signals always indicate 'Stop' and these are lowered only when train comes. But the in automatic coloured signalling is reverse, this type of signal always gives green colour but gives red colour only when that section has been blocked by the entrance of any train. When one train

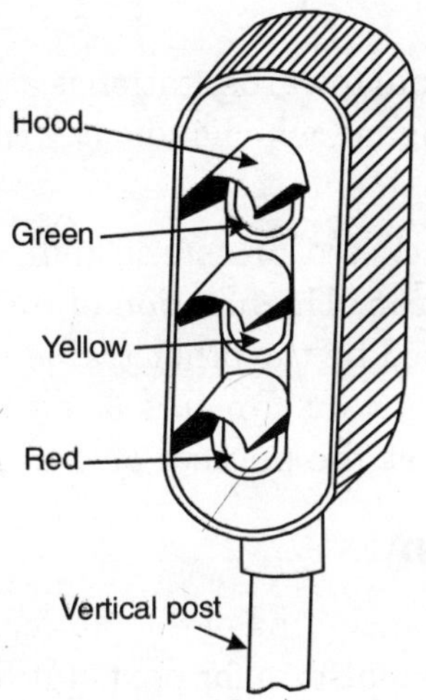

Figure 31.7 *Coloured signals*

enters a block its signal automatically shows red and when the train leaves that section, it automatically shows the green light.

31.10 OUTER SIGNAL

This is a warner signal which is first seen by the driver. When trains are moving at high speed, they require certain distance for bringing them to halt. This distance depends on the speed of the train, its brake power, gradient

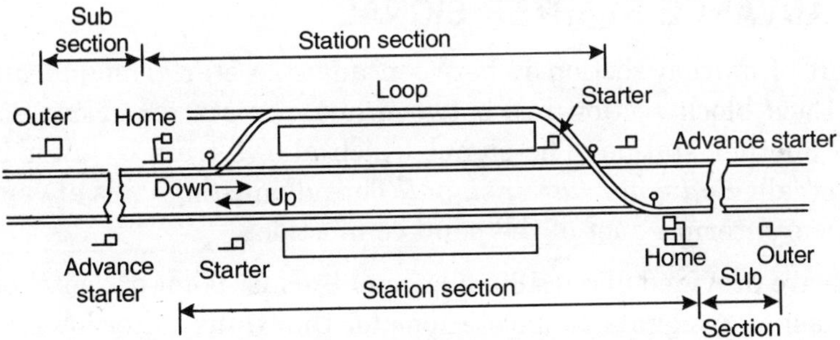

Figure 31.8 *Plan showing position of different types of signals*

and the weight. Therefore the driver should be informed about the position in advance that platforms are clear or not. This signal gives the position of the stop signal ahead. As it is provided at some distance away from the station it is called *distant* or outer signal *and warner signal.*

31.11 HOME SIGNAL

Next signal after outer signal towards station is a stop signal or home signal. It is a simple semaphore signal, which indicates that the platform is clear or not.

These signals have bracketed arms and indicate the line on which the train has to enter the station. The function of the home signal is to protect the sidings already occupied by the other trains. This signal is provided at about 180 metres from the start of points or switches. These signals carry the same number of arms as the number of the diverging tracks.

31.12 STARTER SIGNAL

This is a top signal placed at the end of the platform. It gives the permission to the train for leaving the platform for next station. Any train cannot move until this signal is lowered. This signal indicates to start the train, therefore it is known as *starter signal.*

These signals also mark the limit upto which the trains can stop. Whereas home signals carry as many arms as the number of diverging tracks, the starter signals are provided near each platform. If advance starter has not been provided then in such cases the starter signals are provided outside all the connections of the line. In this way these signals control the movement of the trains. Without getting 'proceed' indicator no train can move from the starter.

31.13 ADVANCE STARTER SIGNAL

The limit of station section is between advance starter signal and home signal. Each block section lies between advance starter of one station to outer of another station. The signal which allows train to enter in block-section is called *advance starter signal.* The advance signal is always placed beyond the outermost set of the point connections.

These are provided about 180 m beyond the last point or switches. These are the last stop signals at the stations for the trains. After lowering these signals the station master conveys the message to the next station that the train has left his station. It is the responsibility of the guard and driver to take their train after getting permission.

31.14 ROUTING SIGNALS

When from one main-line different branches diverge in various directions, it is very difficult to provide one signal for each line at diverging point. At such places various signals for main and branch lines are fixed on the same post. These types of signals are known as *routing signals.* Generally the main

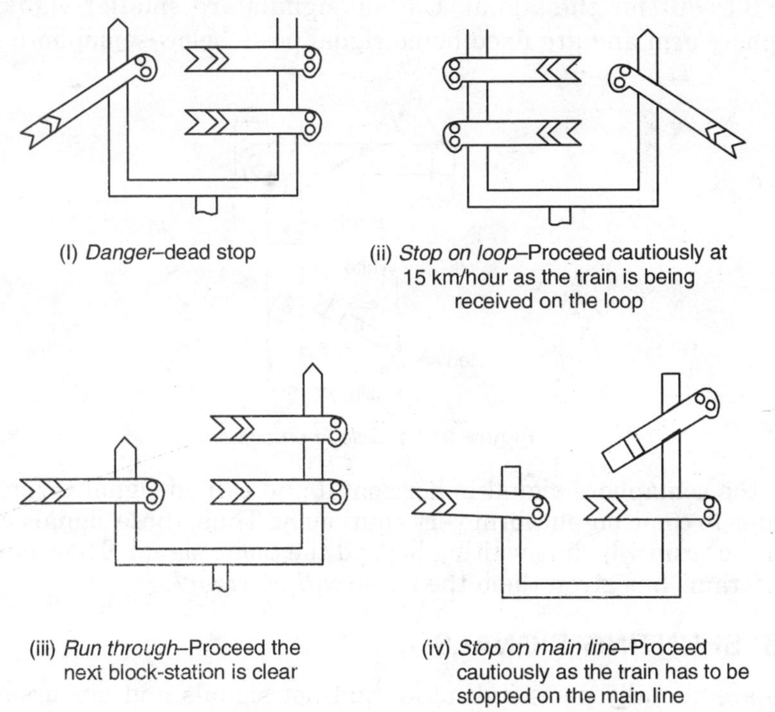

(I) *Danger*–dead stop

(ii) *Stop on loop*–Proceed cautiously at 15 km/hour as the train is being received on the loop

(iii) *Run through*–Proceed the next block-station is clear

(iv) *Stop on main line*–Proceed cautiously as the train has to be stopped on the main line

Figure 31.9 *Home and warner signals combined showing various indications*

line signal is kept at higher level than that for branch lines. Figure 31.10 shows routing signals for some of the branch lines taking off from the main

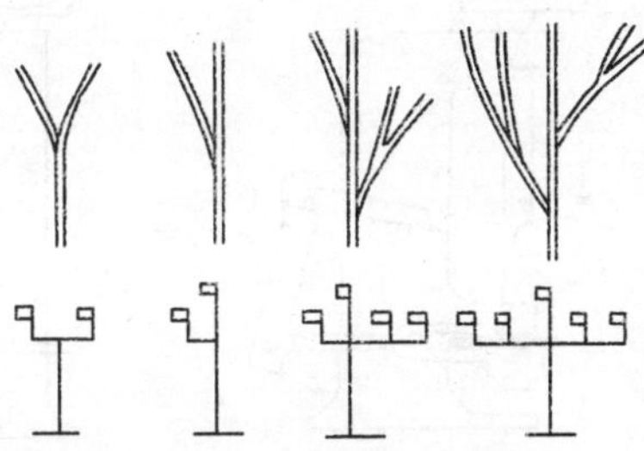

Figure 31.10 *Routing signals*

lines. These signals are placed in order of tracks from left to right and clearly indicate the driver on which line he has to go.

31.15 CALL-ON-SIGNAL

When the train reaches a home signal which is in horizontal position, it will stop and wait for the signal. Call-on-signals are smaller signals then semaphore arm and are fixed home signal posts below semaphore signals.

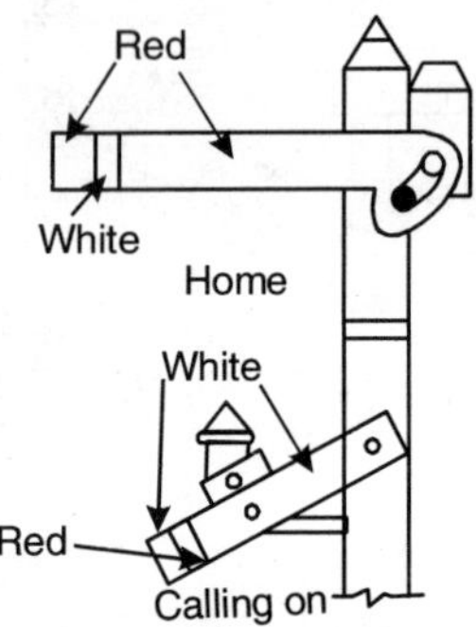

Figure 31.11 *Calling on signal*

When the semaphore signal is horizontal and call-on-signal is inclined, it indicates to come on platform very cautiously. Thus, these signals are used to call the train which is waiting beyond the home signal. Their purpose of calling trains has given them the name *call-on-signal.*

31.16 SHUNTING SIGNALS

These are actually point indicators and not signals and are used during shunting operation to indicate whether points are set for the main lines or

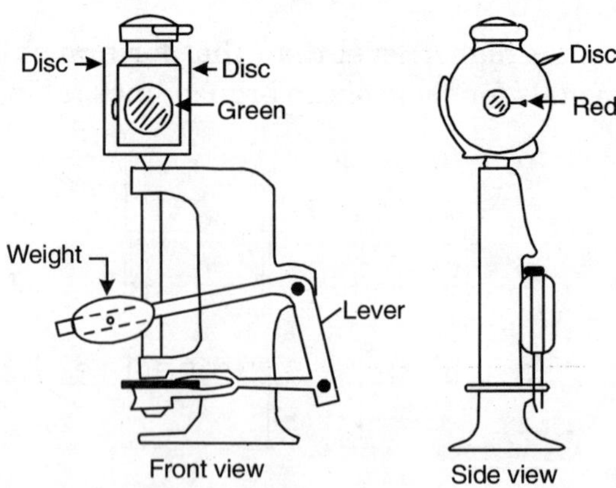

Figure 31.12 *Shunting signals*

not. It essentially consists of a box without a lid. Two circular discs are fixed on the opposite of the box and are painted white. The remaining sides are painted green. This box can be rotated about a vertical axis with the help of a lever. One lamp is fixed inside the box. Four glasses are fixed on the box one on each side. When the points are set for the main track the indicator shows white disc from either side in the day time and white lights in the night. When the box is rotated, it shows green colour from either side in day time and green light in night. The points are interconnected with these point indicators and when points are set these are automatically operated with the help of levers.

31.17 MISCELLANEOUS SIGNALS

In addition to all the above signals, various other types of signals are also required to control the movement of trains such as during relaying of tracks, repairing of bridges and culverts during accidents, etc. For controlling the speed of the trains sometimes boards are fixed in the way on which maximum limit of speed is written. In marshalling yards some special signals are provided for goods sidings. On the boards rings are provided of the capital letter "S" denoting siding is written.

Following types of miscellaneous signals are used in the Indian Railways :

31.17.1 Caution Indicator (signal)

During repairing works of the track the drivers are given caution indicator to reduce the speed of the train and be prepared to stop at any moment. Figure 31.13 shows the caution indicator board which is provided at sufficient distance from the actual repair section.

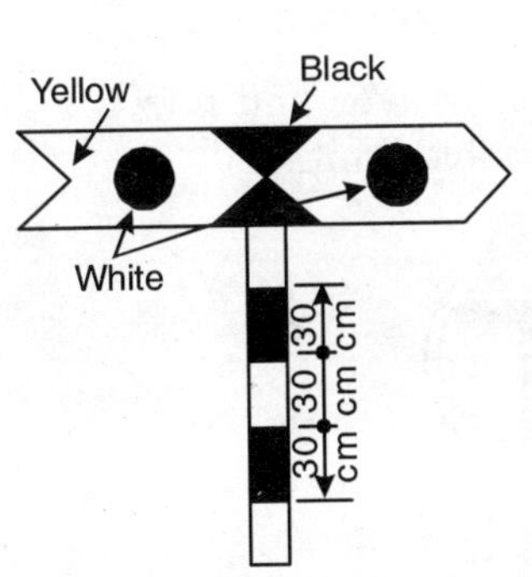

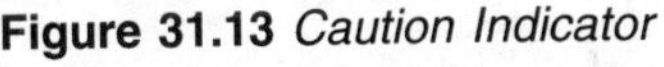
Figure 31.13 *Caution Indicator*

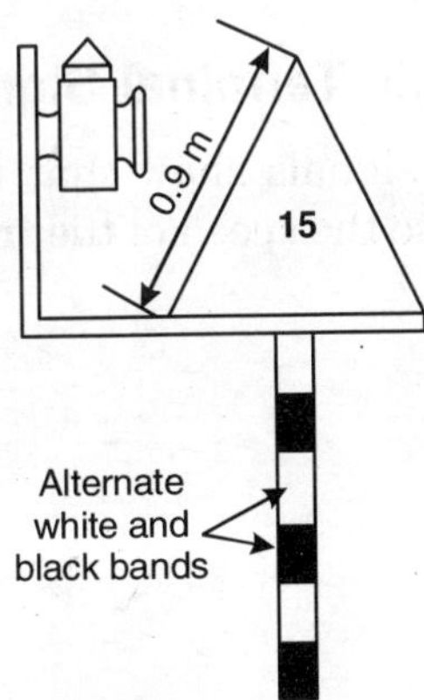

Figure 31.14 *Speed Indicator*

31.17.2 Speed Indicator (signal)

Figure 31.14 shows this type of indicator board which is fixed after the caution indicators. The maximum permissible speed in the section of the track is given on the indicator board.

31.17.3 Stop Indicator (signal)

If the train is required to stop before passing through the section under repair, this type of signal board is provided on the left hand side of the track. Figure 31.15 shows this type of signal.

31.17.4 Shunting Indicator (signal)

These signals denote the limit of the shunting. No vehicle should be taken above this limit during shunting operation. Figure 31.16 shows this type of signal.

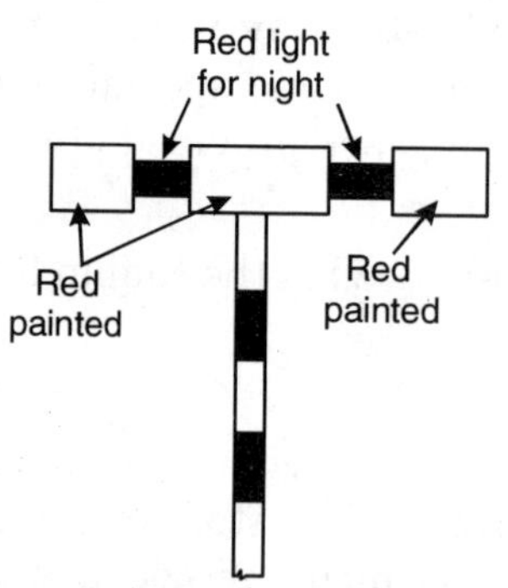

Figure 31.15 *Stop Indicator signal*

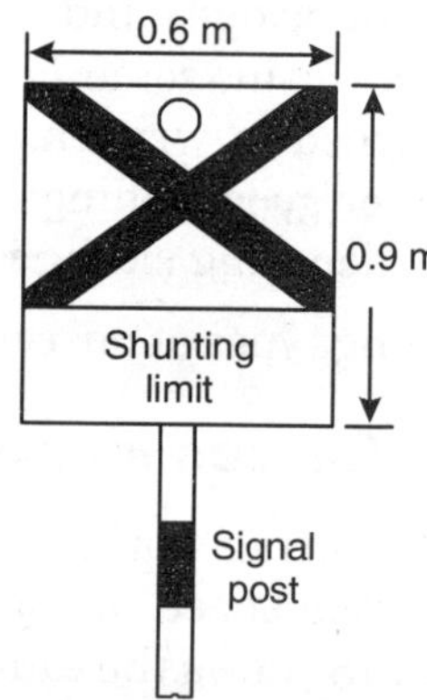

Figure 31.16 *Shunting limit*

31.17.5 Terminal Signal

These signals show that the repair section is over and now the driver can increase the speed of the train to the permissible limit.

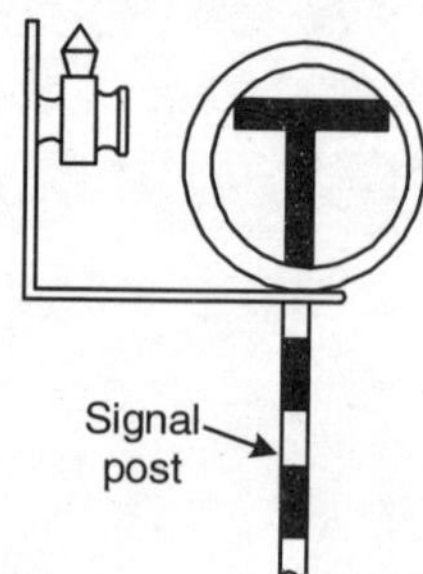

Figure 31.17 *Terminal signals*

All the miscellaneous signals mentioned above are used during the repair and maintenance works. These are temporary signals. These are installed in the direction of the approaching trains in case of double line and on both the sides in case of single line track.

Figures 31.18 and 31.19 show the method of fixing these temporary signals.

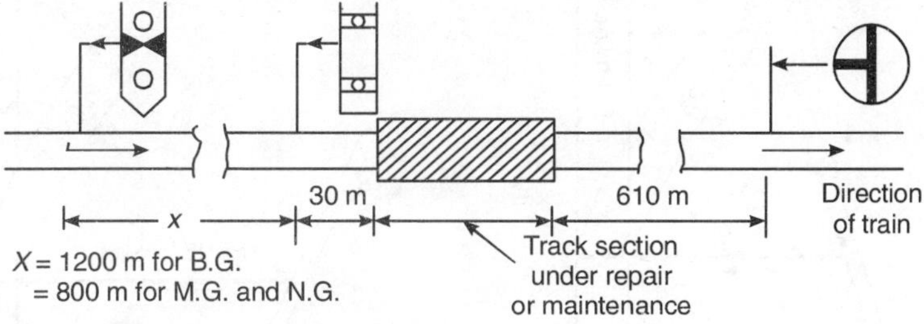

Figure 31.18 *Temporary signalling when the train is required to be stopped*

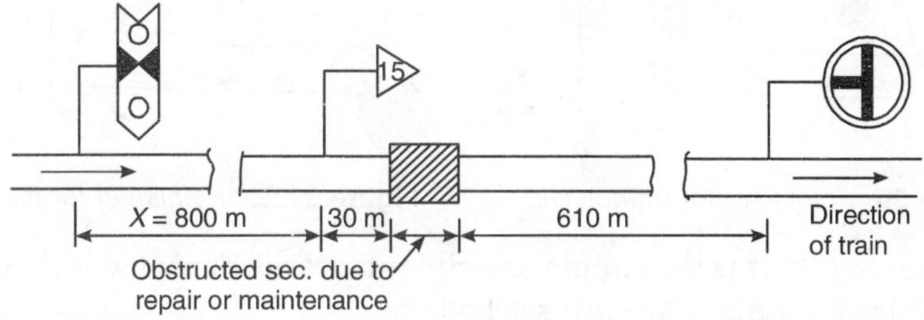

Figure 31.19 *Temporary signalling when the train is not required to stop*

31.18 INTERLOCKING OF SIGNALS AND POINTS

At each station, there are a number of points and signals. These are all operated from one cabin. At the time of operating points, their corresponding signals should be lowered. If the points are set for one line, and signal is lowered for another line, it will cause serious accident. Therefore to avoid such confusions, all these points and signals should be interlocked together in such a way that if points are set for a line, it becomes compulsory to lower the signal of the same line. Generally the lowering of signal of a line becomes only possible when all the points of that line are set. In this way interlocking avoids the possibility of pulling wrong levers and prevents conflicts of lowering signals.

31.18.1 Principles of Interlocking

The principle of interlocking can be easily understood by observing Figs 31.20 and 31.21. When the lever is operated to set the points for main line, the signal for main line is lowered and that of branch line remains at the stop position. The signal for branch line cannot be lowered until the points are set for branch line. In this way, signal and points for each line have been interlocked and it is impossible to lower wrong signal under any

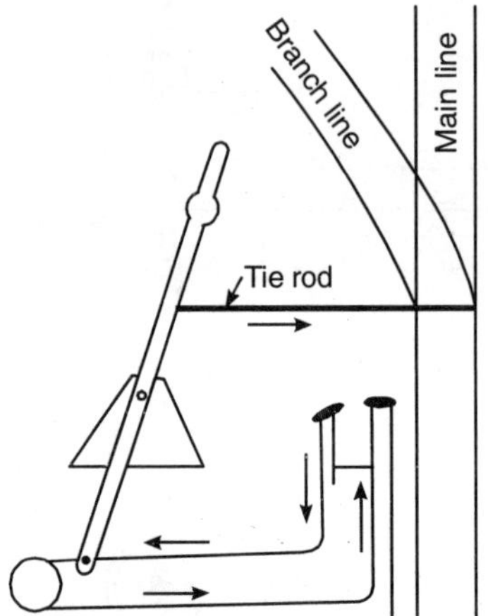

Figure 31.20 *Points set for branch line*

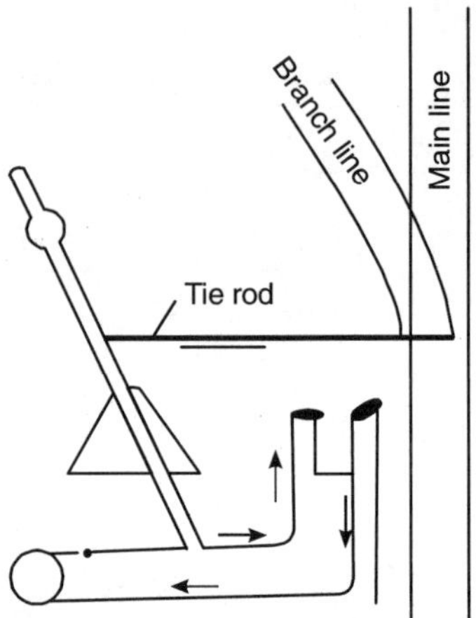

Figure 31.21 *Points set for main line*

circumstances. This is the simple principle of interlocking on which all points and signals of a station are interlocked.

At the time of designing interlocking the following points should be kept in mind :

(a) It should not be possible to lower any two signals that can lead to a collision between trains.

(b) It should not be possible to lower a signal for an approaching train until all the points have been set for the line.

(c) After lowering the signal for an approaching train, it should not be possible to disconnect any point, until the train has passed and the signal has been raised to show stop.

(d) All the points should be interlocked to avoid risk of collisions.

(e) It should not be possible to lower the warner of distant signal unless the home and starting signals have been lowered in advance.

31.18.2 Interlocking Method

In the cabins for the facility in operating correct levers in lesser time it is done by preparing Interlocking Tables which clearly specify which are levers to be pulled and which are to be released for setting required points and lowering the signals. These tables are fixed on the wall in the back of levers and the cabinman can easily operate the required levers for giving signals to the trains. Sometimes for more facility the levers are painted in different colours with its indications. Generally following colours are used:

Blue — Levers for locks.

Green — Levers for warner.

White — Spare levers.

Black — Levers for points.

Red — Levers for stop signal.

Yellow — Levers of crossing gate.

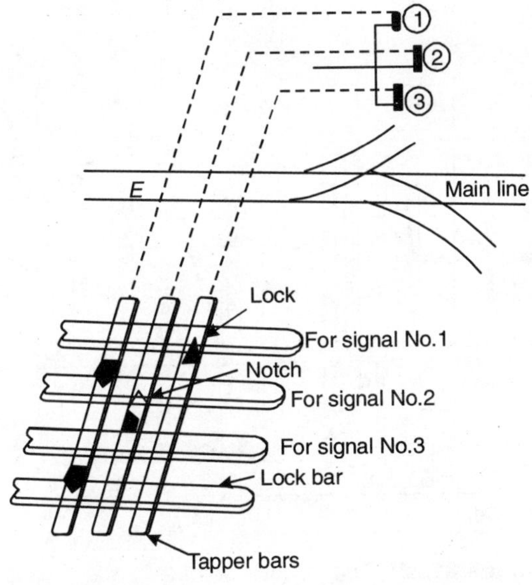

Figure 31.22 *Interlocking arrangements with tappers and locks*

Figure 31.22 shows method of interlocking with tappers and locks. In the tapper-bars, notches are made at suitable places and when lock-bars are stretched, locks fit in the notch and do not allow the movement of tapper-bars until lock-bars have been operated. When the points are set for left hand turnout, its signal is lowered and when these are set for right hand turn-out the corresponding signal only can be lowered.

31.19 MECHANICAL DEVICES FOR INTERLOCKING

The following are the main devices which are required for interlocking :

(a) Point look

(b) Treadle bar

(c) Detectors, and

(d) Roads, cranks and compensators.

Figure 31.23 shows how interlocking is done with these mechanical devices. The toe ends of both the switch rails are connected together by a stretcher bar of flat iron about 40 mm × 15 mm placed on edge. In the middle of this stretcher bar there are holes. The plunger box has a plunger which is installed in the centre of the stretcher-bar. The stretcher-bar moves inside the plunger box. The plunger bar is connected to the lock bar by means of a three-way crank as shown. Now, when lever is operated in the cabin the three-way crank motion to plunger and lock-bar at the same time. The plunger moves at right angle to the stretcher-bar. When the points are set for main line or branch line the required hole comes in front of the tongue of the plunger

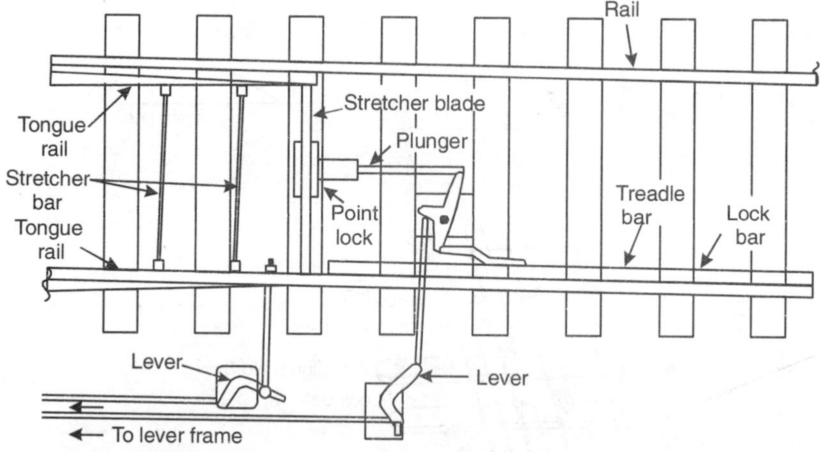

Figure 31.23 *Plunger type point lock and treadle bar*

which enters in the hole and prevents the movement of the stretcher-bar. Now, when the stretcher-bar has been locked, the points are locked and cannot be removed till the tongue of plunger is inside the stretcher-bar.

31.19.1 Point-Lock

Point lock is provided in the middle of the track a little in front of the toes of tongue-rails. The main purpose of the point-lock is to check that the points are set correctly and then lock them till desired. Point locks may be of the following types:

(i) For locking switch rail to the stock rail clamp and pad-locks may be used when the speed does not exceed 16 km/h. Bolts and cotters can also be used.

(ii) Properly designed keys are used for locking the point when the speed limit is between 16 km p.h. to 49 km p.h.

(iii) Plunger type points locks are used when the speed limits are more than 40 km p.h.

Point lock mainly consists of two stretcher blades, connected to each tongue. When operated these blades slide at right angles the track as shown in Fig. 31.23. The plunger moves by a rod connected to the lock bar by means of three-way crank, in the plunger casing. When the levers are operated in the cabin, the crank moves both the lock and plunger. During setting the points for a particular route, the hole in the blade in the blade attached to one of tongues directly comes opposite the plunger rod inside the plunger casing and the rod enters in the slot locking the points and making it impossible to move the switches.

31.19.2 Lock Bar

It is a long piece of steel the length of which is longer than the wheel base of vehicle. Figure 31.23 shows its position in the plan. The main function of it is to lock the movement of three-way crank so that it may not be operated even by the levers from the cabin, when the vehicle wheel is over it. In the beginning when the lever is operated in the cabin, clips revolve round a pin, raise the lock-bar slightly above the rail level and through it again in forward or lock-backward direction in its normal position. Now this lock-bar cannot be given any movement before raising it. In locked position it will not allow the movement of plunger at any cost. When the vehicle is standing over this lock-bar even by operating levers in the cabin we cannot remove switches, because before their removal the plunger is to be moved and for the movement of the plunger lock-bar is to be moved and the lock-bar can be moved only after raising it above the rail, which is impossible, because the vehicle is standing on it. In this way we see that lock-bar locks the switches and does not allow their removal till the vehicle is passing over it.

31.19.3 Detector

It is provided to detect the failure of the locking arrangement of points. If anything has been damaged or broken or some obstruction has come between the main rail and tongue rail, the signal will remain at the danger position. Therefore, in the presence of detectors if the points are not set properly due to any reason, it will be impossible to lower the signals. The signal can be lowered only after removing the defect and setting points properly.

The mechanism of a detector is a box near the points wherein the bars from the signal wires are placed at right angles to the bar from the stretcher bars of point. These bars are known as *'slides'*. Signal slides are placed at right angle the point slide. All the slides are suitably held in their position, and no vertical movement in them is allowed. Number of notches are provided in the point slide, whereas only one notch is provided in the signal slide.

Mechanical detector works as under:

1. During setting of the points for the main track, the notch in the signal slide and the notch in the point slide are face to face and the main signal can be lowered.

2. When the points are reversed, the point slide will be moved to the left and the notch in the slide of another signal 'B' and notch of point slide will remain face to face and the signal 'B' can be lowered.

3. In this way when one signal is lowered the other goes up in 'danger' position.

31.19.4 Compensator

As the distances from cabins to signals and points are very much, these have been connected by wires and rods, which change their length during the change in temperature. If the rods are directly connected to locks and cabins, they may cause pull or push of locks, when their length changes.

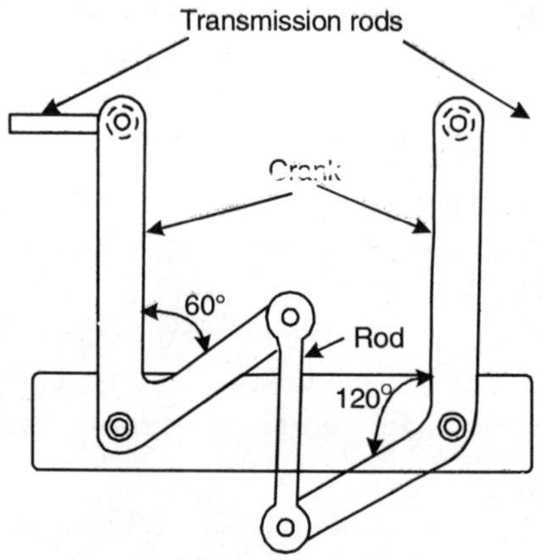

Figure 30.24 *Compensator*

Therefore, to check this change in length, compensator is provided in the way, which compensates this effect of change in length. Figure 30.24 shows a compensator, the working of which is self-explanatory. It is fixed in the middle of bars operating the levers.

Signals are connected to cabins by double wires which neutralize the effect of change in the length.

REVIEW QUESTIONS

31.1. What do you understand by signalling ? What are the main objects of signalling ?

31.2. What are the various signalling systems ? Describe the working of the Absolute block system.

31.3. Write short notes on various types of signalling systems.

31.4. Describe semaphore signal with the help of a neat sketch. What is a warner signal ?

31.5. Write short notes on :

(i) Disc signals (ii) Coloured light signals

(iii) Routing signals (iv) Calling-in-signals

(v) Shunting signals

31.6. Differentiate between the following types of signals with the help of neat sketches :

Outer signal; Home signal; Starter signal and Advance starter signal.

31.7. What is the principle of interlocking ? What points should be considered while designing an interlocking system ?

31.8. What are the various methods of interlocking ? Describe in short.

31.9. Write short notes on mechanical devices for interlocking with the help of a neat sketch.

31.10. Write short notes on :

(a) Lock-bar (b) Detector

(c) Compensator (d) Point-lock

31.11. Write short notes on hand signals.

31.12. When detonating signals are used in railways ? Describe.

31.13. What are the requirements of good signalling systems ?

31.14. Under what circumstances repeater signals are provided. Describe their principle with the help of a suitable sketch.

31.15. Write short note on the temporary signalling used during the maintenance of the track.

32

Types of Railways

GENERAL

In small towns and cities, rickshaws, tongas and cycles are the main vehicles for moving from one place to another. In large towns fast moving vehicles such as auto-rickshaws, taxies and buses are utilized for the movement of traffic from one place to another. In some cities trams are used, but as they occupy the central portion of the roads, they mostly cause obstruction in the movement of other vehicles. Therefore, nowadays trams are not preferred at any cost. In the case of very large cities where the traffic is very heavy and people have to go several kilometres daily in connection with their business and service, all the above types of vehicles are not suitable, because their speed is less and capacity is small. Therefore, in large industrial and commercial cities, where the people come daily for their service from neighbouring villages and towns, the best way is to provide train services, because they have unlimited capacity and can run at very high speeds. That is why in such cities local trains run to take the people from one place to another. It has been seen that about 80% traffic moves on trains in very large cities, because it is the cheapest type of transport as compared with other types of transport systems.

32.1 TYPES OF RAILWAYS

Depending of the traffic, availability of space, obstructions caused to road traffic and safety measures, the following types railways are used in different countries:

1. Surface railways
2. Underground railways
3. Elevated railways
4. Tube railways.

32.1.1 Surface Railways

This is the cheapest type of railway system, in which track is laid directly on the ground. This is the only system which is used in India. But when a new track is to be laid in thickly populated area, it becomes costly, because at the time of acquiring land multi-storeyed buildings are to be pulled down and proportional compensation is to be paid. The construction of surface railways in hills is very costly, because at so many places to avoid steeper gradient, the railway line is to be laid around the hill thus increasing the length. But at certain places tunnelling is to be done, which is also very costly. Similarly while crossing valleys a number of bridges are to be constructed. But even then this is the cheapest type of railway system.

32.1.2 Under ground Railways

This is very common in western countries in thickly populated areas, and is used to avoid the congestion of traffic. When surface railway passes through densely populated business areas, it has to cross a number of roads and lanes and thus interfere with the road traffic. At every level crossing it holds up the road traffic for long times. Secondly, it covers up more costly areas. Therefore to avoid all these things, tunnels are constructed below the ground and tracks are laid in these tunnels over which trains move. The construction of tunnels is very costly, but this type of railway is totally unobstructed by any other type of traffic, and so high speeds can be maintained. In underground railways only electric locomotives are used, because steam or diesel locomotive will produce smoke, which will pollute the atmosphere inside the tunnel. Adequate ventilation is provided in tunnels and pumps are used to drain out the water. At every station access is provided from the ground by means of stairs or lifts. These types of railways are mostly used in London and Berlin.

Following are the main advantages of the underground railways :

1. The passengers and the goods can be easily taken from one place to another at high speed, as there are no obstructions in the track.

2. It greatly helps in the removal to traffic congestion.

3. Large area of the city which would have been used for the surface railways, can be used in other important structures, industries and in reducing the house problems.

4. During war, it provides safety during serial attack and maintain the transportation system, even when the surface transportation may be damaged.

32.1.3 Elevated Railways

This type of railway is constructed on columns of R.C.C. or steel over which suitable deck is laid. The track is laid over the deck. Below the deck between columns, there is continuous viaduct, in which road may be provided. Elevated railways are also used to remove traffic congestion. This is the costliest type of railway systems. This type of system has been provided in Chicago and Berlin. The station buildings are also provided at the elevated level corresponding to the track level. In some places all the three types of railways are provided simultaneously, as in Berlin.

32.1.4 Tube Railways

This is a type of underground railway at greater depth, more than 25 metres. The section of the tunnel in which the railway lines are laid is circular, therefore these are known as tube railways. The main purpose of taking the tube railways so deep it to remove the interference of the sewerage systems, gas, water, oil or other obstructions. These railways are generally laid even below the sea or river beds.

In India the construction of tube railways from Mumbai to Poona and from Dum-Dum Air Port to Kolkata city are under consideration.

Following are the main characteristics of the tube railways :

1. The railway stations of the tube railways are also of the cylindrical shape.

2. Only electric locomotives are used to avoid the smoke and the ventilation problems.

3. Escalators (moving stairs) or lifts are required to reach and for coming out from the railway.

4. Automatic signalling is provided for efficient working.

5. Automatic ticket issuing machines are to be installed on the platforms for quick issue of the tickets without wasting time.

6. The trains of the tube railways have automatic door closing devices, which close all the doors immediately before the start of the train.

7. These trains automatically stop if the signal is at stop-position.

32.2 RAIL-ROAD CROSSING

The surface railways meet with roads at different places. As the (rains move at high speed on restricted band of area, it cannot stop at rail-road crossings. Therefore, for the safety of road traffic, sufficient arrangement should be made at crossings. The rail-road crossing may be of any one of the following types:

1. A level crossing.
2. An under-bridge.
3. An over-bridge.

32.2.1 Level Crossing

When the track and roads cross each other at the same level, it is known as *Level Crossing.* Figure 32.1 shows the plan of a level crossing. It essentially

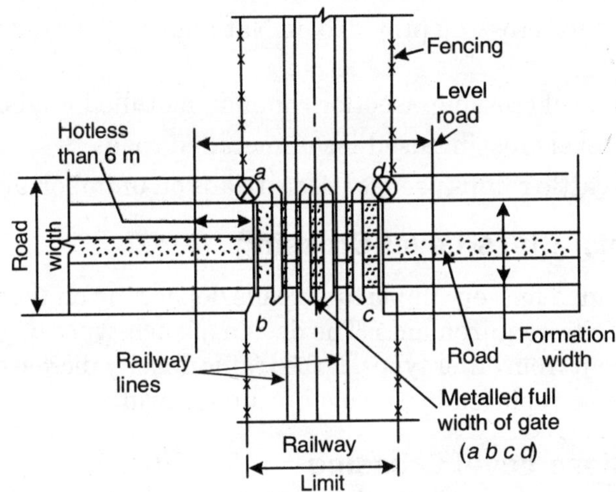

Figure 32.1 *Plan showing Right angle Level Crossing*

consists of two main gates which are closed when the train is passing and are opened for road traffic when the train is not passing. These are provided where the volume of road traffic is less and number of trains passing is small.

When a large number of trains is passing, a level crossing becomes a nuisance for the road traffic, because it holds up the traffic for long times. Two wicket gates are provided in the level crossing so that pedestrians may cross the track, when the main gates are closed. These level-crossings are interlocked with signal in such a way that only after closing the gates, the signals can be lowered. To protect the rails on level crossings and for clear flangeway, check rails are provided. The level of these check rails are kept slightly higher than the main rails, so that road traffic may not damage the main rail. Gate-keepers are provided at each level-crossing who look after them day and night. When gate-keepers receive the intimation about the arrival of a train, they close the gates and after the passing of the train, the gates are opened. As far as possible level-crossing should not be provided in cuttings. These should be located at such a place that road traffic can see it from long distance and move accordingly.

The type of various facilities to be provided at the level crossing mainly depends on:

(a) Classification and nature of the road.

(b) Nature of the traffic moving on the road.

(c) Number of trains passing over the level crossing.

The railway has classified the level-crossings into the following categories:

1. Special class level crossing having exceptional heavy traffic.
2. 'A' class level crossing provided on National Highways or grand trunk roads.
3. 'B' class level crossing used for common metalled roads.
4. 'C class level crossing used for unmetalled roads.
5. 'D' class cattle ramps used for pedestrians on unimportant rural roads.

32.2.2 Special Class Level Crossing

These are provided for very important roads depending on the volume of the traffic and the site-requirements. The design of such types of level crossings is totally different from other types. If the traffic density increases intensively, it may have to be replaced by the over or under bridge.

32.2.3 'A' Class Level Crossing

These crossings must satisfy the following conditions:

1. The approach road on both the sides of the crossing must be straight at least for a distance of 3 m when measured from the centre line of either side of the railway line.
2. As far as possible the crossing angle at the level crossing should be right angle. But if 90° is not possible then in no case it should be less than 45°.
3. In case of acute crossings the angle should be increased by providing reverse curves (S-curves) in the road.
4. The minimum radius of the road curve should be 45 m.
5. The road gradient should not be more than 1 in 30.
6. For preventing the accidents the view of the level crossing should be visible from long distance, so that road vehicles may stop in time, if the gate of the level crossing is closed.
7. On both sides of the level crossing a level length of at lest 7.5 m should be constructed in level for the waiting of the vehicles.
8. On the level crossing minimum road width of 7.3 m should be fully paved.

9. Clear distance of 2.1 m should be maintained from the centre line of the track to the crossing gates.

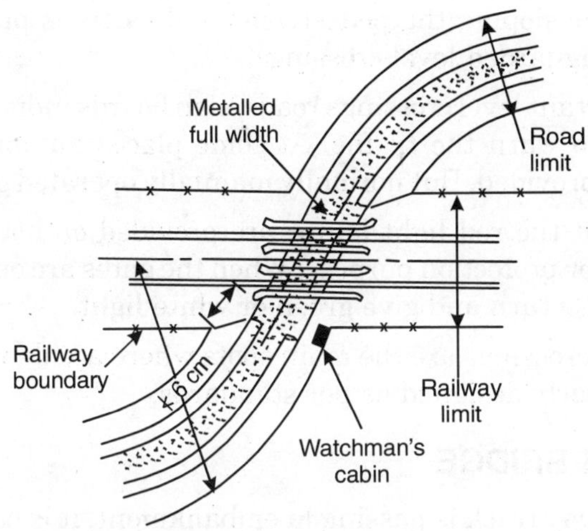

Figure 32.2 *An Acute-angle Level Crossing*

10. The hut of the watchmen should be constructed near the level crossing.

11. At least two watchmen must be employed at each level crossing to take care of the level crossing.

12. If possible telephone facility should be extended to the level crossing watchmen.

32.2.4 'B' Class Level Crossing

All the above conditions of the class 'A' level crossing are applicable except the following to 'B' class level crossing :

1. The radius of curve of the approach roads may be reduced to 30 m instead of 45 m.
2. The minimum gate width may be kept 37 m instead of 7.3 m.
3. In place of two watchmen, only one watchman may be employed.

32.2.5 'C' Class Level Crossing

All the conditions of class 'B' level crossing are applicable to this type of level crossing except the following :

1. The minimum width of the metalled road should be 27 m.
2. The maximum gradient of approach roads should not be more than 1 in 20.

32.2.6 'D' Class Level Crossing

This type of level crossing is provided on unimportant roads. Instead of gates only fencing for sloping the pedestrians and cattle is provided. These are generally of unguarded level crossings.

At the important level crossings roads, sign boards indicating the crossings are provided to warn the traffic. At some places automatic gates (traffic actuated) are provided. But normally manually operated gates are provided.

During night the red light lamps are provided on both the sides of the level crossing for protection purpose. When the gates are open for road traffic, these lamps also turn and give green or white light.

As the level crossings are the main spots where accidents may occur, they should be properly designed as per standards.

32.3 UNDER BRIDGE

When the railway track is passing in embankment, it is better to provide an under bridge, so that road traffic can pass below the railway line. At such places where traffic is very heavy and is held up for long times, an under bridge is provided. The railway track is taken at its alignment without any

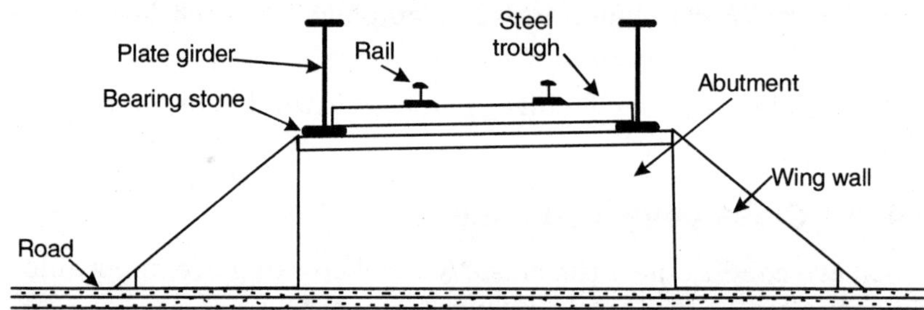

Figure 32.3 *Under bridge* (*cross-sectional view*)

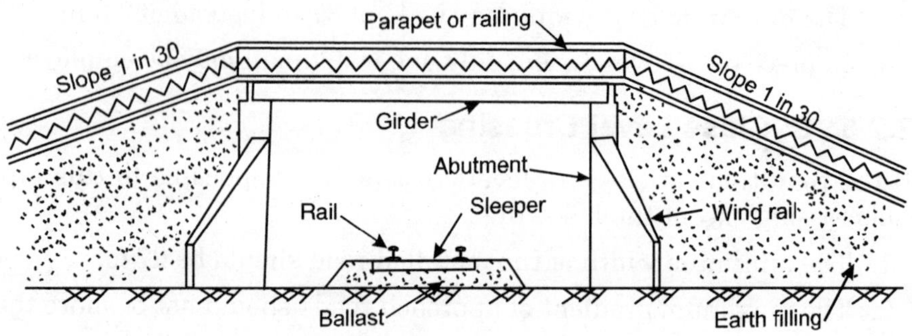

Figure 32.4 *Over bridge* (elevation)

change in level. The road is taken through the cutting with slope from both sides and at the position of track, bridge is provided. Generally the track is directly laid on girders as shown in Fig. 32.3. Check rails are provided in the entire length of the bridge. Such types of bridges generally cause difficulties in the drainage of water. Sometimes pumping is required. If the area is low lying, the water will be filled below the bridge and will cause difficulty in the movement of the road traffic.

32.4 AN OVER-BRIDGE

When the track is passing in slight cutting it is economical to provide an over-bridge. In this case the track passes below the bridge and the road passes above the track on the bridge. When bridges are constructed to cross the track, no gate-keepers are required as the road traffic is not held up at any place. The initial cost of such bridges is very heavy but the maintenance cost is low as compared with level crossing.

Figure 32.4 shows an over-bridge. The detail of such bridges can be seen from the bridge engineering portion of this book.

REVIEW QUESTIONS

32.1. What are the different types of railways? Describe any one of them.

32.2. Write short notes on :

(i) Underground railways, (ii) Elevated railways.

32.3. Describe a level-crossing with the help of a neat sketch.

32.4. Under what circumstances an over-bridge and an under-bridge are provided? With the help of a neat sketch, describe any one of these.

32.5. Write short note on the classification of the level crossings.

32.6. What are the main conditions which class 'A' level crossing must fulfil?

33

Alignment and Survey

GENERAL

Necessity of New Railway-Lines

The following are the main points which create the necessity of new railway-lines.

- To develop the backward areas, so that people may come in contact with modern developed areas.
- To take out the natural minerals from mines and carry it upto industries.
- To increase the trade between developed areas.
- For strategic considerations, so that in case of emergency, defence forces may move from one place to another in less time.
- If one railway route which is very long already exists between two very important cities, a new railway line is laid to connect them directly with minimum possible length.
- To join the ports with trade centres.
- To open the new trade centres. To tape the new sources of oil, mines and other natural wealth and bring them to the factories etc.
- To shorten the distance between the big trade centres. When the old line has been laid in zig-zag way, new line may be required to meet the present requirements.

33.1 ALIGNMENT

It is the position and direction of the centre line of the track. It is decided from the plan and longitudinal section along the centre line.

The following points should be kept in mind while aligning a railway-line :

1. Length of track should be as short as possible.
2. Construction charges should be minimum.
3. It should be cheap in maintenance.
4. It should have easy gradients so that locomotives can haul more load and the transport charges may be minimum.
5. It should have easy curves.
6. It should give maximum safety to the passengers without any chance of accident or derailment.
7. It should pass through important cities and industrial areas.
8. It should pass through aesthetic areas so that passengers may enjoy their journey.

33.1.1 Selection of Good Alignment

Though the shortest route seems to be most economical, but it is not necessary as if in the shortest routes, there are hillocks, rivers, cities or industries, it may be the most costliest route. Taking into consideration all the practical difficulties such as obstructions, steep gradients, construction and maintenance problems, the route of the alignment is decided.

Sometimes an alignment which is economical in initial cost may not be economical in the operation cost or maintenance cost or both.

Following points should be considered while selecting the good alignment of the railway track :

33.1.1.1 Points through which a track must pass

1. The track must pass through the important points connected with social, commercial, political and defence. The importance of the towns, expected volume of the traffic for revenue purpose, expected goods load from and to the area, should be considered.

2. There should be minimum possible bridge constructions on the routes as these are the most costly works.

3. All the deep cutting or heavy banking or tunnelling or viaduct constructions should be avoided as far as possible.

4. As far as possible the existing tunnels, bridges, hill passes or saddle may be used to reduce the overall cost of the project.

5. In case of high and thin ridges, if tunnelling is economical, they should be constructed instead of deep cuttings or increasing the length of the track unnecessarily.

33.1.1.2 Points through which track should not pass

1. Costly land, whose high compensation is to be given.

2. Religious places such as church, mosque, tomb, temple; as the acquisition of these places have been prohibited under law.

3. Water-logged areas.

4. Unsafe areas as mud stone hills, weak grounds, snow-fall areas as cutting of snow have to be done frequently.

5. It should not pass through the rocky soil as its excavation is costly and there is every possibility of the embankment slip.

33.1.1.3 Nature, amount and position of the traffic

1. Areas having good freight earnings should be aligned as the income from goods train is more than the passenger trains.

2. The alignment must pass from the areas having maximum population, as the traffic directly depends on the square of the population.

33.1.1.4 Geometrical standards

1. The track must be aligned confirming to the latest codes of Indian Railways for ruling gradients, minimum permissible radius of curve.

2. To increase the hauling capacity of the locomotives the radius of the curves should be as large as possible.

3. As far as possible the curves should be avoided near the stations and bridges for better visibility of the signals.

4. The alignment of all the major and small bridges must be at right angles at the site of bridge to the river and at least 30 m straight portion must be provided on both the sides of the bridges.

5. As far as possible vertical curves must be avoided as they increase the cost of alignment and the operational cost.

33.1.1.5 Topographical effects on the alignment

1. If the track is to be aligned in the valley, it should be done at a uniform gradient.

2. When the alignment passes througn the various states, it has to cross so many watersheds. In the case of very long routes, it is impossible to align the track at uniform grade. In such cases summit and valley curves are provided at suitable places to reduce the cost of construction.

3. In case of hill areas sufficient care should be taken while doing the alignment, as slight changes may adversely effect the cost of construction. At suitable places tunnelling may be done if the length of the track as well as cost of construction is reduced.

33.1.1.6 Economic considerations

1. Various alternative alignments should be prepared and the cost of construction, maintenance Cost, revenue earned, population benefited etc. should be determined. The best alternative alignment suiting the most of the population and economical in construction and maintenance should be adopted.

2. If possible the track should pass through the proximity of the quarries and villages, from where the stones and labour for the construction and maintenance shall be easily available.

33.1.1.7 Political consideration

Political considerations must be kept in view while aligning the tracks.

33.2 SURVEY

Before the actual execution work of the railway, its project is prepared and got sanctioned. The survey work includes up to the final location of the alignment. The survey work is divided into following parts :

(a) Traffic survey

(b) Reconnaissance survey

(c) Preliminary survey

(d) Location survey

33.2.1 Traffic Survey

Before investing any money in big projects the future earning should also be considered. Therefore, before constructing railways its future should also be through over. The earning directly depends on the traffic, therefore the first survey work is the *Traffic survey.*

The following information should be collected in this survey:

1. Population of the towns, villages and cities within 15 kilometres, on both sides of track.

2. Position of local industries, the nature, destination and the actual weight of goods which will be available for carting to the railway, after laying the new line.

3. Future prospects of development of industries.

4. Approximate number of passengers who will travel by the railway.

5. Religious festivals in different places along the new proposed line, their times and expected visitors who will travel on such occasions.

6. Pilgrimage places on the line.

7. Export materials with their destinations.

8. Import materials with their destinations.

9. Density of population is different places for the location of railway stations.

10. The traffic history of the existing tracks or road transport facilities available or expected to develop the areas proposed to serve by the railway.

After collecting all the above information regarding traffic if it is found that the new railway system can be provided, the next stages of survey work are done. The cost of railway per kilometre can be calculated from the previous experience of providing railway system in other parts. From this the approximate expenditure, the operating cost and the revenue earned can be calculated.

Actually the traffic-survey work is not the work of an engineer. It is the work of an experienced economist, who can do this job in the best way. All the remaining three types of survey works are done when it seems that providing of rail-route will be economical.

33.2.2 Reconnaissance Survey

First of all in this survey work, the general configurations of the land are studied from the available topographical, industrial and agricultural maps. With the help of these maps the approximate alignment is fixed. After it, following datas are collected from the field:

1. Position of rivers, streams, valleys, ridges, etc.

2. Nature of soil at various places and datas about the surface formation of the ground, dip of the existing rocks and the hill-slopes.

3. Sources of water in the areas with their approximate quantity of water and impurities, etc.

4. Physical condition of the country.

5. Climatic conditions.

6. Approximate elevation of different ground surfaces.

After collecting the above datas, various alternatives of the alignment are prepared. Each alternative is considered separately. The success of the reconnaissance survey mostly depends on the experience of the engineer and his personal qualities, and broad sense etc. He should be capable of observing all the salient features of the country and his general knowledge about the whole territory should be very good.

Following are the main objects of the reconnaissance survey :

1. To have a rough idea of the various possible alignments.

2. To locate the various controlling points on the alignment from where the line must pass.

3. To decide the place of maximum gradients and curves for the alignment.

4. To know the geological, geographical physical and climatic details of the areas through which the alignment is to be done.

5. To prepare the rough estimates for the various proposed alternatives.

33.2.3 Priliminary Survey

When the various alternatives have been proposed after the reconnaissance survey, the next stage is the preliminary survey. In this survey the survey work is done with great accuracy along all the alternatives and the total cost of each route is calculated separately.

Generally preliminary survey work is more or less an open traverse survey of 200 to 300 metres wide belt along the centre line of the proposed alternatives. This survey work is done with tacheo-metre, level instrument, prismatic compass and plane table. In this survey work the bearing capacity of soil, position of ground, water-table, soil conditions in river or lake beds, availability of skilled and non-skilled labour are the factors to be determined.

Following informations are also collected during the preliminary survey :

1. Geological information like soil, rocks etc.

2. Details of the existing culverts, bridges, tunnels, irrigation canal works etc.

3. Position and details of the level crossings proposed.

4. Availability of the construction materials, water and labourers at various places along the proposed alignment.

5. H.F.L. and L.W.L. of all the rivers, streams etc. on the proposed alignment.

33.2.4 Location Survey

From the preliminary survey work, the most economical route will be determined. Therefore, in location survey work the full detailed survey work will be done along the most economical route which has been determined by the preliminary survey work.

While carrying out the detailed survey work the following work will be done:

1. At every 30 m along centre line firm pegs will be driven and their chainage will be written at their top.

2. Masonry pillars will be constructed on the centre line at every 600 metres and their top level will be written on their top.

3. At every tangent point of curve, pillars will be constructed.

4. At every tangent point, its bearing will be taken with the help of prismatic compass.

5. From the levels one longitudinal section will be prepared and alignment line will be marked over it.

6. Sufficient number of cross-sections will be drawn.

7. The position of alignment will be clearly marked in the plan.

8. The centre line of bridges, culverts, tunnels, station buildings, yards, signal cabins etc. will be marked on ground as well as on paper.

9. Full datas for the construction of bridges, culverts and tunnels will be determined.

In the final location survey, all the work except of the construction work will be completed. And after the receipt of sufficient funds the execution of the work will be started.

33.3 PROJECT REPORT

After completing all types of surveys the project report is prepared along with all the necessary drawings and details.

Following details are given in the report:

1. Introduction of the project along with history and the whole geographical details.

2. Main requirements and the factors to be kept in view while doing the proposed alignment.

3. Details of the alignment with respect to the proposed gauge, gradients, lengths and levels of the various controlling points.

4. Description about the various other proposed alternative routes.

5. Details of the proposed alignment along with the various datas and the maps.

6. Specifications of the details of the constructional standards.

7. Conclusion and the recommendations for the proposed project of the new railway track.

Following map are generally prepared along with the project report:

1. Over-all map of the portion of the country to a scale on 1 cm = 20 km, which will be served by the new line.

2. Index map of the area to a scale of 1 cm = 2.5 km.

3. Index plane and section to the scale of 1 cm = 640 m (horizontal) and 1 cm = 12.5 m (vertical).

4. Detailed plan of the proposed alignment including longitudinal section.

5. Contour plans and longitudinal sections at the sites of all the bridges and culverts.

6. Plans of all the station yards to a scale of 1 cm = 50 cm.

7. Detailed drawings of all the bridge and buildings to a scale of 1 cm = 1 m.

8. Plan of the level crossings to a scale of 1 cm = 50 m.

9. Details of all the important structures lying within 300 m on either side of the proposed route.

REVIEW QUESTIONS

33.1. What are the main points, which create the necessity of a new railway line ?

33.2. (a) What do you understand by the term "alignment"? What point should be kept in mind while aligning a railway-line ?

(b) What factors govern the alignment ?

33.3. What are the different types of survey-works in railways which are done before its construction? Describe any one in detail.

33.4. (a) What information is collected at the time of doing traffic-surveying?

(b) Write short note on location survey.

33.5. Write short notes on :

(a) Reconnaissance survey.

(b) Preliminary survey.

33.6. Describe what details are given in the project reports.

33.7. Describe what are the main objects of the reconnaissance survey.

33.8. What points should be taken into account while doing the selection of best alignment of the new proposed railway track?

PART 3

Bridge Engineering

Chapter [illegible] : Introduction

Chapter [illegible] : [illegible] Design

Chapter [illegible] : Foundation

Chapter [illegible] : Pier, Abutment and Wing Wall

Chapter [illegible] : [illegible]

Chapter [illegible] : Causeways and Submersible Bridges

Chapter [illegible] : Masonry Bridges and Culverts

Chapter [illegible] : [illegible]

Chapter [illegible] : [illegible] Bridges

Chapter [illegible] : Floating Bridges

Chapter [illegible] : Bearings

Chapter [illegible] : Protective Works

Chapter [illegible] : Loads and Stresses

Chapter [illegible] : [illegible] of Bridges

Chapter [illegible] : Maintenance of Bridges

Introduction

GENERAL

Bridge. It is a structure that affords passage over low ground, water or other obstructions. The communication route over the bridge may be a roadway, footpath, a cycle track, railway track or tramway.

Span. It is the centre to centre distance between two supports. The clear distance between two supports is known as *clear span.*

Pier. It is the intermediate support of a bridge super-structure.

High-level Bridge. It is a bridge, other than a culvert, which allows all the food water to pass below it. All bridges of roads, railways or other obstructions come under this category.

Submersible-Bridge. It is a bridge, which allows normal flood to pass through its vents, and heavy flood water to pass over it. The formation level should be such that during floods the traffic should not be held up for more than three days at a time and not more than six times in a year.

Causeway. It is a 'pucca' dip which allows flood to pass over it. It may have vents for L.W.F.

Culvert. It is a small bridge having maximum span of 6 metres between the abutments.

Apran. Layer of concrete, masonry stone etc. laid like flooring at the entrance or outlet of a culvert to prevent scour.

Clearance. It is the minimum distance between the specified positions on a bridge. Clearance in the case of road or railway bridge is the distance available for traffic. But in case of river or canal, there are two clearances, i.e., Vertical and Horizontal.

The distance between the afflux and the lowest point of the bridge super structure is the vertical clearance.

Deck Bridges. The bridges having their flooring at the top of superstructures.

Linear waterway. The length available in the bridge between extreme edge of a water surface at the highest flood level, measured at right angles to the abutment faces.

Freeboard. It is the difference between the H.F.L. and the level of the crown of the road at its lowest point.

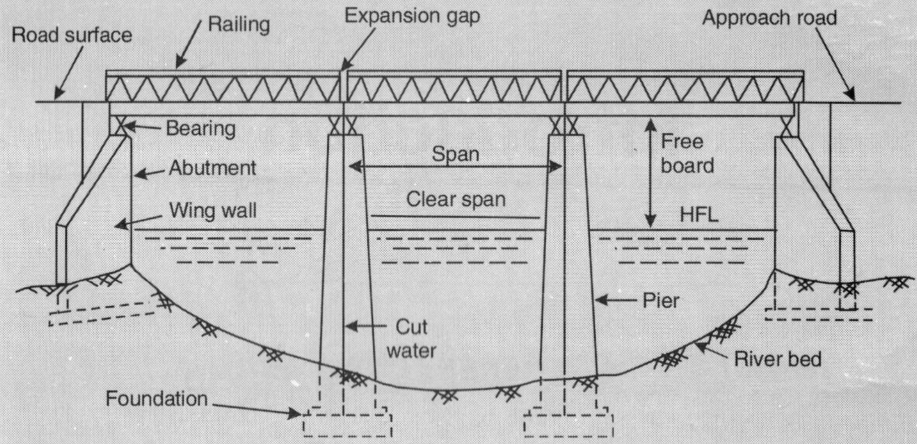

Figure 34.1 *Component parts of a bridge*

Headroom. In a through bridge it is the distance between the highest point of a vehicle and the lowest point of any protruding member of the bridge.

Highest Flood Level (*H.F.L.*). It is the level of the highest flood ever recorded of a stream or river.

Low Water Level (*L.W L.*) It is the minimum water level in the dry weather.

Ordinary Flood Level (*O.F.L.*). It is the flood level which normally occurs every year.

Scour. The vertical cutting of river-bed is known as *scour.*

Afflux. It is the rise in the level of the water surface of a water-course, caused due to the obstruction by the bridge in the flow of water.

Run-off. The portion of the rain-fall on a catchment area, which flows to water-course, is known as *'Run-off.*

Water-way. It is the area of the opening, which should be sufficient to pass the maximum flood discharge that would ever pass under the bridge, without increasing the velocity to a dangerous limit.

Economic-span. The span for which the total cost of bridge will be minimum is known as the *economic span* of the bridge.

Abutment Pier. It is a heavier pier built in the arch bridges, designed to take the horizontal thrust of the arch, in case of bridge failure. Usually every fourth or fifth pier is designed as abutment pier.

Air Lock. It is chamber provided at the top of the caisson shaft leading to the work. Air lock allows the men and material to move from the atmospheric pressure to the compressed air pressure inside the working chamber.

Apron. It is a layer of concrete, masonry or stone etc. which is placed at the entrance or outlet of a culvert or waterway to prevent the scouring.

Boil or Blow. The phenomenon of the soil (quick-sand) of being forced into the caisson, cofferdam or excavated area from under the bottom by upward water pressure. If the water head on the sides is more, this is usually caused while doing excavations for the bridge foundation.

Buffle Wall, Dwarf Wall, Drop Wall or Curtain-Wall. This is a thin wall which is used as a shield or protection against scouring action of the stream. It is usually used with the apron in the stream bed.

Buckle Plate Flooring. It is steel curved plates used for railway bridge decking for supporting the ballast.

Buoyancy. It is the loss in weight of a body when immersed wholly or partly in a fluid, due to the resultant upward pressure exerted on it by the fluid.

Caisson. It is a water-tight box like structure or chamber, made of wood, steel or concrete. It is usually sunk by excavating within it, for the purpose of gaining an access to the bed of the stream for the construction of foundation.

Catchment Area. It is the area from which rainfall flows into a drainage line, out-fall or reservoir etc. The boundary line of this basin or reservoir is called the water-shed.

Catchment Drain. It is the drain excavated on the upper slope of a hill road which intercept and collects water flowing towards the road.

Coffer Dam. It is a temporary enclosure built to exclude the water from the working area on the stream bed and to permit free access to the area within, during the construction of a foundation or other structure that must be undertaken below the water level.

Cribs. It is the temporary pier made in the river bed.

Cut-off Wall. It is a wall, collar or other structure intended to cut-off or reduce percolation of water along the smooth surfaces, or through porous strata.

Cut-water. This is the upstream nose of a bridge pier, shaped for easy and smooth flow of water through it.

Deck-Bridle. The bridge having the carriageway constructed at or near the top level of the main supporting members of the superstructure.

Dike. It is an earthen dam or embankment.

Dolphin. It is the cluster of piles which are driven around the piers to protect them against the floating objects.

Dumb-bell pier. The pier consisting of two R.C.C. columns for taking the load of the superstructure. These piers are connected by thin R.C.C. web for their full or part height.

Ease-water. It is downstream nose of the pier, which is shaped so to promote merging of the water flowing from the bridge openings.

Flood Escape. It is the lowered section of a road specially designed to permit the escape over it of flood water rising over the specified level, without damaging bridge or road structure.

Fender. It is a replaceable device for the protection of the bridge structure against the damage caused by the floating bodies.

Gorge. It is the narrow passage provided in the hills.

Guard Rail. It is a side barrier or protection rail constructed with rail supported on the posts to the traffic.

Kerb-Inlet. It is the apertures formed in a kerb for conveying the storm water to a gulley.

Lattice-Girder. It is an I-beam with some web material cut away. In this girder the flanges are held apart and the shear is totally resisted by diagonal and vertical bracings instead of by the solid web.

Piping. It is the flow of water under or around a structure built on the permeable foundations, which if not prevented or stopped will cause the flow of sub-soil particles which will tend the failure of structure.

Revetment. This is material of stones, concrete blocks or mattresses placed on the bottom or banks of a river for minimizing and controlling the erosion.

Rip-Rap. It is the graded stones arranged from the heaviest size to the coarse sand in such a manner around the piers and abutments for their protection against ocean waves or heavy scouring.

Scuppers. It is a miniature form of causeway which extends across the entire with the formation.

Square Crossing. The bridge centre when made at right angle to the stream centre line.

Sub-structure. The piers, abutments wing walls along with their foundations, which support the superstructure of the bridge.

Through-bridge. It is the bridge having the carriage-way of flooring supported or suspended at the bottom of the main supporting members of the superstructure.

Trestle Bridge. The bridge constructed with the pile bents or towers which carry the bridge-decking.

Trough Flooring. This is steel-plates flooring in the form of troughs used for the railway bridge flooring.

Under Bridge. It is the bridge in which the roadway passes under the bridge.

Viaduct. It is a long continuous structure, carrying the road or railway line, over dry valley (instead or over stream). It essentially consists of a series of spans and constructed over the trestle bents instead of solid piers.

Water Cushion. It is a pool of water constructed on the downstream side of the dam, chute, drop or other spillway structure, and acts as cushion to absorb the impact of falling water.

Water-way. The area through which the water flows under a bridge.

Foundations. These are the structures which distribute the dead load of the superstructure, piers and abutments along with live loads which come on bridge, over a large area of subsoil.

Abutments. The end supports of the superstructure of a bridge, whether flat or arched are called abutments.

Wing-walls. The walls constructed on both the sides of the abutments to retain the embankment of approaches and also to protect them from the wave action of water are called *wing-walls.*

Approaches. These are the construction works which carry the road or railway track upto the bridge. Generally these are done in embankments.

34.1 BRIDGE STRUCTURE

The bridge structure can be divided into two parts :

1. Substructure
2. Superstructure.

The function of the substructure is similar to that of foundations, columns and walls of a building, because it supports the superstructure of the bridge. The substructure consists of foundations, piers, abutments, wing-walls and approaches. They all support the superstructure of the bridge.

The superstructure is that part of the bridge over which the traffic moves with safety. It consists of parapet or railing, roadway and girders, arches or trusses over which the roadway is supported.

34.2 DECK, THROUGH AND SEMI-THROUGH BRIDGES

The flooring of bridge which carries the traffic can be supported at the top, bottom or at some intermediate level of the superstructure or main girders of the bridges. When it is supported at the top, it is called *"Deck Bridge'* and when it is supported at the bottom, it is called *'Through-Bridge'*. But if it is

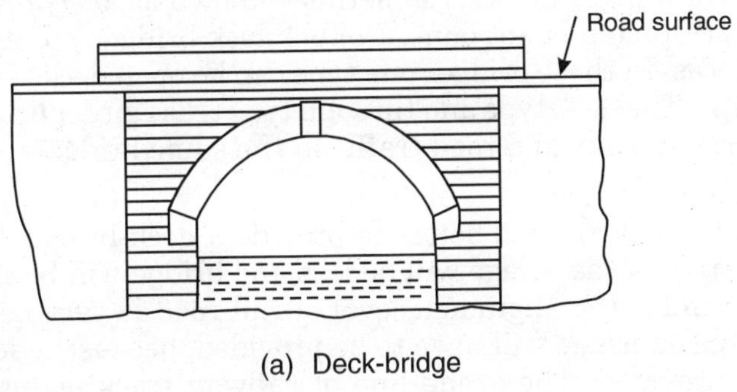

(a) Deck-bridge

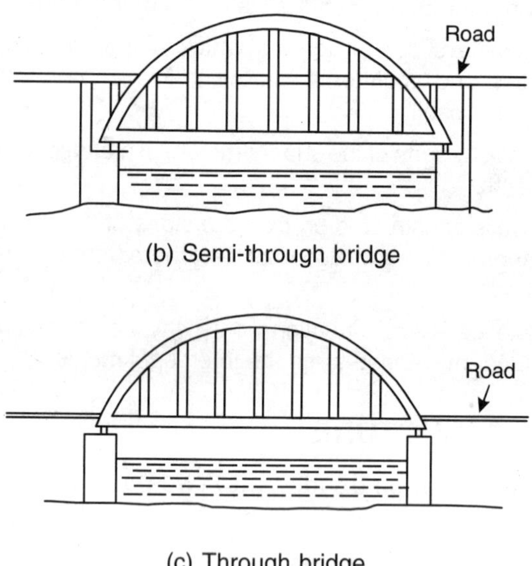

(b) Semi-through bridge

(c) Through bridge

Figure 34.2

supported at some intermediate level, it is known as *'Semi-through Bridge'*. The given sketches show the essential difference between these bridges.

The floor rests on the top of main load carrying members in *case of deck type bridge*. In plate girder bridge of deck type, the floor rests on top of flanges. In truss girder bridge of deck type the floor is placed on the top chords. In such bridges no bracing is done over the top of traffic.

In plate girder bridge of *through type,* the flooring is done on the bottom flanges. In through type trussed girder bridges, the floor rests on the bottom chords or ties. In through type bridges mostly bracing is done on the top of traffic.

In the *semi-through bridges,* the load of floor is directly taken by members placed in between the top and bottom of the main load carrying members. Such types of bridges are also sometimes known as *'pony bridges'*. In these bridges over-bracing is not done. Double deck bridges are also constructed at some places. In these bridges one type can be open deck type and other as through type. The deck type and through type truss girder bridges are mostly used to carry railway and road traffic in the same bridge.

When the alignment of a highway or a railway line is at much higher level than the H.F.L., it is better to provide a deck-bridge, because in this case the length of the bridge will be less and bridge will be cheaper. But on the other hand if the alignment level of rail-road is very near to H.F.L., a through type of bridge will have to be provided, because a deck type bridge cannot be provided. The grade line of railway track or highway and the clearance required under the bridge decide the use of through type or deck type bridge. If sufficient clearance is available under the bridge, the deck type bridges will be economical as compared with through type bridges. In case of deck type bridges the load carrying members may be placed close together than the through type bridges. This causes reduction in the length of the floor beams and lateral movement in the floor system. The height of abutments and piers are also reduced in deck bridges, causing economy in the substructure. At some places where both, road and railway line, have to pass on the same bridge, one may be supported at the top and the other at the bottom. This type of bridge is called *'Deck and Through Bridge'*, because the deck and through slabs are both laid in such bridge. If the level of the alignment of railway line and road is the same, both can be provided side by side at the same level, but some fencing will be required in between them.

34.3 SKEW BRIDGE

When the bridge is constructed at some other angle than 90° to the flow of water, it is called a *'Skew Bridge'*. But these type of bridges should be avoided as far as possible, due to the following reasons :

Let us take a small piece of card-board having the shape of the plan of a Skew bridge. Now if we place a load in the middle of it, we will see that

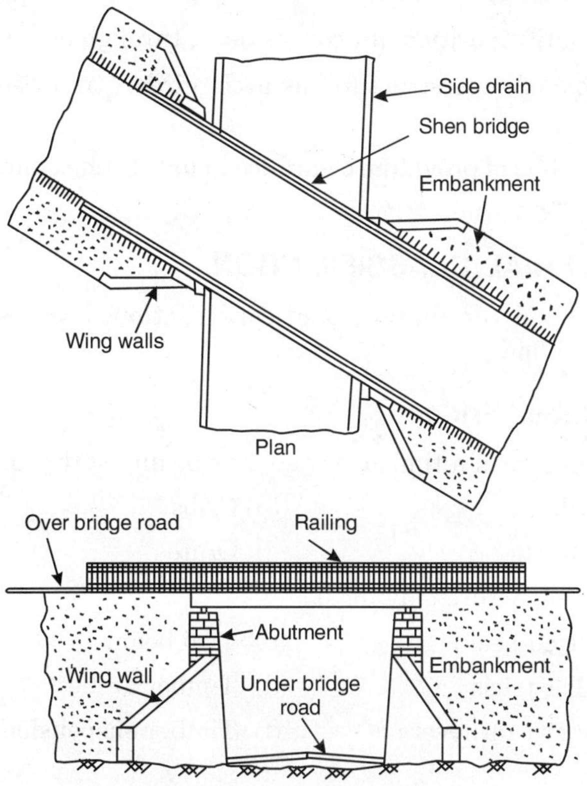

Figure 34.3 *Skew bridge*

acute corners will rise off the support and the load is taken by the other corners only. This clearly shows that if a skew bridge is constructed, the load will be carried only by two corner supports and the other two corners would tend to rise off the support. If piers are constructed at right angle to the bridge, this will cause eddies and scouring in the river-bed, which cannot be allowed in any case.

The alignment of the highway of railway does not play much part in deciding the site of the bridge. If the approaches are given a curve for a right angle crossing, it will be better in place of providing a skew-bridge.

34.4 CLASSIFICATION OF BRIDGES

Bridges can be classified according to :

(i) its function or purpose, as highway, railway, foot bridge, viaduct and aqueduct etc.;

(ii) the materials of construction as timbers, masonry, steel, R.C.C. or prestressed concrete;

(iii) the nature of life, as temporary bridges or permanent bridges;

(iv) the relative position of floor, as deck, through or semi through;

(v) the type of superstructure, as arch, girder, truss, suspension bridge etc.; and

(vi) the method of providing clearance in navigation channels, as movable or transporter.

34.5 MATERIAL CLASSIFICATION

Depending upon the materials of construction, bridges can be mainly classified as follows.

34.5.1 Timber Bridges

The sub-structure of a timber-bridge may be any of the following :

(a) Trestles (b) Cribs

(c) Pile bents (d) Crates

The super-structure may have :

(a) Strutted beam (b) Tied beam

(c) Timber truss (d) Timber arches

(e) Timber cantilevers (f) Timber suspension bridge

34.5.2 Cause-ways

These may be of:

(a) Flush type
(b) Low level type
(c) High level type

34.5.3 Masonry Bridges

These may be of brick-work or stone masonry work and may be of any one of the following types of arches :

(a) Three-hinged arch
(b) Two-hinged arch
(c) Fixed arch
(d) Bowstring-girder arch

All these types of arches may be of :

(i) Open spandrel type; or
(ii) Filled spandrel type

34.5.4 Steel Bridges

These can be further classified as :

(a) Steel-trough plate type
(b) Steel-girder type
(c) Steel-arch type
(d) Bow string girder type
(e) Steel rigid frame type (Fixed bridges)
(f) Steel suspension bridge
(g) Movable bridges

The classification of steel bridges can also be done as :

(i) Riveted bridges
(ii) Welded bridges
(iii) Pin-connected (and bolted) bridges

34.5.5 Reinforced Cement Concrete Bridges

These can be further classified as follows:

(a) Slab type
(b) T-beam and slab type
(c) Rigid frame type
(d) Multiple span portal frame type
(e) Girders with varying moment of inertia type
(f) Double cantilever type
(g) Barrel arch type
(h) Open spandrel arch type

(i) Bow string girder type

(j) Arch ribbed type with partially hung decking

34.5.6 Floating Bridges

These may be of the following types :

(a) Boat bridge (b) Raft-bridge

(c) Pantoon-bridge

Bridges can also be classified according to the level of crossing of highway and railways :

(a) Over-bridges (b) Under-bridges

34.6 TEMPORARY BRIDGES

Out of all the above classifications the following types of bridges come under temporary bridges:

(1) Timber-bridges of all types.

(2) Cause-ways of all types.

(3) Steel-bridges of movable types only.

(4) Floating bridges of all types.

34.7 FACTORS AFFECTING THE SELECTION OF TYPE OF BRIDGE

The following factors generally affect the selection of a type of the bridge :

1. Volume and the nature of the traffic.
2. The nature of the river and its bed soil.
3. Availability of materials and funds.
4. Time-limit, within which the bridge is required to be completed.
5. Physical features of the site.
6. Availability of workers.
7. Whether navigation is done in the river or not.
8. Facilities available during construction.
9. Economic span length of the bridge.
10. Level of H.F.L. and clearance requirements.
11. Facilities available for maintenance.
12. Climatic conditions.
13. Strategic conditions.

14. Hydraulic datas.
15. Foundation condition.
16. Length of the bridge.
17. Width of the bridge.
18. Live loads on the bridge.
19. Appearance.

With the help of the above factors every aspect of the bridge should be considered. The type which is most economical and can give maximum service should be designed.

34.8 SUMMARY OF VARIOUS TYPES OF BRIDGES

Following table summerises the various types of bridges.

Nature of bridge	*Span upto which used*	*Circumstances under which employed*
Timber bridges		
(i) Saturated beam	8 metres	These are used for purely temporary crossings of streams, where the traffic is light and there is no impact on the bridge during the movement of traffic.
(ii) Tied beam	8 metres	These are used for purely temporary crossings of streams, where the traffic is light and there is no impact on the bridge during the movement of traffic.
(iii) Timber trusses	20 metres	These are used for purely temporary crossings of streams, where the traffic is light and there is no impact on the bridge during the movement of traffic.
(iv) Timber arches	60 metres	Where the span is more to reduce the moment, arches are provided. At such places where live load is more as compared to the dead load of the bridge, these types of bridges are provided.
(v) Timber cantilever	20 metres	When timber is available in plenty, but other constructional materials are not available and the banks are firm.

Contd.

Contd.

Nature of bridge	*Span upto which used*	*Circumstances under which employed*
(vi) Timber suspension bridges	60 metres	Where the traffic is very light and constructional materials and skilled labour is not available.
Cause-ways		
(i) Flush cause-ways	No limit	When normal water depth is very small such that traffic can move and the maximum depth of water does not exceed 1.7 metre in flood and total interruption does not exceed 15 days in one year.
(ii) Low-level cause-ways	— do —	When normal water depth is about 30 cm and maximum flood does not interrupt the traffic for more than 72 hours at a time.
(iii) High-level cause-ways	No limit	When the river has small width with good soil near the bed and duration of heavy flood does not exceed 3 days at a time and not for more than six times in a year.
Masonry-bridges		
(i) Filled spandrel arch bridge	60 metres	Where the ratio of rise to span is less. It is provided for small bridges, culverts and where other constructional materials are not easily available.
(ii) Open-spandrel arch bridge	60 metres	Where the ratio of rise to span is more and traffic loads are more as compared with the dead-load of the arch
Steel-bridges		
(i) Trough-plate	5 metres	For small bridges and culverts.
(ii) Girder-bridges	15 metres	For small railway bridges where span is less and loads are very heavy causing impacts and vibrations.
(iii) Arch bridges	250 metres	For very long span bridges, where bed soil is not good and it is not possible to construct intermediate piers, but side banks are firm and can resist the horizontal thrust of the arch.

Contd.

Contd.

Nature of bridge	*Span upto which used*	*Circumstances under which employed*
(iv) Bowstring-bridges	250 metres	Same circumstances as in steel arch-bridge, but when side banks cannot resist the horizontal thrust of an arch, these are provided because horizontal tie takes this thrust.
(v) Rigid-frame bridges	20 metres	For small spans where more clearance is required below the bridges and time required for the construction is less.
(vi) Suspension bridges	1300 metres	For very long spans, where other types of bridges are not economical and to give aesthetic beauty to the bridge.
(vii) Truss-bridges	80 metres	When the live loads are very heavy causing vibrations and impacts.
(viii) Movable bridges	20 metres	In navigation channels where it is not possible to construct bridge super structure at such a higher level that navigation ships can pass below it.
R.C.C Bridges		
(i) Slab type	10 metres	For small span bridge.
(ii) T-beam and slab type	20 metres	For moderate span bridges where the traffic is heavy but causes no sudden impacts on the bridge.
(iii) Rigid frame-type	16 metres	For small span when more clearance is required below the bridge.
(iv) Double cantilever type	70 metres	For very long bridges having many spans.
(v) Barrel arch type	70 metres	For medium traffic load where side banks are firm to resist the horizontal thrust of the arch.
(vi) Open-spandrel arch type	250 metres	For long span where the banks are firm.
(vii) Bowstring girder type	100 metres	For long spans where side banks are not firm and cannot resist the horizontal thrust of arch.

34.9 FEW IMPORTANT INDIAN BRIDGES

Following are the details in brief of few important Indian Bridges.

1. Ganga Bridge at Allahabad (U.P.) on National Highway No. 2.

This bridge has 12 spans of 159.18 m. The 7.5 m wide roadway with 1.52 m wide foot-path on either side has been provided. The three end spans of tied cantilever type in prestressed concrete consisting of 13 units of 159.18 m have been provided. Each superstructure unit has been supported on independent circular well foundation of 5.48 m diameter and average depth of 30 m.

The total length of the bridge is 2084.6 m. The cost of construction of the bridge was Rs. 10,816 per m.

2. Beas Bridge at Mirthal on National Highway No. 1 A. This bridge has 13 spans if 44.8 m and two spans of 46.63 m. The 8.3 m wide roadway with 2 m wide foot-path on either side has been provided. The superstructure mainly consists of pre-stressed girders on each span. The well foundations about 220 m deep have been provided in the bridges.

The total length of the bridge is 673 m. The cost of construction was Rs. 20,089 per metre.

3. Chambal Bridge near Dholpur on National Highway No. 3. This bridge consists of 4 R.C.C. fixed arches of 43.58 m span and 5 R.C.C. fixed arches of 33.52 m in the middle. All these arches are supported on the well foundations taken deep upto the hard rock. On both the sides 26 Nos. R.C.C. units of rigid frames having spans ranging from 13.71 m to 16.45 m have been provided. This is submersible type bridge having collapsible railings, the road level being 8.22 m (average) below H.F.L.

The total length of the bridge is 742.17 m. The cost of construction of the bridge was Rs. 5,915 per metre.

4. Kali Nadi Bridge on National Highway No. 17 (Mysore). This bridge consists of central spans of 121.92 m and four side spans on both sides of 75.34 m each, 2 penultimate spans of 66.49 m each and two end spans of 19.4 m.

All the intermediate spans have prestressed cantilever type beams which are constructed monolithic with the piers and hinged at the centre of span. Well foundations about 24.4 m deep have been provided upto rock bed.

Two lane roadway with 1.52 m foot-path on both sides have been provided. The total length of the bridge is 666 m. The cost of construction was Rs. 21,217 per m.

5. Baitarin Bridge near Akhuapada on National Highway No. 5 (Orissa). This bridge has R.C.C. deck slab type superstructure supported on prestressed concrete girders. The span of the prestressed girders is 40.53 m centre to centre. R.C.C. well foundations 5.75 m diameter and about 21.5 m depth have been provided. R.C.C. hollow cylindrical type piers 9.74 m high connected by diaphragm are used to support the superstructure. The bridge has 7.5 m wide roadway about footpaths.

The length of the bridge is 957 m. The cost of construction of the bridge was Rs. 10,637 per m.

6. Ghogra Bridge at Dhorighat on National Highway No. 23 (Himachal Pradesh). The bridge has 10 spans of 54.86 m and two spans of 51.81 m with end spans of 7.62 m each. The superstructure mainly consists of simply supported prestressed concrete girders. Well foundations of about 43.89 m deep wells are provided. Prestressed concrete circular piers with R.C.C. hammer heads are provided to support the superstructure. The 2 lane roadway of 7.5 width with 1.5 m foot-paths on both sides has been provided.

The total length of the bridge is 647 m. The cost of construction was Rs. 23,184 per metre.

7. Krishna Bridge near Araldinne on National Highway No. 13 (Mysore). This bridge has 30 spans of 26.8 m. It has open foundations on the rock. It has R.C.C. column type piers supporting the prestressed concrete superstructure. The bridge has two-lane road way of 7.5 m width with 1.53 m wide foot-path on each side.

The total length of the bridge is 805 m. The cost of construction was Rs. 6,204 per metre.

8. Bhagao Bridge on National Highway No. 8 A (Gujarat). This bridge has 130 spans of 9.14 m each. It has open type foundations with R.C.C. superstructure. This is a submersible bridge having total length of 1237.8 m. The cost of construction was Rs. 2,834 per m.

9. Sone Bridge at Dehrian Sone on National Highway No. 2 (Bihar). This bridge has 93 spans of 32.9 m each. The superstructure consists of prestressed concrete girders. Well foundations 7 m diameter and about 21.3 m depth have been provided. If has R.C.C. piers. The 7.3 m wide roadway with 1.52 m wide foot-paths on each side have been provided.

The total length of the bridge is 3060 m. The cost of construction was Rs. 8,802 per m.

10. Yamuna Bridge near Humayun's Tomb on National Highway No. 14 (Bye-pass) - (Delhi). This bridge has 11 spans of 42.36 m and two spans of 41.3 m. The superstructure mainly consists of prestressed concrete girders supported on R.C.C. piers. The bridge is constructed in two independent halves except of intermediate verge which is supported on two halves. Circular well foundations about 38.5 m deep have been provided. Two carriageways of 7.9 m each with 1.8 m wide central verge and 3.04 m foot-paths on each side have been provided.

The total length of the bridge is 549 m. The cost of construction was Rs. 10,965 per m.

REVIEW QUESTIONS

34.1. (a) What is a "Skew Bridge"? Explain why it is not usually adopted in bridging major river, even though the alignment of a highway or a rail-road warrants its adoption.

(b) What is the difference between a "Deck" and "Through" type bridge? What are the deciding factors for selecting the former or the latter? Give sketches of both types.

34.2. What are the main classification of bridges? Differentiate between permanent and temporary bridges.

[**Hint**. When the bridges are constructed for short time, such as during construction of Dams, permanent bridges, in crossing rivers during Magh Melas, during survey work for preparing projects, during repair work of permanent bridges, forests for transporting timber from one bank to another, these are known as temporary bridges, because when the object of their construction is fulfilled they are dismantled. But temporary bridges are also constructed where traffic is low and funds do not allow the construction of a permanent bridge.

On the other hand permanent bridges are built to last for centuries. On very important routes, where traffic is heavy and temporary bridges are unsuitable, permanent bridges are constructed to cross the rivers or other such obstructions continuously.]

34.3. What are the factors which affect the selection of the type of a bridge to be provided for crossing rivers?

34.4. Name few important Indian Bridges, with their short description.

General Principles of Design

GENERAL

The general principles of design of bridges depends on so many factors. These may be broadly classified as follows:

1. *Site where to cross the river.* The selection of site is done considering the economy and safety.
2. *Amount of water,* which will pass below the bridge.
3. *Design of bridge,* including foundations, piers, wingwalls, abutment and superstructure is done on the basis of both the above points and so many other factors to be dealt with latter.

35.1 SELECTION OF SITE

The selection of site for a bridge is very important from safety and economy point of view. At some places it is decided by the alignment of the road or railway, because sometimes it is better not to change the alignment even if the bridge is not economical. Generally where bridges are provided to existing Railway Lines or Roads the site will be governed by the alignment of the Road or Railway as the case may be, but for new alignment it may depend on other factors too.

Usually, in the case of major bridges, the site is governed by economic and political considerations, therefore, the engineer has little choice in the selection of the site. The place of site is determined by the position of road or railway and mainly there are three possible crossings of the road or railway line over a river:

1. Simple right-angle stream crossing.
2. Stream crossings on skew angle.
3. Side crossing.

But if the position of railway line or road alignment does not influence the site, the following factors govern the selection of an ideal site for a bridge.

(a) *Straight reach of the stream.* Both the upstream and downstream sides of the river should be reasonably straight, because this will ensure smooth and uniform flow throughout the entire length of the bridge which will avoid eddies and scour. If the site is selected on a curve or loop, the river will have a tendency to change its course by eroding the banks, therefore such site should be avoided.

(b) *Well defined firm banks.* At the site of the bridge, both side banks should be permanent and high, because it will save river training works. These side banks should be free from erosion, collapse or sinking by impact of water in high winds and floods. Firm banks also provide good foundation for the construction of towers of suspension bridges.

(c) *Stream line flow.* At the site of the bridge the river should have stream line flow, because turbulent flow causes scour in the bed. The velocity of water should not be more so as to wash the soil between the piers and undermine the foundations.

(d) *Minimum width and right angle crossing.* This means cheapest bridge. Wherever possible, small streams may be diverted to cross a right angle instead of introducing curves in the railway line making skew-bridge.

(e) *Firm foundations.* The nature of the soil in the bed must provide good foundation. Hard rock or soil should be available near the bed for the economic construction of foundations. If the stream is small and the bed is good the foundation can be laid on hard moorum.

(f) *Dry bed of approach embankments.* The bed of the approach embankments should be dry and hard to bear the weight of the high embankments. Site should be such that approaches do not involve heavy earth work.

(g) *Availability of constructional materials.* The situation of bridge should be such that plenty of good, hard and durable materials, for the construction of bridge, are available near the site.

(h) *Labour.* A site where labourers for the construction of bridge are available in nearby localities will be preferable.

Actually an ideal site is never found in practice. At every site some of the above mentioned ideal conditions are lacking. Therefore in practice while making selection of site the site having least objections should be selected. All these things depend on the experience of the engineer and his judgement.

35.2 SCOUR

When the depth of the water below the bridge increases due to vertical cutting of the bed, it is known as *Scour*. In other words, further deepening of river or cutting of bed due to the action of water is called *Scouring*. Scouring should not be understood as horizontal widening of river, because it is then known as *Erosion*.

If no steps are taken to prevent scouring, the soil below the bridge and around the piers will be washed and it will undermine the foundations. When all the soil around the piers will be washed scouring will be started in the foundation. The bearing capacity of the soil will be reduced and the pier will sink down, causing the failure of the bridge.

For the prevention of scour, the following measures should be taken :

(a) The site of the bridge should have stream-line flow.

(b) The slope of the pier should be designed in such a way that it may not cause eddies and currents in the water.

(c) Sufficient water-way should be provided below the bridge, so that velocity of water may not exceed the limit after which scouring starts or the erodable particles of the bed material do not stand.

(d) At the site of the bridge the particles of the bed should have such a shape, weight and intercohesion property which can stand the high velocity of water.

(e) The river bed on upstream side, downstream side and portion below the bridge should be properly pitched with heavy and long stones.

(f) To prevent scouring piles can be driven in the river bed.

(g) If river bed consists of sand, both on upstream and downstream sides of the bridge, sheet piling should be driven, which will prevent scouring.

There are various methods for the determination of maximum scour depth, but the method which is commonly used in India is based on Lacey's regime equation

$$D = 0.473 \left(\frac{Q}{f}\right)^{1/3} \times R \qquad \text{... (35.1)}$$

where

D = Scour depth in metres

Q = discharge in cu.m/sec.

f = Lacey's silt factor

R
- = 1.27 for straight reach
- = 1.50 for moderate bend
- = 1.75 for sharp bend
- = 2.00 for right-angled bend
- = 2.00 for noses of piers

The value of Lacey's silt factor f is approximately calculated from 1.76 $\sqrt{m}$, where m is the weighed mean-diameter of the river bed particles in mm.

The values of f are generally taken as per Table 35.1.

TABLE 35.1

S.No.	*Types of bed material*	*Mean dia. of particle in mm*	*Value of f*
1	Very fine Silt	0.052	0.400
2	Fine Sill Godavery river (Western India)	0.081	0.500
3	Fine Silt	0.120	0.600
4	Fine Silt Kistna Western Delta type	0.158	0.700
5	Medium Silt	0.233	0.850
6	Standard Silt	0.323	1.000
7	Medium Sand	0.505	1.250
8	Coarse Sand	0.725	1.500
9	Fine bajri Sand	0.988	1.750
10	Large *size* Sand	1.290	2.000

In quasi-alluvial streams having rigid banks and erodable beds, the normal scour depth, in case of streams having large width as compared with having small width, are determined by the following formula :

(a) $$D = \frac{Q}{W.V.} \quad \text{... (35.2)}$$

(b) $$Q = \frac{1.1}{N} W.D^{5/3}\sqrt{S} \quad \text{... (35.3)}$$

(c) $$D = \frac{1.2\ Q^{0.63}}{f^{0.33} \times W^{0.60}} \quad \text{... (35.4)}$$

where

W = The surface width of the stream

V = Velocity of flow

N = Manning's coefficient

S = Slop

f = Silt factor

Normal scour depth in case of quasi-alluvial streams for contracted water-way is determined by Eq. 35.1.

For natural streams the maximum scour depth is not uniform. Table 35.2 gives the maximum scour depth in streams under different conditions.

TABLE 35.2

S.No.	*Conditions of flow in the stream*	*Maximum, Scour depth. Dm.*
1	For straight reach	1.27 D
2	At moderate bend	1.50 D
3	At sharp bend	1.75D
4	At right angle bend	2.00 D
5	At upstream noses of guide banks	2.75D
6	At the noses of bridge piers	2.00 D

It has been seen that maximum scour occurs at the upstream corners in case of abutments, whereas in piers it occurs at the downstream nose end.

Sometimes in the above formula the value of R is determined by

$$R = \frac{w}{L}^{0.61} \quad \text{... (35.5)}$$

where

w = Lacey's regime width of river

L = width of water-way below bridge.

The most effective measure to prevent scouring is to reduce the velocity of the stream.

Example. *Determine the maximum scour depth in case of an alluvial stream carrying maximum discharge of 450 m^3/sec, when the silt factor is*

1.1, in the following conditions :

(a) when the bridge to be constructed has three spans of 45 m each.

(b) when the bridge to be constructed has four spans of 25 m each.

Solution Regime surface width of the stream

$$W = 4.8\sqrt{Q}$$

where

Q = Flood discharge = 450 m^3/sec

$\therefore \quad W = 4.8\sqrt{450} = 101.81$ m

Case 1. Since the proposed bridge consists of three spans of 45 m each

Therefore $\qquad L = 3 \times 45 = 135 > W$

Therefore, Normal scour depth = Regime depth

$$\therefore \qquad D = 0.473\left(\frac{Q}{f}\right)^{1/3} = 0.473\left(\frac{450}{1.1}\right)^{1/3} = 3.5 \text{ m}$$

The maximum scour depth in this case will be at the noses of pier.

$\therefore$ Max. Scour depth $= 2D = 2 \times 3.5 =$ **7.0 Ans.**

Case 2. In this case as the bridge will consist of four spans of 25 m each

$$\therefore \qquad L = 4 \times 25 = 100 \text{ m} < W$$

As the water-way is less than the natural water-way, hence the normal scour depth.

$$D' = D(W/L)^{0.61}$$

$$= 0.473\left(\frac{450}{1.1}\right)^{1/3}\left(\frac{101.81}{100}\right)^{0.61} = 3.5 \text{ m}$$

The maximum depth in this case will also occur at the nose of the piers,

$\therefore$ Max. scour depth $= 2D' = 3.5 \times 2 =$ **7.0 Ans.**

35.3 AFFLUX OR BACKWATER

It is the rise in the level of the water surface of a water-course above the normal on the upstream side of a bridge. This rise is due to the obstruction caused by the bridge abutments and piers in the flow of the water. *Afflux* is taken as the difference of levels of the downstream and upstream water surface of the bridge.

The natural water-way of the rivers or streams is usually greater than the water-way below the bridge, which causes afflux. Usually in plains the

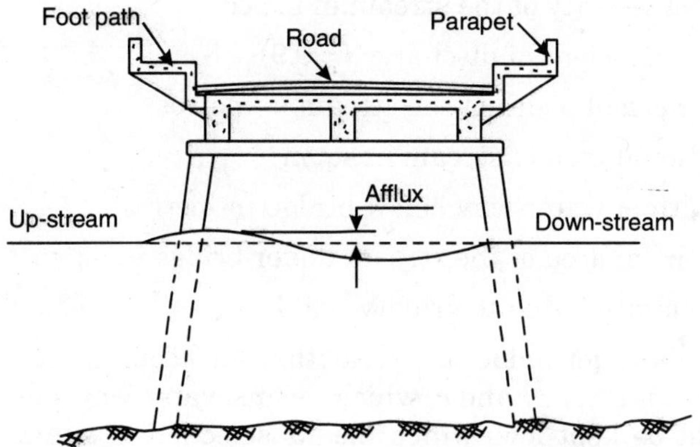

Figure 35.1 *Cross-section of a bridge showing afflux*

natural stream widih is in excess of that required for the regime of the alluvial rivers, therefore, **Water-ways** below the bridges are always kept less for the purpose of stability. Afflux helps in :

(i) Determining the regime of the streams channels or rivers upstream of the bridge.

(ii) Determining the height of protective and river training works.

(iii) Determining the height of the underside of the bridge.

Due to the following reasons afflux should be kept as low as possible.

(a) If the afflux is greater, it will increase the velocity of water in the downstream which will cause more scour and deep foundations will be required.

(b) When afflux is more, the water level will be high and this will submerge more area of land, therefore high and long guide banks will be required for protection works, causing increase in cost.

Calculation of afflux. Generally afflux is calculated by Merriman's formula,

$$x = \frac{50V^2}{g} \frac{l^2}{c^2 l^2} - 1 \qquad \text{... (35.6)}$$

as Molesworth's formula,

$$x = \frac{V^2}{17.8} + 0.015 \quad {\frac{A}{a}}^{3} - 1 \qquad \text{... (35.7)}$$

where

x = afflux in metres

V = mean velocity of the stream in m/sec

c = the coefficient of discharge (= 0.9)

L = the normal width of the stream in metres

A = sectional area of stream in sq. m

l = length of water-way below bridge in metres

a = sectional area of the stream under bridge in sq. m

g = acceleration due to gravity (981)

From the above formulae it is clear that for keeping the value of afflux low, we should increase l and c, which means water-way should be more. V and A should be kept low, which means selection of site should be done where velocity and area of the natural stream are as low as possible.

Afflux may be taken 60 cm in alluvial and deltaic regions, 90 to 120 cm in trough regions and higher in steep reaches of rivers with boulders and rocky beds for rough calculations.

Example 1. *The flood discharge under a bridge is 974 m^3/sec. If the bridge site is on a moderate bend and Lacey's silt factor for river bed is 1.5, determine scour depth.*

Solution By Lacey's regime formula

$$\text{Maximum scour depth, } D = 0.473 \left(\frac{Q}{f}\right)^{1/3} \times R$$

where

D = maximum scour depth in metres

Q = maximum flood discharge in m/sec

f = Lacey's silt factor

R = constant, whose value on moderate curve is 1.5

$$D, \text{ for moderate bend} = 0.473 \left(\frac{Q}{f}\right)^{1/3} \times 1.5$$

$$= 0.473 \left(\frac{974}{1.5}\right)^{1/3} \times 1.5$$

$$= \mathbf{6.14\ metres.\ Ans.}$$

Example 2. *In the above example, if the normal stream width is 481 m and the length of the water-way below the bridge is 400 m, calculate the afflux, if the mean velocity of the approach is 1.5 metre / sec.*

Solution From Merriman's formula

$$\text{Afflux } x = \frac{50V^2}{g} \; \frac{L^2}{c^2 l^2} - 1$$

$$= \frac{50 \times 1.5^2}{g} \; \frac{481^2}{(0.9 \times 400)^3 - 1}$$

$$= 0.069 \text{ m}$$

$$= \mathbf{6.9 \text{ cm. Ans.}}$$

Example 3. *A bridge has a linear water-way of 185 m. If the natural linear water-way is 245 m and the average flood discharge is 1400 m³ / sec, determine the value of afflux under the bridge by Molesworth formula, if the average flood depth is 3.5 m.*

Solution Natural waterway area

$$A = 245 \times 3.5 = 857.5 \text{ sq. m}$$

Contracted water-way $a = 185 \times 3.5 = 647.5$ sq. m

The velocity of approach $V = \dfrac{Q}{A} = \dfrac{1400}{857.5} = 1.63$ m/sec.

The value of afflux by Molesworth formula

$$x = \frac{V^2}{17.9} + 0.025 \quad \frac{A}{a}^{2} - 1$$

$$= \frac{1.63}{17.9} + 0.015 \quad \frac{857.5}{647.5}^{2} - 1$$

$$= (0.091 + 0.015)\,[1.753 - 1]$$

$$= 0.106 \times 0.753$$

$$= 0.08 \text{ m}$$

$$= \mathbf{8 \text{ cm. Ans.}}$$

35.4 RUN-OFF

When the rain falls on the ground, a portion of it is soaked or penetrates in the ground and some portion is evaporated in the atmosphere and the rest flow on the ground. The portion of the rain-fall on a catchment area, which flows to water course is known as *Run-off.*

The run-off mainly depends on the following factors:

(a) Area of the catchment area.

(b) Slope and shape of the catchment area.

(c) The degree of porosity of the soil of the catchment area.

(d) Obstruction in the flow of water as trees, fields, gardens etc.

(e) Initial state of the catchment area with respect to wetness.

(f) Intensity of rain-fall in the catchment area.

(g) Duration of rain-fall.

35.4.1 Determination of Run-off

The run-off which reaches at the site of the bridge can be determined by empirical formulae or by taking actual observation.

The following empirical formulae are commonly used for the calculation of the run-off.

(1) Dicken's formula :

$$Q = CA^{3/4} \qquad \text{... (35.8)}$$

(2) Ryve's formula :

$$Q = CA^{2/3} \qquad \text{... (35.9)}$$

where

Q = discharge in cu.m/sec

A = catchment area in sq. km

C = an empirical constant depending upon the nature of the catchment area which affects the run-off. It has different values for every formula

Value of C in *Dicken's formula*

= 11.3 for Northern India

= 22 for Western Ghats

= 13.8 to 19.3 for Central Provinces.

Value of C in *Ryve's formula*

= 6.7 for areas within 24 km of sea coast

= 8.42 for areas within 24 to 160 km of sea coast

= 10 for areas near hills.

These formulae take into consideration only catchment area, whereas so many factors govern the run-off. Secondly, these formulae can be applied only to restricted zones for which they have been derived, and they give very rough value of the run-off. Therefore these formulae are of very limited used and are used in reconnaissance survey.

(3) Inglis Formula:

This formula is used only in Maharashtra State. Following three different cases are used:

(a) For fan-shaped catchment small area

$$Q = 123.2\sqrt{A} \quad \text{... (35.10)}$$

(b) For larger areas between 160 to 1000 sq. km

$$Q = 123.2\sqrt{A} - 2.62\ (A - 259) \quad \text{... (35.11)}$$

(c) For all types of catchment areas

$$Q = \frac{123A}{\sqrt{A + 10.36}} \quad \text{... (35.12)}$$

(d) Nawab Jang Bahadur's formula

$$Q = C\ \frac{A}{2.59} = (a - b\ .\ \log\ .\ A) \quad \text{... (35.13)}$$

where a, b and c are constants.

$a = 0.993$

$b = 0.0714$

$c = 59.5$ for North India

$= 48.1$ for South India

(4) Khosla's Formula:

This formula depends on the losses, which depend on the temperature.

The run-off by Khosla's formula

$$R = P - 0.482\ T_m \quad \text{... (35.14)}$$

where

R — run-off

P — rainfall

T_m — mean temperature in centigrade.

The above formula is applicable when the atmospheric temperature is above 40° F.

(5) Creager's formula:

$$Q = 46°CA^{(0.894A - 0.043)} \quad ... (35.15)$$

(6) Rain-fall method:

The run-off is also calculated by the formula—

$$Q = 100.\ A.P.I_c.f \quad ... (35.16)$$

where

Q discharge of run-off in cu. m/hour

A catchment area in hectares

P percentage coefficient of run-off

I_c critical intensity of rain-fall in centimetres per hour

f constant depending on the catchment area of the stream

The method is also known as 'Rational Method'. The value of coefficient P varies from place to place, depending on the obstruction caused in the flow of run-off water. Some portion of water percolates in the ground, some portion is filled in natural ponds, lakes and reservoirs etc.

The run-off coefficient is a constant, which when multiplied with the total quantity of rain-fall, gives the quantity of water reaching the site of bridge.

Table 35.3 gives the common values of run-off coefficient which are mostly used for determining the discharge of run-off.

TABLE 35.3 *Run-off coefficient for various surfaces*

S.No.	*Type of surface*	*Value of P*
1	Forest and wooded area	0.01 to 0.20
2	Open grounds, unpaved streets and rail road yards	0.10 to 0.30
3	Parks, lawns, meadows and gardens	0.10 to 0.25
4	Gravel roads and walks	0.15 to 0.30
5	Macadam Roadways	0.25 to 0.60
6	Inferious stone, brick or blocks open joint pavements	0.40 to 0.50
7	Stone, bricks and blocks pavements with joints filled with mortar	0.50 to 0.70
8	Good quality pavements and roads	0.75 to 0.85
9	Asphalt pavements in good condition	0.85 to 0.90
10	Water-tight roof surfaces and other such covered surfaces	0.70 to 0.95

TABLE 35.4 *Value of f*

S. No.	Catchment area of the Stream in hectares	Values of f
1	Nil	1.0
2	4,000	0.85
3	8,000	0.76
4	12,000	0.70
5	16,000	0.67
6	20,000	0.67
7	40,000	0.62
8	80,000	0.60

As every locality consists of different types of surface areas, therefore for calculating the overall run-off coefficient the following formula is used:

Run-off coefficient P (overall)

$$= \frac{A_1P_1 + A_2P_2 + ... + A_nP_n}{A_1 + A_2 + ... + A_n}$$

$$\frac{A.P}{A} \quad \text{... (35.17)}$$

where A_1, A_2, A_3, ... are the different types of areas and P_1, P_2, P_3, ... are their run-off coefficients respectively.

From Eq. 35.16 it is clear that the run-off depends on :

(1) *Catchment area.* If the catchment area is large, run-off will be more and if it is small the run-off will be less.

(2) *Percentage coefficient of the run-off.* This depends on the nature of the soil of the catchment area. If percentage coefficient is more, run-off will be more and *vice-versa*.

(3) *Intensity of rain-fall.* The run-off directly depends on the intensity of rain-fall in the catchment area. The more is intensity of rain-fall, the more will be run-off.

35.5 CRITICAL INTENSITY OF RAIN-FALL

Intensity of rain-fall in centimetres per hour is calculated by the formula :

$$i = \frac{T+1}{t+1} \times \frac{F}{T} \quad \text{... (35.18)}$$

where

i = intensity of rain-fall in centimetres

T = duration of the rain-fall in hours

t = the time in hours for which intensity of rain-fall remains maximum

F = total rain-fall in centimetres

From the above Eq. (35.18) it is clear that if t will be shorter, the intensity of rain-fall will be higher. But if t is short, the time taken for run-off from the distant part of the catchment area will be more and it will not reach at the site of the bridge.

The intensity of rain-fall for which the discharge is maximum, when it remains for concentrated period, is known as the critical intensity of rain-fall.

If I_c be the critical intensity of rain-fall:

$$I_c = \frac{T+1}{T_c+1} \times \frac{F}{T} \qquad \text{... (35.19)}$$

where T_c is the concentration time in hours.

Time of concentration by Richard's formula

$$T_c = \phi \frac{L^3}{H}^{1/3} \qquad \text{... (35.20)}$$

where

T_c = time of concentration

L = the distance from the critical point to the site of bridge or culvert

$$\phi = \frac{9}{4} \frac{5280}{C^2.I.K}$$

where

K = coefficient of runoff

C = Bazin's coefficient

I = intensity of rainfall

The value of ϕ for most of the cases may be taken as 0.89.

Time of Concentration by Danson's formula

$$T_c = \phi \frac{L^3}{H}^{1/3}$$

where $$\phi = \frac{5280}{C\sqrt{S.I.}} \qquad \text{... (35.21)}$$

where

S = slope of catchment.

L and H have got the same meaning as in Eq. 35.20.

The values of ϕ and θ mainly depend on the intensity of rainfall and nature of the soil in the catchment area.

35.6 DETERMINATION OF MAXIMUM DISCHARGE

There are following three methods for determining the maximum discharge under a proposed bridge:

First Method. In this method the catchment area is measured and the rates of rain-fall in catchment area are determined and then the total quantity of water which will reach the site of bridge is estimated.

The following are the steps which are required in this method :

1. A topographical sheet of the whole area is prepared and after marking the site of bridge, the catchment area from which the water will flow to the site of bridge is calculated.

2. With regard to the size, shape, slope, strata, etc. the nature of the catchment area is determined. At one time rain does not fall simultaneously on a wide area. Therefore if the catchment area is very wide at a time, the rain will fall on smaller portion of it and lesser percentage of run-off will be taken. On the other hand if the catchment area is small, we will take more percentage of run-off.

3. The intensity of rain-fall is determined.

4. The percentage coefficient of run-off is determined which generally varies from 20 – 70% depending on the nature of the catchment area.

5. The discharge is calculated by the formula 35.16 after substituting the values of A, P and I in it.

Second Method. In this method the discharge is calculated by formulae 35.8, 35.9, 35.10, 35.11 or 35.12. These are not very accurate methods and give only approximate discharge.

Third Method. In this method following steps are required for the calculation of discharge or run-off.

1. The series of levels are taken along the cross-section of the river at the side of the bridge.

2. From these levels the cross-section of the river is plotted and H.F.L. line is marked on the cross-section.

3. Two cables 30 metres apart are stretched across the cross-section and are divided in equal parts as shown in Fig. 35.2, for determination of the velocity of water.

4. The width of each part is marked on the cross-section and the area of each part is calculated.

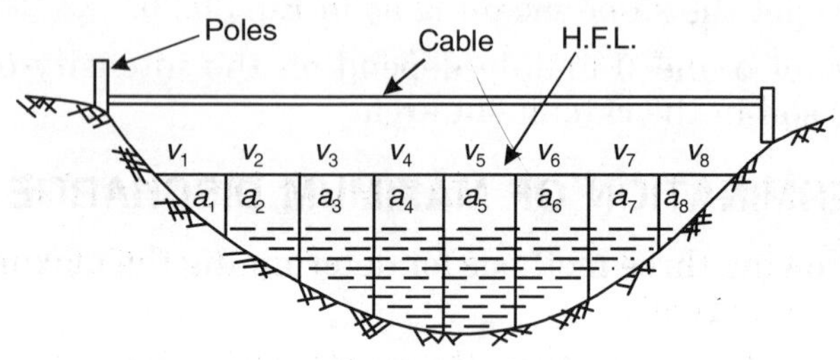

Figure 35.2

5. The mean velocity of water of each part is calculated by means of surface float, floating rod or current meter.

6. Each area is multiplied by the mean velocity of water of its section, and these all are added together which will give the total maximum discharge.

$$Q = v_1a_1 + v_2a_2 + ... + v_na_n. \quad ...(35.22)$$

where

Q = discharge in cu metres per second

a = area of section in sq. metres

v = mean velocity of water of the same area in metres per second.

Example. *The catchment area of a river is of forest and wooded area. The total catchment area is 1,200 hectares, and the length of the catchment is 35 km and the fall in level from the critical point to the site of bridge is 133 m. If the maximum rain-fall recorded the catchment area is 15 cm in three hours, determine the quantity of maximum discharge which will pass through the bridge.*

Solution The intensity of rain-fall in the catchment area per hour from Eq. 35.18.

$$i = \frac{T+1}{t+1} \times \frac{F}{T}$$

$$= \frac{3+1}{1+1} \times \frac{15}{3}$$

$$= 10 \text{ cm/hr}$$

From Eq. 35.20, the time of concentration

$$T = \phi \frac{L^3}{H}^{1/3}$$

Putting

$$\phi = 0.82$$
$$L = 35 \text{ km}$$
$$H = 133 \text{ m}$$

$$\therefore \quad T_c = \frac{0.89 \times 35^3}{133}^{1/3} = 6.59 \text{ hr}$$

From Eq. 35.19, the critical intensity of rain-fall

$$I_c = \frac{T+1}{T_c+1} \times \frac{F}{T}$$

$$= \frac{3+1}{6.59+1} \times \frac{15}{3}$$

$$= 2.635 \text{ cm/hr}$$

From Eq. 35.16 maximum discharge is the run-off which will pass through the bridge

$$Q = 100\, A.P.I_c.f$$

From Table 35.3, value of $P = 0.10$

and from Table 35.4, value of $f = 0.70$

$$\therefore \quad Q = 100 \times 12{,}000 \times 0.10 \times 2.635 \times 0.70 \text{ cu. m/hr}$$

$$= \mathbf{61.483 \text{ cu.m/sec. Ans.}}$$

35.7 WATERWAY

For deciding the waterway the first thing to be determined is the maximum expected flood discharge which will pass under the bridge. *The water way is the area of the opening, which should be sufficient to pass the maximum flood discharge that would ever pass under the bridge, without increasing the velocity to a dangerous limit.* The maximum velocity depends upon the nature of the bed of the river. But in ordinary cases 3 m/sec is the maximum permissible velocity.

The velocity of river below bridge should not be more. The maximum permissible limit depends on the nature of soil in the bed. Table 35.5 gives the permissible limit of river velocity in various types of soils.

TABLE 35.5 *Permissible river velocity in different soils*

S.No.	*Type of soil*	*Permissible velocity m/sec*
1	Clayey soil	1.5 to 2.0
2	Sandy clay	1.00 to 1.50
3	Fine sand	0.60 to 0.90
4	Soft stone or gravel	1.5 to 1.7
5	Medium stratified stones	1.8 to 3.0
6	Hard stone bed	3.0 to 6.60

When the maximum flood discharge is calculated the water way can be calculated by the formula

$$Q = V \times A \quad \text{... (35.23)}$$

where

Q = maximum flood discharge

V = permissible velocity under the bridge

A = area of the waterway

If h and ha be the head causing the velocity of normal stream and afflux respectively

$$Q = \text{Length} \times \text{Depth} \times \sqrt{2G(h + ha)} \quad \text{... (35.24)}$$

The length is divided into required numbers of spans. If the spans are more, the cost of the bridge will be more. Total numbers of spans is calculated by considering the economical design of the bridge.

When the waterway is restricted to such an extent that the resultant afflux will cause the stream to discharge at erosive velocities, proper protection against damage by scour should be done by providing deep foundations, curtain or cut-off walls, rip-rap, bed pavement, bearing piles, sheet piles or some other suitable works. Similarly embankment slopes adjacent to all structures subject to erosion should be adequately protected by pitching revetment walls or other suitable construction.

35.8 ECONOMIC SPAN

When the site for a bridge is fixed with reference to clear waterway, flow of water, alignment of road or railway etc., the length of each span will have to be decided keeping in view the depth of the foundation, height of pier and the nature of the river bed at the site of the bridge. The span for which the total cost of the bridge will be minimum, is known as the *economic span* of the bridge.

The length of each span depends on :

(a) the nature of river bed

(b) the depth and type of foundations required, and

(c) the height of piers required.

In navigation canals and rivers, the span, height of piers and clearance depends on the size of biggest vessels passing under the bridge.

Derivation of formula. Before deriving following assumptions are made :

(i) Generally the width of the road or railway bridge is fixed, therefore the cost per unit run-off dock slab and parapet walls will be a fixed sum, which will vary proportionally to the span.

(ii) While designing main girders or trusses bending moment is taken as $\frac{wl^2}{8}$ where w is the road per unit length and l is the span. Therefore the cost of girder will vary according to the square of the span.

(iii) As the dimensions of every supporting system e.g., pier and foundation is constant, its cost will also be constant.

(iv) Approximately the cost of side abutments with foundations is constant.

If length of bridge $= L$

Total number of spans $= N$

Cost of one pier $= C_p$

Cost of one abutment $= C_a$

Cost of both main girders $= C_g = Kl^2$ (where K is a constant)

Cost of roadway slab $= C_s$

$$\therefore \quad l = \frac{L}{N} \text{ or } N = \frac{L}{l}$$

Now the total cost of the bridge :

$$C = 2C_a + (N-1)\,C_p + N.C_g + C_s$$

$$= 2C_a + \frac{L}{l} - 1\; C_p + \frac{L}{l}C_g + C_s$$

$$= \frac{L}{l}\left(C_p + Kl^2\right) + 2C_a - C_p + C_s \text{ (Putting } C_g = Kl^2\text{)}$$

Differentiating with respect to l,

$$\frac{Cd}{dl} = L \frac{-C_p}{l^2} + K$$

For the cost to be minimum

$$\frac{Cd}{dl} = 0$$

$$\therefore \quad \frac{C_p}{l^2} = K \text{ or } C_p = Kl^2 = C_g \qquad \text{... (35.25)}$$

Hence for the total cost of the bridge to be minimum, the cost one pier should be equal to the cost of main girders or trusses whichever may be the case.

Equation 35.25 is suitable for steel girder bridges, truss bridges and arch bridges.

Indian Road Congress have recommended the following formulae for determining the economic spans for small road bridges and culverts.

(a) For R.C.C. slab bridges

$$L = 1.5\,H \qquad \text{... (35.26)}$$

(b) For masonry arch bridges

$$L = 2.0\,H$$

where

L economic span

H the height of pier above the foundation in case of R.C.C. Bridge

the height of peir above the foundation upto springing line in case of masonry bridge.

Now we know that cost of a bridge will increase with the number of spans because the number of piers have so increased. On the other hand if the span is increased the cost of main girders or trusses will increase which will also increase the total cost of the bridge. From our above derivation, it has been proved that the cost of the bridge will be minimum when the cost of the pier and main girder is equal. This derivation holds good for arch and girder bridges only.

Following approximate 'thumb rules' are generally taken as a guide for determining the length of span in small bridges, where open foundation are constructed.

S. No	Description	Span Length
(i)	R.C.C. slabs on masonry piers	1.5 H
(ii)	R.C.C. beam and slab on masonry piers	1.75 H
(iii)	Masonry arches	$2H$ or more
(iv)	R.C.C. slab on pile bents	0.75 H – 1.0 H
(v)	Steel truss spans on masonry piers	3 H

where, H is the total height of the abutment or pier, from the undersides of its foundation to its top, and for arches to the intrados of the key-stone.

The above thumb rules are not applicable when the steel, cement, aggregates etc. are to be transported from large distances.

Example. *While estimating the cost of a bridge for various spans the following calculations where made. Determine the most economic span length for the bridge.*

Span in metres	*Cost of superstructure per span in Rs.*	*Cost of one pier in Rs.*
10	5,000	16,000
15	8,500	17,700
20	16,700	18,300
25	29,500	20,000

Solution As the cost of superstructure varies directly according to or with the span of bridge.

$$C_g = Kl^2 \text{ or } K = \frac{C_s}{l^2}$$

The value of constant K,

$$\text{For 10 m span} = \frac{5{,}000}{100} = 50$$

$$\text{For 15 m span} = \frac{8{,}500}{225} = 37.8$$

$$\text{For 20 m span} = \frac{16{,}700}{400} = 41.75$$

$$\text{For 25 m span} = \frac{29{,}500}{625} = 47.2$$

$$\text{Average value of } K = \frac{50 + 37.8 + 41.75 + 47.2}{4} = 44.2$$

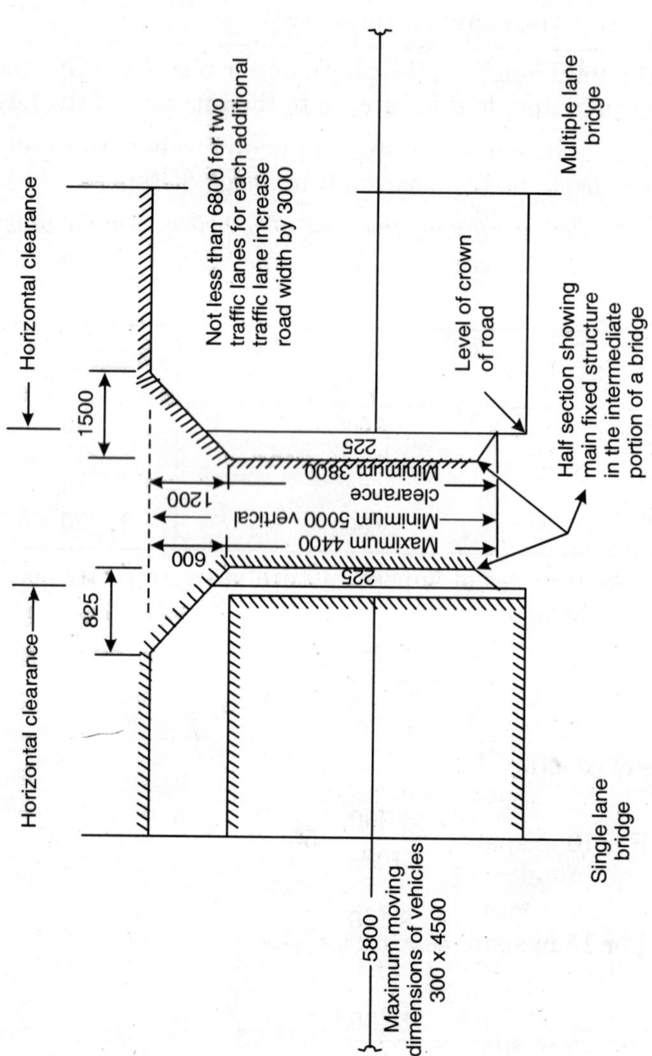

Figure 35.3 *Minimum clearances for road bridges*

$$\text{Average cost of pier} = \frac{16{,}000 + 17{,}000 + 18{,}300 + 20{,}000}{4}$$

$$= 17825$$

$$\therefore \quad \text{Economic span} = \frac{\sqrt{17825}}{44.2} = 20.1 \text{ m (say 20 m)} \textbf{ Ans.}$$

35.9 CLEARANCES

For the free passage of the vehicles in the bridge, horizontal and vertical clearances are provided to protect the vehicles from striking the bridge structure.

In case of bridge constructed on a horizontal curve with superelevated road surface the horizontal clearance must be increased on the inner kerb side, by an amount equal to 5.0 m multiplied by the value of super-elevalion. The minimum vertical clearance distance should be measured from the super-elevated side of the road.

Figure 35.3 clearly shows the recommended minimum clearances.

35.10 DETERMINATION OF LENGTH OF BRIDGE

After determination of waterway and economic span length, the length of bridge is determined. The pier is designed and its dimensions are noted. Then the total length of the bridge will be equal to the width of all the piers plus linear waterway under the bridge.

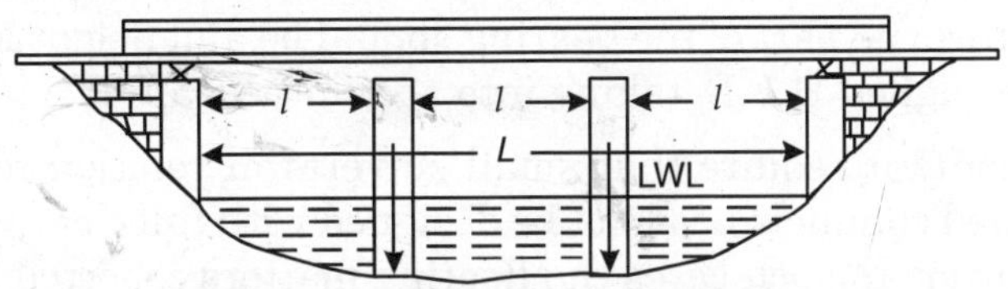

Figure 35.4 *Length of bridge*

If there are N spans of length l and all piers are of width b, the length of bridge is,

$$L = Nl + (N - 1)b \quad \text{... (35.27)}$$

$$= \text{waterway + width of all piers}$$

35.11 FREE BOARD OR VERTICAL CLEARANCE

Free board is the vertical difference between the designed highest flood level, allowing for afflux, and the lowest part of the bridge structure.

Free board is provided for passing floating debris, fallen trees, trunks and boats etc. from one side of the bridge to the other side.

Following values of the free boards are recommended in various types of bridges:

(a) Arch bridge 300 mm

(b) High level bridge 600 mm

(c) Girder bridge 600 to 1000 mm

(d) Navigational stream bridge 2500 to 3000 mm

For arched openings of high level bridges, the clearance below the crown of the intrados of the arch should not be less than l/10th of the maximum depth of water plus l/3rd the rise of the intrados. In small bridges the springing line should be above the high flood level.

As per IRCS-1970 following minimum vertical clearances are recommended.

TABLE 35.6

S. No.	*Discharge in cu. m/sec*	*Minimum Vertical Clearance*
1	Upto 0.3	15 cm
2	Between 0.3 – 3.0	45 cm
3	Between 3.0 – 30	60 cm
4	Between 30 – 300	90 cm
5	Between 300 – 3000	120 cm
6	Above 3000	150 cm

In case of bridges provided with metallic bearings to support the superstructure, no part of the bearing should be at a height less than 50 cm above the designed H.F.L. taking into account the afflux.

For economy it require that small culverts in relation to large bridges across defined channels need not be designed with full clearance for passing floating matter. In such cases the floating matters collected at the culverts are removed by the permanent gangs kept for the maintenance of the canals.

REVIEW QUESTIONS

35.1. Outline and explain the factors which govern the selection of an ideal site of a railway bridge over an alluvial river.

35.2. Discuss briefly the points you will consider for finally selecting the site of a major railway bridge.

35.3. (a) What is "Scour"? What will happen if no steps are taken to prevent scouring, and what measures are adopted for prevention of scour?

(b) What is afflux? Why and how is it kept as low as possible?

The flood discharge under a bridge is 4,764 m^3/sec. If the normal width and waterway are 954 m, and 900 m respectively, determine scour depth and afflux. Lacey's silt factor is 1.5 and bridge site is on straight reach.

35.4. What do you understand by "Run-off of a catchment area"? What are the different empirical formulae adopted in India for calculating the run-off of big rivers?

Also explain why these are regarded as of very limited use, discussing the factors on which run-off largely depends.

35.5. Write short note on critical intensity of rain-fall.

35.6. Describe the major factors that will affect the run-off from the catchment area and state the procedure that you will follow in fixing the water way to be provided for a major bridge.

35.7. Explain the principles on which waterways for bridges are determined. Give details of field data required and computations necessary for deciding upon a suitable waterway for a river crossing a moderate width.

35.8. How is the length of a bridge determined? What factors are considered in determining the depth of piers for a large bridge.

35.9. What do you understand by the economical span length for a bridge? Derive the formula for it and show how it can be applied for different types of bridges?

The approximate costs of one superstructure span including flooring, girders, railing etc. and one pier of multiple span bridge are tabuled below for various lengths of span. Determine the economic span.

Span in Metres	10	15	20	25	30
Superstructure cost in Rs.	4,000	11,000	16,600	23,500	39,700
Cost of one pier in Rs.	17,500	18,700	19,500	20,000	21,800

35.10. (a) Explain step by step, how you will finally determine the expected maximum discharge under a proposed bridge?

(b) On what factors does the intensity of "run-off" from a catchment area depend?

35.11. (a) Write short note on the various methods for determining the run-off from a catchment area.

(b) Calculate the flood discharge from a catchment of 65 sq. km, when the rainfall during a storm was 15 cm in two hours. The time of concentration is 20 hours and the run-off coefficient for the catchment is 0.35.

35.12. Explain briefly the different methods used in estimating the maximum flood discharge of streams. Which method, according to you, is accurate? Give reasons.

35.13. The catchment area of a stream is of sandy soil with light vegetation cover and the area of the catchment is 12000 hectares. The length of the catchment is 25 km and the fall in level from the critical point to the bridge site is 480 m. Calculate the peak run-off designing the bridge, if the severest storm as recorded yielded 18 cm of rain in 4 hours. Assume value of area factor = 0.70 and coefficient to account for lossed due to absorption = 0.20.

35.14. (a) Discuss briefly the available methods for estimating the flood discharge and methods of determining the linear waterway for a bridge.

(b) If the designed maximum discharge in a large natural stream of an alluvial bed is 2500 cu. sq./m, calculate the linear waterway in metres.

Take the value of C in the formula $L - C\sqrt{Q}$ (I.R.C. recommendation) as 4.8. Also determine the maximum depth of scour below H.F.L. around piers constructed on the above stream. Assume Lacey's silt factors as 0.70.

35.15. (a) Explain how would you determine the discharge of a river by any suitable method.

(b) A bridge constructed across a stream has a linear waterway of 400 sq. m and unobstructed water is 500 sq. m. The flood creates an afflux of 0.3 m. Calculate the flood discharge.

35.16. "For maximum economy the cost of one pier including foundation should be equal to the cost of one span of superstructure exclusive of the floor system". Explain how the statement has been arrived at and how far the assumptions involved are true.

35.17. (a) Define 'Normal Depth of Scour'. How will you find the maximum depth of scour?

(b) Calculate the maximum depth of scour and the depth of foundation with the following data :

Linear waterway provided	3 spans of 30.5 m (100 ft) each
Width of stream	106.75 m (350 ft)
Average area of cross-section	465.0 m (5000 sq. ft)
Mean velocity of flow	1.22 m/sec (3 ft/sec)
Silt factor	1.1.

35.18. (a) How is the maximum scour depth estimated in :

(i) An alluvial stream?

(ii) A quasi-alluvial stream having relatively hard banks compared to its bed?

(b) How is the bridge site selected over a meandering river?

(c) What is meant by the term 'free board'? How much free board is provided in the case of various bridges?

35.19. What do you mean by 'Economical span length' in bridges? How is it arrived at? Has it anything to do with the type of structures, you may propose for different situations?

35.20. What are the merits and demerits of this method as compared to the methods of determining discharge, by using :

(i) Empirical formulae and

(ii) The slope and cross-sectional area of the stream?

35.21. Write short note on the free board in bridges.

Foundations

GENERAL

It has been said that he who designs and builds important and difficult foundations does not sleep well at night. The planning and designing of foundations involve some of the greatest engineering skill, though it is inconspicuous and unnoticed by the public.

If a steel girder is too weak and flexible, it can be strengthened; if a column is found to be inadequate, it can be assisted by intermediate columns; if flooring of a bridge is found weak it can be changed or reinforced; but if the foundation yields unevenly and causes settlement, the buildings, piers, abutments, howsoever strong it may be, cannot stand and will tilt badly causing complete failure of the structure. It is not possible to rebuild an inadequate foundation and repair it without incurring much expenditure. Therefore planners and designers of foundations must feel their responsibility.

36.1 FUNCTIONS OF FOUNDATIONS

The following are the functions of the foundation :

(a) To distribute the load of the structure over a large area of the subsoil, so that the intensity of pressure may not cross the safe bearing capacity of the soil.

(b) To prevent tilting and overturning of the piers and abutments.

(c) To prevent the lateral escape of the supporting material of the riverbed, so that piers may not sink and cause the failure of the bridge.

(d) To provide a levelled base for the construction of piers and abutments.

(e) To prevent unequal settlement of sub-soil and super-structure, by transferring the load uniformly over the sub-soil.

36.2 SOIL PRESSURE

The maximum permissible pressure on the sub-soil below the foundation depends on the ultimate bearing capacity of the soil. The bearing capacity i.e., ultimate bearing capacity of the soil can be determined either from various formulae (given in the books of Soil Mechanics) or by plate load test. Plate load test is a common test adopted in determining the bearing capacity of the soil. In this method the load is given on a rigid plate kept at the foundation level and the settlement of the soil corresponding to each load increment is recorded. The ultimate bearing capacity is then determined by drawing graph. The load at which the settlement of soil starts at a rapid rate is generally the bearing capacity of the soil.

Bearing capacity of the soil mainly depends on the nature of soil and its degree of compaction. The size and shape of the foundation also affect it. The permissible bearing capacity of the soil is obtained by dividing the ultimate bearing capacity by suitable factor of safety, which is generally 3. In case of bridges the value of live load generally varies from 5% to 9% of the dead load depending on the span and type of bridge.

If in any bridge foundation the ratio of dead load to maximum live load is $(1 - N) : N$, and the maximum live load = $K \times$ Normal live load;

Then factor of safety over load

$$= F \; 1 - \frac{N}{K}(K-1)$$

The value of F (Factor of Safety) should be assumed as 3 for clay and 2 for sandy soil. As a normal practice the normal live load is taken between 100% to 50% of the maximum live load depending on the length of the span.

36.3 DEPTH OF BRIDGE FOUNDATIONS

The depth of bridge foundations is determined by considering the safe bearing capacity of the soil, and effect of scour. The bridge foundations should be taken below the scour level at the place having minimum variation in the moisture content of the sub soil.

In case of erodible soils, if D is the anticipated maximum scour depth below the designed highest flood level, the minimum foundation depth d should be 1.33 D, subject to minimum of 2.0 below the scour line for arched bridges and 1.2 m for other types of bridges. If the bridge piers and abutments are directly resting on the rocks, the minimum depth should be 0.3 m into rock. In case of very hard soil the foundation must be carried at least 0.6 m inside such hard soil. In case the rock taking the bridge foundation is sloping, the masonry should be anchored to the rock by means of dowel bars by embedding them about 30 cm.

In non-erodible channels or streams for small bridges of moderate height on dry land, 1.5 to 2.0 m depth of foundations may be provided. Depth of foundation on lands not exposed to the erosive action the depth calculated by the Rankine's formula should be used.

When the river bed materials are protected against the scouring by providing bed floor or pavement with certain walls (drop walls) or sheet piles etc., the shallow deep foundations may be provided.

When the waterways has been restricted as well as the velocity is not allowed to exceed 2.0 m/sec, the block or loose stone pitching should be done extending upto the end of the wing walls.

When the waterway has been restricted as well as the velocity is allowed to exceed 2.0 m/sec, for constructing culverts, the bed floor should be paved and a paved apron should be constructed extending upto the end of the wind walls.

If the waterway is not restricted and the velocity is not expected to increase during floods beyond the critical velocity, there is no necessity of providing pavement in the bed.

The width of the foundations of the abutments and the wing walls may not be extended more than 1.0 m parallel to the bridge length.

The design of the pavement should be done keeping in view the upward pressure resulting from the head of water of the stream. The top of the pavement floor should be kept about 30 cm below the natural bed of the stream. To protect the bed paving dwarf walls should be provided on the upstream and downstream sides. The depth of the dwarf wall depends on the velocity of the flow and the erodibility of the bed materials or scour depth, but it should not be less than the depth of the footing.

In channels having provision of flow control the pavements and dwarf walls are not provided unless so designed. In some such cases sometimes a single drop wall on the downstream side is provided which is sufficient. If the outfall of the stream is free it must be provided with a suitable cistern with drop wall.

While calculating the pressure at the base of the bridge sub-structure, no reduction due to skin friction on the sides of the sub-structure should be made.

36.4 TYPES OF FOUNDATIONS

The design of bridge foundation mainly depends on the depth of the foundation at which suitable soil for foundation is available. In case the foundations are required to be carried to a considerable depth, number of piers should be reduced as far as possible. If good rocks are available at shallow depths in hills and the valley is narrow and deep, single span bridge shall be economical. If the valley is shallow, the bridge may have more spans supported on the intermediate piers.

In the case of deep and wide valley, it is better to construct arch bridge, because number of spans shall be reduced as well as maximum rise will be available in these bridges.

The selection of suitable foundation for the bridge mainly depends on the following :

(a) Nature of the subsoil in the bed of the stream.

(b) Presence of boulder, rock kanker, buried free trunks etc. which may cause difficulties during excavation.

(c) L.W.L. and the depth of water at the site.

(d) Availability of the various equipments during construction.

(e) Availability of the skilled workers and the raw materials for the construction.

Broadly, foundations may be classified as :

(i) Foundations constructed on dry land, called land foundations.

(ii) Foundations constructed in soils having water after some depth.

(iii)Foundations constructed under water.

The construction of the first two types is much easier than of the third type, because while constructing under-water foundations, special methods to exclude the water from the area are required, which are called *coffer dams*.

The following are the different types of foundations, which are generally used for the transfer of loads from piers and abutments in bridges —

(a) Spread Foundation or Open Foundation

(b) Raft Foundation

(c) Pile Foundation

(d) Well Foundation

(e) Caisson Foundation

36.5 SELECTION OF FOUNDATION

The selection of the type of a foundation depends on so many factors. The factors which effect the selection are, the nature of the soil, type of the bridge, velocity of water and superimposed load on the bridge. Different types of foundations are provided under different conditions which are as follows :

(i) *Spread Foundation*. If the good and hard soil is available within 2.3 metres below the bed level of the river, spread foundations are provided. This type of foundation is best suited at such places where scouring is minimum. If the bed contains sand the scouring can be prevented by sheet piling in upstream and downstream and by pitching of the bed floor. This type of foundation can also be provided under piers, which are constructed on dry land having good soil.

(ii) *Raft Foundation*. When the river bed consists of different types of soils having less bearing capacities, raft foundation is provided. This type of foundation is provided when bed contains soft clay, and good soil is not available within a reasonable depth of 1.5 to 2.5 m.

(iii) *Pile Foundation*. If the soil at the site of bridge is very soft and good soil is at a greater depth, pile foundations are provided, because under such circumstances these are economical. When the load on a foundation is excessive and heavy, and other types of foundations are uneconomical, the pile foundation is the only alternative.

(iv) *Well Foundation*. When the good soil is available at nearly 3 metres below the bed level and bed consists of sand, well foundations are provided. These are used in soft soil or sandy beds where scouring is liable to occur. Well foundations are provided to distribute excessive load of the bridge.

(v) *Caisson Foundation*. When the depth of the water in the river is more and good soil is available near the river bed, caisson foundations are provided.

36.6 SPREAD FOUNDATIONS

In this type of foundation the load from the pier is distributed on wider area by providing steps. For the construction of such foundation cement or lime concrete in sufficient thickness is laid over the required area of the sub-soil, which is calculated by dividing the total load by the permissible bearing capacity of the soil. If the soil is bad, reinforced cement concrete is laid. Over this concrete after making the centre line of the pier, masonry work is started, the area of which is gradually reduced to the pier area at the top by providing steps or giving a batter as shown in Fig. 36.1.

Sometimes instead of giving steps, the pier is taken with the same slope and R.C.C. column footing is provided. Spread foundations are also known as 'Open foundations' because these are constructed in open excavations.

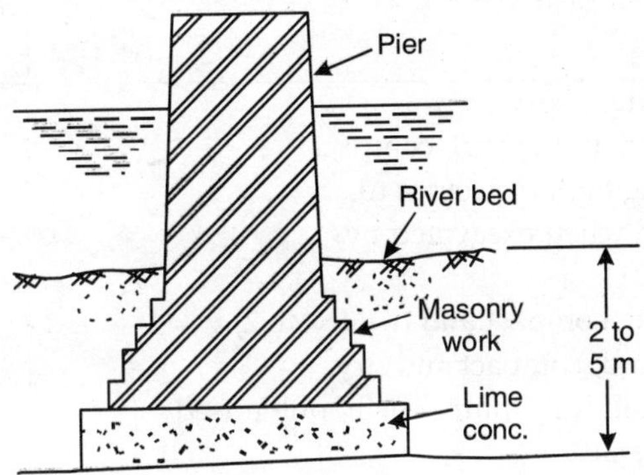

Figure 36.1 *Spread foundation*

If the foundation soil is not stiff and there is sufficient quantity of water coming out in the excavated area, like springs, the sides should be protected by shoring. A small sump is made in the corner of the excavated trench and water is continuously pumped to keep the trench dry. The main aim of the shoring is to permit the excavation in the minimum required area. Sometimes shoring, sheathing with timber planks supported by walls and struts are provided to protect the excavated trench sides from collapse or sliding.

The depth of the foundation is determined by considering the safe bearing capacity of the soil, angle of repose and the density of soil. The safe bearing capacity for each pier is determined by conducting practical tests. The bore holes are also driven to determine the thickness of the sub-soil and the thickness of various layers in the bed.

The minimum depth of the spread foundation is determined by the formula

$$d = \frac{P}{w}\left(\frac{1-\sin\phi}{1+\sin\phi}\right)^2$$

where

d = Minimum depth of foundation is metres

P = The bearing capacity of the sub-soil in kg/m^2

w = Density of the sub-soil in kg/m^3

ϕ = Angle of repose or internal friction of the sub-soil

Table 36.1 gives the safe bearing capacity of the soils according to IS : 1904 – 1961.

TABLE 36.1 *Safe bearing capacity of soils*

S.No.	*Description of the soil*	*Safe bearing capacity (t/m^3)*
	Cohesionless soil	
1.	Gravel, sand and gravel, compact and offering high resistance to penetration when excavated by tools (Note 1).	45
2.	Coarse sand, compact and dry (Note 2).	45
3.	Medium sand, compact and dry.	25
4.	Fine sand, silt (dry lumps easily pulverised by the fingers).	15
5.	Loose gravel or sand mixture, loose, coarse to medium sand dry (Note 1).	25
6.	Fine Sand, loose and dry.	10
	Cohesive Soils	
1.	Soft shale, hard or stiff clay in deep bed, dry.	45
2.	Medium clay readily indented with a thumb soil.	25
3.	Moist clay and sand clay mixture which can be indented with strong thumb pressure.	15
4.	Soft clay indented with moderate thumb pressure.	10
5.	Very soft clay which can get penetrated several cm with the thumb.	5
6.	Black cotton soil or other shrinkable or expensive clay in dry condition (50% saturated).	15

Note 1. Compactness or looseness cohesionless materials may be determined by driving a wooden picket of 5 × 5 × 70 cm size with a sharp point. The pocket shall be pushed vertically into the soil by the full weight of a person at least 70 kg. If the penetration of the picket exceeds 20 cm, the loose state shall be assumed to exist.

Note 2. The bearing capacity of peat, fills or made-up-ground should be determined after investigation.

Note 3. Increase or decrease in the safe bearing capacity should be done as follows :

1. The safe bearing capacity may be increased by an amount equal to the weight of the material removed from above the bearing level i.e., the base of the foundation.

2. For cohesionless soil the safe bearing capacity should be reduced by 50% if the water-table is above or near the soil. If the water-table is below the bearing surface of bearing surface of the soil at a distance at least equal

to the width of the foundation, no such reduction should apply. For intermediate depth of the water-table, proportional reduction of the safe bearing capacity may be made.

36.7 RAFT FOUNDATION

It is also known as floating foundation or mat foundation. In this type of foundation if the bed consists of very soft clay, a thick reinforced concrete slab is laid in the bed across the whole length of the bridge and over this slab piers and abutments are constructed at required distance. It is better to provide this raft slab with inverted T-beams for the purpose of economy as shown in Fig. 36.2. The T-beam slab may consist of primary and secondary

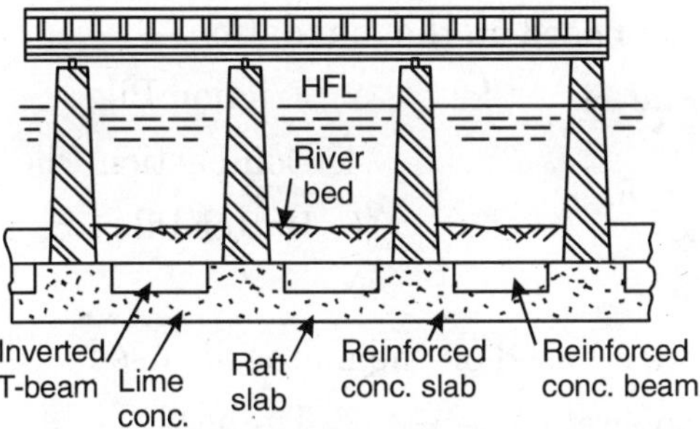

Figure 36.2

beams. For the construction of such foundation first of all, lime concrete is laid over the entire area over which reinforced slab with T-beams is to be laid. These T-beams and secondary beams are casted at the time of casting slab, so that all these may act as one monolithic mass. At the position of piers cement concrete paddings are provided over which piers rest as shown in Fig. 36.2.

These foundations are provided in the situations where the soil contains compressible pockets causing differential settlement which is difficult to control. The raft acts as a bridge over the erratic deposits and eliminates the possibilities of the differential settlement. If the loads coming through the piers are too excessive, the raft foundations are provided to reduce the settlement by distributing the load over wider area. Even if the whole raft foundation may undergo large settlements, it will not cause harmful differential settlement or crocks in the bridge.

36.8 PILE FOUNDATIONS

Pile foundations are similar to spread foundations in some respects. In spread foundations, the loads are transmitted directly to the soil, or possible on hardpan, by the foundation, but in plies which, in turn, transmit them to the soil, hardpan or rock. The upper end of a pile is called the head or butt, and the lower end the point or tip. The piles are part of the substructure. These are good foundations, but piles are no better than the materials to which they transmit their loads. In pile foundations, the soil under and around the piles is the thing which should be studied carefully before deciding the type of foundations to be used and its designing. Of course the piles should be long, strong, durable and suitable for particular conditions where they are to be used.

Whenever piles fail, it is always due to settlement, therefore, it should be designed carefully.

Piles are broadly classified in the following two groups :

(A) Classification based on the foundation —

1. Bearing Pile
2. Friction Pile
3. Screw Pile
4. Compaction Pile
5. Uplift Pile
6. Batter Pile
7. Sheet Pile

(B) Classification based on the materials used —

1. Cement Concrete Pile
2. Timber Pile
3. Sand Pile
4. Composite Pile
5. Steel Pile

The following are the circumstances under which the pile foundations are used :

1. when it is not economical to provide spread foundations and hard soil is at greater depth;
2. when it is much expensive to provide raft or grillage foundation;
3. when heavy concentrated loads are to be taken by the foundation;
4. when the top soil is of made-up type and of compressible nature;
5. when the scouring is very much in the bed of the river where the foundation is to be constructed;
6. when there are chances of construction irrigation canals in the area in near future.

36.8.1 Pile Driving Methods

The methods which force the pile into the ground, upto the required depth are known as *pile driving methods*. For driving the piles, hammers, leads and winches, and pile frames are required.

36.8.1.1 Pile frame

It is steel structure frame whose height varies from 10 to 25 m. At its bottom it has platform for supporting the engine, boiler, winches and space for standing driver. It has intermediate platforms for standing the workers while driving the pile.

36.8.1.2 Pile hammer

The hammers which give blow at the top of the piles for driving them are known as *pile hammers*. These may be of the following types :

(a) Drop hammers

(b) Single acting steam hammers

(c) Double acting steam hammers

(d) Diesel hammers

(e) Differential acting steam hammers

(f) Vibrators

(a) *Drop hammers* are simple weights varying from 1.0 to 4.0 tons in weight. These are lifted by means of manual labour through winches and are allowed to fall under gravitational force from a height of 1.5 m to 6 m directly on the pile through guides.

(b) *Single acting steam hammers* are raised by means of steam or compressed air and, then are allowed to fall under force of gravity on the top of the pile. Such hammers may give about 60 blows per minute.

(c) *Double acting steam hammers.* In these hammers the lifting and dropping both are done by means of steam or compressed air force, hence are known as double acting steam hammers. These types of hammers are most commonly used. The weight of hammer is about 500 kg, but due to dropping under steam and gravitational force, its effective weight becomes about 3 tons. Such hammers can give from 100 – 200 blows per minute.

(d) *Diesel hammer* is a small, light weight, self-contained hammer, which is self-activating type. It requires small diesel engine for its operation. Some times these are fitted on the chasis of the diesel trucks and are operated by the diesel engine of the truck. These are mobile and can be easily taken from one site to another, one place to another.

(e) *Differential acting steam hammers* are hammers having advantages of both the single acting and double acting steam hammers. The weight and height of fall the hammers are that of single acting steam hammers, whereas the number of blows are same as that of double acting steam engine.

(f) *Vibrating hammers are mainly used for driving sheet piles.* These give vibrations at very high rate to the sheet piles and the same are driven in less than 30 seconds upto a depth of about 20 m.

36.8.1.3 Leads

These are used for guiding the hammer and piles while driving, to keep them in exact vertical movement. These are also movable and can be adjusted according to the need.

36.8.1.4 Winches

These essentially consist of steel drum, wire ropes and pulleys fitted at the top of the pile frame. These are used in lifting and lowering the piles and hammers for keeping them in position.

36.9 PILE DRIVING FORMULAE

The bearing power of pile is the static longitudinal load that pile can support safely as a part of the permanent substructure. It can be determined by :

(i) Dynamic pile driving formula

(ii) Static formula

(iii) Actual loading test

(i) **Dynamic pile driving formula :**

The following are the dynamic pile driving formula :

Engineering News formula :

$$R = \frac{16.65\, WH}{S + 0.254(P/W)} \qquad \text{... (36.1)}$$

Engineering News formula steam hammers :

$$R = \frac{16.65\, WH}{S + 0.254} \qquad \text{... (36.2)}$$

Engineering News formula for drop hammers :

$$R = \frac{16.65\, WH}{S + 2.54} \qquad \text{... (36.3)}$$

where

B = Allowable static safe-load on pile in kg (or tonnes)

W = Weight of striking parts of hammers in kg (or tonnes)

H = Height of fall of hammer in metres

S = Penetration of pile per blow in cm (average of last five blows)

P = Weight of pile as driven in kg (or tonnes)

Hiley formula. This formula is also commonly used in calculating the safe load on piles,

$$R = \frac{eWH}{S + \frac{C}{2}} \times \frac{W + K^2 W_1}{W + W_1}$$

where

R Maximum or ultimate load carrying capacity of the pile in kg

W_1 Weight of pile including shoe and cap in kg

W Weight of the striking part of hammer in kg

H Height of fall of hammer in cm

S Final settlement of pile in cm (to be taken as average of last five blows)

e Efficiency of hammer

1.00 for drop hammers with free fall

0.65 to 0.90 for single acting, double acting and differential acting steam hammers

K Coefficient of restitution, varying from 0 to 1 and depending on the type of hammer, material of pile and its cap

$$C = C_1 + C_2 + C_3$$

C_1 Temporary compression of the pile head and cap in cm

C_2 Temporary compression of the pile in cm

C_3 Temporary compression of the ground in cm

(ii) Static formula

$$R = a.p + A.f \qquad \text{... (36.4)}$$

where

a Cross-sectional area of pile at bottom in m^2

p Bearing capacity of soil at pile point kg/m^2

A Surface area of pile in m^2

f Frictional resistance of soil at pile surface in kg/m^2

R Allowable static safe-load on pile in kg

This formula equate the soil resistance to the load carrying capacity of the pile. The static resistance of the pile is total of the point resistance and side friction developed between the pile and the surrounding soil. This formula also give fairly accurate results.

(iii) **In actual loading test:** In actual loading test, the load which a pile can bear is determined by placing loads over it and then testing it.

Load carrying capacity of a group of piles are determined by the formula

$$P = p.n.r \; 1 - \frac{\phi}{90} \; 2 - \frac{1}{n} - \frac{1}{r}$$

where

P Total load carrying capacity of r rows of pile, with n number of piles in each row

p load carrying capacity of one pile

ϕ Angle in degree such that $\tan \phi = D/S$

where

D Diameter of the side of pile or actual diameter in case of circular pile

S Facing of pile from centre to centre

36.10 TYPES OF PILES

Several types of load bearing piles are used, but the most common types are described below.

36.10.1 Wood or Timber Piles

These are tree trunks 20 cm to 50 cm in diameter with branches and barks trimmed off. Mostly they are driven with small end down, but sometimes

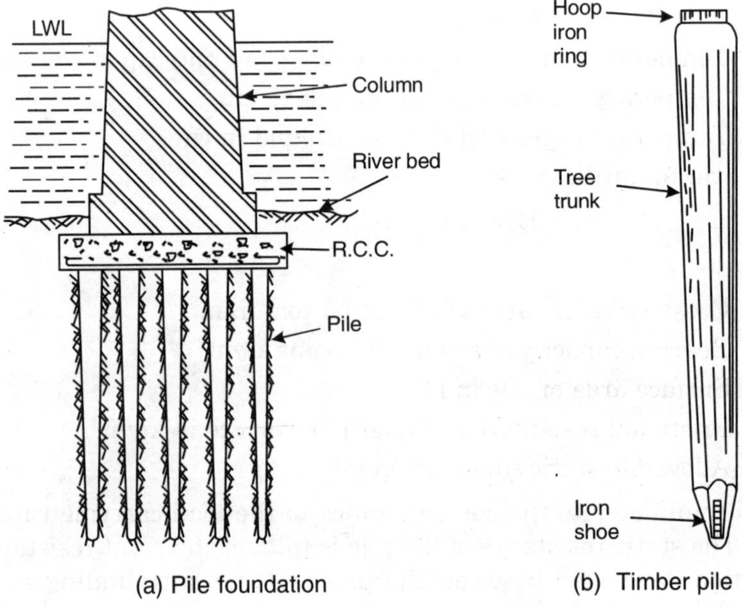

(a) Pile foundation (b) Timber pile

this end may be square or provided with pointed metal shoe. When the soil is soft, unpointed piles are used, but if the ground contains boulders, metal points should be used. The top end of a wooden pile is provided with an iron ring or band so that it may not split under the blows of hammer.

Timber piles are economical and can be driven rapidly without heavy machinery and much technical supervision.

Long timber piles cannot be used and are liable to decay by white ants and salty water. If these piles are not driven below the permanent water-table, they will decay due to fungi and insects.

36.10.2 Precast Concrete Piles

R.C.C. precast piles having 15 cm to 60 cm circular, square or octagonal cross-section and lengths upto 30 metres are used. The reinforcement consists

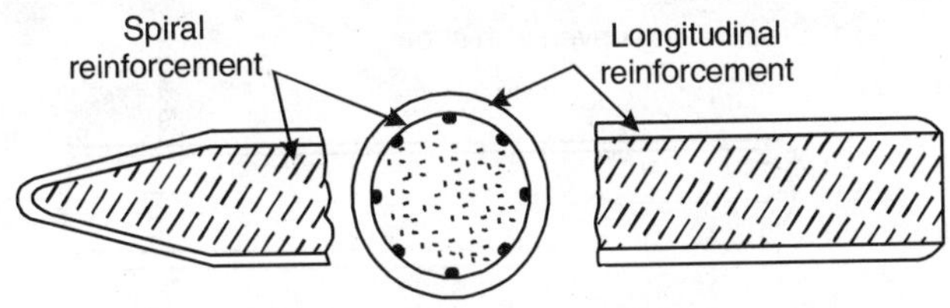

Figure 36.4 *Precast concerete pile*

of longitudinal bars with hoop or spirits as shown in Fig. 36.4. In plastic clay-soil blunt toe is provided, while long tapering point type toe is used for driving in sand or gravel.

The present concrete piles can be cast early before starting the foundation work so that the execution of work can be done very quickly. Reinforcement remains at required position and any defect of the casting can be removed. Unlike timber piles these can be used above the ground-water table.

These piles can be driven under water easily. There is proper control over the composition and design of these piles. All defects such as hollows, in casting, wrong position of reinforcement etc. can be checked and removed. These piles have good resistance to biological and chemical actions of the ground. Good quality concrete of M 200 grade is generally used. The lifting points for piles must be clearly indicated/marked on the piles. These lifting points are such that maximum bending moment due to handling are kept at the lowest possible magnitude.

But these piles are very heavy in transpiration and there are chances of their damage. If a pile is found too short or long, at the time of driving, it causes much difficulty in increasing or cutting its length.

36.10.3 Cast in Situ Piles

These piles are constructed by making a bore (hole) in the ground by driving a casing, which is then filled with cement concrete after placing the reinforcement (if any). These piles may be further cast-in-situ concrete piles, if the casing is allowed to remain with pile or uncased cast-in-situ concrete piles, if the casing is taken out.

In the former case, a sheet-shell or a casing closed at the bottom point is driven into the ground and is filled with concrete. This may be of Raymond or Union pile type.

In shell-less pile, a shell with inserted core is driven into the ground and after withdrawing the core, concrete is filled and then shell is taken out. This may be of Simplex or Pedestal pile type.

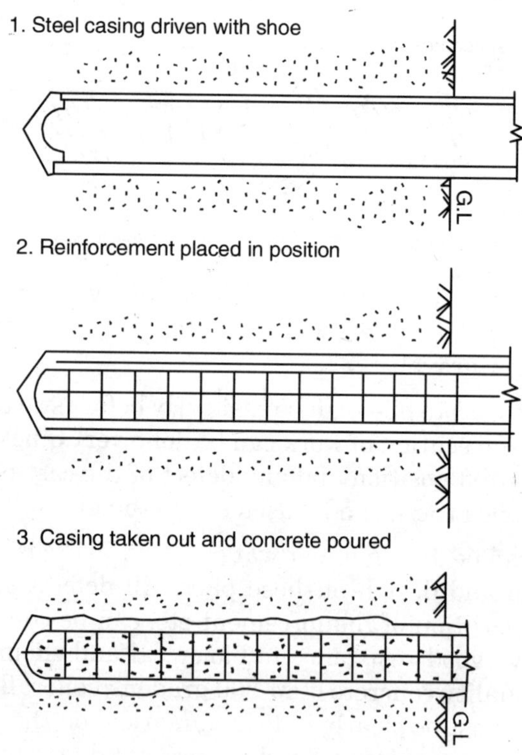

Figure 36.5 *Method of simplex pile driving*

Cast-in-situ piles are casted directly at site, therefore, there is no pile transportation. These are sound in construction as they have not to bear hammer blows. These are casted in exact lengths and there is no wastage like precast piles. These are easy in handling and driving. These piles do not require any extra reinforcement to resist the stresses developed during handling and driving operations. The extra cost of transporting the pile is saved.

Following are the disadvantages of cast-in-situ piles :

(a) It is not possible to cast these piles under water.

(b) It is difficult to keep the reinforcement at correct position.

(c) It is very difficult to have proper control over the composition and design of these piles.

Nowadays prestressed concrete piles are also used, because these are light in weight and easy to handle. These piles are generally of hollow circular section of length ranging upto 60 cm.

36.10.4 Structural Steel Piles

Rolled steel section are very advantageous for using as bearing piles on sound rocks. The steel-piles can be driven in soil containing boulders and thin strata of hard pan. But this type of piles cannot be used as friction piles because they are expensive and full area is not available for frictional resistance. (See Fig. 36.6).

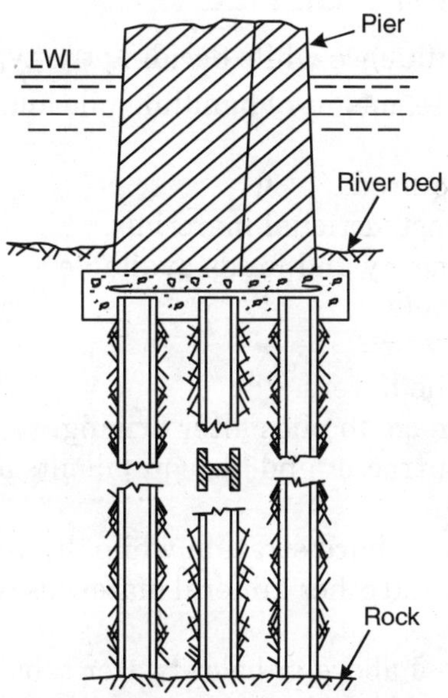

Figure 36.6 *Structural steel pile foundation*

36.10.5 Composite Piles

These piles are very suitable for foundations constructed on soil charged with water. In this type timber piles driven up to the ground-water table

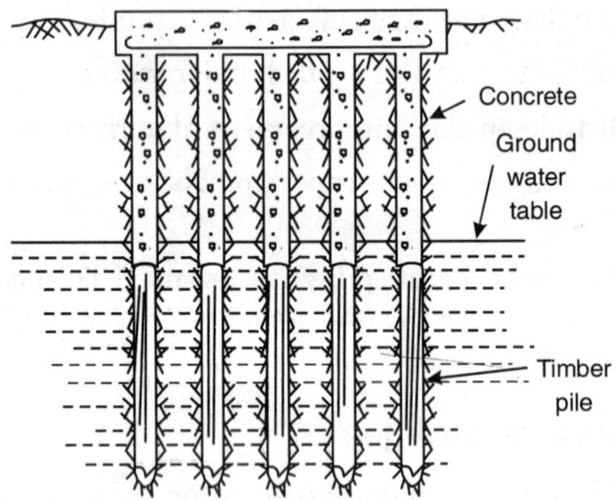

Figure 36.7 *Composite pile foundation*

and above it concrete is used. Thus both materials can be used with advantage and also economy is obtained as timber is cheaper than concrete. (Fig. 36.7).

36.11 CHOICE OF TYPE OF PILE

The following factors influence while deciding the type of pile to be used :

(a) Accessibility of site, means of handling pile and bearing value desired per pile.

(b) Length of pile required.

(c) Availability of constructional materials.

(d) Degree of permanency and funds available.

(e) Depth of water table.

(f) Time available.

(g) Presence of marine bovers.

(h) Possibility of damage to piles after driving.

(i) Availability of contractor and his equipments able to install a certain type of pile.

(j) Resistance of pile to hard stratum while driving.

(k) Presence of acids and other mineral materials which may cause injury to pile materials.

(l) Driving is required above or below water table.

(m) The presence of acids and other mineral materials causes bad effect on piles.

(n) The pile is to be driven on dry land or in deep water.

36.12 WELL FOUNDATION

In India there are two systems of well foundation :

(a) Punjab system — in which deep well foundations are provided without flooring for protection against scouring.

(b) Madras system — in which shallow well foundations are provided with continuous flooring in upstream and downstream against scouring.

In India well foundations are constructed with brickwork of masonry work, while in western countries iron is used for construction of foundations, which are known as Tubular foundations.

36.12.1 Method of Sinking Well Foundations

If the water depth is shallow at the site, an earthen coffer-dam is erected in a ring form enclosing the area over which foundation is to be laid. After it

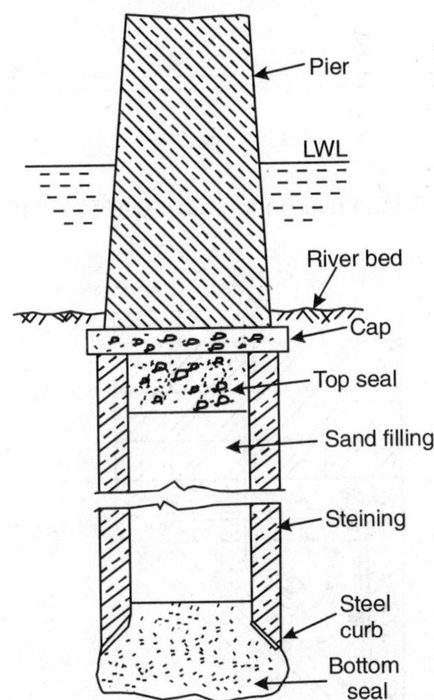

(a) Section showing bridge pier on well foundation.

the inside water is pumped out. But in case the water depth is more, sheet-piling or caissons are used for enclosing the area and then the water is pumped out.

When all the water has been pumped out, a curb of R.C.C. or iron is laid on the bed and excavation is started inside the curb. When it reaches about one metre depth, steining of brick-masonry, concrete block or cement concrete

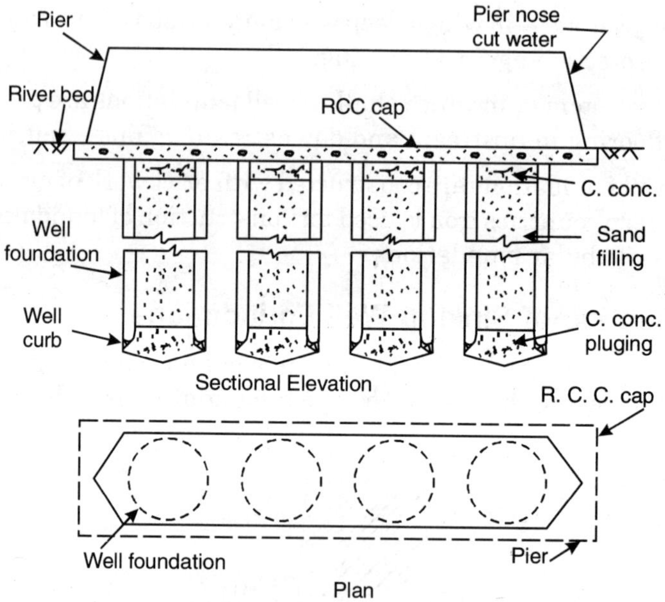

(b) Bridge pier on four nos. well foundation.

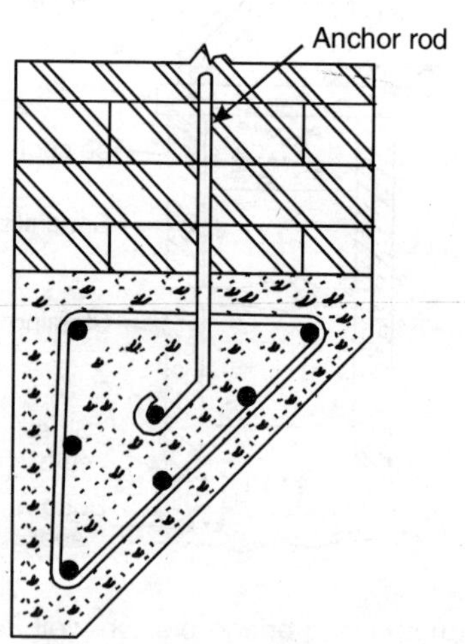

(c) Bridge pier on four nos. well foundation.

Figure 36.8

is laid over curb upto about one metre above the bed level. Again excavation is done and some weights are kept on steining for sinking it. This process is repeated many times, till steining has been sunk to the required depth. If during sinking or excavation process water starts entering in the foundation, it is taken out by continuous pumping.

When steining reaches upto the desired depth, the bottom is dredged to uniform level and all the materials are cleared under the cutting edge of the curb. After it a thick layer of concrete is laid and the bottom is sealed. This is known as *plugging*. After this again all the water from the inside is pumped out and the inside of the wall is filled with sand as shown in Fig. 36.8 (a). Then a concrete layer is laid on it upto the level of the masonry-work of steining. Now a R.C.C. concrete cap is provided at the top cover the group of wells under one pier which acts as base for the construction of pier.

36.13 DIFFICULTIES DURING SINKING OF WELLS

During well sinking, in the construction of bridge foundations, various types of difficulties arise causing great difficulties. Following are the main difficulties:

(a) Sand blowing

(b) Shifting of wells

36.13.1 Sand Blowing

During the process of dewatering in sandy stratum, this difficulty takes place while well sinking. While dewatering the sand particles also come with the water, causing hollowness inside the ground. This causes wide cracks in the ground the well under sinking operation. During well sinking the sand falls around the ground and is filled suddenly inside the well-known as *sand blowing*. The fall of sand in the caisson is very quick and it amounts to a depth of about 5 to 15 metres. This sand blowing may cause fatal accidents which may result in burying the workers and plants inside the sand.

During well sinking, if slight indications of sand blowings are seen the dewatering should be discontinued and all workers and equipments should be taken out from the well. Now the bundles of grass and other such materials should be laid all around the steining of the well which will control the possibilities of sand blowing.

After ascertaining the safety and no possibilities of sand blowing, the work of well sinking should be started further.

36.13.1.1 Tilting of Wells

During sinking of wells, the main common difficulty is to sink the well true in plumb line or exactly vertical. But during sinking operation sometimes the wells get tilted towards one side. This tilting is mainly due to non-uniform excavation and due to difference in the bearing power of soils which come across while excavating in various soil stratums.

Following methods are generally adopted for bringing in vertical position, to the tilted wells.

(a) *Eccentric loading*. In this method extra loads are placed on the well side which has sunk less than the other side. The extra loads bring the steining in true vertical position during further sinking.

(b) *By water setting*. In this method the friction between the soil and the well steining towards the higher side are reduced by forcing water jets. Due to reduction in the friction by water jet the higher side sinks more than the lower side and this causes both the sides to come in true vertical position after some further sinking.

(c) *By Jacks*. In this method the tilted well is brought in true vertical position by use of Hydraulic or Screw Jacks.

(d) *By use of temporary obstacle*. In this method temporary obstacles are provided below the cutting cope which has sunk more and when the higher side comes down and the well comes in true vertical position, the temporary obstacle is removed and the further well sinking is continued.

(e) *Deposition of earth*. In this method the excaved earth is deposited along the well steining towards the tilted side at ground level. This causes increase in the friction between the well steining and the soil on the more sunk side. Now during further well sinking operation the side towards which the soil is deposited sinks less as compared with the higher side, thereby bringing the well in true vertical position.

(f) *Control in dredging*. In this method the excavation or dredging operation is controlled. The excavation is done below the cutting edge or well curb on the higher side. But the excavation should be done carefully otherwise the well may sink quickly causing accidents.

36.13.2 Shifting of Wells

Sometimes during well sinking and during bringing the tilted wells in true vertical position the centre line of the wells is displaced from the actual centre line. This causes the shifting of the well on either side. The shifting of wells will cause eccentric loads on the wells as well as change in the bridge span lengths.

As the main source of shift of wells is their tilting, full care should be taken to avoid even a slight tilt during sinking operation. The outer surface of the well curb and steining should be constructed uniformly to avoid difference in friction resistance along the difference. The radius of well curb should be 2 to 4 cm larger than that of steining. The thickness and sharpness of the cutting edge should be uniform. The excavation or dredging should be done uniformly.

36.14 CAISSON FOUNDATION

The term caisson is derived from the French word *caisse*, meaning box. There are three forms of caissons used in constructing foundations underwater, which are :

1. Box caissons 2. Open caissons 3. Pneumatic caissons

36.14.1 Box Caisson

These are used in constructing bridge foundations under water where little excavation is required except for preparing the side to give good bearing for the caisson. The caisson, which may be constructed of timber, reinforced concrete or steel, is towed into position; filled with concrete

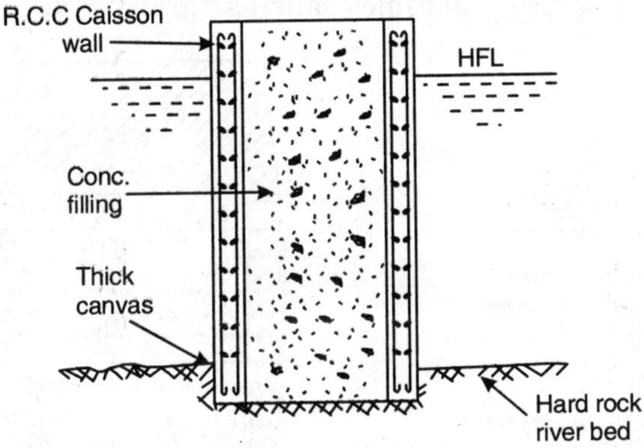

Figure 36.9 *Section through R.C.C. box-caisson*

or stone masonry and sunk until it resist on the river bottom, which has been prepared to receive it or on a pile cluster, to form the lower part of a bridge pier.

36.14.2 Open Caisson

As the name implies, an open caisson is one that has no top or bottom during its use as a protection for the excavation process. It is like a box with vertical sides only. Sometimes it is like a honeycomb structure with open ends. In plan, it may be square, rectangular, circular, oval or whatever shape best suits the situation. Small caissons consist of one opening or well, but large size contains series or group of wells.

Caissons are made of steel plates riverted or welded together or made of reinforced cement concrete. The thickness of the steel plates depends on the size of the caisson, on the material to be penetrated, on the method of advancing and on other factors. The thickness varies from 6 mm to maximum of about 18 mm. One cutting edge or curb of steel is provided in the bottom of the caisson. A steel caisson may be a single unit for the entire length or its length may be increased by adding sections as the sinking progresses. The length of a reinforced-concrete caisson is normally increased by providing new sections, in lifts, as the sinking progresses. To maintain the caisson in a vertical position, it can be propped against the side of the pit, but generally steel ropes are used for this purpose. At some places towers are erected for the purpose.

The following methods are used for removing the soil from the interior of the caisson at the time of sinking—

(a) *Hand methods*. Using the pick and shovel to place the soil in buckets, which hoist soil to the surface.

(b) By orange-peal and small clam-shell buckets which can be operated through water. These are sometimes called as grab buckets.

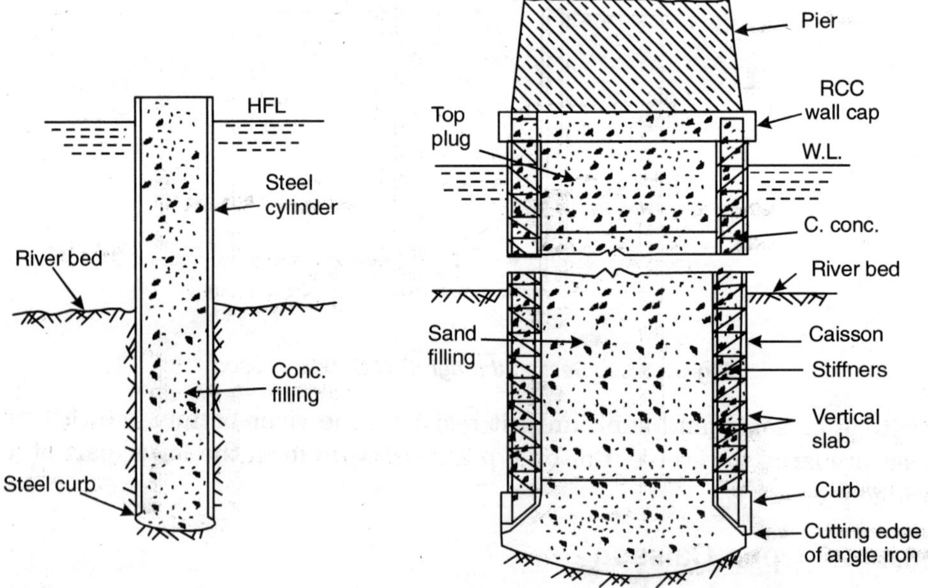

(a) Section through steel open-caisson (b) Section showing pier on cassion foundation

Figure 3.10

(c) By interior water jets which loosen the soil and raise the water level inside the caisson high enough to allow soil particles to flow outward under the cutting edge and upward along the outside of the caisson to the surface.

After excavating the soil from the interior of the caisson, these are advanced so that its cutting edge is at or below the bottom of the excavation. This process is continued until the hard stratum is reached for construction of foundation.

When the advancing of the caisson is completed, all the water and soil particles from the interior are excluded by pumping, the excavated space is prepared to receive concrete. Now the foundation bed is thoroughly examined before starting the concrete-work. Concreting is started and all the interior space is filled with concrete to form the pier. The steel sheets above the river bed, acting as form-work of the pier are removed and are used in the construction of next piers.

36.14.3 Pneumatic Caisson

A pneumatic caisson is one with a permanent or temporary roof near the bottom arranged in such a way that the inside pressure can be increased and men can work in the lower part. The essential parts of a pneumatic caisson are the working chamber, the shaft and the air locks as shown in Figs. 36.11 and 36.12. The working chamber may be constructed of timber, steel or reinforced concrete. The shaft is usually constructed of steel.

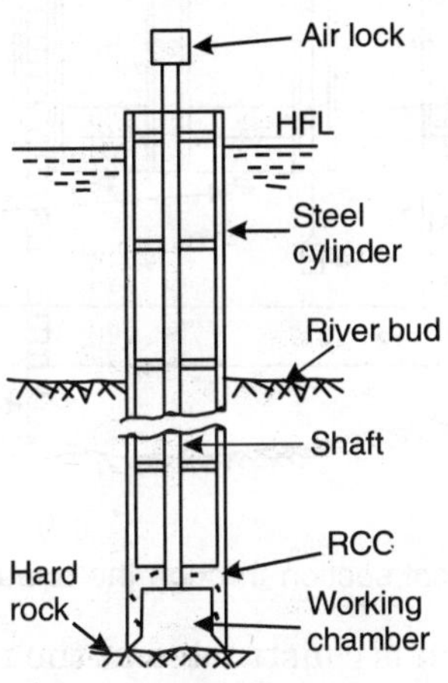

Figure 3.11

The pneumatic caisson is composed of an inner and outer skin plate of steel, with steel trusses or girders as horizontal supports to form a box-like structure. In the bottom cutting steel edges known as well curb are provided for easy penetration in the soil. A working chamber of about 3.4 metres deep is formed by providing air-tight roof. Access in the working chamber is through shafts and air locks. If a man wants to enter the working chamber, he first enters the air lock. Then the door of air lock is closed and the pressure is increased equal to the pressure of the working chamber. Now the door of the shaft is opened and the man enters the working chamber. After this, the door of working chamber is closed. Similarly, if a person wishes to come out of the working chamber the process is reversed.

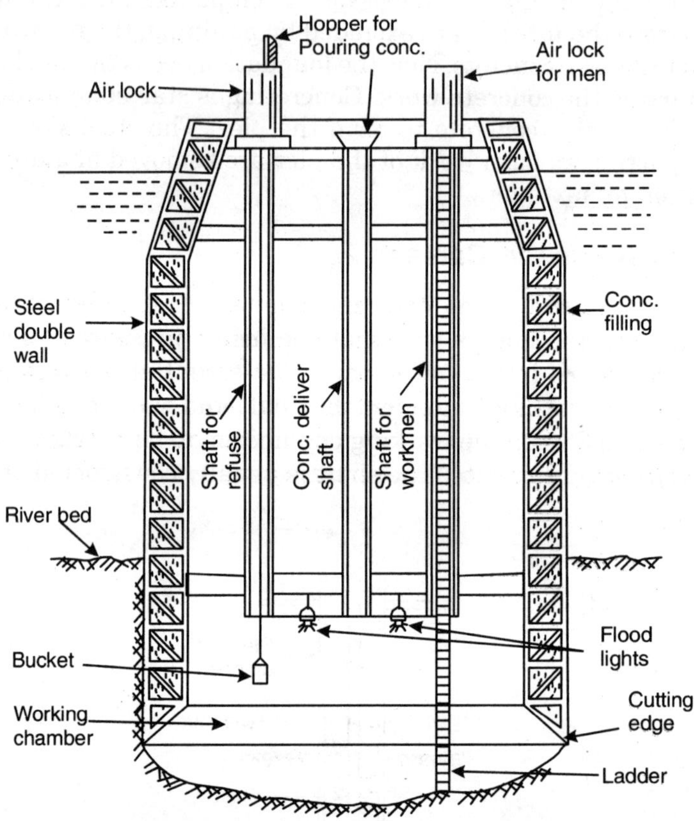

(a) Typical section through the pneumatic caisson.

The pneumatic caisson is constructed at the river bank, then towed out to the place where it is required. At the site of the pier, concrete is filled between its double walls and it is sunk into the river bed. Three shafts—one for refuse, second for workers and the third for concrete delivery are provided as shown in Fig. 36.12. When the caisson is sunk on the river bed compressed air is forced in the working chamber to exclude all the water from the working chamber. After it, excavation is started in the river-bed and the caisson is allowed to sink. As the caisson sinks the pressure of the compressed air is increased to balance the pressure of the water outside, so that it may not enter in the working chamber. When the caisson has reached upto desired depth the bottom is prepared to receive the concrete and the concreting is started through the concrete shaft and a thick layer of concrete is laid which acts as a seal. Now the air pressure is released and air locks and shafts are removed from the caisson. The remaining portion of the caisson is filled with concrete under atmosphere pressure. Care must be taken to avoid air

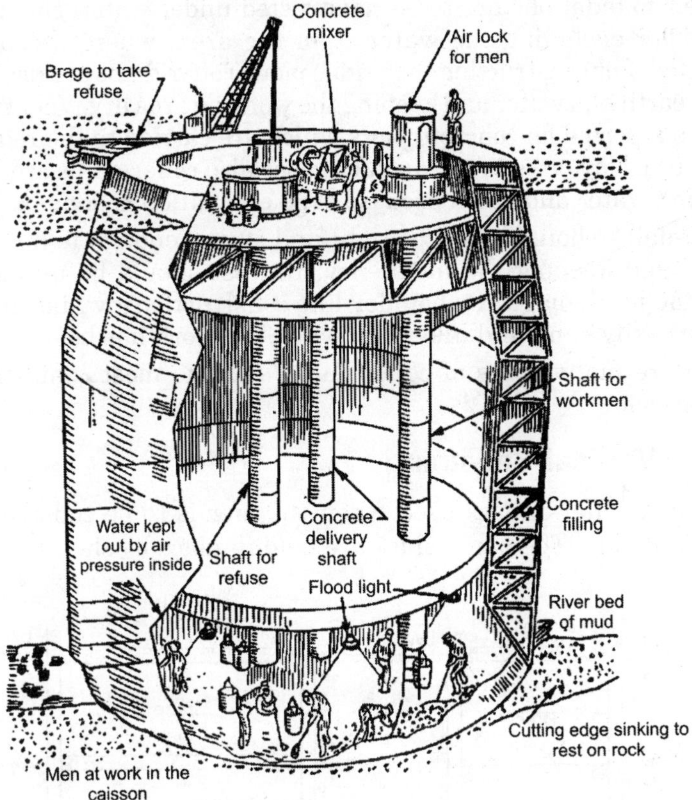

(b) Pictorial view showing works of pneumatic caisson.

Figure 36.12

pockets and to over come the effect of the shrinkage which occurs in the concrete while it is setting.

36.15 COFFER-DAMS

There are various types of coffer-dams, but the following are the most commonly used :

1. Vertical sheathing
2. Poling board
3. Sheet piling
4. Cellular coffer-dam
5. Puddle coffer-dam
6. Rock-fill coffer-dam
7. Rock-fill crib coffer-dam

Coffer-dam may be defined as a temporary structure built to exclude water and earth from a given area, so that work may be carried out in that area under atmospheric or dry conditions. If there is some leakage of water, the enclosed area can be kept dry by using pumps.

The superstructure of a bridge rests on piers and foundations. As the piers with their foundations are to be constructed under water, some devices are required for excluding the water from the area, where these have to be constructed. For constructing the bridge piers, coffer-dams are used for holding both the earth and water and keeping the working area dry. Very tall buildings are also supported on concrete piers which are carried to the required depth to reach film clay or bed rock. In such cases also coffer-dams are used to hold the ground water and earth because deep excavation is done.

Coffer-dams should be water-tight and strong enough to hold the water-pressure and keeping the inside dry. Its cost should be minimum and it should not get damaged by boulder-like constructions, while sinking down. It can be easily sunk and taken out when the work has been completed.

There are many types of coffer-dams, but the most common types are described below.

36.15.1 Vertical Sheathing

This type of coffer-dam is used when the pier is to be constructed upto a small depth only. The earth-sides are held in place by sheathing of wooden

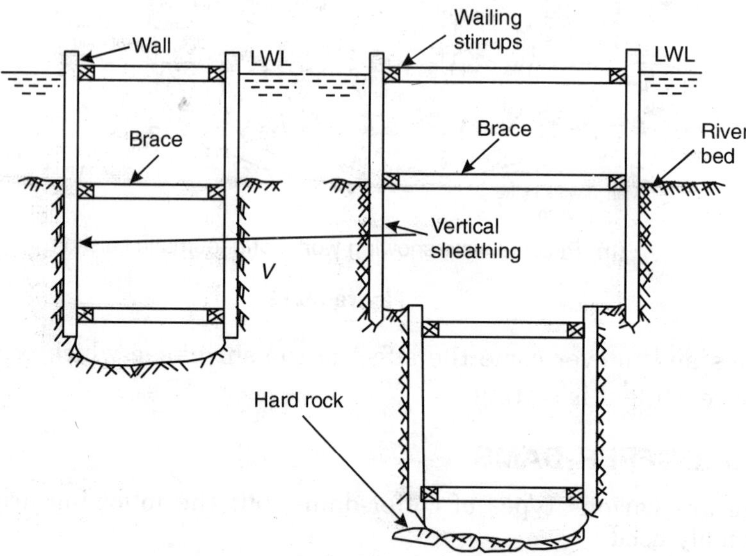

Figure 36.13 *Vertical sheathing coffer-dam.*

planks which are placed vertically and are supported by wood frames called wales or rangers and braces. One set of sheathing is drived upto 3.5 metres only and for more deeper excavations more than one set of sheathings are used as shown in Fig. 36.13.

36.15.2 Poling Board Method

Instead of driving the sheaths upto 5 metres as described above, a set of 60 cm to 100 cm height, sheathing set, the boards of which are held in place by metal rings is first sunk in pre-excavated trench of some depth and lateral

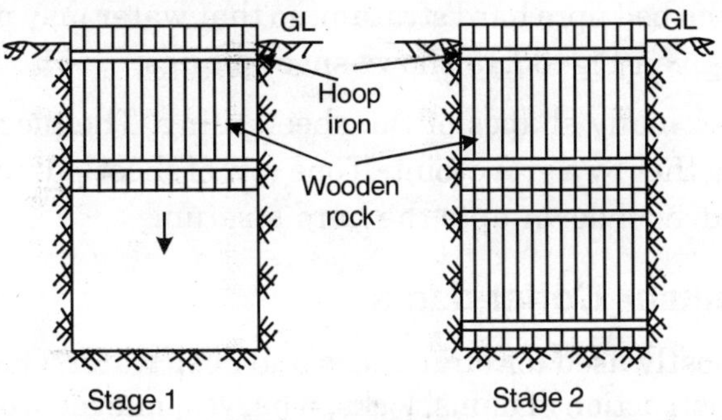

Figure 36.14 *Poling board coffer-dam.*

dimensions. Then the excavation is started below and when it reaches about 1 metre, the first sheathing is lowered down and other set is kept over it and this process is repeated until the required excavation is completed. The sheathings have tongue and grove joints with each other. When the excavation is carried below ground water level some pumping may be necessary to keep the water out. It is also necessary to supply fresh air to the workers working in the well.

36.15.3 Sheet Piling

At some places when the river-bed is sandy, if we construct coffer-dams and take out the water from the enclosed area, the water of the river will reach

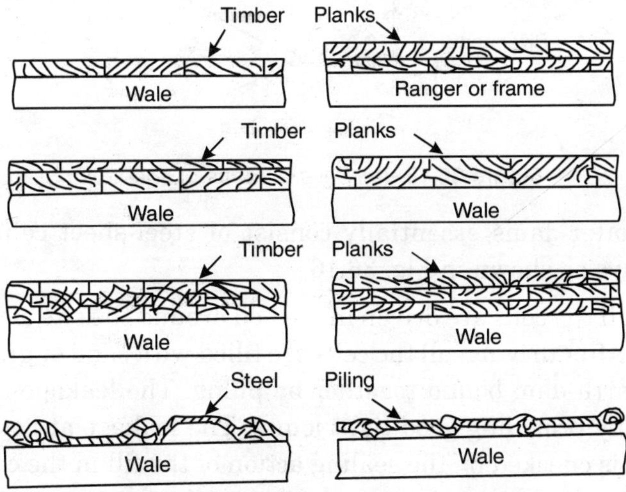

Figure 36.15 (a) *Several forms of sheet-piling.*

in the enclosed area due to seepage from sandy bed and may cause difficulties in keeping the area dry. If we rapidly pump it, the sand and silt of the adjacent strata will go out with pumped water and will create cavities in the bed.

Due to the above difficulties the site is surrounded by sheet piles which are driven in the bed upto hard stratum, so that water may not come through seepage. The given Fig. 36.15 shows sheet pilings.

There are so many shapes of the sheet-piling. The pieces of sheet-piles are jointed together by a grove joint. If one row of sheet-piling is not sufficient, two rows are driven down upto the hard stratum.

36.15.4 Cellular Coffer-dams

This type is mostly used for large areas and deep water. Generally it is used during the construction of dams, locks, wharves, marine works and so many other large water-front structures.

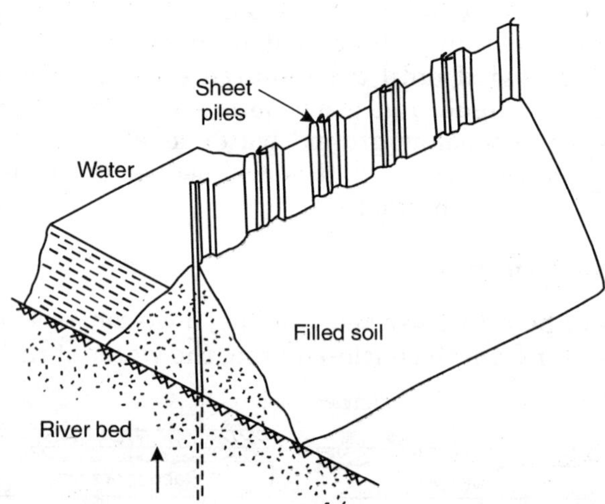

Figure 36.15 (b) *Several forms of sheet-pilling.*

These coffer-dams essentially consist of steel-sheet cells arranged in various ways as shown in Fig. 36.16.

First of all the cells are driven in the bed around the site of pier to form an enclosure. After driving, all the cells are filled with sand or gravel and act as a sort of earth-dam bound together by piling. The leakage at cell joints is prevented by providing good tight joints. The leakage at bottom in case of rocky soils is checked by the sealing action of the fill in the cells. The top of

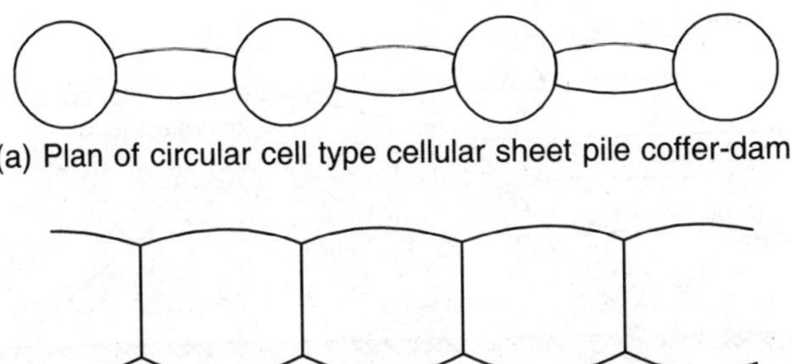
(a) Plan of circular cell type cellular sheet pile coffer-dam.

(b) Plan of diagram type of cellular sheet pile coffer-dam.

Figure 36.16

all the cells is covered with riprap or 10–15 cm concrete slab. The top of cellular coffer-dam if provided with thick concrete slab can be used for the movement of excavating plants and for storing other plants and equipments.

36.15.5 Puddle Coffer-dams

These are temporary structures constructed for excluding the water from a portion of the river-bed for inspection and construction of piers. The section of a puddle coffer-dam is shown in Fig. 36.17.

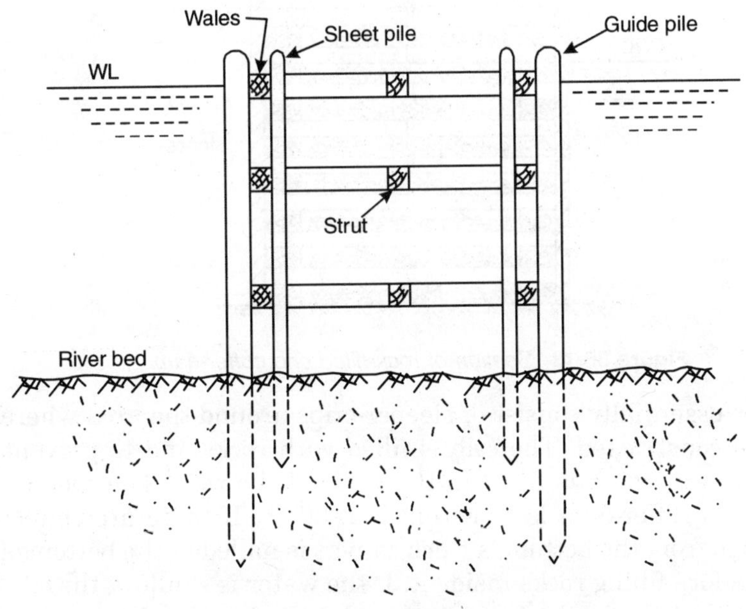

Figure 36.17 (a) *Section through a puddle coffer-dam.*

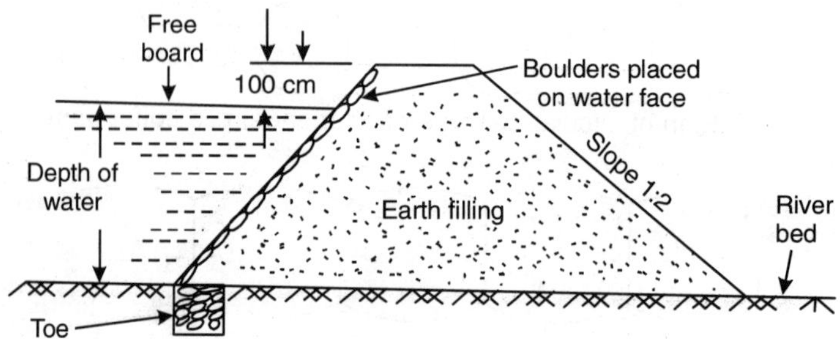

Figure 36.17 (b) *Earthen coffer-dam.*

It consists of two rows of wood or steel piling which are vertically driven in the bed, with puddle filled in between the sheet pilings to form a water-tight wall. Every guide pile is connected by a horizontal string or wall piece as shown in the figure, to prevent spreading of the piles due to earth pressure. Interior string pieces and ribbon pieces keep the sheeting piles in position. These coffer-dams are used upto 6 metre height.

(vi) *Rock-filled crib coffer-dams*. When the bed of the river is uneven and the current of water has much velocity, rock-filled crib coffer-dams are best

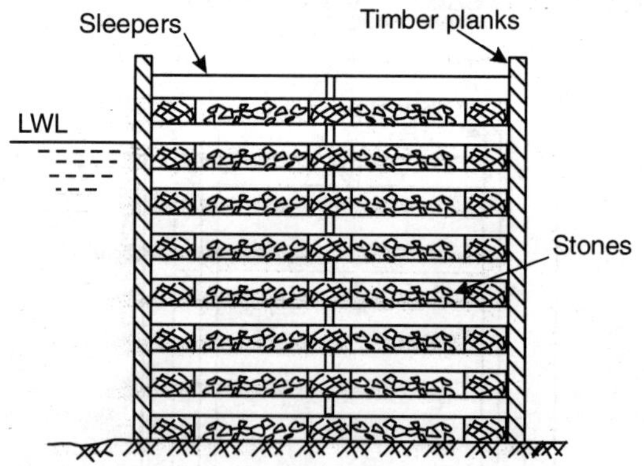

Figure 36.18 *Section of rock-filled crib coffer-dam.*

suited. They essentially consist of sleeper-cribs around the area where the pier is to be constructed. The crib is filled with rocks and to prevent the passage of water, earth is filled in the voids of the rock. The section of a rock-filled crib coffer-dam is shown in Fig. 36.18. If there are chances of water leakage from the bottom, a thick canvas is nailed in the bottom of the whole crib before filling rocks inside it. If the water is shallow, the ribs can be constructed at the bank of the river and floated to the site and lowered thereby filling rocks and earth in it. On the other hand if the depth is considerable it can be constructed in sections and lowered one by one.

36.15.6 Rock-fill Coffer-dams

The given Fig. 36.19 illustrates a cross-section through a rock-fill coffer-dam. It is constructed by rocks and boulders with side slopes. The water

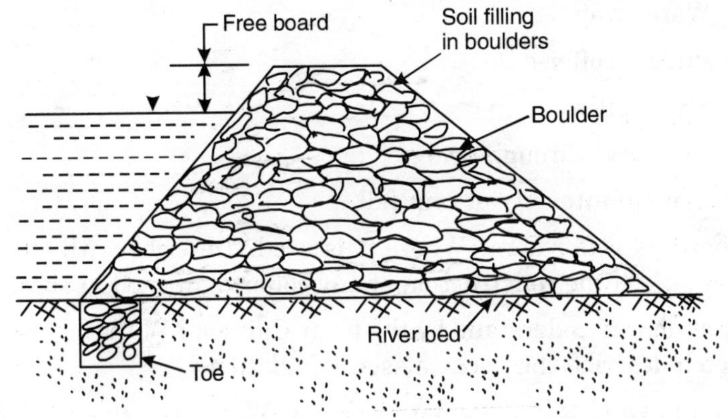

Figure 36.19 *Section of a rock-fill coffer-dam.*

face is made water-tight by laying layers of timber planks on earthen mat. But these types of coffer-dams are not completely watertight and pumping is required during construction of piers after some interval of time to keep the inside without water. As the materials of construction for such coffer-dams are easily available at the site, these are very cheap, but continuous pumping for keeping inside dry increases its overall cost.

REVIEW QUESTIONS

36.1. (a) How will you determine what types of foundations will be required for the piers and abutments of a proposed major bridge?

(b) Give a typical cross-section through a 6 m high abutment on deep well foundation.

36.2. State in what type of soils, you would adopt "well foundation" for founding the sub-structure of a major bridge? Describe briefly with sketches the construction of these foundations upto the stage from where the pier masonry can be conveniently started.

36.3. State, under what circumstances well foundations for bridges are provided. Describe the method of sinking masonry well in shallow rivers.

36.4. What are the main types of load bearing piles? When are they used for foundation of bridges? What do you understand by the bearing power of a pile? How it is calculated?

36.5. What factors influence while deciding the choice of type of a pile? Describe different types of piles which are used for the foundation of bridges?

36.6. Write short notes on the following :

(i) Water-way

(ii) Puddle coffer-dam

(iii) Sheet-piles

(iv) Deck and through bridge

(v) Critical intensity of rain-fall

36.7. What is a coffer-dam? Where it is used? Describe with the help of a neat sketch the construction of any coffer-dam used in shallow water.

36.8. What do you understand by the term 'Caisson'? With the help of a neat sketch describe the open caisson foundation.

36.9. What is air lock in pneumatic caisson? What is its function? How does it work?

36.10. Describe pneumatic-caisson with the help of a neat sketch.

36.11. Write short notes on cellular coffer-dam.

36.12. Write short notes on bridge foundation.

36.13. (a) When would you suggest the use of pneumatic caisson foundation?

(b) Describe stage by stage of construction of pneumatic caisson foundation.

(c) Difficulties encountered in sinking wells and methods to overcome them.

36.14. Describe in detail the different types of piles and the uses to which they are put in bridge construction.

What are the considerations involved in using and designing pile foundations of a bridge.

36.15. Discuss the comparative merits and demerits of the different shapes of wells commonly used for bridge foundations.

36.16. (a) Describe in detail the different types of piles and the uses to which they are put in bridge foundation.

(b) What are the consideration involved in using and designing pile foundation of a bridge.

36.17. (a) Describe a well foundation, giving a neat sketch showing various components. How will you correct the tilt of well foundation while sinking?

(b) Write short notes on the following :

(i) Various formulae for determining the load carrying capacity of a pile.

(ii) Sheet piling.

36.18. Describe the methods adopted for constructing well foundation in :

(i) Shallow water,

(ii) Deep water, and

(iii) Where foreign matter like tree trunks, boulders etc. are expected to be met in the bed.

36.19. (a) When would you suggest the use of pneumatic caisson foundation?

(b) Describe stage by stage the construction of pneumatic caisson foundation.

(c) Difficulties encountered in sinking wells and methods to overcome them.

37

Piers, Abutment and Wing-Wall

GENERAL

The substructure of the bridge is that part of it which supports the superstructure. It includes foundations, piers, abutments and wing-walls. Piers are the intermediate supports for the superstructure. Abutments are the end supports for the superstructure. Wing-walls are the walls constructed on both sides of the abutments to retain the embankment of approaches and protect them from the wave action of water.

Piers and abutments are usually constructed with masonry, plain cement concrete or R.C.C. The masonry used for piers and abutments is usually stone masonry in 1 : 4 cement sand mortar. If plain, cement concrete as mass concrete is used, and it corresponds to M100 or 1 : 3 : 6 mix by volume. The size of the stone ballast used is about 40 mm. In case of the large size piers and abutments bigger size stone plums of 200 to 150 mm of about 20% volume of concrete may be used for obtaining economy.

The concrete used for R.C.C. piers and abutments is of M I50 grade or 1 : 2 : 4 by volume.

37.1 ALLOWABLE STRESSES

Table 37.1 gives the allowable stresses for masonry and mass concrete.

TABLE 37.1 *Allowable stresses*

S. No.	*Material*	*Allowable maximum compressive stress kg/cm²*	*Allowable maximum tensile stress in bending kg/cm²*
1.	Brick masonry in lime mortar	60.0	12.0
2.	Brick masonry in 1 : 4 cement sand mortar	100.0	20.0
3.	Coarsed rubble (granite) masonry in 1 : 4 cement sand mortar	150.0	30.0
4.	Mass concrete work 1 : 3 : 6 by volume	200.0	25.0

37.2 PIERS

Piers are the intermediate supports for the superstructure for multispan bridges. These are intended to transmit the loads from the superstructure

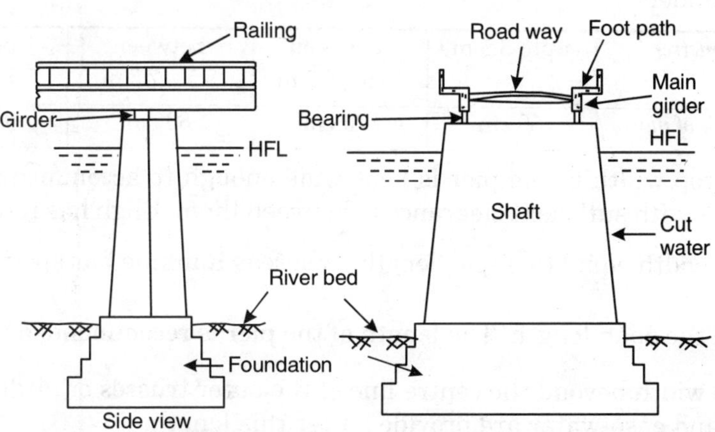

Figure 37.1 *Side view* **Figure 37.2** *Pier longitudinal view*

of the bridge to the foundations. A pier essentially consists of a column or shaft (simple or multiple) and a foundation (simple or multiple). When this is constructed on dry river-bed, it is made up of these parts alone. But in water it is sometimes provided with projections called cutwaters, the function of which is hydrodynamic. In plan, a pier has a shape extended transversely to the direction of a bridge at right angles if the bridge is straight, and it is slanting if the bridge is on the skew. When viewed in elevation parallel to its longest dimension in plan, it generally takes the shape as shown in Fig. 37.2. The vertical shaft is provided with batter and the foundation forms a widened footing at the base of the pier.

Generally, the type of pier to be adopted depends on the superstructure. At certain places the type of pier is decided by soil conditions and construction procedures. Every pier should have sufficient area at its top to receive the bearings. It should appear strong rather than weak or flimsy. Pier should be capable of taking the lateral and longitudinal loads including vertical loads. It should be attractive, practicable, durable and require minimum maintenance. The height of pier is decided by the alignment of road or railway and H.F.L.

The top of the pier is generally kept 1– 1.5 m above the H.F.L. to prevent the bridge girders, arches or other bottom structures from damage by the floating objects. The height of the pier is measured from the bridge bearing support or the springing point of the arch to the foundation top. The sides of the piers in case of short piers are kept vertical, but in the case of piers having considerable height the sides are given a batter of 1 in 12 to 1 in 24 to improve the appearance as well as to reduce the load on the foundation.

Top widths of piers for IRC heavy loading piers for spans other than arches are as under:

Span in metres	upto 3.5 m	between 3.5 – 5.5 m	between 5.5 – 7.5 m	% between 7.5 – 1.0 m
Top widih of pier	60 cm	70 cm	80 cm	95 cm

The top width of the pier is kept wide enough to accommodate the two bearings with sufficient clearance in between them. Bligh has recommended the top width equal to $\sqrt{\text{span length}}$, whereas Rankine has specified it l/6th to l/7th the span length. The length of the pier is recommended as $1\frac{1}{2}$ times the top width beyond the centre line of the outer trusses or girder. The cut-water and ease-water are provided after this length.

The piers are usually provided a batter of 1 in 30 for brick masonry in 1 : 4 cement sand mortar and 1 in 24 for brick masonry in lime mortar. Maximum batter of 1 in 12 can be provided in high piers. No batter is necessary in case of short piers. The width of the pier at its base should not be less than $\frac{1}{3}$rd of its total height.

To receive the load of the superstructure of the bridge, the tops of the piers are provided with R.C.C. caps. These caps are projected about 10 cm all around the pier tops to improve the appearance as well as to act as drip for dripping down the rain water away from the pier sides. The R.C.C. cap is generally provided with normal reinforcement of 12 mm ϕ @ 20 cm c/c spacing along the longer side and 12 mm ϕ @ 15 cm c/c spacing. The thickness of the pier cap is kept between 45 to 60 cm and of M 150 grade. The thickness of the cap is reduced to 30 cm at the ends.

37.2 DIFFERENT TYPES OF PIERS

The following types of piers are mostly used in the bridges :

1. Abutment pier
2. Dumb-bell pier
3. Solid pier
4. Column pier
5. Trestle pier
6. Cylindrical pier
7. Pile pier

37.2.1 Abutment Pier

In an arch-bridge, it is not necessary to design intermediate piers to resist horizontal thrusts of arches, because the thrust of two half arches on each side of the pier act in opposite directions and counteract their effect. But if the bridge is very long, then it is desirable to provide some intermediate piers of heavier sections, so that they can bear the horizontal thrust if required. Generally in practice every fourth or fifth pier is provided for taking horizontal thrust during construction. These piers which are designed for horizontal thrust are known as *Abutment piers.*

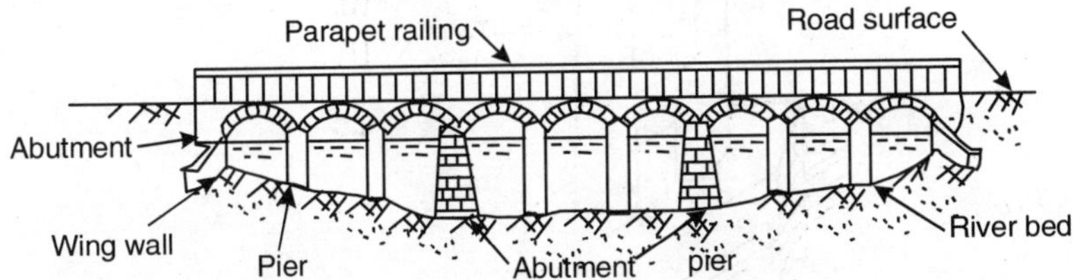

Figure 37.3 *Abutment pier*

These abutment piers are very useful and save centring charges, because if no such pier is provided, the centring for the whole length of the bridge is to be done at one time. But if such piers are provided the centring can be done in sections and construction work can also be done in sections. Also the partial damage caused due to floods and earthquakes to the bridge will not extend further beyond the first abutment pier. Thus the repairs can be done easily.

37.2.2 Dumb-bell Pier

If the superstructure of the bridge has twin-girders, the central portion of the upper part of the wall type pier is completely useless for transferring loads to the foundation. If the bridge is longer, the pier will be higher and wall type pier will not be satisfactory in appearance. In addition to this the mass of masonry or concrete is expensive and gives unnecessary load on the foundation. Therefore if twin column shafts are provided just below the girders, it will be economical and will also give good appearance. But if this type of piers in the full height are not sufficiently strong to offer resistance

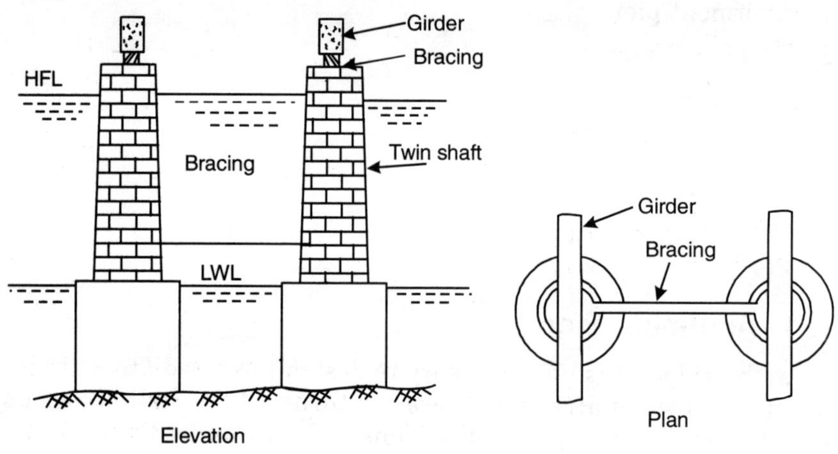

(a) Dumb-bell pier

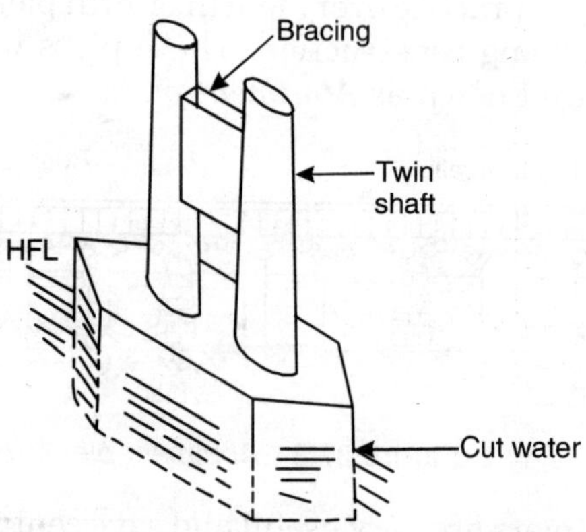

(b) Dumb-bell pier (with solid portion upto H.F.L.)

Figure 37.4

to the impact of floating bodies. The wall type pier is taken upto the H.F.L. and above this level twin piers are constructed. At some places where there is no danger of impact of floating bodies and the bearing capacity of the soil is very good, two separate foundations should be provided for economy purpose. There are so many shapes of such types of piers which are also known as *Dumb-bell piers.* Some sketches of Dumb-bell piers are shown in Fig. 37.4 (a) and 37.4 (b).

37.2.3 Solid Piers

In masonry bridges the shafts of piers are almost always shaped like a wall set parallel to the current. This shape is one of the most commonly used for

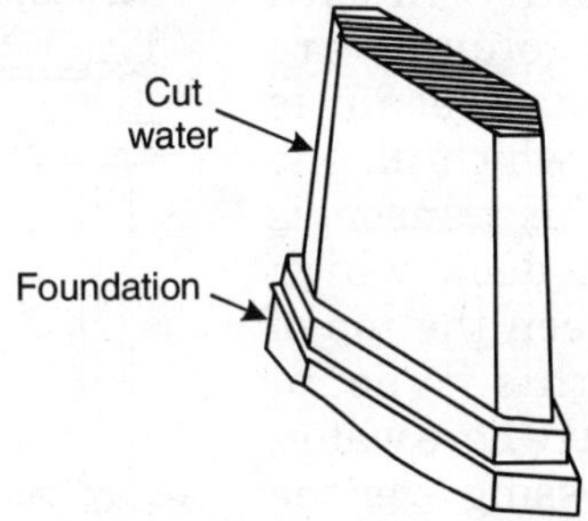

Figure 37.5 *Solid pier*

the upper portion of the piers. These types of piers are known as *solid-piers* and they provided excellent resistance against floating bodies and can be used for any type of superstructure of the bridge. The minimum top width 1.20 metre is required in these piers.

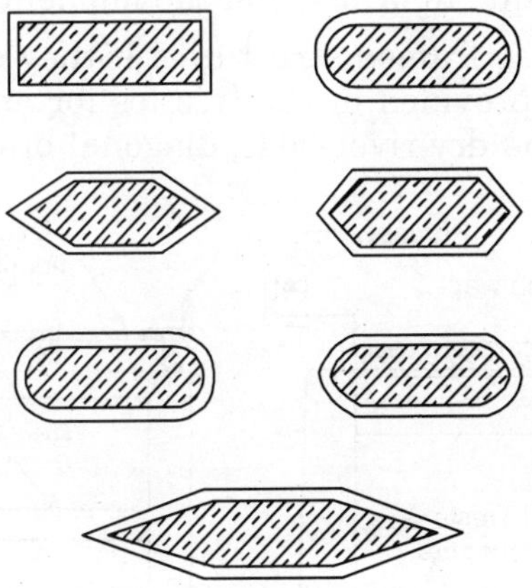

Figure 37.6 *Shapes of bridge-piers*

Modern trend is towards the slender or thin piers, therefore these are designed to resist the vertical load in addition to the impact of floating bodies. Different shapes of cutwaters are provided in these piers as shown in Fig. 37.5. The edges of the cutwaters and the side faces of the piers are usually given a tapering of 1 in 50 to 1 in 40, for the improvement in its appearance. Modern trend is towards the slight tapering because more taperings are out of date.

These piers may be constructed with stone-masonry or concrete. Sufficient arrangements are done for the putting of bearings on the pier.

37.2.4 Column Piers

When the beam of girder of the superstructure of a bridge is very close together, it is impossible to construct a separate pier for each girder. In such cases small number of columns are designed and a transom or beam is provided at their top on which all the girders rest. Sometimes this transom is provided in the bridge deck which increases the space between the top of pier and deck. But this type of arrangement is very useful where water, oil or sewage pipes are passing over the bridge, because they can be laid in the open space.

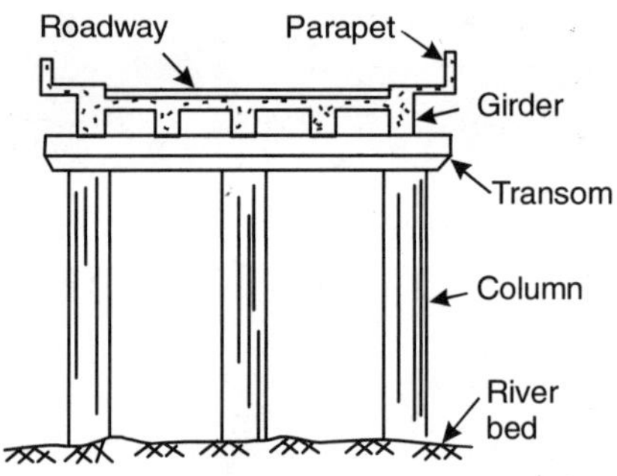

Figure 37.7 *Column piers*

37.2.5 Trestle Piers

These are mostly constructed on land sites or on dry river bed portions. Now as there is no fear of shock by the impact of floating bodies, these are designed for compressive forces due to bridge and the moment caused by wind.

These are designed as framed structures. If they carry sub-way road or railway, openings are provided in the trestles for another carriageway. If these are constructed on dry river-beds, diagonal bracing may be provided

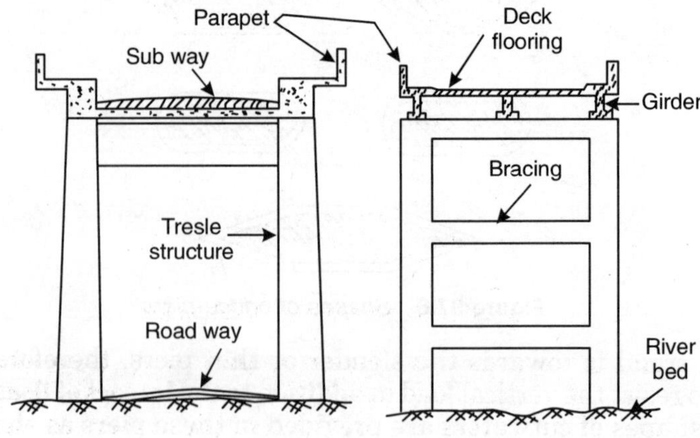

Figure 37.8 *Trestle-piers*

Fig. 37.8. Trestles may be of R.C.C. or steel depending upon the choice and design.

37.2.6 Cylindrical Piers

These are generally provided on steel cylinder caisson foundations. Mild steel girders are extended from the foundation and are connected with each

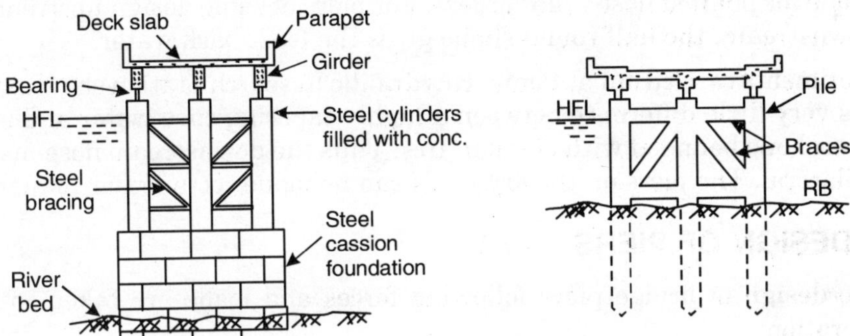

Figure 37.9 *Cylindrical piers*

Figure 37.10 *Pile-piers*

other by horizontal and diagonal steel bracing as shown in Fig. 37.9. Concrete is poured around the vertical steel girders in the shape of thin cylinders, in which girders act as reinforcement. Bridge girders are directly laid over these concrete columns.

37.2.7 Pile Piers

In shallow waters the piles can be used to support the main girders of the bridge directly over their caps. But in such cases to give more lateral rigidity to the piles, these are braced together as shown in Fig. 37.10. These types of piers are cheap and easy to construct as piles can be driven in the river-bed easily.

37.3 SHAPE OF PIERS

Various common shapes of the piers in the plan are given in Fig. 37.6.

37.3.1 Cut-water

It is the shape given to the side of the upstream nose of a bridge pier for smooth flow of water past it. Cut-waters need not be very long. Best form of cut-water is half round or semi-elliptical.

37.3.2 Ease-water

It is the downstream nose of the bridge pier shaped to promote merging of water flowing out of the adjacent opening of a bridge. The best form of ease-water is half round or pointed.

With certain shapes of the piers the force of the current is reduced, but they are uneconomical. The cut-water shape shall be carried upto the pier base.

The best practical form of nose is either the half round or the semi-elliptical, these being better than the pointed nose. Back water may be appreciably reduced by using an efficient tail or ease-water. A 90° nose is not satisfactory best angle for pointed nose is 45° or less. For piers of same design upstream and downstream, the half round shape gives the least back water.

Experiments carried out at Ponna Hhydraulic Research Station show that there is very little difference between a highly tapering ease-water and an equilateral one bounded with circular arcs. Thus the downstream nose may be equilateral. The piers on the dry lands can be made rectangular in plan.

37.4 DESIGN OF PIERS

For the design of bridge piers following forces and loads are taken into consideration:

37.4.1 The Self-Weight of Pier

If the foundations of the piers are pervious, the weight of that portion of the pier below water level should be reduced by 1000 kg/m^3.

37.4.2 Load of Superstructure

The end loads of the two adjacent bridge spans (including all dead, live and impact loads) coming on the piers. Eccentricity of the pier loading should also be taken into account, if the two loads coming over it from the adjacent spans differ widely.

37.4.3 Reactions of the Foundation

The pier should be safe and stable against the following :

1. Its sliding parallel to the axis of the bridge due to thrust of the arch.
2. Its sliding towards downstream side due to thrust of the current of water.
3. Its overturning about the downstream toe.
4. Its overturning in the direction of the axis of the bridge.

The maximum pressure at the toe of the downstream side should not exceed the allowable masonry and foundation soil bearing capacity pressure.

37.4.4 Buoyancy or Uplift Force

In the situation where the water can enter under the pier or abutment foundations, the consideration of the buoyancy or uplift force should also be done. In case of submersible bridges, it should be designed assuming that the fill behind the abutment has been removed. If the bridge is constructed on the impermeable or firm strata, no allowance for the uplift force should be done. For the allowance of the buoyancy a reduction in the gross weight of the pier or abutment equal to the weight of a volume of water of the submerged portion is made.

37.4.5 Impact Load

The impact effect is produced by sudden application load, blows or shocks. It is greater on the short span than on the long span bridges. It increases within certain limits with the speed of the vehicle. The impact effect on the railway bridge is much more and mainly depends on the weight of the engine, coaches and bogies and the speed of the train.

In case of road bridges the impact effect is reduced to a considerable effect, if earthen cushion is provided over the road slab or arch crown. If earth cushion of more than 60 cm is provided, the impact load shall be half as calculated for non-cushion roads. If the cushion of more than 1.0 m is provided, the effect of impact on the bridge shall be zero.

37.4.6 Traction Forces

The traction forces produce longitudinal thrust due to braking and acceleration act of the vehicle engines. This force is parallel to the centre line of the bridge and tend to overturn the pier and abutment in the plane of the force. The traction force is about 20% of the live load on the span. Proper accounting of this force is to be done.

37.4.7 Centrifugal Force

This force is caused only on those bridges which are situated on the curves. All large and major bridges are not provided on the curves. Only small bridges are provided on the curves, in case the change of the alignment of the bridge is uneconomical and other reasons prevent its change. Due allowance for it should also be made.

37.4.8 Horizontal Force Due to Water Current

All the piers and the abutment submerged in the stream water shall be designed for horizontal force due to water current. While designing the full H.F.L. of the water and the maximum velocity of the current is taken into account.

On the piers parallel to the direction of the water current, the intensity of pressure is calculated from the following formula :

$$P = 52\,KV \qquad \text{... (37.1)}$$

where

P the intensity of pressure due to the water current in kg/sq. m

V velocity of the current at the point where the pressure intensity is being calculated in m/sec

K coefficient, whose value depends on the shape of the nose of the pier as shown in Table 37.2.

TABLE 37.2 *Values of Coefficient 'K'*

S. No.	*Description of pier shape*	*Value of coefficient 'K'*
1.	Piers with cut and ease waters of equilateral of circles	0.43
2.	Piers with arcs of cut waters intersecting at 90°	0.47
3.	Piers 5—6 times longer than their breadth with triangular cut and ease waters subtending an angle of 30° or less	0.50
4.	Circular piers or with semi-circular ends	0.66
5.	Piers with triangular cut ends having 60°	0.70
6.	Piers with triangular cut ends having 90°	0.90
7.	Piers with square ends (rectangular piers)	1 .50

Following formulae are also used for determining the force due to water current:

1. Horizontal force due to static head

$$= \frac{WH^2}{2} \times \text{area of pier face (or } 2 \times \text{area of cut-water face)} \qquad \text{... (37.2)}$$

2. Pressure due to impact of water and debris

$$= \frac{WV^2}{2g} \; \frac{a}{A-a} \qquad \text{... (37.3)}$$

3. Pressure due to eddies

$$= W \cdot \frac{(V_1 - V)^2}{2g} \qquad \text{... (37.4)}$$

where

W Weight of water

H [Depth of water (height of pier)

+ (afflux) + head due to velocity of approach]

V Velocity of approach (velocity of stream)

A Area of flow just upstream of site

a Area of obstruction in the flow due to bridge

V_1 Velocity through the opening (under bridge)

At the sites where the pier is at an angle to the current, the velocity should be resolved into two components, one parallel and the other normal to the pier. As the velocity of the water current is maximum near the surface and minimum near the bed, the point of application should be taken as $H/3$ from the top. A force of 20% of the pressure of water parallel to the pier, should be taken as acting at right angle to the current and normal to the pier.

In case of a bridge having a pucca bed flooring the effect of cross-current should not be taken into account less than that of a static force due to a difference of head of 25 cm between the opposite faces of the piers.

All the piers should be located approximately parallel to the direction of the current to prevent the shift in the river channel, erosion of the foundation bed and unnecessary obstruction in the flow of stream. The skew spans should be avoided as far as possible.

37.4.9 Temperature Forces

These are caused due to the expansion and contraction of the superstructure of the bridge. The force produced in a member or support restrained from expansion or contraction is equal to the stresses in the restrained member multiplied with its cross-sectional area.

Following range of temperature is usually made in the design :

1. *Concrete bridge* : 9 mm per 10.0 m of exposed length.
2. *Steel bridge* in Moderate climate : from – 18°C to 50°C.
3. *Steel bridge* in extreme climate : from – 35°C to 50°C.

The magnitude of the forces should be taken from the latest I.S.I. Codes of Practice and I.R.C. Standard Specification and Code of Practice.

37.5 ABUTMENTS

The end supports of the superstructure of a bridge, whether flat or arched are called *abutments.* Its function is to transmit the load from the bridge superstructure, give final formation to the bridge, and to retain the earth-work of the embankment of the approaches. It serves both as a pier and retaining wall. It tan be of brick-masonry, stone-masonry, precast concrete blocks or of R.C.C. Generally, weep holes are provided in abutments and wing-walls for draining the water of the soil of embankment. The height of abutment is equal to that of piers and length is equal to the length of pier excluding cut-water edges.

The top surface of the abutment is kept flat when the superstructure of the bridge consists of girders, trusses or semi-circular arch. But if the superstructure consists of segmental or other types of arches the top of the abutment is given a slope to receive the arch.

For draining off the sub-soil water entered in the filled soil behind the abutment, weep holes are provided at suitable intervals in the face of the abutment. The cap is provided at the top of the abutment in the same way as on piers.

The face of the abutment towards water is kept vertical, but it may also be provided with a batter of 1 in 1, 2 to 1 in 25. The face retaining the earth fill may be provided with steps or a batter of 1 in 6.

37.6 TYPES OF ABUTMENTS

37.6.1 Abutment for Masonry-arch Bridge

As the arch exerts thrust on the abutment, it is designed for it. The waterside face may be given a batter or slope if desired. Embankment side face may be

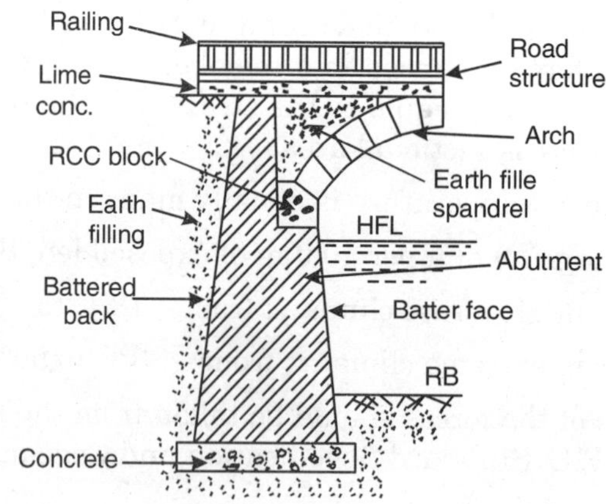

Figure 37.11 *Abutment for masonry arch-bridge*

inclined or vertical. Generally, it is given an inclination as shown in Fig. 37.11.

37.6.2 Abutment for R.C.C. Arch

In this case the abutment is specially designed for the thrust and things of R.C.C. arch as shown in Fig. 37.12. The abutment may be constructed with

R.C.C. as counterfort type of retaining wall, which can be inclined to counteract the thrust of the arch.

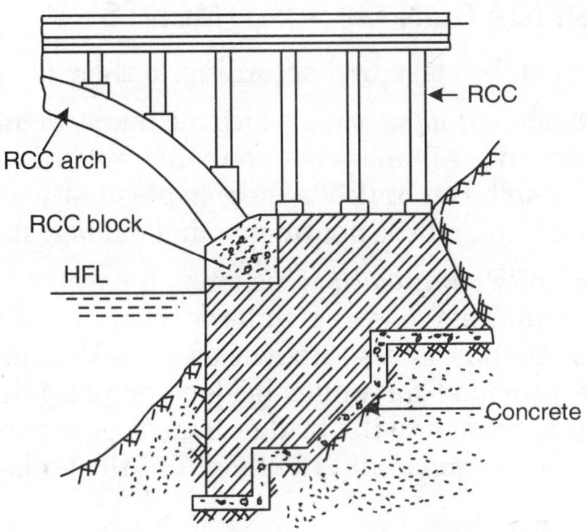

Figure 37.12 *Abutment for R.C.C. arch bridge*

37.6.3 Abutment for Girder-bridge

Abutments for girder-bridge are designed to take the vertical loads of the superstructure. These are constructed with masonry or R.C.C. Abutments for girder bridges are usually subjected to tilting forward due to eccentric

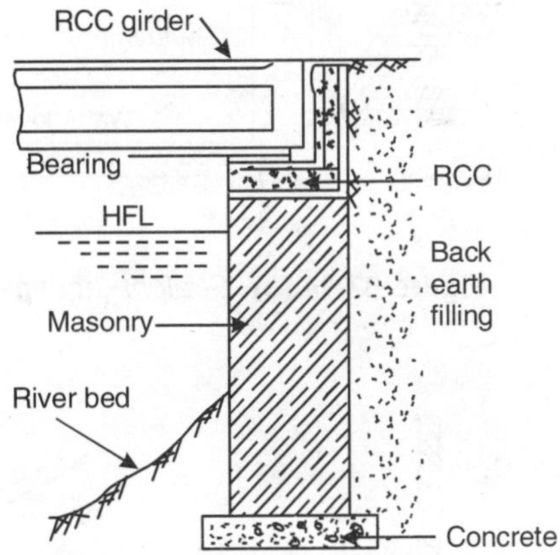

Figure 37.13 *Typical abutment of R.C.C. girder-bridge*

load acting on it and also due to the deflection caused in the girder. Hence when they are loaded there is a tendency in them to move away from the soil bearing against them at the rear. Therefore, at the time of designing, care is taken for it also. Figure 37.13 shows an abutment used for a girder-bridge.

37.7 CLASSIFICATION OF ABUTMENTS

The abutments can be classified according to their layout in plans.

(*a*) *Abutment without-wing walls* which include straight and Tee-abutments. This type of abutment is suitable when the banks are firm and height of approach embankment is small. But in floods there is possibility that flow of water may wash the earth-fill around the abutments and damage it.

(*b*) *Abutment with wing-walls* which may have straight wing walls, wing-walls curved in plan, splayed wing-walls or return wing-walls. The abutment with return wing-walls is also known as Box-Abutment. Sometimes box-abutments are provided for under-bridges or providing another sub-lane under the main approach to the bridge. This type is provided at such places where scouring is too much and the height of the embankment is more.

Figure 37.14 (a) *Straight abutment*

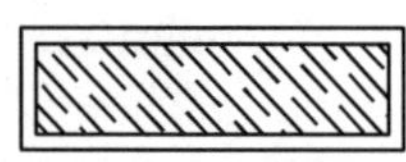

Figure 37.14 (b) *Tee-abutments*

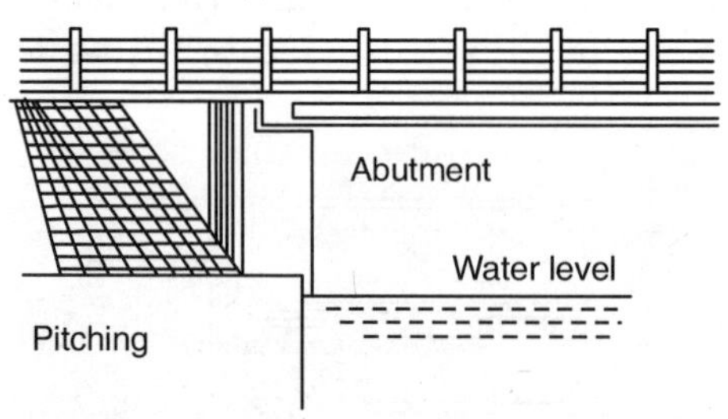

Figure 37.14 (c) *Straight abutment*

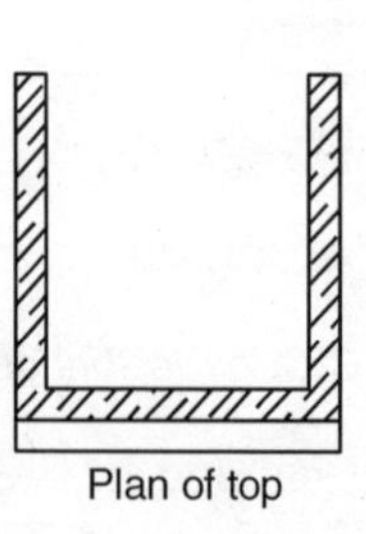

Figure 37.15 (a) *Plan of Top*

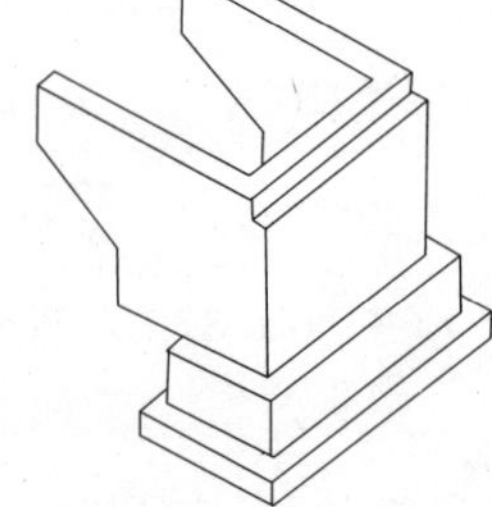

Figure 37.15 (b) *Box abutment (Isometric view)*

Abutment with straight wing-walls is suitable for railway bridge crossing a roadway or under bridge. It is also suitable when one road crosses another road having lower level. But this type is not suitable for bridges on rivers, because water may get access in the back of wing-walls and damage it.

Abutment with curved wing-walls is sued for smooth flow of water. The wing-walls may be of convex or concave shape facing the water.

Abutment with splayed wing-walls is mostly used. Its angle may be 45° or

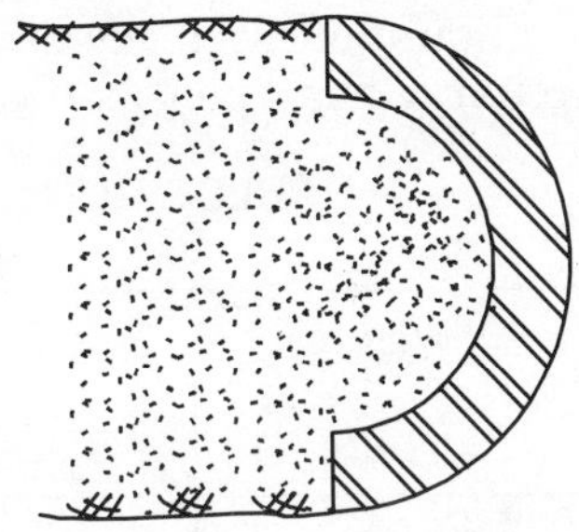

Figure 37.16 *Hollow abutment*

30° with the face of the abutment. Splayed wing-walls prevent damage to embankments of approaches.

(c) *Buried abutment.* At certain places when the channels are narrow, the abutments are constructed inside the banks and do not cause nay obstruction to the flow of water. Such abutments are called *buried abutments.* In this case the earth pressure in the back of abutment is balanced by the fill in the front, therefore cheaper abutments will serve the purpose.

(d) Hollow or Arch abutment. These abutments are curved in plan and are used only in rail and road crossings on land. These are mainly provided for economy in certain conditions. The seats of high inclined skewback arch thrusts are provided only in rocks. These are oftenly used for spans over gorges. Sometimes these abutments are provided with curved wing-walls.

37.8 DESIGN OF ABUTMENTS

The forces which are considered and taken into account for the design of the piers are also taken into account for the design of abutments.

The forces acting on the abutment are the load of the end of the bridge span with any which it may carry, the pressure from the earth-fill, the surcharge pressure due to live load, the traction force of the vehicles, centrifugal force, wind force, water pressure, reaction of the foundation and the buoyancy force. The abutment must be safe and stable against sliding and overturning. It should be safe against its self-structure also. It should also be stable with and without the bridge in place. Abutments are also designed against saturated earth pressure caused due to back fill.

The length of the abutment at the top is kept usually equal to the formation width. The abutments are also effected by the impact force caused by the moving vehicles, but it is supposed that entire impact is dissipated through the embankment. While determining the pressure on the abutment equivalent live load surcharge in addition to the usual earth pressure is also considered. In case firm approaches are provided for 3.0 m length on either side, no live load surcharge is taken into account. With the increase in the height of the abutment, the effect of live load is reduced due to earth cushion over it.

Table 37.3 shows the amount of earth cushion, which if provided over the culvert, no live load surcharge shall be taken into account.

TABLE 37.3

S.No.	*Height of abutment in m*	*Minimum earth cushion height to be provided for eliminating live load surcharge*
1.	upto 3.0	1.20 m
2.	between 3.0 – 4.5	1.10 m
3.	between 4.5 – 6.0	0.90 m

Note. In Table 37.3, the earth cushion height means the height between the top of abutment and the road level.

While constructing the abutments, care should be taken not to fill the back earth until the deck is constructed over it and it is capable of taking load.

The filling of earth in the back of the abutment should be done in layers not exceeding 30 cm in depth. The slope of each laver should be kept 45° with the ground. The ramming of this filling should be done at optimum moisture content. Sand and sandy soil or such sound materials should be used for back filling.

While doing the earth filling in the approaches of the bridge or the spandrels of the arches, the filling should be done simultaneously with the wing-walls also.

In the abutments for preventing any accumulation of water weep holes or drain pipes are provided at about one metre spacing in both the directions in the face.

37.9 SIZE OF ABUTMENTS

Usually 70 cm top width of the abutment is provided upto 3.0 m span. For the spans above 3.0 m the abutment top width of about 80 cm is provided.

Indian Road Congress special publication No. 13 recommends the following widths of the abutments.

TABLE 37.4 *Abutment widths at ground level*

S. No.	*Height of abutment from ground level to top of bed stone (cm)*	*Width of abument for about 22 t/m² maximum pressure on the sub-soil (cm)*	*Width of abutment for about 44 t/m³ maximum pressure on the sub-soil (cm)*
1.	150	158	128
2.	180	173	143
3.	210	190	152
4.	240	207	160
5.	270	233	170
6.	300	265	185
7.	360	330	218

Note. The dimensions of width given in Table 37.4 are for spans upto 3.0 m, for span above 3.0 m the top width should be increased by 8.0 cm.

The face of the abutments may be kept vertical or with a batter of 1 : 12. The ratio of the height of the abutment or pier to its base width should not exceed 6.

Minimum bottom width of 138 cm is provided including 24 cm foundation concrete all around.

In case of ordinary soils the minimum depth of 1.85 m (1.40 m for masonry and 45 cm for concrete) is provided. This depth is for IRC heavy loading, where no approach slab is provided.

Usually the safe bearing capacity of the soil in the foundation increased by 0.5 t/m² for each additional 30 cm depth below one metre depth.

Table 37.5 gives the equivalent heights of surcharge of earth for live loads as per IRC : 6.

TABLE 37.5 *Equivalent heights of surcharge of earth for live loads*

S.No.	*Depth of abutment below road level in metres*	*I.R.C. class 'AA' and class 70 R loading*		*I.R.C. class 'A' loading*		*I.R.C. class 'B' loading*	
		Single lane bridge	*Multiple lane bridge*	*Single lane bridge*	*Multiple lane bridge*	*Single lane bridge*	*Multiple lane bridge*
1.	0.2	26.0	15.4	14.3	17.2	8.3	10.0
2.	1.0	15.0	9.1	8.5	10.0	5.1	5.8
3.	2.0	8.0	5.5	5.1	6.1	3.0	3.7

Contd.

Table 37.5 Contd.

S.No.	Depth of abutment below road level in metres	Single lane bridge	Multiple lane bridge	Single lane bridge	Multiple lane bridge	Single lane bridge	Multiple lane bridge
4.	3.0	6.8	4.1	3.8	4.6	2.3	2.7
5.	4.0	5.5	3.3	3.0	3.5	1.8	2.1
6.	6.0	3.8	2.3	2.2	2.6	1.3	1.5
7.	8.0	3.0	1.8	1.7	2.0	1.0	1.2
8.	10.0	2.6	1.5	1.4	1.7	0.9	1.0

Note. Following figures are used in the calculations, the values of Tables 37.5 :

(i) Length of abutment – 4.5 m for single lane bridge and 7.6 m for multiple lane bridge.

(ii) Angle or repose of soil – 30°.

(iii) Density of back fill material – 600 kg/cu m.

(iv) The resultant earth pressure acts in horizontal direction.

When the abutments are designed to retain the earth behind them, but no protection is made in the front, the foundation should be designed to withstand, the earth pressure and the horizontal forces for conditions of maximum scour of 0.63 D. If proper protection in the front of the abutments is provided, it should be taken into account.

Table 37.6 gives the top width of piers and abutments of slab and girder bridges just below the caps.

TABLE 37.6 *Top widths just below caps*

S.No.	Span of the bridge in m	Top width of piers carrying simply supported span in cm	Top width of piers and abutments carrying continuous spans in cm	
1.	3 .0	50	40	The top width of wing walls and return-walls should not be less than 50 cm
2.	6.0	100	75	
3.	12.0	120	100	
4.	24.0	160	130	
5.	40.0	200	170	
6.	60.0	220	190	

37.10 WING WALLS

Normally an abutment has one front wall on which the deck or girder of the bridge is supported, and side walls perpendicular to front wall which are also known as *return walls,* when they are parallel to the bridge, and *wing-walls* when they are curved in plan or at any other angle.

The design of wing-walls is independent. For straight wing-walls it may be assumed that the equilibrium of each of their cross-sections is comparable with that of an imaginary horizontal wall of indefinite length of the same cross-section holding up an embankment.

Generally, the wing-walls have steadily decreasing cross-section. Their mean thickness at any given section is one-third of their height. The design of the wing walls mainly depends upon the nature of the soil in the embankment.

37.11 CLASSIFICATION OF WING-WALLS

Depending on the material of construction the wing-walls may be classified as follows.

37.11.1 Masonry Wing-Walls

These are generally used with masonry abutments. These may be splayed or return parallel to the road way. The face towards the water may be kept vertical or battered. The face towards the earth fill may be stepped or provided with a slope or batter. Mostly the top of the wing-walls is kept slopy outward, but may also be kept horizontal.

37.11.2 Reinforced Concrete or R.C.C. Wing-wall

In case of R.C.C. or plain cement concrete abutments, the wing-walls may be constructed with R.C.C. or plain cement concrete. These are designed as cantilever type or counter-fort type retaining walls. The top of these may also be kept slopy outward or horizontal as in the case of masonry wing-walls.

The wing-walls may also be classified on their lay-out in plan. On this basis wing-walls are classified as follows :

(a) *Straight wing-walls.* These are used in small bridges and culverts meant for crossing drains. These are cheaper as compared with rcturn-walls.

(b) *Splayed wing-walls.* These provide smooth entry of the water under the bridge. This type of wing-wall is also very useful when road approaches are curved in plan as shown in the figure. Splayed wing-wall can be used on small as well as high bridges.

(c) *Curved wing-walls.* Curved wing-walls are generally used in bridges of canals or irrigation channels, because they provide smooth entry of water under the bridge. The convex or concave face may be towards the flow of water. If convex face is towards water, the entry and exit is smooth, but when concave face is facing water, it forms eddies.

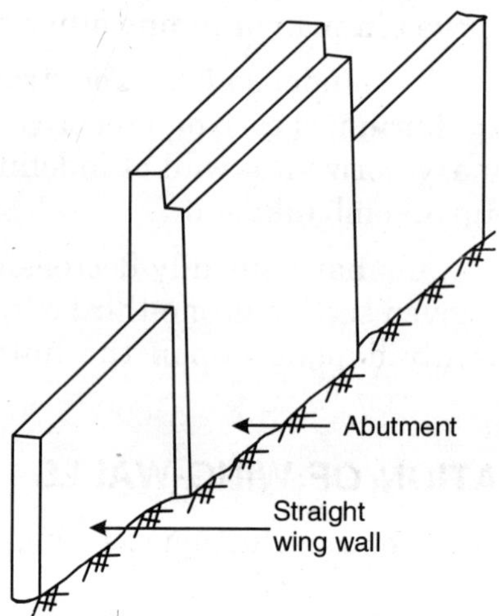

Figure 37.17 *Abutment with straight wing-walls*

(d) *Return wing-walls.* In this case the wind-walls are at right angles to the abutment and parallel to the roadway. The wing-walls run into the approach embankments. This type of wing-walls are suitable when the embankment is very high and banks are firm.

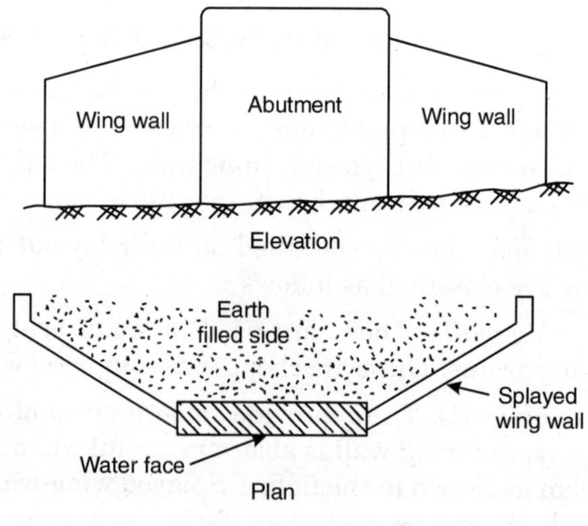

Figure 37.18

37.12 DESIGN OF WING-WALLS AND RETURN WALLS

The wing-walls are the splayed extension of an abutment of a bridge on a culvert. The main function of the wing-walls is to retain the embankment soil and side slope of the embankment. The wing-walls also serve the purpose of the guide for allowing the water to pass through the desired openings. The top of the wing-wall is kept level with the shoulder of the embankment and then, it is constructed at the natural angle of repose of the earth fill material to a height of 0.6 to 1.2 m above the ground level or water level.

The height of the lower end of the wing-wall is usually kept lower than 60 cm. If the height of the embankment is less than 240 cm, the height of the lower end of the wing-wall is kept usually half the height of the embankment.

Usually upto 3.0 m span culverts, the wing-walls have vertical faces with batter at the back of 1 : 4 and 23 cm minimum top width as shown in Fig. 37.19.

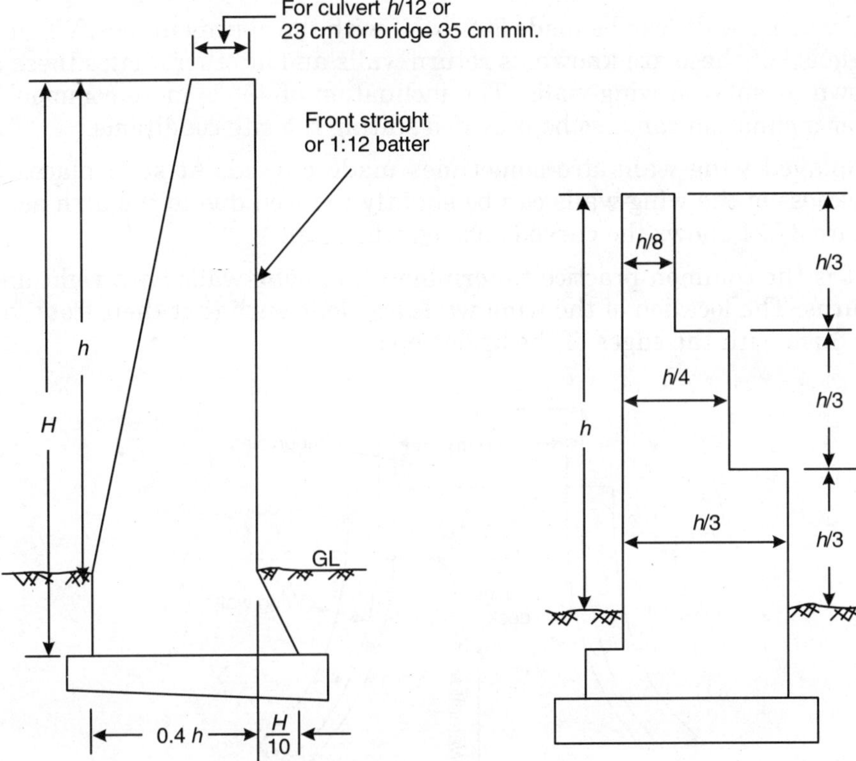

Figure 37.19 *Section through wing-wall of small culvert*

Figure 37.20 *Cross-section of a stepped wing-wall*

For bridges having span between 4.5 m to 9.0 m the wing-walls are provided with 1 : 12 batter in the face and 1 : 6 batter in the back constructed in steps, with minimum top width of 46 cm. Usually the thickness of wing-wall is also reduced as its height reduces towards the outer edge.

When the soil of the backfill is of poor quality, the wing-walls are extended to the ground level. The length of the wing-walls and the return walls are so provided, so as to prevent the falling of the backfill soil into the water-way.

The wing-walls are designed as retaining walls. The main difference between them and the abutment design is the superimposed vertical deck load of the superstructure which is taken into account in the case of abutments. The load is the main important fact or which causes difference in the unequal settlement-between the wing-wall and the abutment. For this reason a small gap between the abutment and the wing-walls is to be provided, and should be filled with bitumen and sand for making it water-tight.

Figure 37.20 shows the cross-section of a stepped wing-wall, which is commonly used in railways.

The wing-walls can be made 30° to 90° with the abutment face. When the angle is 90° these are known as return walls and at other angles these are known as splayed wing-walls. The inclination of 45° is more common but other inclination can also be provided suiting the site conditions.

Splayed wing-walls are sometimes made curved. At such places the thickness of the wing-walls can be slightly reduced due to the arch action. Figure 37.21 shows the curved splayed wing-wall.

It is the common practice to terminate the wing-walls with right angle returns. The location of the wing-walls are done such that their front faces are flush with the edges of the bridge opening :

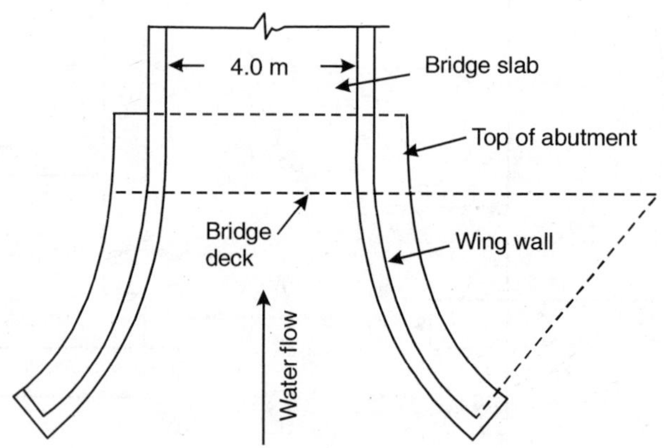

Figure 37.21 *Curved splayed wing-wall*

The return wing-walls (U-abutment) are provided at such places where the banks of the river are steep and the hard soil is not subjected to erosion and the stream water does not require guiding due to large width of the stream. The return wing-walls are taken sufficient inside so that the earth slope along them terminates outside the water way. The return wing-wall increases the strength of the approaches.

It has been seen that wing-walls splayed at 30° used with pipe-culverts increases the capacity of the culvert over that obtained with straight end walls.

37.13 APPROACHES

The portion of the road constructed to reach the bridge from their general route or height is known as approach of the bridge. The alignment and the level of the approaches mainly depend on the design and layout of the bridge. Indian Roads Congress recommended that a minimum of 15 m length on either side of the bridge should be kept straight. This length may be increased depending on the minimum sight distance in case of fast moving traffic. The width of the approach should not be less than the roadway width on the bridge.

If horizontal curves are required on the approaches to meet the original aligned road, they should be provided after the straight length of the road as specified on both sides of the bridge. The gradient, radius of horizontal or vertical curve should be provided as per latest I.R.C. recommendations for roads.

For high level culverts and bridges the approaches are constructed in embankments, whereas for low-level submersible or causeways these are constructed in cutting. In steep or hilly areas where the streams generally flow in deep valleys or gorge, the length of the approaches are less. Even during floods the water flows at a greater vertical distance from the bridge, of course with great velocity. The flood water do not spread over to vast areas, nor there is any possibility of flood water to damage the abutment or approaches.

In case of plain areas as the ground has a very gentle slope, the velocity of the stream water is low. Therefore, during floods large area of the ground on both the sides of the stream is flooded with water. In most of the major bridges the flood water is also filled on both the sides of the approaches. This requires stable approaches. In some cases vents or relief culverts are provided in the approaches for allowing excessive flood water from up-stream side to down-stream side.

In some cases the bridge is extended into the approach banks for some distance to provide stability to the approaches. The design of this extended portion generally differs from the main bridge and may consist of simple piers with deck slab. In case of arch bridge, it is more economical to construct approaches on series of small arches constructed on small walls. In the cities or at such places where the land is costly the approaches may be constructed with retaining walls filled with the earth inside them.

REVIEW QUESTIONS

37.1. Write short notes on the following :

(a) Abutment-pier;

(b) Dumb-bell pier;

(c) Box Abutment;

(d) Splayed wing-walls.

37.2. Give sketches of :

(i) A typical masonry pier, and

(ii) A complete masonry abutment used for a bridge.

Name the various parts and indicate their approximate proportion.

37.3. Give sketches of various types of abutments which are adopted in bridges, mentioning the usages to which each type is put.

[**Hint**. (a) *Abutment with straight wing-walls*. It is used where velocity is more and greater waterway is required at straight reach.

(b) *Abutment with splayed wing-walls*. These are used where approaches are higher than the banks. Therefore to retain soil of the embankment these are provided.

(c) *Abutment with return wing-walls*. These are provided where approaches are high and there is danger of erosion.

(d) *Straight abutment without wing-walls*. These are provided where side banks are firm and flow of water is less and no danger of side erosion exists.

(e) *The abutment without wing-walls*. These are used at such places where approaches are not high and there is no danger of erosion.]

37.4. (a) Explain with sketches (i) Abutment (ii) Piers (iii) Wing-walls for a Railway girder bridge.

(b) State the condition of stability of the abutment of a Bridge.

37.5. Sketch the various types of abutments used in the construction of bridge. Describe briefly the advantages and disadvantages of each type.

37.6. Give sketch of (i) A typical masonry pier, (ii) A complete masonry abutment for a bridge. Name the various parts and indicate their approximate proportions.

37.7. What loads are taken into consideration for the design of (i) Abutment, (ii) Pier, (iii) Return wing-wall.

37.8. What is an approach road for a bridge? Explain with the aid of nearly drawn sketches the details of a car ferry-approach suitable for an alluvial river.

37.9. (a) Describe the different types of bridge piers.

(b) What loads and forces are considered in the design of a bridge abutment?

37.10. Describe with neat sketches the different types of wing-walls indicating their advantages and disadvantages.

37.11. What is an approach road for a bridge? Explain with the aid of nearly drawn sketches the details of a car ferry-approach suitable for an alluvial river.

37.12. Enumerate the forces to be taken into account in designing a pier and abutment of a railway bridge.

37.13. What do you know about the construction of approaches for a road bridge.

37.14. Describe with sketches the types of piers used for R.C.C. bridges. What are the advantages of a hollow pier? Why are cut-waters used in a pier? Sketch suitable types of cut-waters.

37.15. Write short notes on :

(i) Dumb-bell Pier

(ii) Abutment Pier

(iii) Approach Slab

(iv) Pedestal Pier

(v) Wing-walls

37.16. Give sketches of :

(i) A typical masonry pier.

(ii) A complete masonry abutment used for a bridge.

Name the various parts and indicate their approximate proportions.

37.17. (a) Discuss different types of bridge piers. Give neat sketches.

(b) What are return-walls and wing-walls. Discuss the purpose for which these are provided.

Timber Bridges

GENERAL

The bridges which are constructed by timber have 15–20 years life, therefore, these are classified as temporary bridges. These bridges are very common and are suitable for hills, because they can be constructed with local materials and labour very quickly and economically. These bridges are constructed on unimportant roads or provisionally on important roads pending the construction of permanent bridges. The timber should be properly seasoned before constructing such bridges, and as far as possible large size timber with more elasticity should be used. If possible, creosote should be applied to the timber before using it, to make it safe against fungis, etc.

In bridges timber can be used for constructing (a) sub-structure, (b) trusses of frame for supporting a roadway, and (c) roadways. The construction of all these three types of structures will be dealt in the following article.

38.1 SUB-STRUCTURE

The design of sub-structure for timber bridges mainly depends on the span, load to be carried and the width of the bridge. There are many types of timber piers and abutments as detailed below :

38.1.1 Trestles

These can be constructed of round or square timbers. If heavy sections are required, tree-trunks are used but for light sections salwood *ballis* are best suited. Round timbers are fixed by rope or steel wires at their joints, but spikes and nails are used for fastening square sections. It is always economical to construct trestles at the site in shallow water or dry beds. But if it is difficult to construct them at site, they are constructed on the banks and then taken to site on boats or rafts and then lowered down.

Trestles are used when the bed of the stream is hard and the depth is not more. These may be two-legged, three-legged or four-legged as shown in Fig. 38.1 (a), (b), (c). Two-legged trestles are mostly used. Three-legged trestles are used when the river bed consists of mud; but these are not suitable for uneven bottoms and gives difficulty in placing them in position. Four-legged trestles are very suitable for placing at intervals in a long bridge of two-legged trestles to give it longitudinal stability. But generally these being of very heavy structures and difficult in placing are not used.

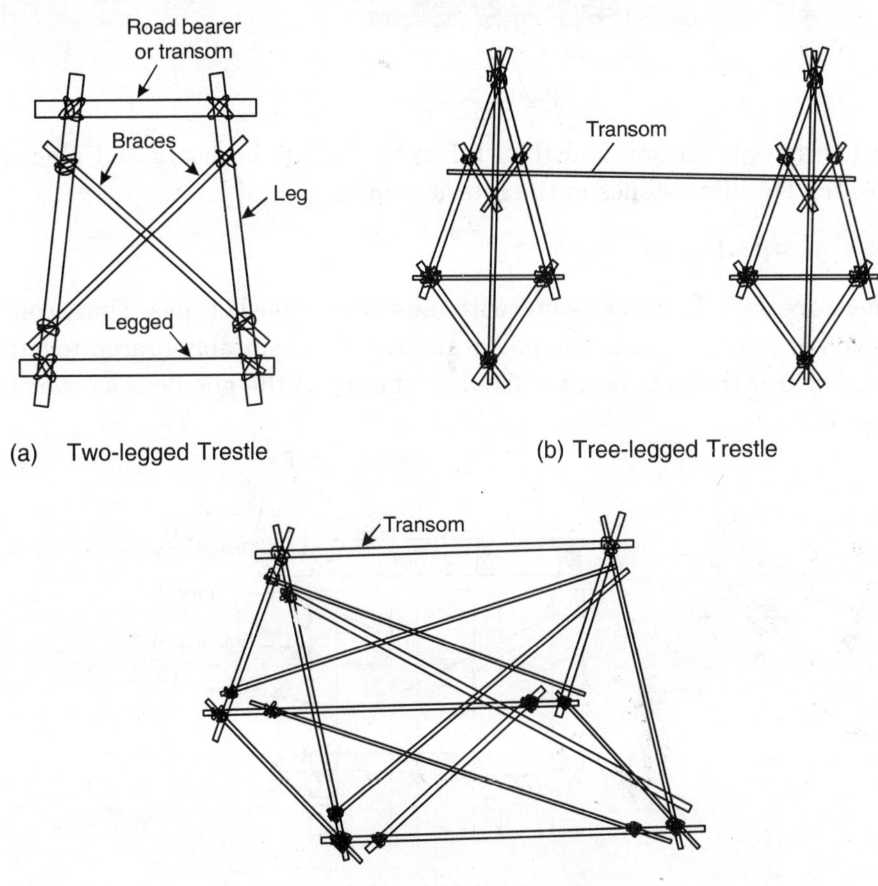

(a) Two-legged Trestle

(b) Tree-legged Trestle

(c) Four-legged Trestle

Figure 38.1

38.1.2 Cribs

Figure 38.2 shows crib-pier which essentially consists of wooden sleepers placed one above the other. These are used for temporary deviation of railway line or for taking heavy loads. These cribs are constructed at river banks

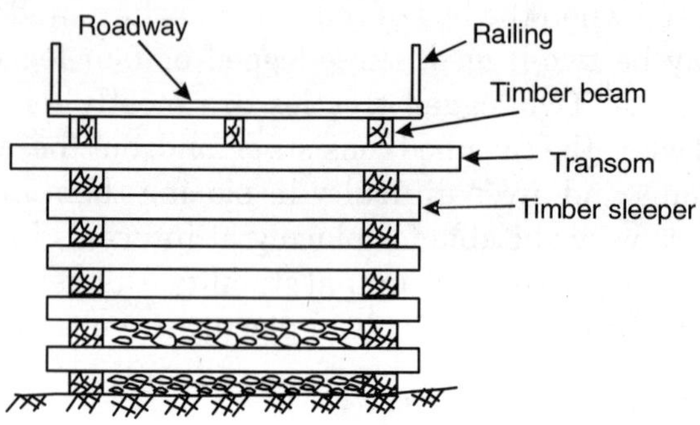

Figure 38.2 *A crib-pier*

with tray-type bottom and then taken to the site on boats and placed in position by filling stones in the interior space.

38.1.3 Pile-bents

These are used in deep streams with muddy erodible bottoms. These consist of timber piles driven in the bed of the river and suitably braced together. Road bearer transom beam is fixed on the top of the pile-bent as shown in Fig. 38.3.

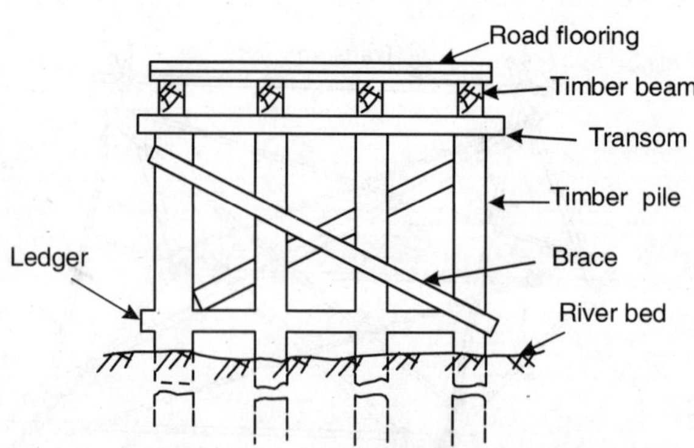

Figure 38.3 *Timber pile-bents*

38.1.4 Crates

Figure 38.4 shows the framework of a crate, which consists of four strong upright poles with horizontals. One each at the top and bottom to keep in upright poles at fixed distance and to carry the road-beams at their top. This crate is construcied on the bank of the river and wire netting is done on all sides and bottom to form a tray for filling brush-wood and stones to keep

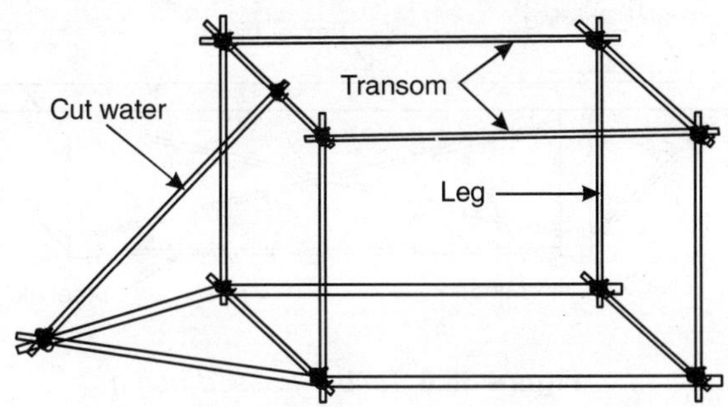

Figure 38.4 *Crate-pier*

the crate in position. Sometimes one inclined pole is fixed to the crate and kept towards upstream side, which acts as a cut-edge for water.

38.2 SUPER-STRUCTURES, TRUSSES OR FRAMES

The flooring of timber bridges can be constructed on any of the following types of super-structure:

(a) Strutted Beam
(b) Tied Beam
(c) Timber Trusses
(d) Timber Arches
(e) Timber Cantilever Bridge
(f) Timber Suspension Bridge

38.2.1 Strutted Beam

A simple type of strutted beam is shown in Fig. 38.5 which can carry ordinary road traffic over a span of about 8 metres. The main beam and diagonal

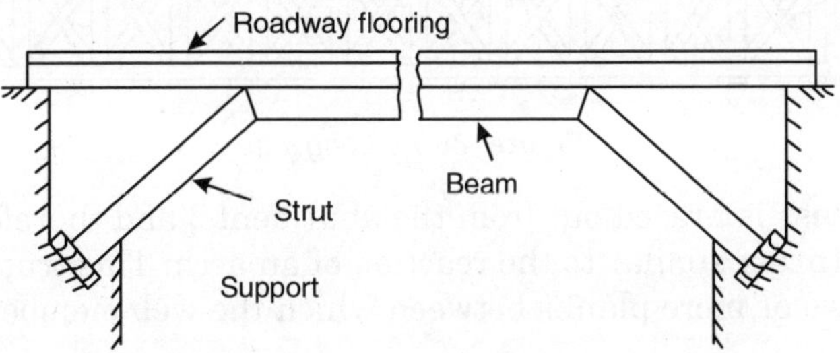

Figure 38.5 *Strutted-beam*

members have about 23 × 23 cm section, and all the vertical members are 23 × 15 cm section.

38.2.2 Tied-Beam

This type of truss is also suitable upto 8 metre span. The timber is used to take the compressive force while steel roads are used for taking tension.

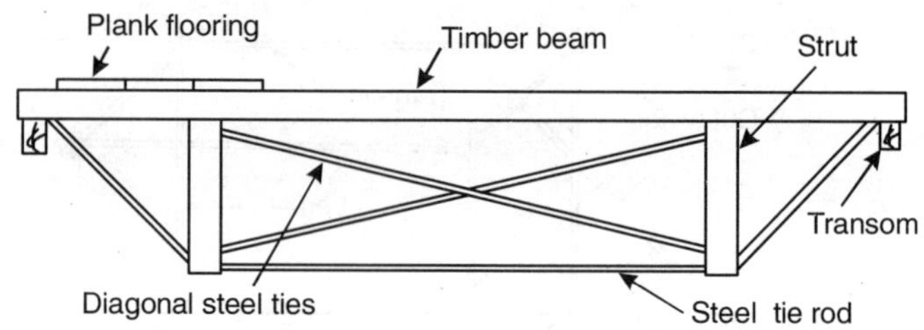

Figure 38.6 *Timber-trussed beam*

Generally, at the time of construction a camber of about 1 in 50 to 1 in 60 is provided initially so that when loaded the top beam will become horizontal.

38.2.3 Timber-trusses

The idea of the truss was used by an Italian architect, Andrea Palladio (1518–1580), in the middle of the sixteenth century. Ulric Grubenmann, a Swiss carpenter, constructed a wooden truss bridge of 63 metre span over the river Rhine in 1788. This was followed by other structures and finally a great wooden bridge of 130 metre span was constructed by Ulric and Jean Grubenmann—which is said to be the longest span wooden bridge ever built. These are used for long span timber bridges. In these, compression members are designed of timber, whereas steel rods are employed for taking tension. These may be of Long's, Howe or Pratt type as shown in Fig. 38.7.

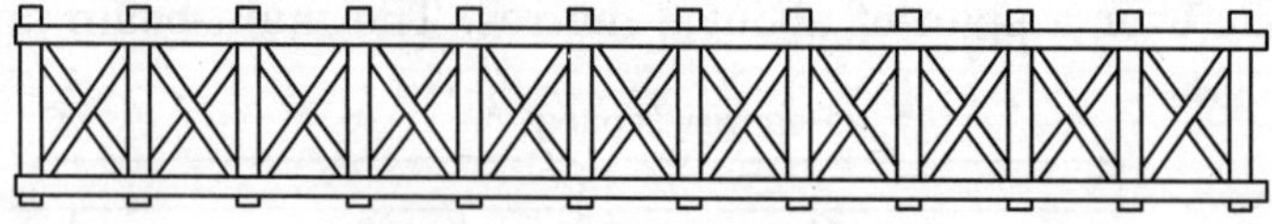

Figure 38.7 *Long's truss*

Long's truss is braced out from the abutments, and therefore it exerts a horizontal thrust similar to the reaction of an arch. This truss is built with chords of two or more planks between which the web members are placed.

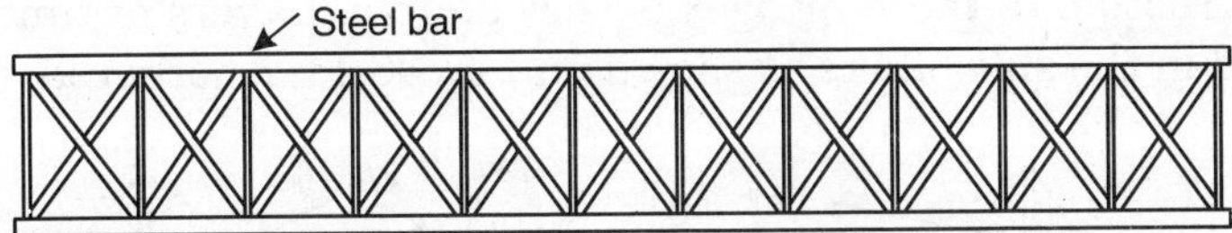

Figure 38.8 (a) *The Howe truss*

Pratt and Howe trusses are improvements over Long's truss. Howe employed vertical iron rods for web members whereas Pratt used diagonal members of steel.

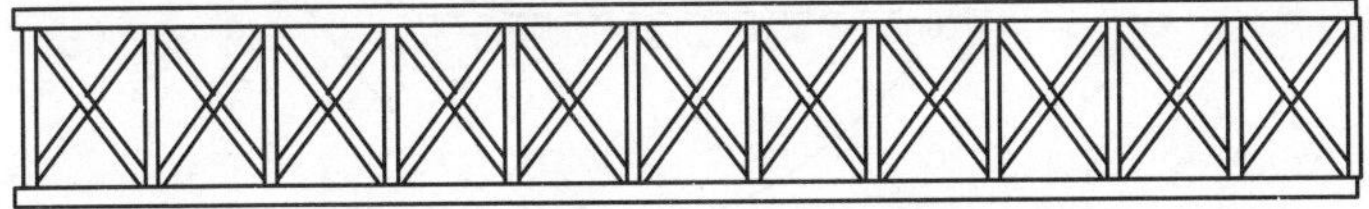

Figure 38.8 (b) *The Pratt truss*

38.2.4 Timber-Arches

Timber-arches can be constructed for spanning in the bridges. In 1806 one bridge was constructed over Schuylkill river at Philadelphia which has timber arches. Side two arches were of 50 metre span, and the central arch was of 65 metre span.

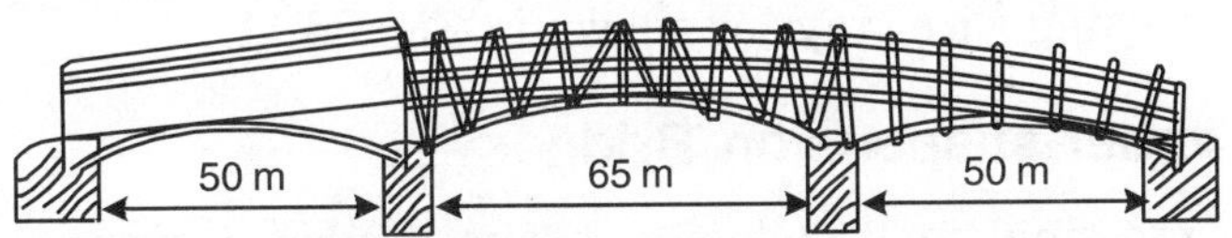

Figure 38.9 *Timber-Arch bridge*

One timber arch bridge of 113 metre span over Schuylkill river at Philadelphia was constructed in 1812. The given Figs. (38.9 and 38.10) show these two types of timber arch bridges.

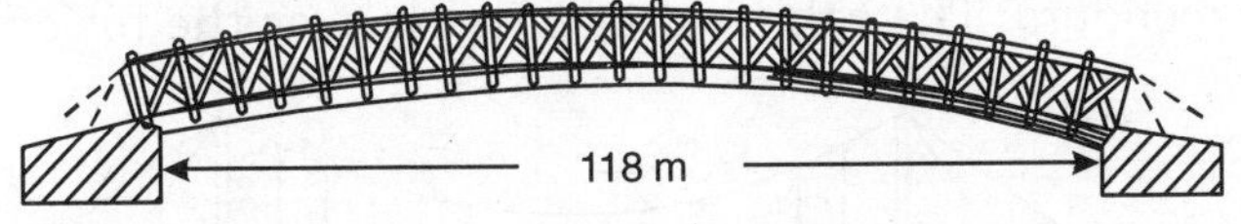

Figure 38.10 *Colossus Bridge-arch*

Timber-arch bridges for small spans are generally constructed in hills.

38.2.5 Timber-Cantilever-Bridge

In hilly areas where timber is available in plenty but other constructional materials are not easily available, timber cantilever bridges are very suitable.

The construction of these bridges is very simple. Layers of timber logs are embedded in the side banks as shown in Fig. 38.11. Another layer of timber

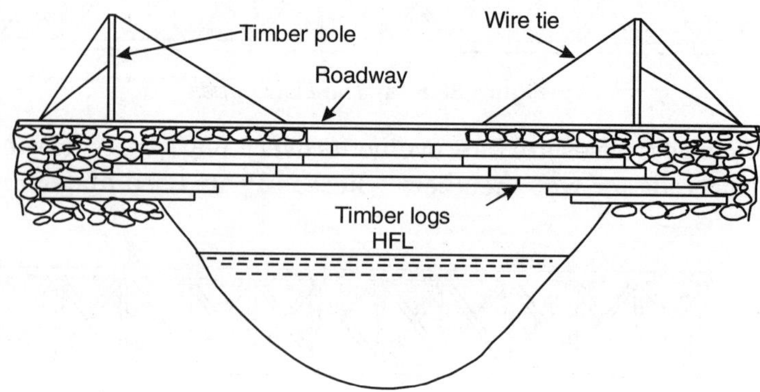

Figure 38.11 *Timber Cantilever-Bridge*

logs projecting between 2-3 metres from the lower layer is laid on it. In this way layers over layers are laid with projections from the bottom layers till 5–7 metre gap is left in the middle. Road bearers are laid over this gap and the bridge is completed by levelling the surface by earth and stone-ballast and the top is given a smooth touch. For light traffic these types of cantilever bridges are very suitable. The load bearing capacity of these bridges can be increased by using wire stays as shown in Fig. 38.11.

38.2.6 Timber-suspension Bridges

In hills for crossing small valley or stream, timber suspension bridges are very useful, because these can be constructed very cheaply and in a very short time. These bridges are generally not stiffened and are suitable only for very light traffic. They are of the following types:

(i) Sling bridge. This is the common type of temporary suspension bridge. It essentially consists of cables or ropes which bear the load of the roadway through suspenders. The cables after passing over the towers are fixed on

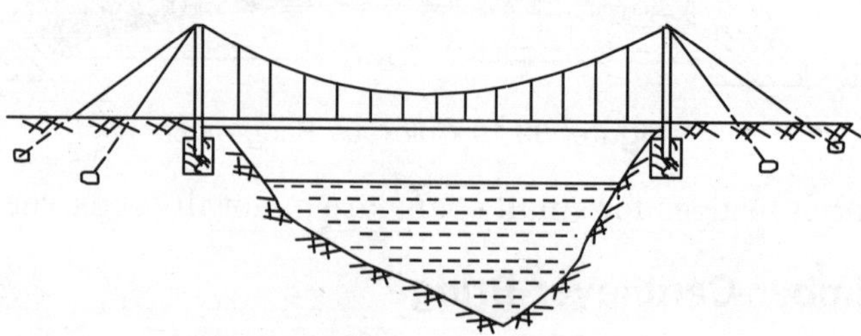

Figure 38.12 *Sling-Bridge*

the other side. The roadway is provided with wooden planks which are connected at ends to the wooden transom beams, hanged by suspenders. Bridge railing is fixed to the suspenders.

(ii) Ramp bridge. In this case the roadway is directly laid on cables for most of its length. This is called ramp-suspension bridge, because ramp joins

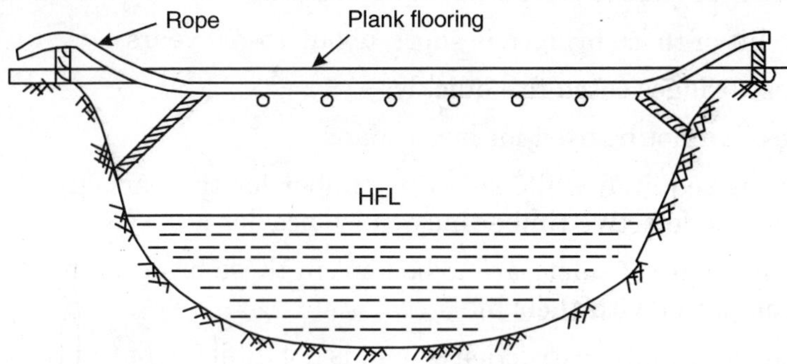

Figure 38.13 *A Ramp-Bridge*

the bank approaches to the bridge floor. This is a very cheap type of bridge, because it requires small quantity of material and can be completed in a very short time. Timber posts are fixed to the flooring, and chain or ropes are passed through them, which serve the purpose of railing. As the traffic moves on this bridge, there is distortion in it, because there is no stiffness in it, and this is the only disadvantage of this bridge. Figure 38.13 shows a ramp-bridge.

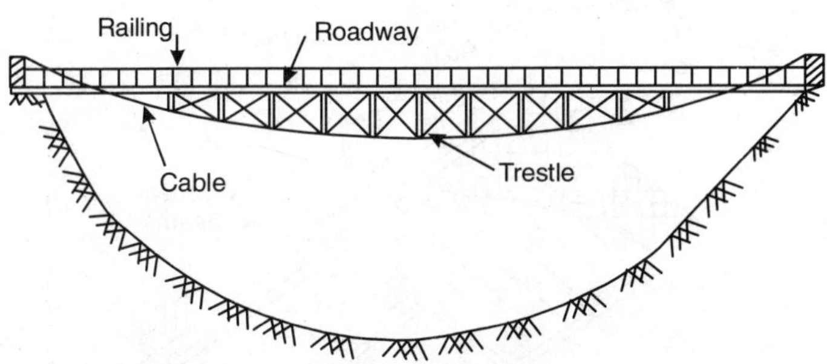

Figure 38.14 *A Trestle-suspension Bridge*

(iii) Trestle suspension bridge. In this case trestles are supported on cables and the roadway is supported on them as shown in Fig. 38.14. This type of bridge has the disadvantages of a very heavy superstructure, but the height of tower is reduced and so there is no distortion in the bridge when traffic moves on it; which is the main advantage of this bridge.

38.3 DISADVANTAGES OF TIMBER-BRIDGES

These bridges are mainly used for road bridges in certain areas in and around forests.

Following are the main disadvantages of these bridges :

1. These get deterioration due to atmospheric effects of the wind, sun and rain.
2. The lift of these bridges is short, about 15–20 years.
3. These bridges catch fire quickly.
4. These cannot be used for large spans.
5. Care is required while selecting timber for the particular member of the bridge, as defective timber may cause accidents.
6. More factors of safety are to be kept in these bridges, increasing their cost as compared with their life.
7. The design and construction of joint is not difficult, but as the strength of timber is not uniform in all directions, it causes difficulties while using timber.

The designs of these bridges are done for I.R.C. Class B loading with no impact allowances.

38.4 ROADWAYS

The flooring of superstructure consists of beams and planks laid on road bearers, wooden trusses, wooden arches or tied beams. The railings of roadways may be of timber posts with iron chains or ropes pushing in their holes.

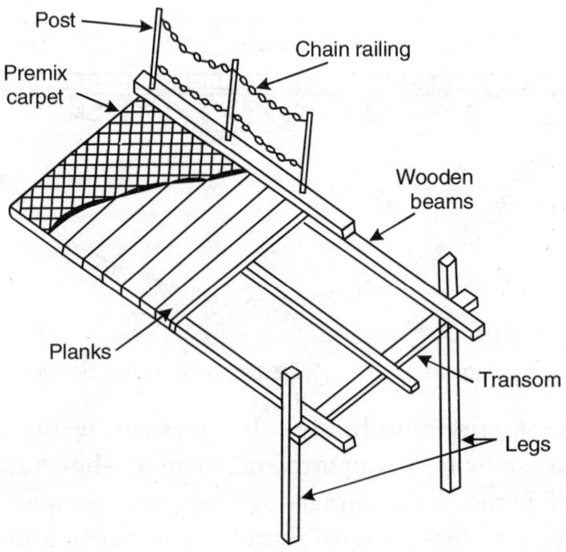

Figure 38.15 *Super-structure with road bearer*

As the timber is not strong against abrasion, therefore 5 cm thick premix carpet should be laid on the planks of the roadway for safety against the wearing due to traffic. The planks of the floor should be sufficiently strong to bear the load of the traffic under worst conditions and the impact caused by them.

REVIEW QUESTIONS

38.1. Write a short note on timber-bridges.

38.2. Describe with the help of neat sketches 'trestles' and circumstances under which each type is suitable.

38.3. With the help of a neat sketch show how the roadway of a timber-bridge is provided. If the timber planks of soft wood are used, what you will do to make it strong against abrasion due to movement of traffic?

38.4. In hilly areas what types of bridges will be most useful for the crossing of valleys and streams by the light traffic? Describe each type with the help of neat sketch. What is the main speciality of each type?

38.5. Write short notes on the following :

(a) Timber-cantilever-bridge

(b) Timber arches

(c) Timber trusses

(d) Crates

(e) Pile bents

38.6. What are cribs? Under what conditions these are provided?

38.7. Describe the various types of temporary bridges. What type of I.R.C. loading is used on a wooden bridge?

38.8. Write short note on timber bridges.

38.9. Write explanatory short note on Trestle bridge.

39

Causeways and Submersible Bridges

GENERAL

Causeways are also known as *Irish Bridge* or *Dip*. These are used for crossing over roads which are of minor importance. In un-important routes in the beginning a causeway is constructed and when in due course of time traffic increases and held up by the streams then high level crossings or bridges are constructed. Thus causeways are temporary crossings which in course of lime will have to be replaced by permanent bridges.

Causeways and *submersible* bridges are types of crossings which are used for minimising the cost of a project, but at the same time the requirements of traffic are not ignored.

The 'Causeway' denotes a submersible road bridge across a stream which is designed bad built in such a way that the normal dry weather flow of river passes entirely through the vents below the roadway and occasional floods pass through the vents and over the roadways, thus entailing temporary movement of the traffic.

39.1 CAUSEWAYS

Causeways are provided under the following circumstances :

1. When the depth of water in the stream is very small and seasonal flow is less.

2. When heavy discharge in streams of water courses come only for small durations, generally not exceeding 12 hours.

3. When funds are not available for the construction of a high level-bridge.

4. In hill roads at concave curves, where a number of small streams flow in wide bed width.

5. In hill roads, when rubbish or shingle flow in streams, which may choke road culverts.

Causeways differ from other types of cross-drainage works in the following respects:

1. There is no foundation, pier and abutment in a causeway.

2. Water flows over the bridge or sometimes through vents.

3. Causeways are very cheap and are constructed by providing a slab and approaches on both sides only.

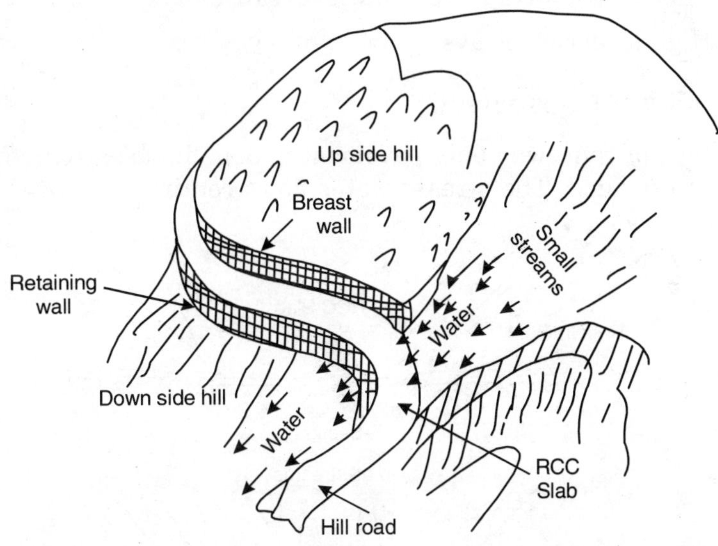

Figure 39.1 *Position of causeway on Hill-road at concave curve*

39.2 LIMITATIONS IN USE OF CAUSEWAYS

(a) The Conference of Chief Engineers held at Nagpur, in 1943, permitted the use of causeways, if their effective submersion due to floods did not exceed 12 hours and not more than six times in a year.

(b) Causeways are unsuitable where the maximum depth of scour exceeds 2.0 m.

(c) Submersible bridges are unsuitable where the floods bring down large boulders or lugs, unless sufficient protective works are constructed to prevent them from damage.

(d) The causeways are unsuitable in heavily silting river beds. These are also unsuitable for torrential *nala* crossings in hills.

(e) Causeways cause obstruction of traffic, according to the design. The main reason for adoption of this type must, therefore be less cost and in every case it has to be decided whether the saving in first cost is such as is worth while considering the convenience. In any case the causeways should not be provided if their cost with approaches is more than $\frac{2}{3}$ rd the cost of a bridge with approaches.

39.3 CLASSIFICATION OF CAUSEWAYS

Causeways can be classified as follows :

(a) Flush causeways (b) Low level causeways

(c) High level causeways

39.3.1 Flush Causeways

In this type of causeway only pavement is done in the stream bed and no vents are provided. The stream water flows continuously over the firmly

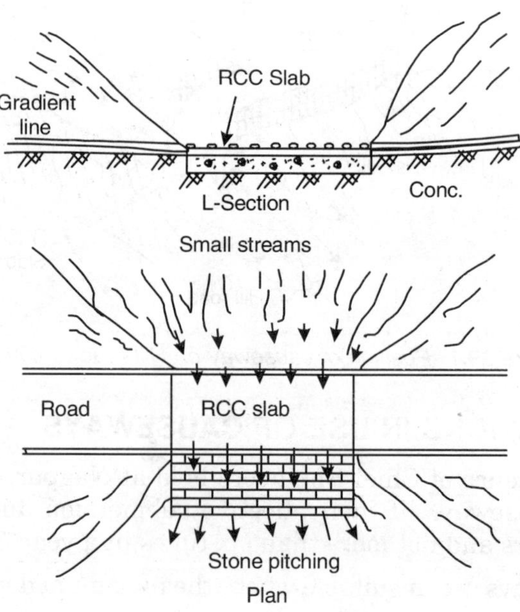

Figure 39.2 *Flush-causeways*

paved bed throughout the year. Sometimes R.C.C. slab is provided in the bed for giving smooth surface. A curtain wall is provided on the downstream side to protect the flooring.

Flush-causeways are provided in hilly roads, when the maximum depth of water does not exceed 1.7 metre in floods and the total interruption does not exceed 15 days in the whole period of one year.

39.3.2 Low Level Causeways

In some streams the depth of water generally remains about 30 cm for most of the period of the year, and the heavy discharge comes only in rainy season for a few hours only. In such cases low level causeways are very useful for the traffic. Small openings of about 30–35 cm are provided below the roadway slab, so that the winter and summer discharges can pass through these vents without disturbing the traffic. Thus for most of the period traffic can pass over the causeway without any interruption, but in the monsoon time just after rains heavy discharge comes which flows over the causeway and the traffic can stay in the rest-houses constructed on both sides for this purpose. When the flood is over, traffic starts crossing the stream.

Curtain walls are constructed on both sides of the road and an apron is provided on the downstream side as shown in Fig. 39.3. Vents should have

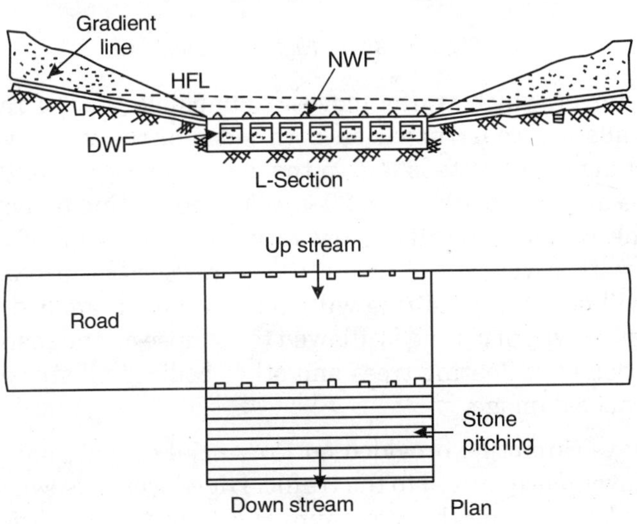

Figure 39.3 *Low-level causeway*

bell mouth entrance for better flow below the road slabs and also should have sufficient waterway to pass the fair weather discharge. The paving should have slight inclination towards downstream side or may have the inclination of the bed of the stream.

39.3.3 High Level Causeways

This is also called as *Submersible bridge.* This may be defined as a bridge which allows normal flood to pass through its vents, and heavy flood water to pass over it. Submersible bridges are provided in the following circumstances:

1 When the river has small width with straight reach.

2. When the floor is sandy, but the good soil is available at lower depth.

3. When the road is less important and the traffic is small.

4. When the duration of heavy flood does not exceed three days at a time and more than six times in a year.

Figure 39.4 shows longitudinal section and cross-section of a submersible bridge. These may be constructed on firm, rocky or loose soil beds. Thick cement concrete is first laid in the bed, over which vents of required section and shape are constructed. At the top of the vent opening arches or R.C.C.

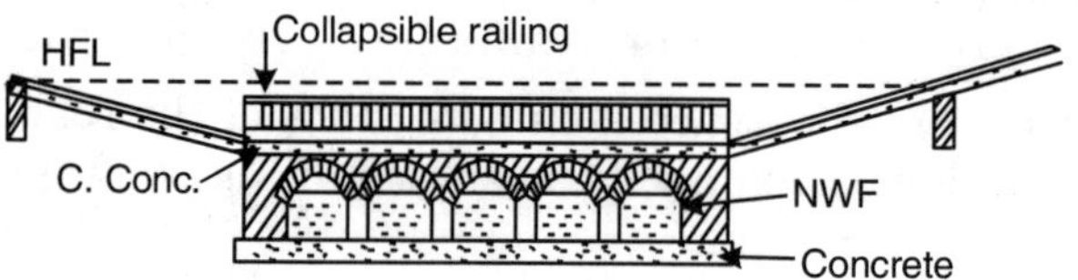

Figure 39.4 *High level-causeway*

slab is laid over which traffic moves. Collapsible type of railing or small parapet walls are constructed on both sides of the road over the bridge. Stone pitching or concrete is laid in the downstream side of the bridge. The approaches are provided at 1 in 20 gradient in cutting on both sides of the bridge. The vents generally allow to pass about 40% of the maximum discharge with a velocity of about 10 metres/sec because at this velocity scouring will not start or silting will occur in the stream bed. During floods the collapsible type of railing is allowed to collapse on the road floor, because if it is not done the floating trees and other bodies will strike against them and may cause damage.

Causeways cannot be provided on Provincial or National Highways due to the inconvenience caused to the traffic. High level causeways are provided after carefully checking the maximum H.F.L. of about 5–6 years.

39.4 REQUIREMENTS OF CAUSEWAYS

39.4.1 Selection of Site

The selection of site for a submersible bridge or causeway should be done at the place, where

(i) the width of the stream is minimum,

(ii) it has well defined high banks, and

(iii) the stream has straight reach for a considerable length.

39.4.2 Traffic Requirement

The causeway should have sufficient width for at least two lanes of traffic, 6.7 m clear width between the kerbs or posts fixed to body or wall.

39.4.3 Foundation

A monolithic base is provided over the entire length and width, with sloped aprons and cut-off walls or dwarf walls on both the upstream and downstream sides. Sometimes a raft of cement concrete is laid over the entire length. To protect against the scour and undermining dwarf walls are provided. For further protection boulder pitching encased in wire netting is provided in the bed, both upstream and downstream, away from cut-off walls.

The abutments of the causeway are made of solid massive construction. The approach roads which usually follow the bank slopes are also paved and are extended 45 cm above H.F.L., and are made in the form of scuppers with dwarf walls both on up-stream and downstream. Pitching if required is also done on both the sides.

In the causeways usually wheel guards are provided instead of railings.

39.5 DESIGN DATA

In addition to the usual data about foundation etc., following datas are collected for the design of causeways:

(i) Flood data for about 10–15 years giving complete flood record as possible viz.. H.F.S., the number of their occurrence and the duration are collected.

(ii) Importance of road and nature of traffic.

(iii) Funds available for the proposed crossing.

(iv) The nature of the banks and strata composing the bed. The depth of the solid strata should be noted.

(v) Contour survey of the area upto 300 m on U/S and D/S side to show the course of the river in the proximity of the bridge site.

(vi) Longitudinal section of the river bed or of the water in its bed for about 1000 m on either side of the crossing and a cross-section at each end and at the proposed crossing. H.F.L. and N.F.L. should be marked on these sections.

(vii) If any bridge is existing over the stream near the site proposed, the waterway and effect of the current should be noted and recorded in the report.

(viii) Specific gravity of sand shingle of the river bed at the sue of the proposed crossing.

(ix) Specific gravity of stone boulders to be used for pitching.

(x) In rivers with deep sandy beds, the depth of sand must be as certained as raft foundations are generally considered suitable for causeway constructed in rivers with deep sandy beds.

From the above data and the requirements of traffic, the ordinary flood level are fixed, at which the causeway should be possible. After determining the N.F.L. at which the causeway should be possible, further design work of the causeway is undertaken.

39.6 WATERWAY DESIGN

Causeway waterway design is governed by the periods during which it is permissible to have the road closed to traffic. There are three main designs.

(a) The modest low causeway with a few openings, to pass the flow in the cold and hot weathers.

(b) Somewhat higher causeway with more openings, passing not merely cold and hot season discharges, but also the flow during the end of the monsoon period.

(c) The high level causeways or submersible bridges passing all ordinary floods.

39.7 PRINCIPLES OF CONSTRUCTION OF CAUSEWAYS

The main principles in the design of a causeway are that it must not contract the stream, that it must maintain a firm roadway against damage by flood, erosion or movement of nala bed and that it should be set at right angles to the current, to avoid scouring along the upstream walls.

The embankment if liable to erosion by flood must be protected by dry stone walling or pitching carried at least 1.0 m, below the nala bed level and 0.45 m above the observed H.F.L.

In case of wide staggling *nalas* with two or more streams, these should when practicable be trained through one causeway of length equal to the aggregate width of all the streams at observed H.F.L.

The design of causeways on sandy beds should be based on the same principles as govern the design of weirs. The true measure of security of a weir on a permeable bed is the distance through the solid which a current of water would have to travel before it could rise up below the weir and that it is of little consequence : whether masonry is laid horizontally on the weir bed or sunk vertically below it, so long as the currents passing through the solid below the structure are exposed to the friction of the same length of passage. The weir must be protected from horizontal scour on the face of the toe. It should have sufficient weight to resist the horizontal pressure of the head it supports and also sufficient weight to oppose the upward pressure in the base of the foundations which that head may produce.

The actual design of causeway on rocky beds do not present special difficulty. The design of causeways in sandy bed rivers require sufficient experience and care.

39.8 DESIGN OF CAUSEWAYS

39.8.1 Lauds

The live loads on the causeways are the same as for other bridges. It has also to carry dead loads depending on its design and in addition to the horizontal thrust and lift due to flood waters, when live loads are absent.

The forces acting on causeways, when the level of floods is just at formation level are:

(a) Static head due to afflux.

(b) Horizontal loads due to water currents.

(c) Friction of water against piers and soffit.

(d) Uplift on the slab or arch-head from water surface to soffit.

(e) Floatation due to submergence—an upward force decreases the pier weights.

39.8.2 Waterway Calculation

After determining the accurate bed fall of the river and the cross-section of the natural stream at the various levels viz., at the flood level at which the causeway is intended to be made passable and at the H.F.L., the natural velocities at the various levels may be found out by Kutler's or Mannings formulae. While doing waterway calculations of the causeway the aim should be to see that the following conditions are satisfied:

(i) It must present the minimum obstruction to all the discharge of the river above the O.F.L.

(ii) The causeway is so designed as to allow the whole of the discharge of the river to pass with the least possible afflux through its opening, at the O.F.L. at which it is intended to be made for dable and that the flood is not allowed to rise higher than that level.

39.8.3 Foundation

The foundation of a causeway requires adequate protection against scour and undermining. For this purpose the piers should be taken below scour depth or if founded on a raft the latter must be provided with cut-offs both up-and down-stream and the flank between the two cut-offs should also be properly guarded. This can be achieved by providing curtain walls between the two cutoffs extending down to their lower levels so as to completely box up the foundations of the causeway. The common practice is to construct the piers on a raft foundation over the sandy bed. In older types of causeways the top of the raft is in just level with the bed, but nowadays, it has been taken below the true bed level to eliminate chances of obstruction by it to the flow of water. In the case of foundations on permeable soil not taken below the scour depth, an impervious flow is an absolute necessity. The flow of the'causeway has the following functions.:

(i) To restrain the pressure gradient against an outburst so that it does not sufficiently lose its potentiality.

(ii) To distribute the load transmitted to it by the piers on to sandy bed, so that pressure per unit area is safe for the soil.

If the floor is not reinforced, there is a change of its cracking under tension if piers are further apart than necessary according to modulus of rupture of the material used to keep the whole flow in compression. With 1.0 m thick raft, the span of the causeway is kept about 2.0 m, with the above requirements. This does not rule out independent foundations of piers on sandy soil, but economy must be worked out for selecting the type of foundation. The span on raft foundation can be increased to 3.0 m to prevent silting, for providing openings, to allow small brushwood to be floated down safely and also for economy, but in this case inverted arches should be provided. By providing the inverted arches, the thickness of the raft can be reduced, thereby increasing the waterway also.

39.8.4 Upstream Cut-off

The upstream cut-off must be taken below the scour depth. If this is done, a sufficient length of boulder pitching should be provided so that it settles down evenly and acts as an ultimate cut-off. For effective function as such the pitching should have a layer of spalls and small round boulders at the bottom with sufficiently heavy ones on top which cannot be disturbed by the heaviest flood. As a thumb rule the length of the upstream apron should not be less than three times the scour depth. The diameter of the boulders required for pitching should be calculated from Cheilly's formula. The thickness of the wall should be $\frac{1}{3}$rd of the height of wall.

39.8.5 Down Stream Cut-off

What applies to the upstream cut-off holds equally good for the downstream side. Since the scour depth is likely to be more than U/S, the cut-off should be proportioned accordingly. The pitching stones should also be heavier in the ratio of the velocity. Foundation should be $(1.5d + 1.3)$ metre, where d is the depth of water above causeway. In sandy beds these depths are increased. The actual depth and length of the cut-offs should be determined by the formulae.

39.8.6 Uplift Pressure

$$\text{The uplift pressure} = W \times h - \frac{V_1^2 - V_2^2}{2g} \qquad \text{... (39.1)}$$

where

W density of water in kg/m^3.

h the uplift head under the decking, which is equal to the thickness of road slab including wearing coat and afflux.

V_1 Velocity under the opening.

V_2 Original velocity of stream before reaching the causeway.

g acceleration due to gravity (= 9.8 m/sec^2).

39.8.7 Pressure due to Eddies

$$\text{The pressure due to eddies} = \frac{W\,(V_v - V)^2}{2g} \qquad \text{... (39.2)}$$

where

V_u Velocity through the openings

V Velocity of approach.

The force is acting in the horizontal direction.

39.8.8 Friction of Water on Surface in Contact with Water

All the horizontal forces which are acting on the causeway are added together.

The forces are assumed to act at $\frac{1}{3}$ rd height.

Figure 39.5 shows the cross-section of typical causeway.

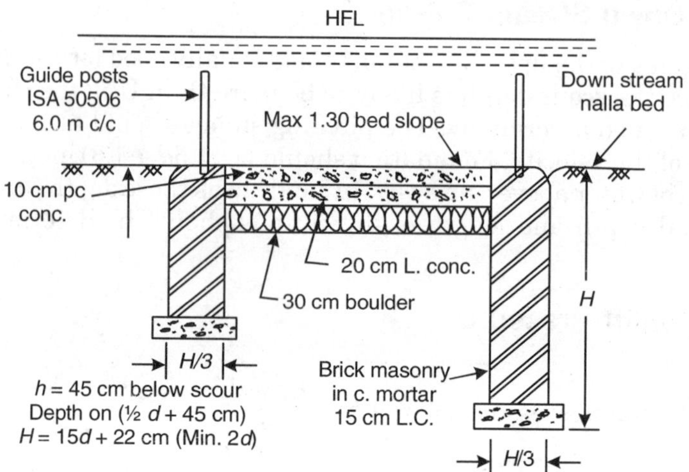

Figure 39.5 *Cross-section of a typical causeway*

39.8.9 Thickness of Floor

The thickness of the floor of the causeway is determined by the formula,

$$T = \frac{4}{3} \times \frac{p}{g-1} \qquad \text{... (39.3)}$$

where

- T floor thickness
- p the residual pressure head at any point
- g acceleration due to gravity.

30.8.10 Protection Mattresses U/S and D/S

Where after practical experience of floods, protection mattresses prove necessary, mattresses in wire case should be fixed securely to the eye bolts already embedded in the D/S wall extending down stream to distance of H, which should not be less than $2d$ or the observed length of scour if greater.

where

- H depth of D.O. below the bed
- d depth of water above causeway during H.F.L.

The length of D.S. apron should not be less than three times the afflux expected during floods.

The slopes or ramps at the end of causeway should follow the bank with a maximum gradient of 1 in 20 and be continued to 50 cm above the H.F.L The method of correct designing of the approaches will naturally vary with the circumstances of each individual causeway. The approaches of causeways are generally in deep cuttings and the alignment of the approaches should be so selected as to prevent, to the greatest possible extent, the accumulation of silt, which is the tendency which occurs when H.F.L. subsides. Vertical curves at changes of gradient should be as per I.R.C. recommendations.

39.8.11 Size of Pitching Stone Boulders

The diameter of the boulders required for pitching can be determined by the following Cheilly's formulae :

$$V = 5067\sqrt{G.D.} \qquad \text{... (39.4)}$$

where

V velocity of stream required to just move the boulder of dia. D

G specific gravity of boulder

D dia. of the stone

A factor of safety of 20 must be taken for straight reaches and of 2.5 of 3.0 on bends depending upon the sharpness of the curves. Ordinarily the depth of pitching should be 1.0 to 1.3 m. A wire mattress should be used to encase the boulders if sufficiently large sizes are not available at reasonable cost.

39.8.12 Railings

Only collapsible type railing may be provided in the causeways. It should be low and of a movable type so as not to cause obstruction to the Hoods and as such could be quickly removed just before the monsoon and refixed thereafter. Reinforced concrete guard stones with proper holes in them can be made to serve. One line of gas or water piping may be passed through the holes to serve as rail and kept locked in position during fair weather to prevent theft or unauthorized removal.

39.8.13 Roadway Over the Causeways should have One-way Camber

Experiments have shown that one way-camber towards the D/S have two advantages :

(a) The efflux is considerably reduced and there is no chance of a centre standing wave being formed as in case of two-way camber.

(b) As this arrangement raiser, the level of the U/S edge, this contributes towards keeping the roadway dry up to a higher depth of causeway upstream. The road surface over the causeway should be preferably of 7.5 cm cc over soling and intermediate cost of stone metal.

39.9 SCUPPERS

It is a small form of a causeway and extends across the entire formation width. The scupper is usually laid on the curves in the direction of the road alignment, convex at the ends and cancave in the middle. The required cross slope of 1 in 12 in hill areas and 1 in 30 in flat areas is provided. These are oftenly used instead of small culverts in the hills. There are provided upto 60 cm span.

REVIEW QUESTIONS

39.1. What are causeways? How do they differ from other types of cross-drainage works?

Draw neat sketches of causeway on a sandy soil showing its component parts and the principles of construction.

39.2. Under what circumstances causeways are provided? Describe an Irish Bridge with the help of a neat sketch.

39.3. What are different types of causeway? Under what circumstances each type is provided. Describe a high-level causeway with the help of a neat sketch.

39.4. Differentiate between a flush-causeway and low-level causeway. Describe the latter with the help of a neat sketch.

39.5. How does a flush-causeway differ from other types of causeways? Describe it with the help of a neat sketch.

39.6. Write short notes on causeways.

39.7. Draw a neat sketch of high level causeway, showing its component parts with their functions. How is the waterway determined in this case?

39.8. Indicate the basic principles of design and the construction features for a submersible bridge.

39.9. State usual types of low cost bridges and list the conditions in which each type would be applicable.

39.10. Sketch and describe a high level causeway for carrying a state highway across a stream 6.1 m (200 ft) wide. In what circumstances a low level causeway is preferred?

39.11. Write what do you know about utility of causeway and submersible bridges?

39.12. What are causeways? How are they different from other types of cross-drainage works? Draw neat sketches of a causeway on a sandy soil showing its component parts and the principles of construction.

40

Masonry Bridges and Culverts

GENERAL

The superstructure of masonry bridges consists of masonry arch over which the roadway is constructed. This arch rests on piers and abutments, which are designed for this purpose. In old times masonry was the only alternative for the construction of arches, but nowadays generally bridges are constructed of R.C.C., steel or prestressed concrete. It must be a matter of universal regret that the masonry arch is passing away, if it has not already passed from the 'practical politics' of the engineer. It is largely a matter of cost. Masonry arches are provided for small bridges and culverts upto a maximum span of 60 metres.

Masonry arches may be of brickwork, masonry or concrete. These are constructed for small span bridges where more headway is required during floods for the passing of boats below the bridge. These are simple to construct, having long life and materials for their construction are cheaply available near the site of the bridge.

40.1 CLASSIFICATION OF ARCHES

Depending on the shape of the arch, the masonry arches of bridges can be classified as under :

1. Semi-circular arches 2. Segmental arches 3. Elliptical arches

40.1.1 Semi-circular Arches

These are provided at such places where alignment of the road or railway is very high, because semi-circular arches give greatest rise to the bridge. These are very strong and do not develop any horizontal pressure on the abutment and pier. They transmit the load vertically on the pier or abutment.

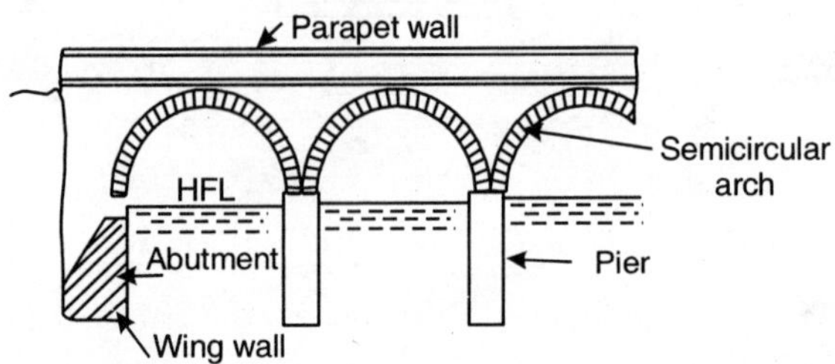

Figure 40.1 *Semi-circular Arch-bridge*

If the alignment of the road is not high, these involve heavy and very expensive approaches.

40.1.2 Segmental Arches

These arches are very easy in construction and are generally employed. The proportion of rise to span generally varies from $\frac{1}{4}$th to $\frac{1}{7}$th, which increases

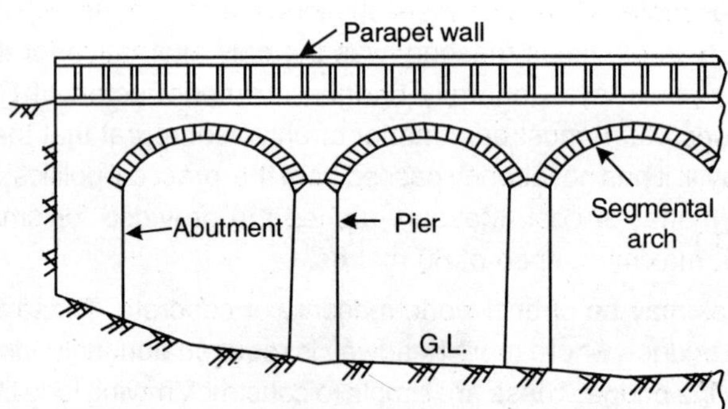

Figure 40.2 *Segmental Arch-bridge*

with the increase in the span. These arches exert great thrust on abutments and supporting piers, which are to be designed for it.

40.1.3 Elliptical Arches

These arches provide the maximum waterway. These are light and very graceful in appearance, but their construction is very difficult. These are less strong than semi-circular arches. Elliptical arches give little horizontal thrust on piers and abutments.

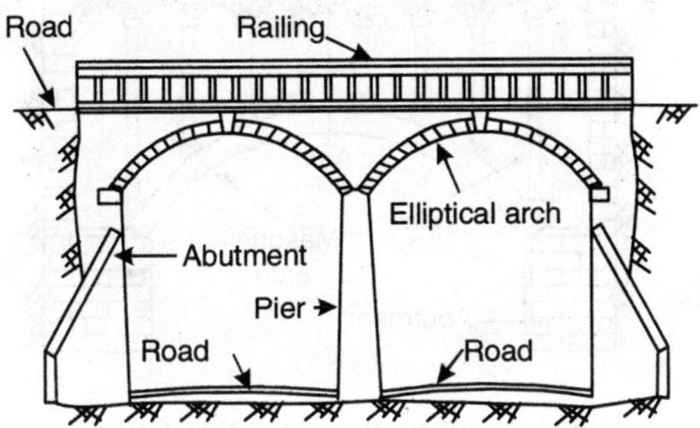

Figure 40.3 *Elliptical Arch-bridge*

Depending on the way, how these arches support the roadway above their top, these may be classified as follows :

40.1.4 Filled Spandrel Arches

These types are used for small ratio of rise to span. Side walls are constructed along the sides of the arches and the space between them is filled with soil or any suitable material to bring the horizontal level at the top over which the roadway is laid.

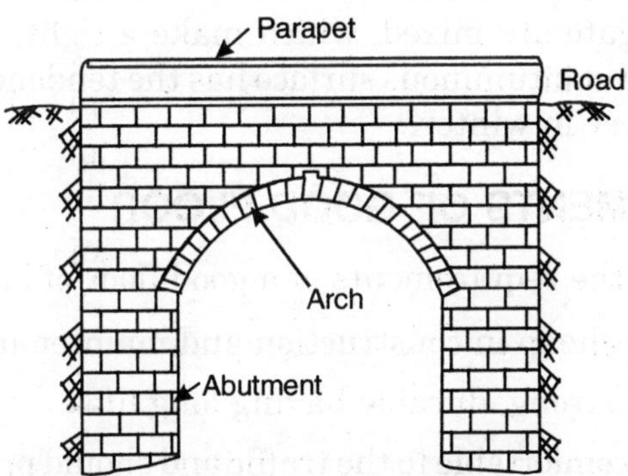

Figure 40.4 *Filled Spandrel Arch*

40.1.5 Open Spandrel Arches

When the ratio of rise to span is very large, the spandrel is not filled, because this increases unnecessary load on the arch, which consequently increases its designed cost. In such cases vertical walls are constructed on the arch rib over which the roadway is laid on arches or R.C.C. slab. This type of open

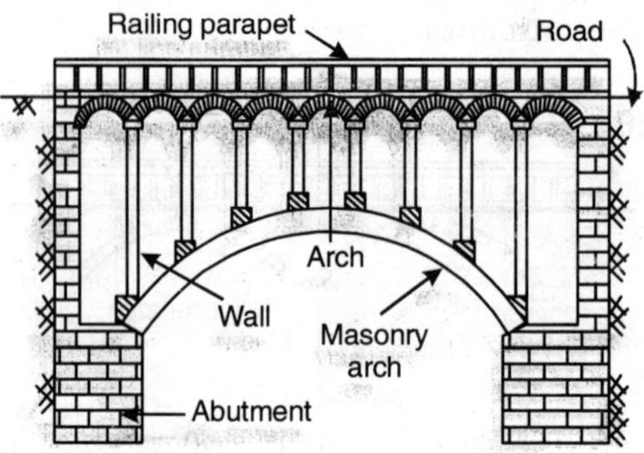

Figure 40.5 *Open Scandrel Arch*

scandrel arches look very good in an appearance and also decrease the load on the arch, thus making the design economical.

40.2 ROADWAY

After the construction of arches, the spandrel filling is done with soil. Over soil filling a thick layer of lime concrete is laid and the top is brought to a uniform level. When lime concrete sets, road finish is done over it. In case of concrete roads, smooth surface is undesirable, because it may cause splipping of vehicles while applying brakes. Coarse or sand paper surface which has a uniform level is best. In case of bituminous roads only required quantity of binder and aggregate are mixed, which make a tight, dense and unclosed textured surface. But bituminous surface has the tendency to flow in summer and become slippery in winter.

40.3 REQUIREMENTS OF GOOD FLOOR

The following are the requirements of a good floor of bridges.

1. It should be cheap in construction and maintenance.
2. It should be strong, durable having long life.
3. It should be comfortable to the traffic and should provide smooth riding.
4. It should be noiseless when traffic pass over it.
5. It should be fire proof and should be able to bear the vibrations and impact caused by the traffic.
6. It should be non-absorbent in nature, dustless, non-slippery during rains and produce a sense of security in the mind of traffic.
7. It should be such that can be easily drained.
8. It should not give glare to the traffic.

40.4 CULVERT

It is a small bridge used for carrying water from one side to another side in the embankment of road or railway. A culvert may have one, two or three spans. Mostly culverts having one or single span are commonly used. In case of the road culverts the maximum span is limited to 5.0 m, whereas culverts upto 6.0 span are used in the railway.

Following four types of culverts are commonly used in practice :

(a) Arch culvert
(b) Slab culvert
(c) Pipe culvert
(d) Box culvert
(e) Steel girder culvert

40.4.1 Arch Culvert

Brick or stone masonry arch culverts were very popular in the earlier times, but nowadays these are very uncommon. These culverts mainly consist of foundation, abutments, wing-walls, arch and the parapet as shown in Fig. 40.6. If the bed soil is good, there is no necessity of providing floor and curtain walls in the bed. But if the soil is poor and there is likelihood of scouring, floor and curtain walls are provided. Spandrel filling is done with lime concrete.

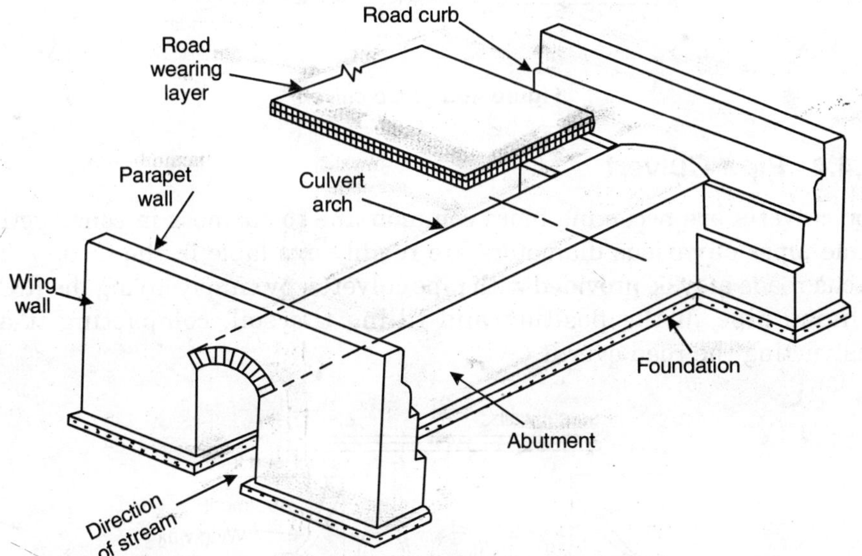

Figure 40.6 *Masonry Arch Culvert*

The parapet or railing provided on both the sides may be solid type or hollow type.

40.4.2 Slab Culvert

Masonry culverts with R.C.C. slab are very common these days. Due to difficulty in centring, shuttering and less life and possibilities of crack, more dead loads due to spandrel fillings etc., the masonry arches in culverts have been replaced by simple R.C.C. slab construction. Slab in these culverts may be of R.C.C., stone, timber sleepers (for temporary culverts) or steel girders.

The design of deck slab of culverts is done for the worst possible effect of I.R.C. loadings. The culverts may be designed for class 'AA' or 'BB' loadings depending on the type of road.

Road wing of the Ministry of Transport and Shipping has prepared standard design and drawings of culverts from 2.0 m span to 6.0 m span, for adopting by the various road departments.

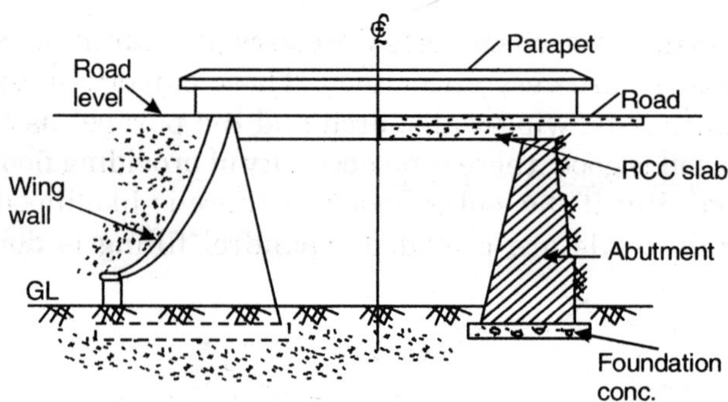

Figure 40.7 *Slab culvert*

40.4.4 Pipe Culvert

Pipe culverts are becoming more common due to easiness in construction. Hume pipes of various diameters are readily available in the factory. The kutcha roads may be provided with pipe culverts, by simply laying the R.C.C. or hume pipe in the position and filling the soil, compacting it and constructing the road over it.

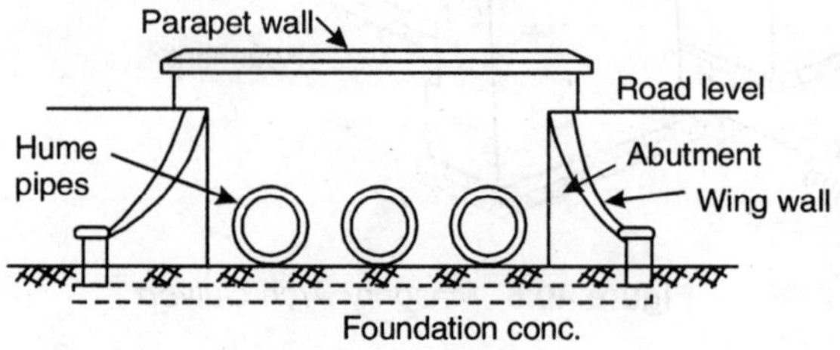

Figure 40.8 *Pipe culvert*

The exact number and the diameter of the pipes are determined by the maximum discharge which will pass through this culvert and the height of the embankment of the road. Pipe may be of masonry, cement concrete, iron or steel or stone ware. But more common is hume pipe of R.C.C.

40.4.4 Box Culvert

These culverts mainly consist of one or more number of square or rectangular openings for passing the water from one side to another. In soft soils where there is possibility of scouring and bearing capacity of the soil is poor, these culverts are commonly used.

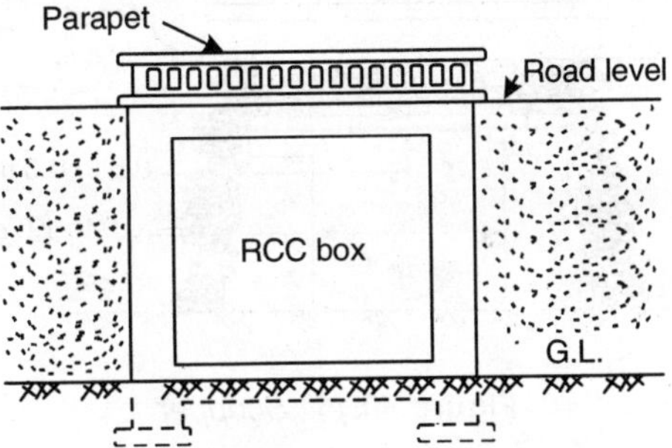

Figure 40.9 *Box culvert*

Figure 40.9 shows this type of culvert. The span of each way is not more than 3.0 m in this case.

40.4.5 Steel Girder Culvert

This type of culvert is only provided in railways. Two main girders are laid just below the rails. Wooden sleepers are provided between these girders

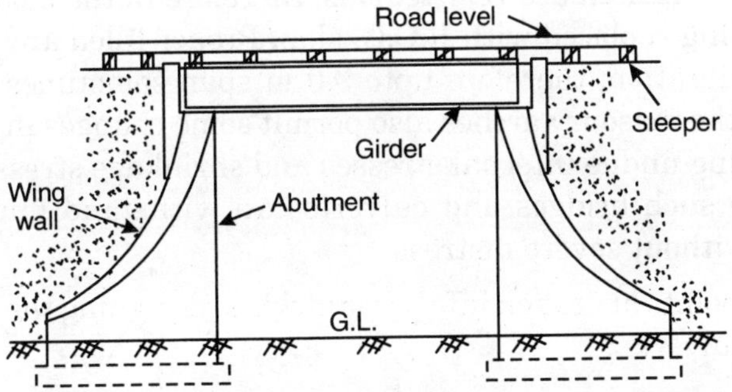

Figure 40.10 *Steel Girder culvert*

and the rails. Therefore sometimes these culverts as also known as *open deck culverts*.

40.4.6 Scupper

This is a cheap type of culvert or cross drainage work, used at such place where small quantity of water is required to pass below it.

Figure 40.11 shows this type of culvert. The width of water passage is about 90–100 cm. Retaining walls of coarsed rubble dry masonry are constructed on both the sides for the purpose of abutments. The width at the top of the opening is reduced by corbelling the masonry from both the sides as shown in Fig. 40.11.

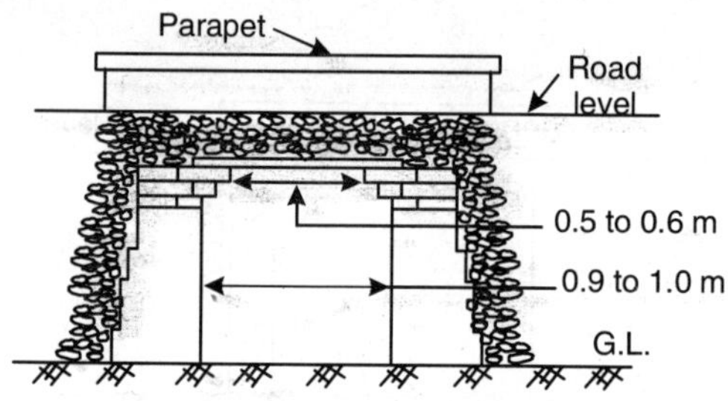

Figure 40.11 *Scupper*

Stone slab or R.C.C. slab is laid over the reduced opening at top. Now dry stones are hand packed over the slab and the road is constructed. Retaining walls are provided on both the ends of the scupper. Scupper is very cheap and is used on unimportant roads.

40.5 DESIGN OF MASONRY ARCH BRIDGES AND CULVERTS

Arches of stone and brick with good rise, thick ring, good foundation and with proper maintenance can carry very heavy loads. But exact strength calculations of the arch are very tedious, therefore in the modern times the arches are being replaced with R.C.C. slab. Proper filled arches reduce the impact and vibration, therefore upto 9.0 m span sometimes are provided. The joints of the masonry arches also permit some changes in temperatures without causing undue internal stresses and shrinkage stresses are not set up, therefore such bridges and culverts can withstand minor abutment movements without severe distress.

Following points are taken into account while designing the masonry arch bridges and culverts :

40.5.1 Rise to Span Ratio

The rise to span ratio should be kept as high as possible. As semicircular arches do not exert any horizontal thrust, therefore the rise should be kept between $\frac{1}{2}$ to $\frac{1}{4}$ of the span. While constructing skew arches, care should be taken that the courses are everywhere at right angle to the line of thrust.

40.5.2 Arch Ring Thickness

For segmental arches the thickness of the arch ring should not be less than $0.45\ \sqrt{R}$, where R is the radius of arch intrados.

For culverts under high fill the above arch thickness is increased by 50%.

40.5.3 Thickness of Abutment

When the height of the abutment h is less than one and a half the span, the thickness of the abutment is calculated by the formula,

$$E = \frac{R}{5} + \frac{r}{10} = 0.6 \text{ m} \qquad \text{... (40.1)}$$

where

E Thickness of the abutment at springing

R radius of the arch intrados

r rise of the arch

The thickness of the abutment at any depth can be calculated by the formula

$$b = E = \frac{EZ}{24.r} \qquad \text{... (40.2)}$$

The back batter of the abutment is determined by the formula

$$\text{Abutment back batter} = 1 \text{ in } \frac{24 \times r}{\text{span}} \qquad \text{... (40.3)}$$

The radius of the arch intrados is determined by the formula

$$R = \frac{a^2 + r^2}{2r} \qquad \text{... (40.4)}$$

where

b thickness of abutment at any depth Z below springing

t thickness of arch ring at crown

S span

a half clear span

The thickness of the abutment at its base R should not be less than $\frac{2}{3}$ h, where h is the height of the abutment.

40.5.4 Thickness of Pier

It should be sufficient to resist the thrust resulting from one of the two arches it supports, when it is covered with the designed load, while the other remains unloaded.

The pier thickness at the top should be adequate to accommodate the skew backs on both sides. Short piers have its sides as vertical or batter of 1 in 24 to 1 in 30. A batter of 1 in 12 is provided in case of long piers.

The thickness of the pier at its base k should not be less than ($h/3$ + 23 cm). It should be 30 cm more than PJ may be 30 cm to 75 cm.

Every 4th or 5th pier should be an abutment pier having the same top thickness as the abutment. A batter of 1 in 12 to 1 in 24 is provided in the abutment pier.

40.5.5 Haunch Filling

The depth of the haunch filling is determined by the formula,

$$D = \frac{r+1}{2} \qquad \text{... (40.5)}$$

where

D depth of haunch filling

r rise of arch

I thickness of the arch ring

40.5.6 Earth Cushion

The depth of the earth cushion over the arch crown C is provided between 45 to 75 cm depending upon the span of the bridge.

40.5.7 Base Width

The base width of the abutments and the piers depends on the bearing capacity of the soil. The pressure at the toe of the abutment should be determined to ensure that the soil is not over-stressed. In the site having scourable beds the length of the pier should be kept minimum.

Table 40.1 gives the ready data for Masonry Arches for culverts and small bridges of I.R.C. class 'A' loading (two laries) or class '70 - R' loading (one lane).

Figure 40.12 illustrates the dimensions of the culverts and small bridges corresponding to Table 40.1.

TABLE 40.1 *Size of masonry arch culverts and small bridges (Ref. Fig. 40.12)*

S.No.	Clear span	Thickness of the arch ring	Earth cushion	Haunch filling	Width of stone masonry (1 : 3) abutment at top	Width of brick masonry (1 : 3) abutment at top	Width of stone masonry (1 : 3) pier at top	Width of brick masonry (1 : 3) pier at top	Depth of foundation concrete cement (1 : 3 : 6)	Depth of lime concrete in foundation
	m	cm	cm	cm	cm	cm	cm	cm	cm	cm
		t	C	D	E_1	E_2	P_1	P_2	G_1	G_2
1.	1.00	25	45	25	65	70	40	50	30	45
2.	1.25	30	45	30	75	90	50	60	30	45
3.	1.50	30	45	35	85	90	55	70	40	55
4.	2.00	35	50	45	90	95	60	70	40	55
5.	2.50	35	60	50	100	110	70	80	45	60
6.	3.00	35	60	60	110	115	75	80	45	60
7.	4.50	40	60	70	120	140	80	90	55	75
8.	5.00	45	60	90	130	150	85	90	55	80
9.	6.00	55	60	100	155	160	100	110	60	85
10.	7.50	60	70	120	175	190	110	130	75	90
11.	9.00	65	75	145	200	210	130	140	75	100
12.	10.5	70	75	160	220	230	150	160	75	110
13.	12.5	80	75	190	245	250	175	180	75	115

40.5.8 General Specifications

The top of all the abutments and piers should be in 1: 4 cement sand mortar for at least 60 cm.

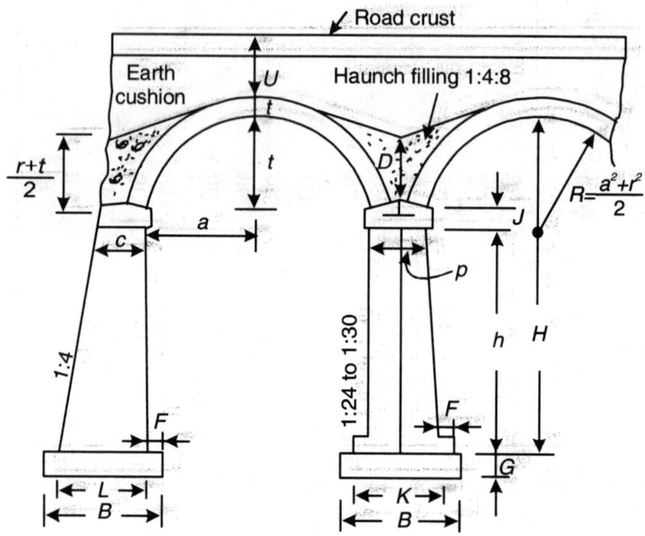

Figure 40.12

The masonry of the arch ring should be either of concrete blocks (1 : 2 : 5 : 5 or 1 : 3 : 6) or dressed stones or bricks in 1: 3 cement-sand mortar.

The crushing strength of the bricks, stone or concrete used in the construction should not be less than 106 kg/cm^2. Only coarsed rubble masonry or ashlar masonry should be used for arch ring.

REVIEW QUESTIONS

40.1. Write short notes on masonry bridge.

40.2. What are the different types of arches which are used in masonry bridges? Under what circumstances each type is suitable?

40.3. With the help of neat sketches differentiate between an 'open-spandrel arch' and 'filled spandrel arch'. What are the circumstances under which each type is suitable?

40.4. How is the roadway constructed in case of masonry bridges? What are the requirements of a good bridge flooring?

40.5. Write short note on the culverts.

40.6. What are the various types of culverts? Describe any one with the help of neat sketch.

40.7. Write short notes on any three of the following:

(a) Timber bridges (b) Box culverts

(c) Causeway (d) Afflux

40.8. (a) Distinguish between :

(i) Culvert (ii) Causeway

(iii) Submersible bridge.

(b) Illustrate the three types of the cross drainage structures with diagrammatic sketches.

Steel Bridges

GENERAL

After the invention of cast iron and ascertainment of its properties, a cast iron arch bridge of span 30.6 metre was first constructed in 1776 at Coalbrokdal England. Theodore Cooper was the first man to make use of iron truss in 1840 by erecting Trumball bridge at Frankfort, New York. In this bridge cast iron was used for compression-members and wrought iron for tension-members. The construction of iron bridges in England was started by four famous English egineers — John Rennie, Thomas Telford, George and Robert Stephenson. Nowadays much advancement has been done in steel bridges and mostly long railway bridges are constructed with steel.

41.1 CLASSIFICATION

Steel bridges can be classified as under:

(1) Steel trough plate bridges
(2) Steel girder bridges
(3) Steel truss bridges
(4) Steel arch bridges
(5) Bow string bridges
(6) Steel rigid frame bridges
(7) Suspension bridges
(8) Movable bridges.

41.1.1 Steel Trough Plate Bridges

Steel trough plae can be used for connstructing small bridges upto 5 metres span. Steel troughs of sufficient sections are laid in the required width of the road and are tied together with steel bars so as to remain in position. On the bottom of both ends, bearing plates are fixed to these troughs, so that they can transmit loads uniformly on abutments. The bearing plate is fixed to the abutment as shown in Fig. 41.1. After placing steel troughs in position, lime concrete is filled in the trough and over them, as shown in Fig. 41.1. Lime concrete is brought to a uniform level and over it road wearing-surface of cement concrete or bitumen is laid. Parapet of steel plates is fixed to the end troughs as shown in cross-section sketch of Fig. 41.1.

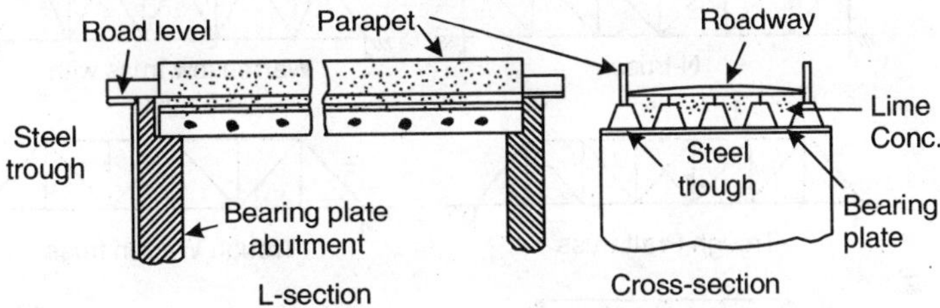

Figure 41.1 *Steel Trough-plate Bridge*

41.1.2 Steel Girder Bridges

These are used for small railway bridges. The track is directly laid over the girders, which are braced together. Steel bearings are provided below girders which transfer the load to abutments. Depending on the span and the load

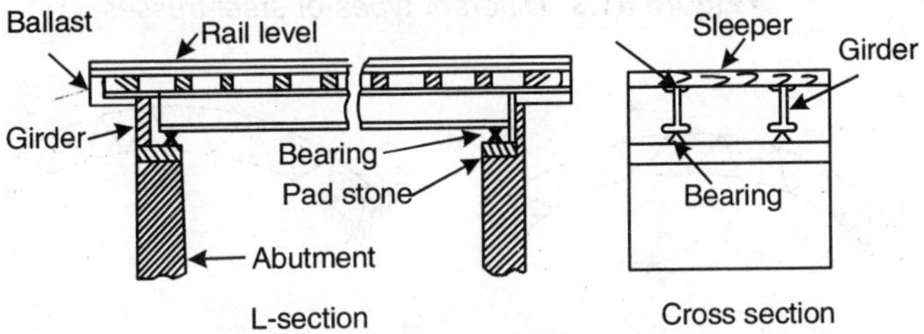

Figure 41.2 *Steel Girder Bridge*

of the traffic, girders may be of rolled joists, built-up girders or box girder type. Sometimes, rails are directly laid on girders, without providing sleepers between them. But when the span is large or long, sleepers should be provided.

41.1.3 Steel Truss Bridges

For very long bridges of railway, generally trussed girders of steel are provided, because these may have roadways on the top or bottom or on both at the same level. Sometimes these are used for combined road and railway bridges. This is a stiffer form of structure and less affected by wind pressure.

Trussed girders are provided upto 80 metres span. Four principal types of trusses are used which are : Nor Linville, Howe, Pratt and Warren. These have been designated on the names of their originators.

These trusses have generally depth of about $\frac{1}{6}$ of the span. The economical length of a panel is more or less independent of the truss depth.

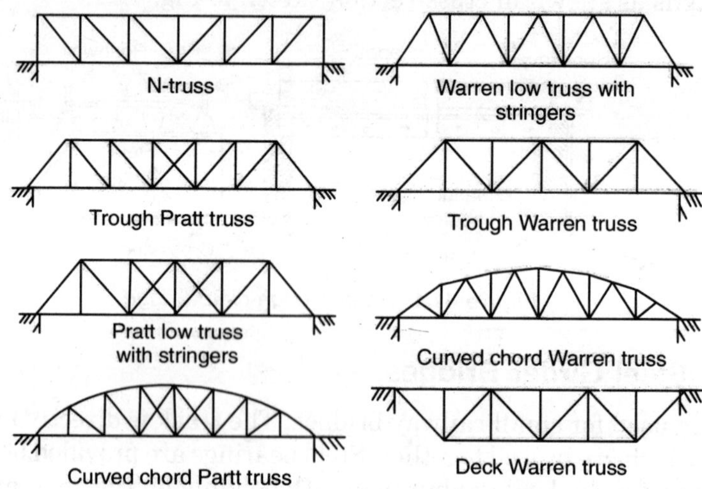

Figure 41.3 *Different types of steel-trusses*

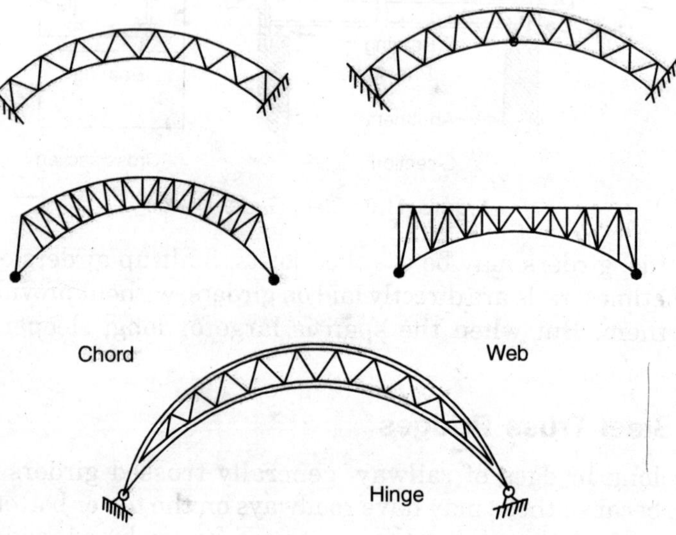

Figure 41.4 *Steel arches*

41.1.4 Steel Arch Bridges

Bridges of very long single spans are constructed with steel-arches where it is not possible to construct intermediate pier. Steel arches, may be of two-hinged or three hinged type which in turn may be either of the spandrel-braced or trussed-arch type. Two-hinged type steel arches are more rigid and economical. Spandrel-braced arch looks pleasing in appearance and can be constructed like two cantilever from both sides of the river bank.

41.1.5 Bow String Bridges

In steel-arch bridges the abutments are designed to carry the great thrust caused by the arches on them. But in the case of bow string bridges steel

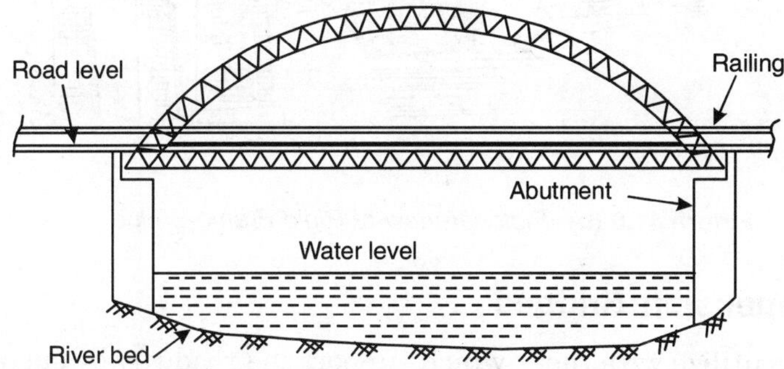

Figure 41.5 *Bow string Girder Bridge*

girder steel tie is provided joining the two ends of an arch to bear the horizontal thrust. Sometimes the roadway is directly laid over these ties and these ties are connected to the suspenders. But in some cases the roadway is suspended from the arch-ribs by suspenders.

41.1.6 Steel Rigid Frame Bridges

In this type, steel portal frames are used for the construction of bridges. The roadway is provided on the top of the portal frames, the corners of which are

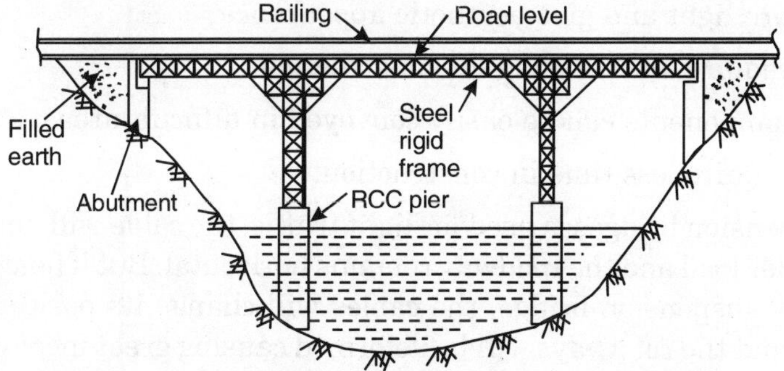

Figure 41.6 (a) *Steel Rigid Frame Bridge*

stiffened for rigidity. The bottom feet of the vertical legs may be hinged or fixed to the foundation.

Steel rigid frame bridges can be constructed in a short time and very economically. It provides more clearance below the bridge and not heavy abutments are required. It does not require bearings and fixtures over the piers.

Figure 41.6 (b) *Pictorial View of Rigid Frame Bridge*

41.1.7 Suspension Bridges

These bridges utilize wire ropes which support the roadway by suspenders. These are required for long spans. The wire-rope or cables are carried over the piers and are anchored to jumpers left in good rocks. The cables rest at the top of the pier in special casing known as saddle which is bolted or made to rest on roller at piers. The suspension bridges are suitable due to the following reasons:

(i) They are economical, because the stresses pass directly through cables and suspenders, to the point of support.

(ii) They require high tensile ropes which are cheaper than mild steel trusses.

(iii) They are light and give aesthetic appearance.

(iv) They require no centering for construction.

(v) Their components can be easily conveyed in difficult areas.

(vi) They require less time in construction.

If the suspension bridges is used for light traffic, the cable will not change its shape under load and the roadway remains horizontal. But if heavy traffic moves on the suspension bridge, the cables will change its position under heavy loads and the roadways will be deformed causing great inconvenience to the traffic. In this case stiffening of the roadway will be required. It can be achieved by providing stiffening trusses at the roadway or by bracing the chains. The given Fig. 41.7 shows the braced cable suspension bridge.

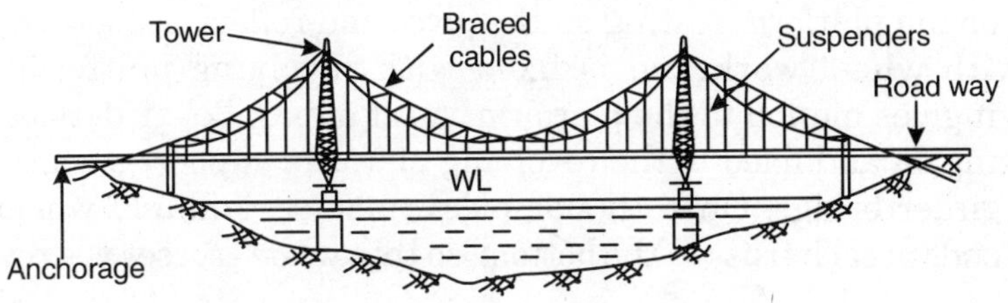

Figure 41.7 *Suspension Bridge*

41.1.8 Movable Steel Bridges

These temporary bridges are also called 'Opening bridges'. In the navigation canals, it is not possible to construct the bridge-superstructure at such a level that navigation ships can pass below the bridge even in high floods. This will require much more clearance which cannot be provided. In long bridges the whole bridge is constructed at the required level, but the design of bridge between one or two spans is done in such a way that the bridge superstructure in that portion can be moved so as to allow the necessary width and clearance for the passing of ships. These types of bridges are called temporary bridges, only because they can be put out of use whenever required, by moving them out of the position. The life of these bridges is not less than any other permanent-steel bridges.

Movable bridges are of the following types :

(a) Swing bridges

(b) Bascule bridges

(c) Traversing bridges

(d) Transporter bridges

(e) Vertical lift-bridges

(f) Flying bridges

(g) Cut-boat bridges.

41.1.8.1 Swing Bridges

In these bridges one pier is provided in the channel. One disc bearing is provided at the top of the pier, on which two continuous stresses are provided, which can revolve about a vertical axis at the centre. These trusses may carry the flooring at their top or bottom booms depending on the alignment of the road. Sometimes a circular rail track is provided round the circumference of a box plate on the central pier as in a railway turn-table. A roller frame with suitable conical rollers is provided, the rollers of which rest on the track. This roller frame turns about the central pivot. There is a circular revolving platform resting on the pivot and rollers. A toothed arc is provided with wheel work and is fixed with revolving platform. This arrangement gives motion to the platform. A set of parallel girders or cross girders resting on and fixed to the revolving platform support a continuous steel truss girder bridge. On both side-piers rails are laid and wheels are provided in ends of each truss in the bottom, so that when trusses are revolved

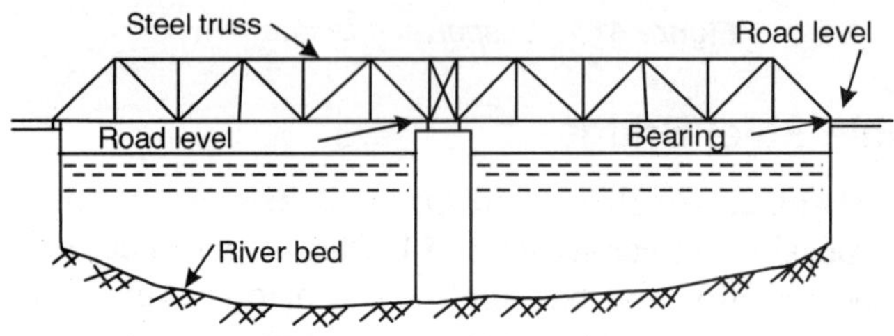

Figure 41.8 *Swing-bridge*

to come in line with the main bridge, the wheels rest on the rails as shown in Fig. 41.8. When the bridge is brought in position with the other part of the bridge, it is locked so that it may not change its position under the impacts of traffic.

41.8.1.2 Bascule Bridges

This is also known as *draw bridge* and can turn about a horizontal axis and in a vertical direction. The bridge is opened in the vertical direction by means of a pinion driving a toothed sector. Mostly cables are used for raising thebridge. These bridges may have *single bascule* or *double bascule*. Figure 41.9

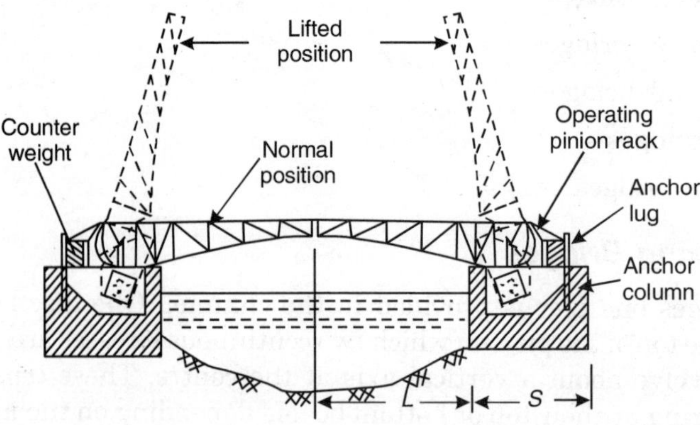

Figure 41.9 *Bascule Bridge*

shows a double bascule bridge. In single bascule bridge the bridge is rotated only about one axis from one side of bank only. These bridges are suitable for small spans only, because for longer spans they will be very costly both in construction as well as maintenance. These are of two types, (i) Rolling lift type, and (ii) Trunnion type. Rolling lift type is better than Trunnion type because it requires less counter-weight; offers less friction and is cheap in maintenance. Bascule-bridges consist of hinged-trusses which can be revolved about a horizontal axis on trunnion and possess vertical position when lifted. Bascule-bridges can be opened more quickly than swing bridge. If heavy wind is blowing it becomes very difficult to open the bridge. The main girders of the trusses are taken continuously at the back of the span to form upturned segmental horn. When the bridge is lifted, these horns roll along the rails having the same radius of curvature. Counter-weights are provided in the trusses for easy lifting. When these are lifted they form 70°–80° with the horizontal position. Bascule bridges may have single or double leaf depending on the width of the channel.

41.8.1.3 Traversing Bridges

These are also called "Draw bridges" and can be rolled forward or backward across the opening. These are provided with rollers on the shore, or moderate spans. These may be operated manually or electrically. Generally these are not provided.

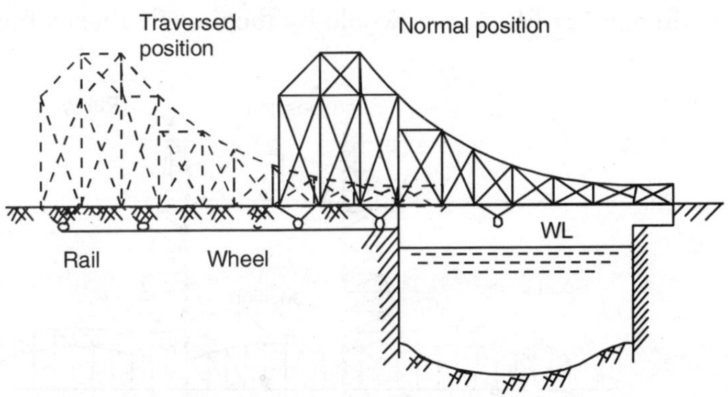

Figure 41.10 *Traversing Bridge*

41.8.1.4 Transporter Bridges

These consist of a travelling car, cradle or cage suspended by cables from an overhead truss resting on two towers. This travelling car moves from one side of the bank to the other side in suspended position. The materials or

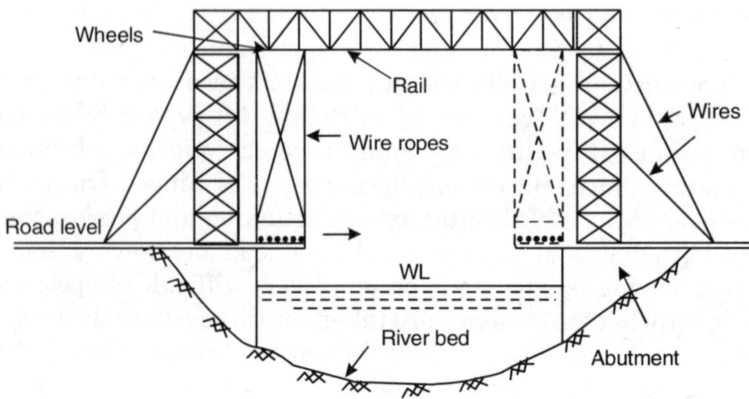

Figure 41.11 *Transporter Bridge*

persons are loaded in the cage and are transported from one bank to the other bank of the river. The navigation ships can easily pass below the bridge when the car is near the bank.

41.8.1.5 Vertical Lift Bridge

In wider navigation canals or rivers where Bascule or Swing bridge is not economical. Lift-bridge is provided. When the navigation boat or ship comes, the main span is lifted as a whole by means of cables which pass over the

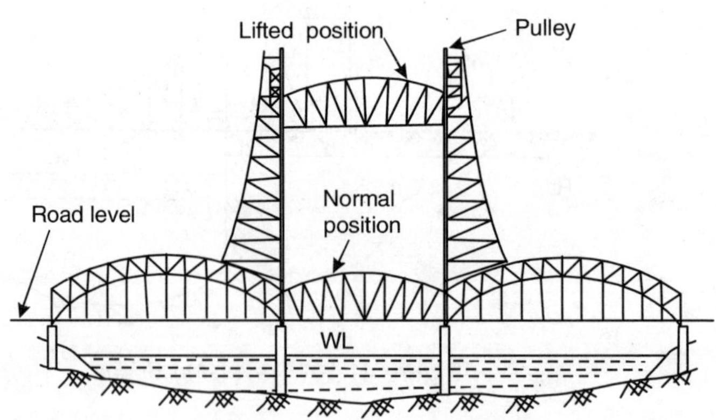

Figure 41.12 (a) *Lift-Bridge*

sheaves fixed on the tower. Two or four towers are constructed which are connected by rigid trusses across the river as shown in Fig. 41.12. These bridges are operated by hydraulic lifts or electrical motors. In the

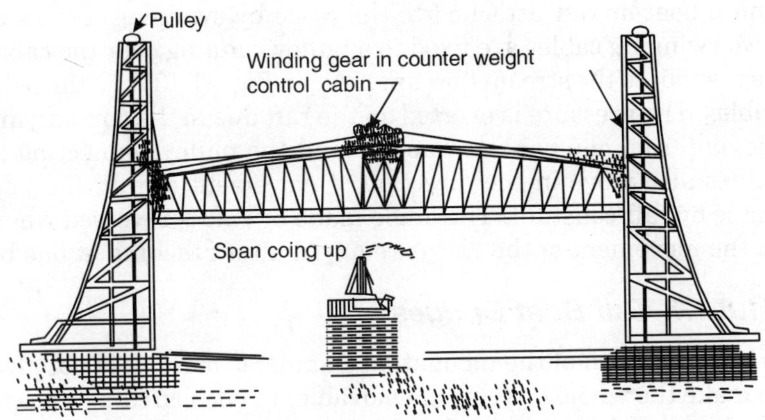

Figure 41.12 (b) *Another type of Lift-Bridge*

normal position the traffic can pass over the bridge, but when any vessel comes it is lifted up and after the passing of the vessel it is again brought in the normal position.

41.8.1.6 Flying Bridge

These are purely temporary bridges, used for crossing streams. It essentially consists of a suspended cable stretched across the stream above the H.F.L.,

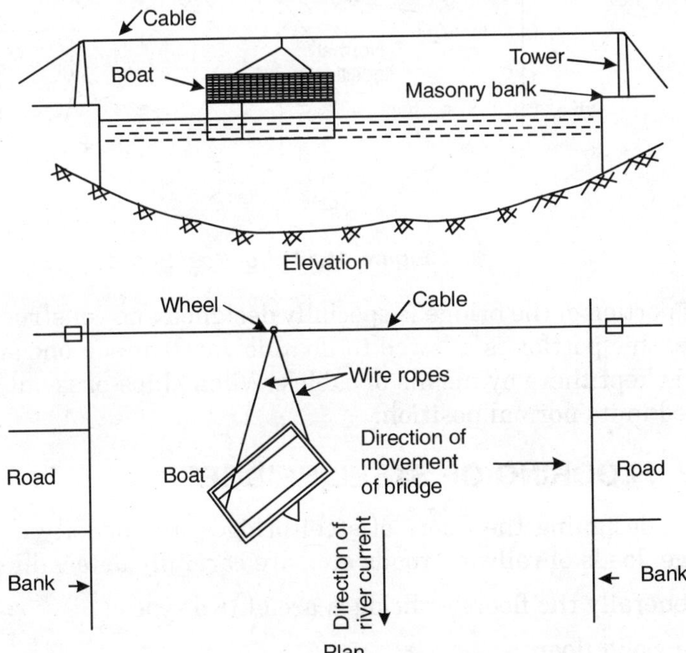

Figure 41.13 *Flying Bridge*

and a boat or raft attached to the cable by swinging cables and anchors. Two swinging cables are fixed to a pulley running over the cable. The raft is kept at 55° to the stream flow as shown in Fig. 41.13 with the help of swinging cables. The pressure is exerted on the raft due to the current, and raft starts moving from one bank to another and the pulley connecting the swinging cables also move in the direction of movement of raft and keep the angle made by raft constant. The angle made by raft is reversed when the change in the movement of the raft is required after reaching at one bank.

41.8.1.7 Cut Boat Bridges

When the width of the navigation channel is more, it is not always possible to make the whole boat bridge movable. In such circumstances only a small portion sufficient to pass the ships and navigation boats is made movable.

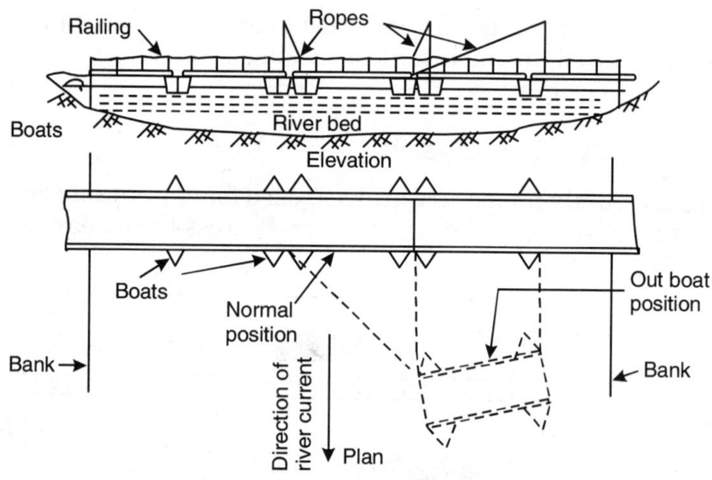

Figure 41.14 *Cut-Boat Bridge*

This portion of the bridge is specially designed and constructed. When ships come, this portion is allowed to float downstream on one side of the bridge and is kept there by means of cables. When ships pass out, it is pulled and placed in its normal position.

41.2 FLOORING OF STEEL BRIDGES

While designing the floors of steel bridges, the density of traffic width of bridge, loads of railway, roads etc., are carefully determined.

Generally the floors or bridges are of two types :

(a) Solid floors

(b) Open floors

41.2.1 Solid Floors

This type of floor is also known as continuous floor or ballasted deck type floor. This system mainly consists of the following types:

1. R.C.C. floors
2. Prestressed concrete floors
3. Precast concrete floors
4. Box-plate floors
5. Trough plate floors
6. Timber floors

Figures 41.15 to 41.19 show various types of floors.

For the construction of R.C.C. floors the main girders are braced together and then the casting of R.C.C. slab is done over them. This slab rests on the cross girders and main girders.

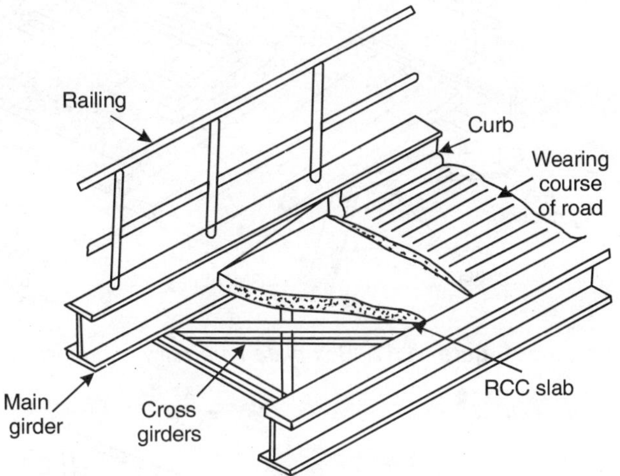

Figure 41.15 *R.C.C. slab type road bridge flooring*

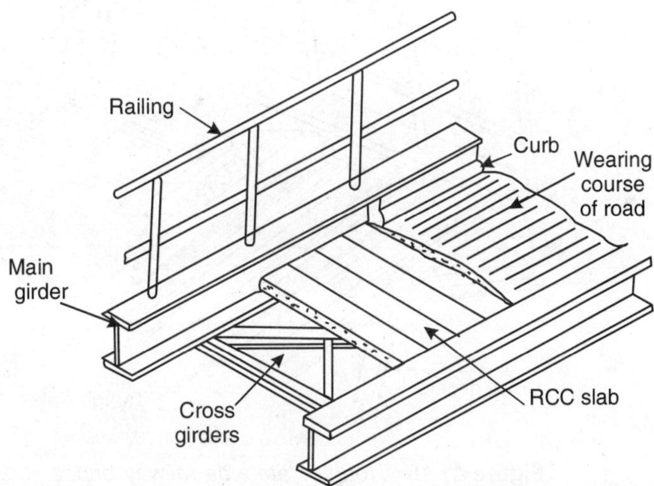

Figure 41.16 *Precast concrete slab type road bridge flooring*

In case of precast R.C.C. floors, the required planks of floors are precasted on the river bank or in factories and are then laid on the main girders as shown in Fig. 41.16. If these planks are light in weight, these are laid by workers by means of ropes. Heavy precast piece are laid by means of cranes.

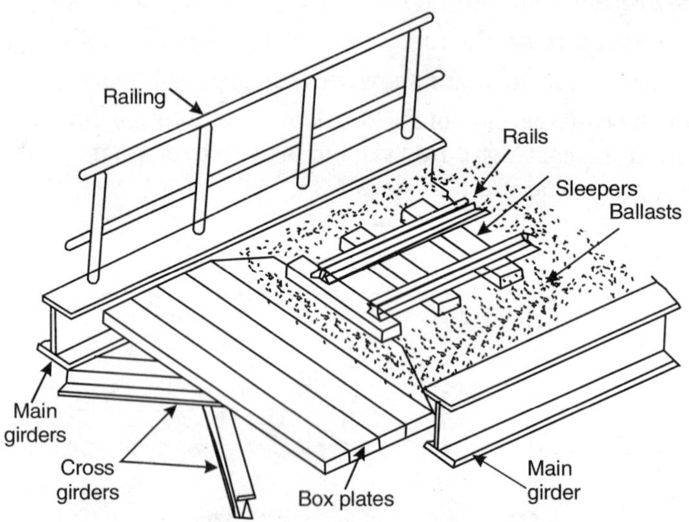

Figure 41.17 *Box plate type railway bridge flooring*

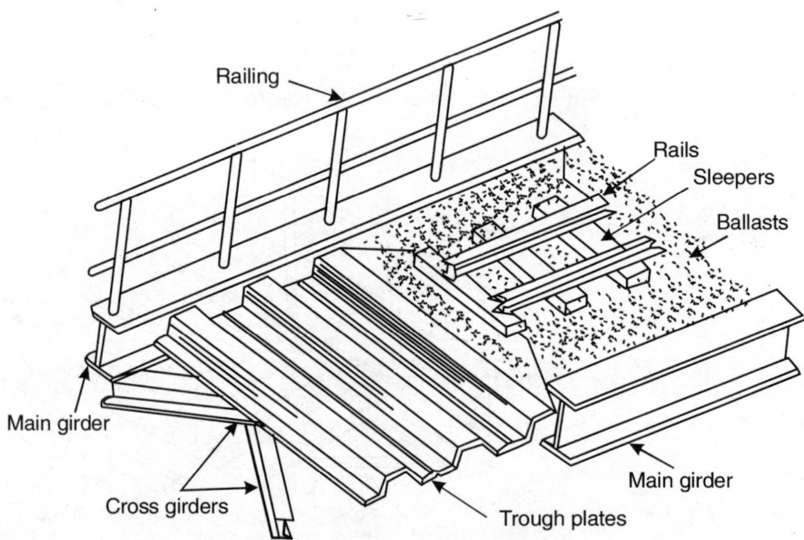

Figure 41.18 *Trough plate type railway bridge flooring*

The construcuon of solid floor with trough plates is shown in Fig. 41.18. In such types of floors troughs are laid on the flanges of main girders and are joined together. After this concrete is filled in case of road bridges and ballast is packed in case of railway bridges.

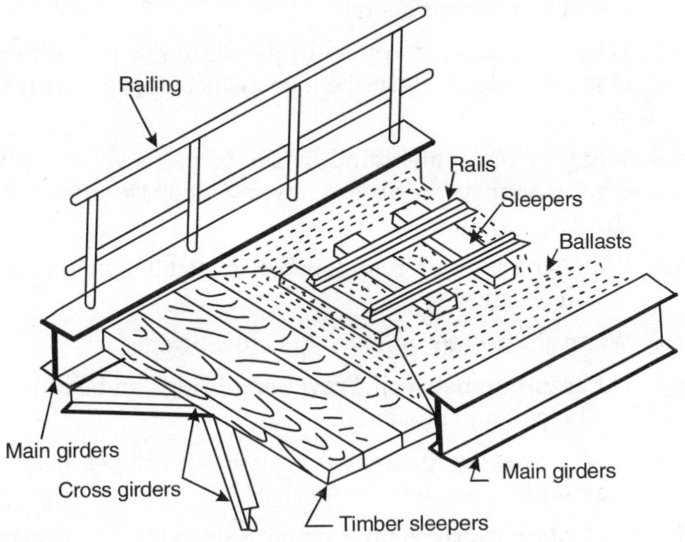

Figure 41.19 *Timber sleeper type railway bridge flooring*

Figure 41.17 shows the method of construction of solid floor with box plates.

Timber planks of sleepers are used for bridges of light-traffic or for foot bridges.

Open floors. These types of floors are used only in railways. In such type floors the main trusses or girders are joined together by means of cross-girders or bracings. The railway line alongwith sleepers is directly laid over it as shown in Fig. 41.20.

Figure 41.20 *Open floor type railway bridge*

REVIEW QUESTIONS

41.1. Write a short note on the history of steel-bridges.

41.2. What are the different types of steel-bridges? Describe any one with the help of a neat sketch.

41.3. What are suspension steel-bridges? How do they differ from other types of steel-bridges? Describe a suspension-bridge with the help of a neat sketch.

41.4. What do you understand by the term 'movable steel-bridges'? Under what circumstances these are provided? Describe Bascule-bridge with the help of a neat sketch.

41.5. What are the different types of movable steel-bridges? Describe any one in detail.

41.6. Write short notes on movable steel-bridges.

41.7. Differentiate between Traversing-bridge and Transporter bridge with the help of a neat sketch.

41.8. Under what circumstances Vertical lift-bridges are provided? How does they differ from Bascule-bridge?

41.9. Describe with the help of a neat sketch the function of cut-boat bridge. How do flying bridges differ from it?

41.10. Describe vertical lift-bridge with the help of a neat sketch.

41.11. Discuss the conditions under which you would recommend the adoption of :

(a) An arched bridge

(b) A suspension bridge

(c) A girder bridge

41.12. (a) Briefly discuss the factors governing the choice of a bridge superstructure in a certain situation.

(b) Compare the use, function and performance of the following :

(i) Balanced cantilever bridge

(ii) Suspension bridge

(iii) Arch bridge

41.13. Draw an elevation and cross-section (line diagram) of a steel truss bridge of about 60.0 m span carrying a single line B.G. track and name the various component parts.

41.14. Describe the various types of flow systems and flow surfaces adopted for R.C.C. and steel bridges.

41.15. Under what condition would a rigid frame bridge be prepared? Briefly comment upon the economics of this type of bridge.

41.16. Discuss briefly the factors that affect the choice between

(i) Arches

(ii) Balanced cantilevers

(iii) Prestressed concrete beams

(iv) Steel girders for a bridge over a river like Ganga in plains say near Kanpur

41.17. Discuss briefly the merits and demerits of a Bow-string Girder Bridge as compared to an Arch Bridge.

41.18. Describe with the help of sketches the various movable bridges.

R.C.C. Bridges

GENERAL

Lambot, a French, was the first man to start R.C.C. construction in 1850, by constructing a R.C.C. boat for Paris exhibition. But Monier was the first man who started actual construction of R.C.C. structures. Edwin Thacher, an American, was the first man who built R.C.C. bridge. After the invention of R.C.C. it was found that it is the best material for the construction of Engineering structures. Concrete is very suitable for taking compressive loads and steel is more suitable for tensile loads, thus the combination of these two will be ideal for structures. R.C.C. is better than steel because steel is neither fire proof nor incorrodible, whereas R.C.C. is safe against these drawbacks.

42.1 TYPES OF R.C.C. BRIDGES

The following are the different types of R.C.C. bridges, which are commonly used under different conditions :

42.1.1 Slab Type

This consists of simply a slab of uniform thickness laid on two abutments. It is used for small culverts and bridges upto a maximum of 18 metre span. It is simple to design and construct and if bearings are properly arranged it offers a certain degree of "flexibility" in the event of disturbances. The piers and abutments are constructed with stone or brick masonry. These are mostly suitable for pedestrians and light weight vehicles.

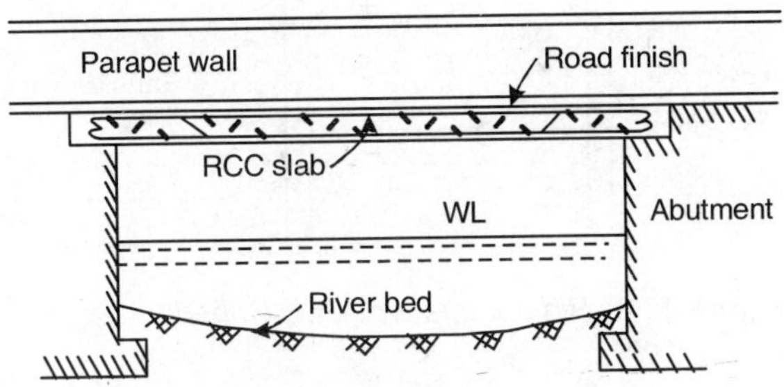

Figure 42.1 *Sub type R.C.C. bridge*

42.1.2 T-beam and Slab Type

It is used for spans upto 20 metres. The T-beam is simply supported over abutments or piers. It is cheaper than slab type and requires less quantity of materials. Over the concrete another wearing layer is laid on which traffic moves. When this wears or layer is worn out it is replaced by another layer. Generally in R.C.C. bridges parapet or railing is also provided of R.C.C. Abutments and piers in these types of bridges are also constructed with brick or stone masonry. Due to long life mostly over rivers and on over bridges stone masonry is more common in abutments and piers.

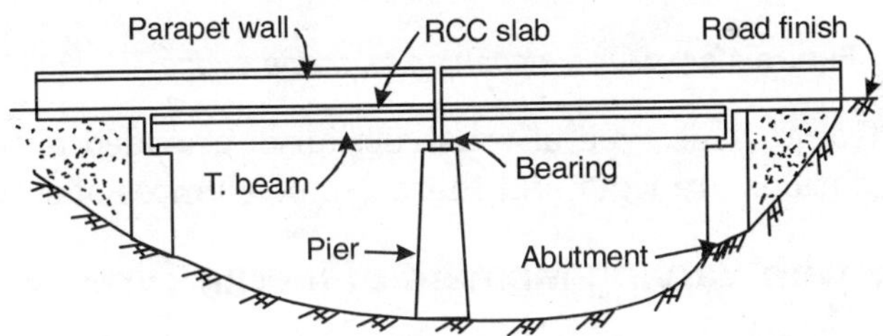

Figure 42.2 *T-beam and slab type R.C.C. bridges*

If the height of the piers is more, they may be constructed with R.C.C., other types and shapes are as described in chapter four. These types of bridges are more commonly used nowadays for road bridges of canals and railway lines.

42.1.3 Rigid Frame Type

In this type of construction horizontal deck slab is made monolithic with the vertical abutment walls. This type of bridges can be provided upto 20 metre span. It is not suitable for larger spans and for spans less than 10 m. A portal frame bridge is statically indeterminate structure. The portal frames have high stability against lateral forces such as wind and earthquake. These types of bridges are not very common in India.

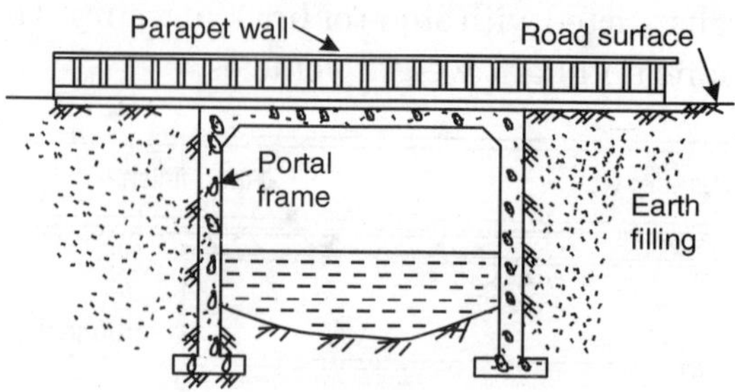

Figure 42.3 *Rigid frame type R.C.C. bridge*

42.1.4 Multiple Span Portal Frame Type

It consists of continuous spans in which the superstructure is monolithic with the supporting abutments and piers. In this type the span limit is 16

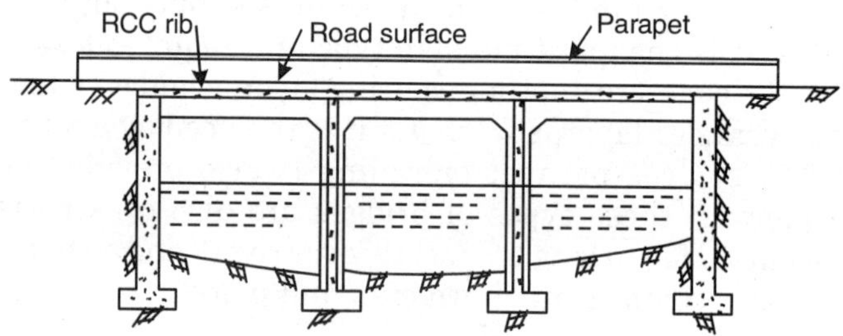

Figure 42.4 *Multiple span portal frame bridge*

metres for each frame. This type of construction is provided only in such cases where foundations are rigid and there are no chances of settlement.

42.1.5 Girders with Varying Moment of Inertia Type

This type consists of continuous girder spans having varying moment of inertia along the span. This type of construction is suitable to much longer

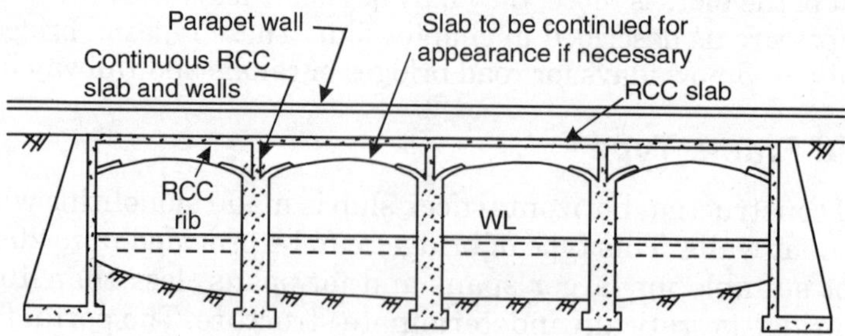

Figure 42.5 *Girders with varying moment of inertia*

spans and the span limit about 50 metres. The super-structure may be continuous with the piers and abutments or may have free supports on the top of abutments and piers.

These are also known as continuous bridges and are used for large spans where unyielding foundations are available, because high stresses will be induced even at slight settlement of foundations. The end spans are generally made 15–20% smaller than the intermediate spans. The deck of these bridges may be slab type, T-beam type or box type. Due to continuous girder over the piers, the magnitude of the bending moments at the supports (on piers and abutments) is more than the mid spans, hence the depth of the girder at the supports is more than the mid-span depths.

These types of bridges have the following advantages :

1. Less quantity of steel and concrete is required leading to over-all economy.

2. Due to less depth of the girder in the middle more head room is available in case of over bridges.

3. Due to continuous girder over 2–3 piers, the number of expansion joints is reduced, leading to low maintenance cost.

4. Due to continuous girder, it requires only one number of bearings at the piers, as compared in other bridges where two bearings are required for each girder at the end.

5. These bridges suffer with less vibrations and deformations even under heavy loads.

Following are the main advantages of these bridges :

1. Their design is more difficult than simply supported girder or slab bridges.

2. More detailing of reinforcement is required.

3. Slight settlement in the foundation may adversely effect the bridge structure.

4. For the construction of these bridges more skilled scheduling is required during placing of concrete and removing the form work.

42.1.5 Double Cantilever Type

It is suitable where the bridge has many spans. The main feature of it is the construction of alternate spans with projecting cantilevers, the ends of

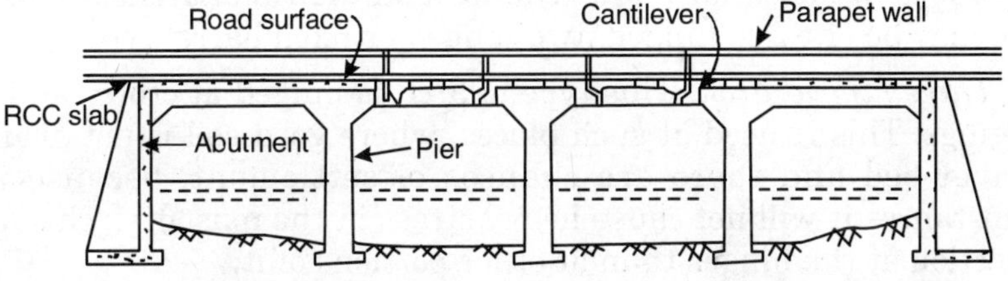

Figure 42.6 *Double cantilever type bridge*

which are used as supports for freely-supported spans constructed between them. This type of construction can be used upto 70 metres span.

These bridges are also known as *Balanced Cantilever Bridges*. These are used in the conditions where foundations are expensive and construction of small bridge shall be uneconomical. Large span bridges are constructed by using balanced cantilever superstructure. These may also be provided over deep gorges when they are to be crossed by single span. The junction between the suspended span and the edge of the cantilever is called as *articulation*. The bearings at articulations should be of fixed and expansion type alternatively. The length of the cantilever is about 20 to 25% of the supported span.

The design of the suspended portion is done as simply supported at the articulation ends. For designing the main span, the maximum negative moments at the supports are determined with the cantilever and suspended spans assuming to be fully loaded with no live load on the supported span. The maximum positive moments at the mid span are determined with full live load on the main span and no live load on the cantilever and suspended span. The cross-section of the balanced cantilever bridge may be of T-beam or Hollow Girder type. As the moments at the supports or piers are more than the mid span, the depth of the section at the support is more.

These bridges also have the following advantages :

1. Similar to girder with varying of moment of inertia, these bridges also requires less quantity of steel and concrete as compared with simply supported girder bridges.
2. Due to less reactions at piers these bridges require lighter and economical piers, than simply supported girder bridges.
3. Only one bearing is required at every pier due to continuity of girder at the piers.
4. Due to less number of expansion joints, the maintenance cost is low.

42.1.7 Barrel Arch Type

In this type the arch is constructed monolithic with the abutments or piers and is provided with closed spandrels. It is suitable for use on spans up to 70 metres. It is not used for more dead load on the rib of arches. This type of bridge may be of three hinged, two hinged or fixed barrel type.

(a) *Three hinged arch*. This type of arch is hinged at crown and at both springings. This is used at such places, where good soil is not available in the river-bed and there are chances of settlement, because in such circumstances it will not cause heavy stress in the masonry. This arch has less section at the hinges than at other portions of it.

(b) *Two hinged arch.* This arch is hinged only at springing positions. Therefore it transmits only reactions at the supports. No bending moments are produced at the springing place. Thus this type of arch gives no horizontal thrust on the piers or abutments.

(c) *Fixed arches.* These arches have no hinges at any place, therefore these are simple in construction than any other type of arch. These arches may be of open-spandrel or of filled spandrel type.

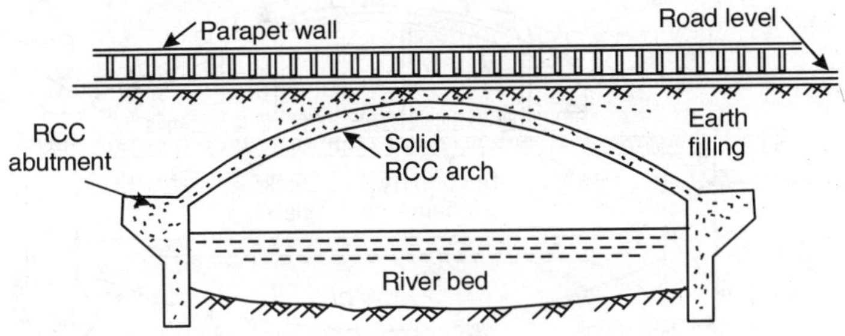

Figure 42.7 *Fixed barrel arch bridge*

42.1.8 Open Spandrel Arch Type

It is just like Barrel arch type but open spandrel is provided in place of filled spandrel. The deck is supported on columns, beam and slab construction

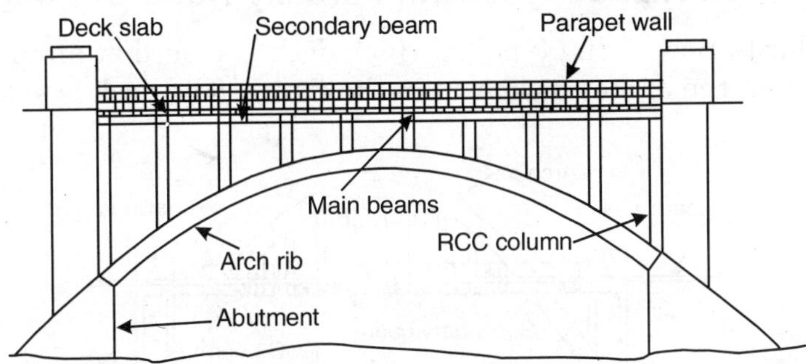

Figure 42.8 *Open-spandrel R.C.C. arch bridge*

or cross-wall and slabs. These types of bridges can be used for much longer spans upto 250 metres.

42.1.9 Bowstring Girder Type

In this case the arch ribs are constructed above the deck level of the bridge. Horizontal ties are provided to resist the horizontal thrust caused by the arches. Usually these horizontal ties are suspended from the arch-rib at suitable intervals and deck is supported on them directly. This type can be used for long spans upto 100 metres.

These types of bridges are more economical where more head room is required under the bridge. The main supporting system is known as bow-string girder, due to resemblance of its arch with the bow and the tie beam with that of string of the bow. There is no effect of slight vertical or angular displacement of the abutments. The bow-string girders being rigid are not affected even by slight horizontal displacement of the abutments.

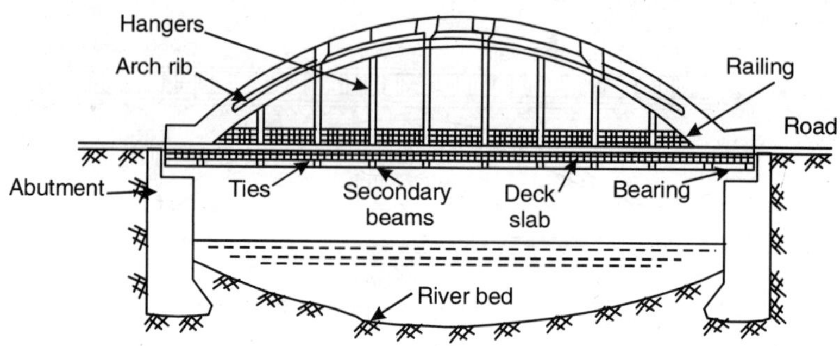

Figure 42.9 *Bow-string girder type bridge*

42.1.10 Arch Ribbed type with Partially Hung Decking

It is similar to Bow-string type, in which deck is placed at an intermediate level between the arches as shown in the figure. The deck is suspended by ties from the arch rib for the major portion in the middle, but at the ends it is supported on columns or cross-wall placed over the rib.

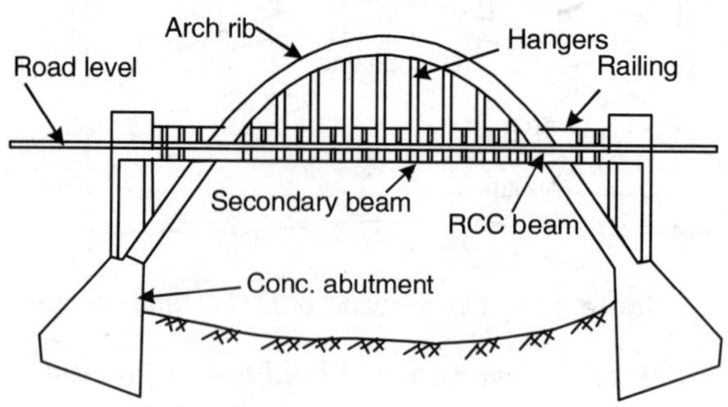

Figure 42.10 *Arch-rib type with partially hung decking*

42.1.11 Hollow Girder Bridges

These bridges have closed box type cross-section of their girder and are economical for spans between 25 to 30 m. The main girders are constructed with multi-cellular rectangular or trapezoidal-shaped hollow cells. The torsional stiffness of these bridges is much more than T-beam bridges. These bridges are normally curved in plan and are most suitable for grade separations.

42.2 ARRANGEMENT IN CROSS-SECTION

The different types of bridges which are described above, are also subjected to variations in the arrangement of the cross-section. The following are the different arrangements, which are possible in cross-section :

1. In the first type there may be a slab of uniform thickness throughout the cross-section. It is found in slab-type and short span portal-frames and parapet girder bridges (Fig. 42.11).

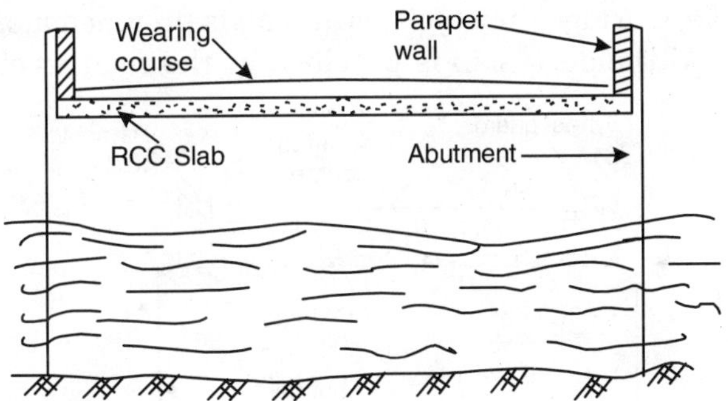

Figure 42.11 *Simple-slab type cross-section*

2. Deck-slab and beam ribs (underside) are found in slab and T-beam bridges (Fig. 42.12).

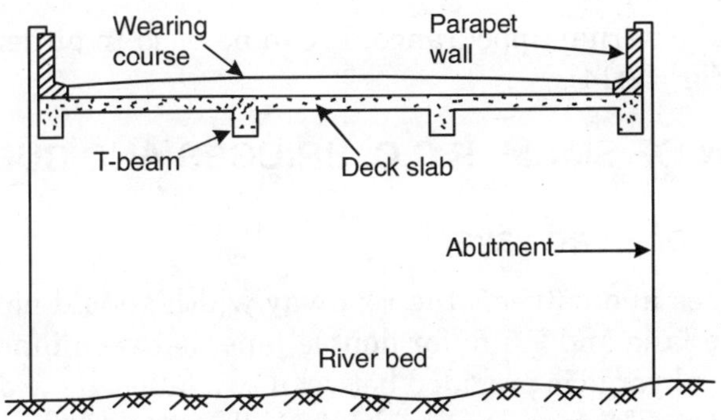

Figure 42.12 *Beam and slab type cross-section*

3. Deck-slab and beam ribs (upperside) are found in parapet type bridges (Fig. 42.13).

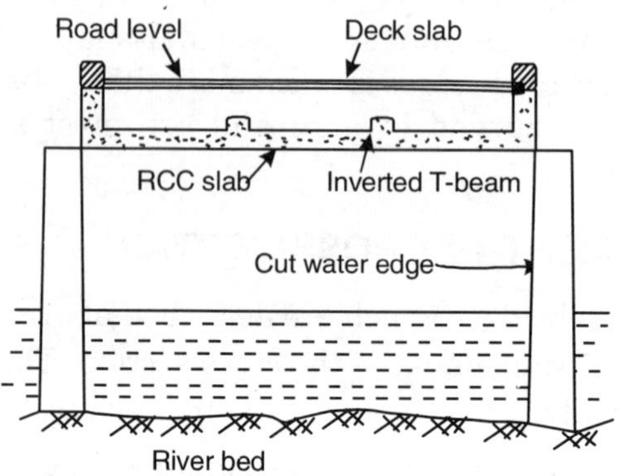

Figure 42.13 *Deck-slab and Beam ribs on upperside Type Cross-section*

4. Cellular type bridge, in which soffit slab is provided along the bottom of the girders and the deck slab is provided at the road level. This type of bridge has solid external appearance. It can be used in girder type or arch type bridges (Fig. 42.14).

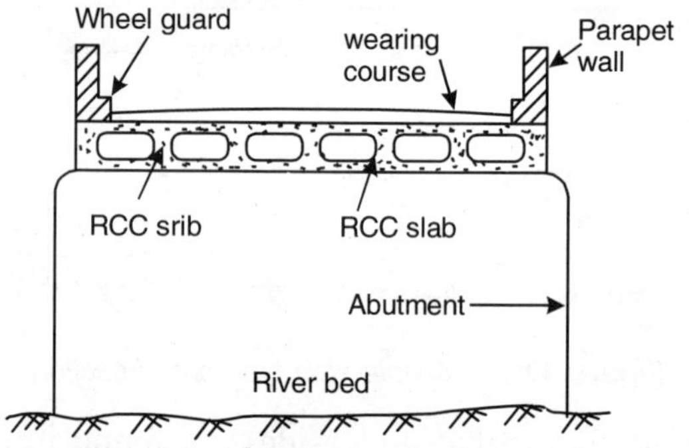

Figure 42.14 *Cellular type cross-section*

42.3 DESIGN OF SMALL R.C.C. BRIDGES AND CULVERTS

42.3.1 Width of Roadways

For small bridges and culverts the roadway width should not be less than 4.5 m for single lane and 7.5 m for double lane, between inner faces of the kerbs. If a central verge is provided between two lanes, its width should not be less than 1.2 m. The formation width for village road bridge is kept 6.0 m and for arterial and district road as 9.75 m.

42.3.2 Kerbs or Wheel Guards

Solid kerb sections are provided on both the sides of the carriageway on the bridges. Figure 42.15 shows the cross-section of the wheel guard usually provided.

On the roadway safety kerbs are also provided, which are designed as foot-paths. The safety kerbs have the same outlines as that of kerb, the only difference being in the top width, which is not less than 60 cm.

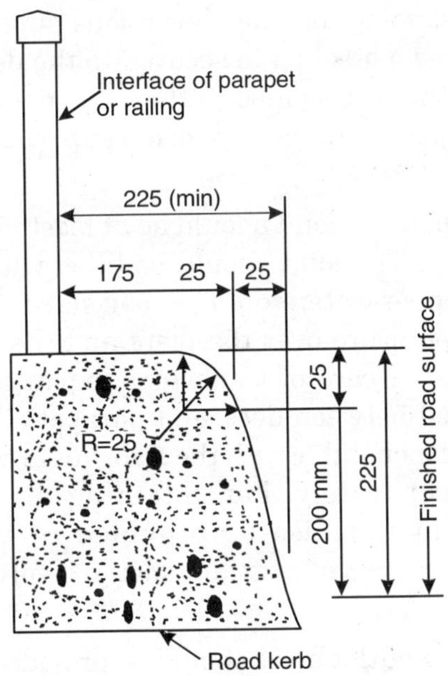

Figure 42.15 *Kerb or wheel guard section*

42.3.3 Footways on Bridges and Pedestrian or Foot Bridges

A minimum width of 1.22 m is provided for footpaths or footways on the bridges. (I.R.C. has recommended 1.52 m as minimum width). It is better to provide 1.83 m wide footway where possible.

Live load of 400 kg/sq. m is taken for design purpose upto 7.60 m span. For other spans upto 30.0 m, the value of live load is taken as 300 kg/sq. m where the live load is less on all footways on bridges and bridges used by the animals. On the bridges where crowd loads occur, the live load of 600 kg/sq. m is used. The approximate dead load on timber planks and decking is taken as 500 kg/sq. m on the girder bridges.

42.3.4 Vertical Clearance

A minimum head room of 5.0 m above the crown of the road surface should be provided. In case of footways and cycle track the value of head room may be 2.25 m.

42.3.5 Joints in R.C.C. Road Slabs

Free movement of the slab part at the expansion joint should be allowed. The places where the wearing surface is to be carried continuously over the joints in bridge decks, some form of continuous support over the joint is provided. In such places a steel plate secured to the deck at one side of the joint and free on the other is provided. If the movement is small, a standard T-section which is dropped into the expansion gap, without attaching to the structure, can be used.

The width of the expansion joint should be at least four times the amount obtained by calculation. The joint should be filled with bitumen and sand and provided with a copper waterstop. The road slabs should be free to move due to change in temperature over the piers and the abutments, over the sliding plates. For taking care of transverse expansions, adequate joints should also be provided between deck and parapet. The kerbs of the slab bridge should be casted monolithic with the slab, but in case of T-beam bridge, it should be casted separately. For obtaining discontinuity joints at appropriate places should also be provided in the railings. Small reinforcement in the direction of restraint should be provided in contraction joints.

50 to 75 mm diameter outlet holes should be provided on both the sides of the road through parapet at about 2.5 to 3.0 intervals. Adequate camber should be given in the road surface for quick draining off the rain-water.

42.3.6 Road Surfacing over Concrete Slabs

Following procedure is followed for road surfacing over concrete slabs :

1. The concrete surface is painted with a coat of asphalt and is covered with 40 mm of earth cushion. Over the earth cushion 120 to 150 mm metalling is provided.

2. Now 50 mm thik asphaltic concrete is laid. Sometimes a small amount of cement can be added to mix as filler for improving the concrete. The ramming is done with wooden rammers or with light roller.

3. Over the slab concrete 75 mm cement concrete can be laid as wearing surface directly as in case of cement concrete roads.

42.4 REINFORCEMENT DETAILS FOR R.C.C SLABS FOR CULVERTS

Table 42.1 gives the reinforcement details for R.C.C. slabs for culverts from 1.0 to 6.0 m span. The reinforcements given are calculated on the basis of I.R.C. 'Guidelines for Design of Small Bridges and Culvert'.

Following points are to be considered while using Table 42.1.

(a) I.R.C. loading class 'A' for two lanes or class 70 'R' for one lane roadway have been used.

(b) The clear width of the roadway is taken as 11.20 m, with 0.40 m parapet wall on both the sides. Thus the total width of the slab becomes 12.0 m.

(c) Ordinary cement concrete of 1 : 2 : 4 ratio by volume or M 150 grade has been assumed while designing.

(d) The stresses of M.S. grade I tested steel reinforcement have been used.

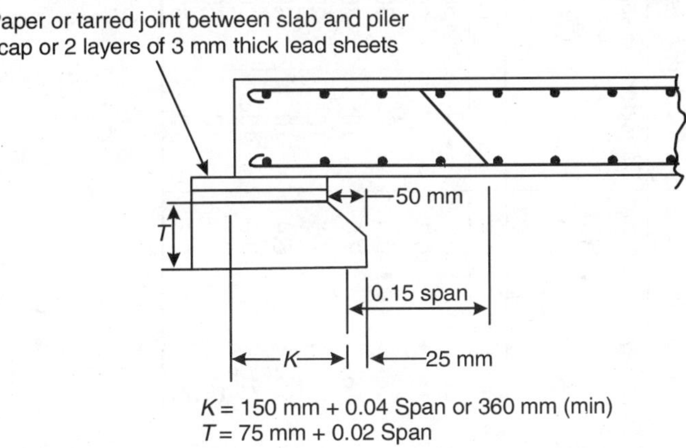

Figure 42.16 *Detail of R.C.C. culvert slab and abutment*

(e) Every alternate longitudinal bottom bar has been bent up at 0.15 span from the edge of the abutments at 45° and taken up for taking the negative bending moments at the ends. Proper hooks are provided at the end of the bars.

(f) Transverse reinforcement given in Table 42.1 may be reduced towards the ends.

(g) In case full length reinforcement bars are not available, they are placed at joint for lap joints for 45 diameters length.

(h) 25 mm minimum clear cover is provided to the reinforcement.

(i) For larger spans more than 30 m longitudinal camber of span/240 should be provided for taking into account the sag at the middle.

(j) 75 mm thick asphaltic concrete has been assumed on the concrete slab.

TABLE 42.1 *Reinforcement details for R.C.C. slabs for culverts*

Clear span	Length of bearing at each end	Overall length of slab	Overall thickness of slab	Details of reinforcement bars							
				Longitudinal bars				Transverse bars			
				Bottom bars		Top of bars		Bottom bars		Top of bars	
				Diameter	c/c spacing	Diameter	c/c spacing	Diameter	c/c spacing	Diameter	c/c spacing
m	*cm*	*m*	*cm*	*mm*	*cm*	*mm*	*cm*	*mm*	*cm*	*mm*	*cm*
1.0	30	1.6	19	16	15	10	30	16	15	10	30
1.5	30	2.1	22	16	10.5	10	30	16	15	10	30
2.0	30	2.6	24	16	10.0	10	30	16	12.5	10	30
2.3	30	3.1	27	16	9.0	10	30	16	12.5	10	30
3.0	30	3.6	30.5	16	7.5	10	30	16	12.5	10	30
3.5	30	4.1	34	16	6.0	10	30	16	12.5	10	30
4.0	30	4.6	37	20	10.0	10	30	16	12.5	10	30
5.0	40	5.8	42	20	9.0	10	30	16	12.5	10	30
6.0	40	6.8	47.5	20	7.5	10	30	16	12.5	10	30

(k) Tar paper is to be provided at the top of the piers and abutments for free movement of the slab.

(l) 6 mm space is to be provided between the end of the slab and the abutment for expansion. This gap is to be filled with bitumen impregnated felt.

REVIEW QUESTIONS

42.1. Write a short note on the history of R.C.C. Bridges.

42.2. What are the different types of R.C.C. Bridges? Sketch any two types and describe in brief.

42.3. What are different types of R.C.C. arches? Describe with the help of neat sketches.

42.4. What is the difference between Barrel arch bridge and Bowstring bridge?

42.5. Write short note on the arrangement in cross-section of R.C.C. Bridge.

42.6. Draw longitudinal section and cross-section of a T-beam and Slab-type R.C.C. Bridge.

42.7. Discuss the points in favour and against the selection of the following types of superstructure for a major bridge :

(a) R.C.C. simply supported slab and beam construction.

(b) R.C.C. slab and portal frame construction.

(c) R.C.C. slab and continuous beams with varying depths.

(d) R.C.C. (balanced) Cantilever type.

(e) R.C.C. Bow-string Girder type.

42.8. State briefly the advantages and disadvantages of the following types of R.C.C. constructions :

(a) Rigid frame bridges

(b) Continuous bridges

(c) Balanced cantilever bridges.

42.9. Discuss the different types of R.C.C. bridges, giving the main features of each type. Why are R.C.C. bridges not preferred for Railways?

42.10. List the various types of superstructures possible for a R.C.C. type of major bridges. Compare their advantages and disadvantages.

42.11. Describe the various types of flow systems and flow surfaces adopted for R.C.C. and steel Bridges.

42.12. State different types of bridges that are normally adopted by highway engineers in India. Discuss the relative merits of each type and suggest a suitable type for a 14 m span bridge.

42.13. Describe the different types of bridge flows and draw neat sketches to show how they are supported. State the essential requirements of bridge flows to be kept in view while adopting flows for bridges.

43

Floating Bridges

GENERAL

These are temporary bridges, which are used at the following occasions :

- When it is very costly to provide permanent bridge floating bridges are provided for crossing rivers all the year round except during monsoon, when these are dismantled.
- In the case of Kumbh Melas, floating bridges are provided for the period of mela, because these are very cheap for such times.
- Military forces use floating-bridges at the time of war for crossing rivers, nallahs, lakes etc.
- At the time of construction of permanent bridges or dams, floating-bridges are used for carrying materials and workers from the river banks to the sites or from one bank to another.

43.1 TYPES OF FLOATING BRIDGES

Generally, there are three types of floating bridges which are commonly used :

1. Boat Bridge
2. Raft Bridge
3. Pontoon Bridge

43.1.1 Boat Bridge

One boat can be used for crossing the river. But to avoid delay in loading and unloading of the boat, a bridge of boats is constructed to provide quick crossing of the river without any difficulty. Steel-boat bridges are used by military because these are lighter than Pontoons. The boats are designed to

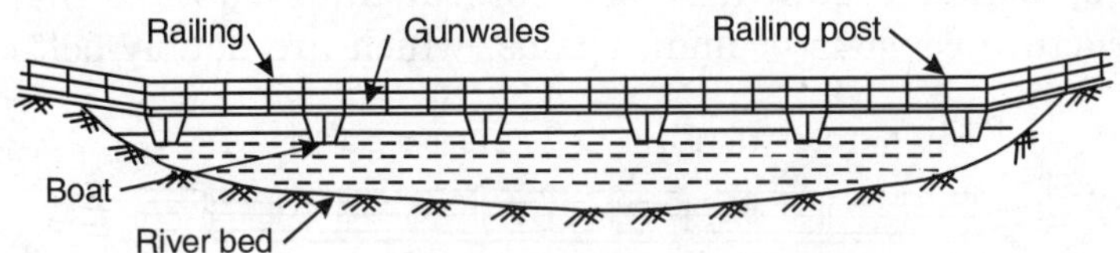

Figure 43.1 *Boat bridge*

form the substructure of the bridges. The superstructure of it consists of longitudinal beams of simple girder or trussed-beam type, over which planks are laid, to serve the purpose of the roadway of bridge-flooring. These beams are known as gunwales and rest on gunwale pieces fixed on the outer edge of the boat. Cross-beams are provided between two gunwale pieces of the same boat, over which the same type of flooring as on gunwales, is done. All the boats are stiffened together by cables and each boat is provided with two anchors, one in upstream and the other in downstream side to hold the boat in position. Parapet of wooden or steel poles with rope or steel chain passing through them is provided on both sides of the roadway.

43.1.2 Raft Bridge

The given Fig. 43.2 shows this type of bridge. It consists of barrels or casks fixed to gunwales at their tops, over which road-bearer or trussed beams are fixed. The substructure of raft-bridge mainly consists of floating piers made of barrels or casks lashed together in lines of 7 or 8 by means of long spans, laid across to them at the top. The superstructure of raft bridge is similar to boat bridge. Wooden planks are fixed to the road-bearer which serve as a floor of the bridge. Parapet of boat-bridge type is provided in this bridge.

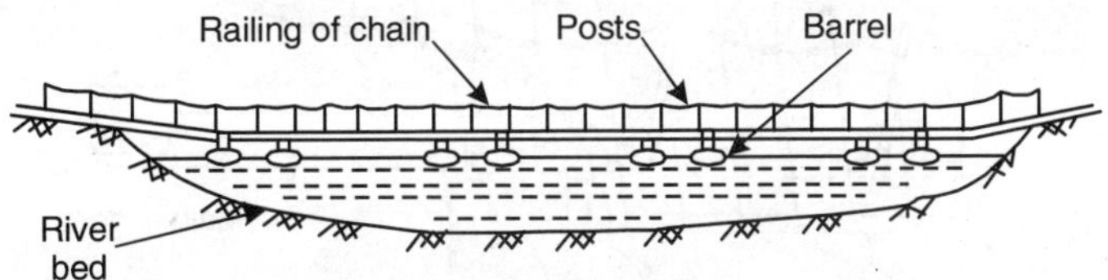

Figure 43.2 *Raft bridge*

43.1.3 Pontoon Bridge

This type of bridge is generally provided at Kumbh-Melas. These are heavier than boat-bridges and can take more load than boat-bridges. Excepting the monsoon, these are also used for crossing rivers when the current of water is more, because this type of bridge can resist it. Figure 43.3 shows this type of bridge, which essentially consists of substructure of pontoons. Superstructure consists of main-beams, which are doubly-bolted at their

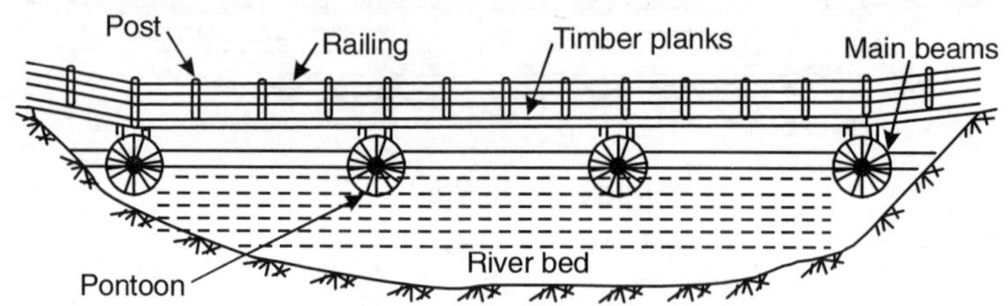

Figure 43.3 *Pontoon bridge*

ends to the pontoon. Timber-planks are laid across the main beams which serve the purpose of the bridge flooring.

In developed countries, the modern trend is to use inflatable rubber pontoons instead of wooden or metal boats. The bailey bridge components which were standardised during second world war are also extensively used nowadays by the army in crossing streams. Bailey bridge components are panels, bracing frames, stringers, bearings, transoms, links, pins etc. which are used as superstructure in the pontoon bridge.

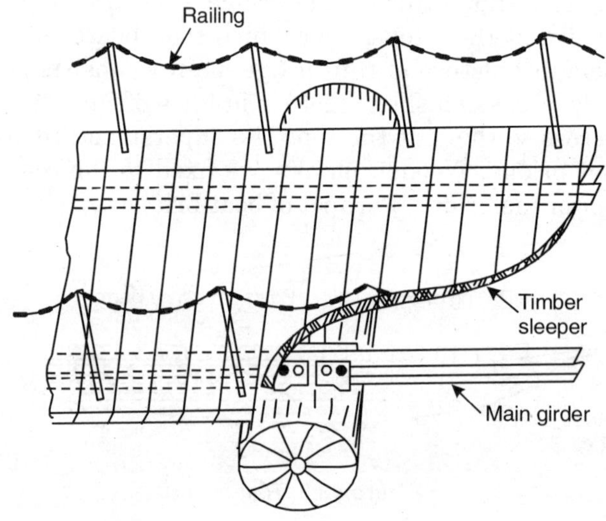

Figure 43.4 *Parts of Pontoon bridge*

Indian army has started use of 'Krupp-Man bridges' which are made of aluminium alloy sections. These bridges parts are very light—as compared with old steel sections and are used for transporting heavy military vehicle across streams or rivers. Govt. Ordinance Factory of Ambajhari (Maharashtra) has started manufacturing of the required bridge sections of aluminium alloy for the military purpose.

REVIEW QUESTIONS

43.1. Under what circumstances floating bridges are provided? What are different types of floating bridges? Describe only one in detail.

43.2. Differentiate between Boat, Raft and Pontoon bridges in detail.

43.3. Write short notes on floating bridges.

43.4. Describe a Boat bridge with the help of a neat sketch. How the roadway of such a bridge is constructed?

43.5. Write explanatory short notes of the following :

(a) Trestle Bridge

(b) Pontoon Bridge

Bearings

GENERAL

In long span bridges, there are some changes in the main girders due to deflection and change in temperature, which give rise to heavy stresses endangering the entire bridge. This expansion or contraction, movement and deflection which occur due to movement of traffic on bridge or wind loads when acts on the superstructure of the bridge, are transferred to the piers through bearings, which allow such movements quickly and safely. Hence bearings are the bridge components which transfer the loads of the superstructure to the piers. The design of the piers mainly depends on the types of the bearings and subsequently they affect the design of the foundation. Therefore the bearings are the most important components of the bridges and require scientific design, skill in their construction and maintenance. In major bridges the cost of bearings may vary between 10 m to 15% the total cost of the bridge. Bearings are provided in the bridges for the following reasons :

- To distribute the load on sufficiently large area and bringing the intensity of compressive pressure within safe limits of the materials of support.
- To allow the girder of the bridge to take free angular movement at the ends when loaded, as also to allow it to rock. The simply supported girders deflect under loads causing their ends to take angular movements over the support. The bearings allow both the ends of the girder to assume a position tangent to the deflected bottom shape of the girder.
- Variation in temperature causes expansion or contraction in the length of the girder but the bearings allow the free movement of the girder without any difficulty and damage to the masonry work. If one end of a simply supported member is

kept free to move longitudinal and the other end is kept hinged by the bearing, the changes in temperature will not cause any stresses on the sub-structure and will make the bridge safe. In case of R.C.C. or prestressed concrete bridges, the shrinkage starts from the moment the concrete is laid in position during construction and it continues for three to four months. In these bridges provision of safety against longitudinal variation is also taken by providing suitable bearings.

- To prevent the damage to the bridge, if any pier or foundation sinks, the bearing allow free end movement as well as angular movement and prevent the superstructure from crack and damages.

44.1 TYPES OF BEARINGS

The following are the different types of bearings, which are commonly used in bridges :

1. Cement mortar pad
2. Laminated rubber fixed bearing
3. Fixed-end bearing
4. Rocker bearing
5. Rocker roller bearing
6. Rocker bearing with curve bottom
7. Knuckle bearing
8. Sole plate on curved bed-plate bearing
9. Sliding bearing

44.1.1 Cement Mortar Pad

In case of girder bridges for short span, the cement sand mortar (1 : 1) is the cheapest type of bearing for supporting the ends of the girder. Dowel bars are embedded in the bed block at one end and their other end passes through the slots kept in the end of the girder plate. These dowel bars are designed to take up the longitudinal force through shear.

44.1.2 Laminated Rubber Fixed Bearing

Laminated rubber fixed bearing consists of reduced thickness of laminated rubber sheets. The reduction in thickness is kept such that, after the deflection the strain in the rubber sheets should be less than the allowable limit. The maximum allowable compressive strain in the laminated rubber bearing, should not be more than 10% the thickness of the sheet including compression due to the rotation of the girder.

44.1.3 Fixed-end Bearing

This type of bearing does not permit any longitudinal movement of the girders. A thick bearing plate is attached by means of counter sunk rivers to the bottom of the bridge girder, which is fixed on cast iron bed plate, fixed to the pier by anchor bolts as shown in Fig. 44.1. Sometimes the lower plate is faced with a phosphor bronze wear surface to prevent erosion and for reducing the coefficient of friction. This type of bearing is recommended upto span of 12.0 m. Instead of the steel plate sometimes a deep cast iron base is fixed to the lower flange of the girder. Deep-seated bearings are provided over 12.0 m span to avoid the concentration of the reactions at the inner edge of the bearing. Sometimes also honey-combed cast iron blocks are made for reducing weight.

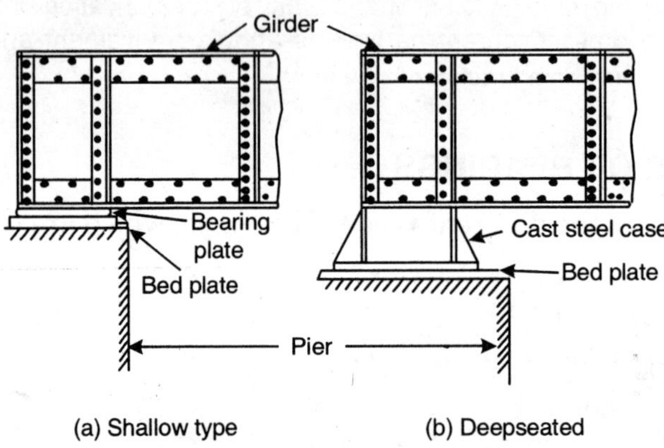

Figure 44.1 *Fixed-end bearing*

44.1.4 Rocker Bearing

This type of end bearings are suitable for more than 25 metres span where there are more chances of deflection in the girder of the bridge. It has as

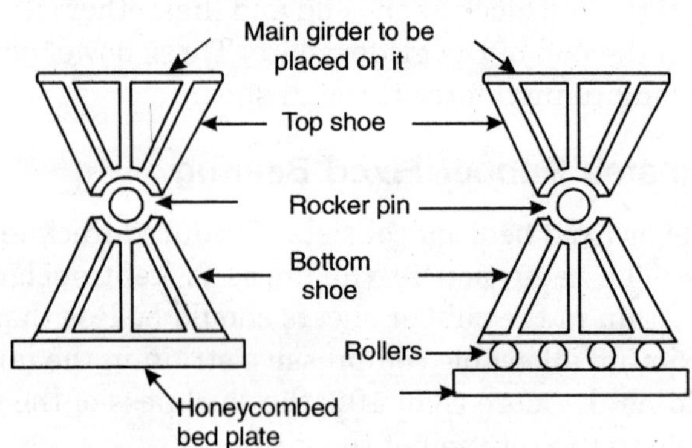

Figure 44.2 *Rocker Bearing*   **Figure 44.3** *Rocker Roller Bearing*

20 cm radius circular rocker pin, placed between top shoe and bottom shoe, which also have the same radius of machined bottom and top respectively.

Usually both these shoes and saddle covers with $\frac{2}{3}$ circumference of the rockers are shown in Fig. 44.2. The girder of the bridge is connected to the top shoe. When the girder deflects due to variation in length, due to loads, the top shoe rotates over the pin. This bearing gives free angular movements of the main girder connected to the top shoe.

44.1.5 Rocker Roller Bearings

This is similar to the rocker-bearing with the only difference that bottom shoe rests on cylindrical rollers which are free to roll on bottom steel plate. This bearing provides free longitudinal as well as angular movement of the main girder of the bridge connected to the top shoe. Figure 44.3 shows this type of bearing.

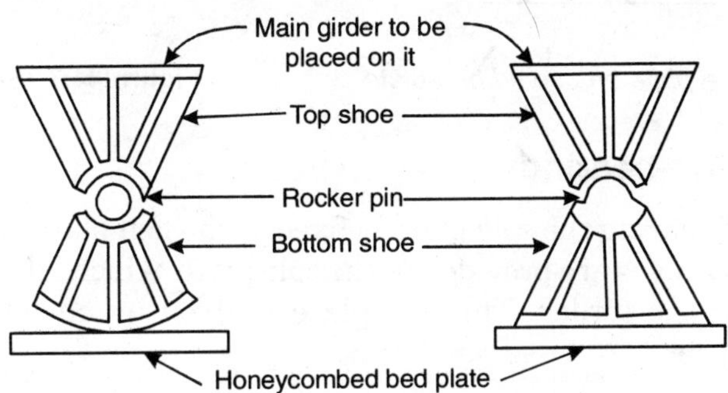

Figure 44.4 *Rocker-bearing with curved bottom* **Figure 44.5** *Knuckle-bearing*

44.1.6 Rocker-bearing with Curve Bottom

The upper portion of this bearing is also similar to the rocker-bearing. The bottom shoe has curved bottom surface as shown in Fig. 44.4 which rests on steel plate. The bearing allows free longitudinal as well as angular movement to the bridge girder.

44.1.7 Knuckle-bearing

Figure 44.5 shows this type of bearing which has semi-circular concave surface of the top shoe. The bottom shoe has semi-circular surface of convex type having the same radius as that of the top shoe. This bearing allows only angular movement of the girder fixed to the top-shoe.

44.1.8 Sole Plate on Curved Bed Plate

In this bearing one sole plate is fixed in the girder of the bridge, which rests on a curved bed plate fixed in the masonry. When the main-girder deflects due to loads the sole-plate takes angular movement on the bed plate. Thus this type of bearing is useful where angular movement occurs in the bridge girder.

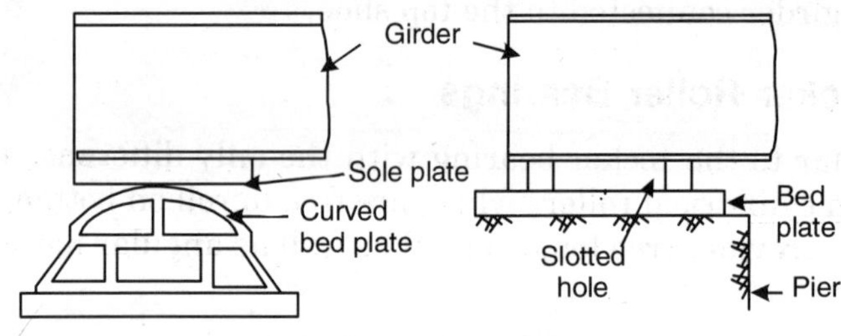

Figure 44.6 *Sole plate on curved bed-plate*

Figure 44.7 *Sliding-bearing*

44.1.9 Sliding-bearing

In this type of bearing the sole plate is fixed to the bed plate and bolted to it. Slotted types of holes are provided in the sole plate, which allow longitudinal movements of the girder. The bed plate is fixed by anchor bolts to the masonry.

44.2 BEARINGS FOR STEEL ARCHES

Special types of bearings are used for the support of steel arches at the pier or abutments. Two types of bearings are most commonly adopted. One is

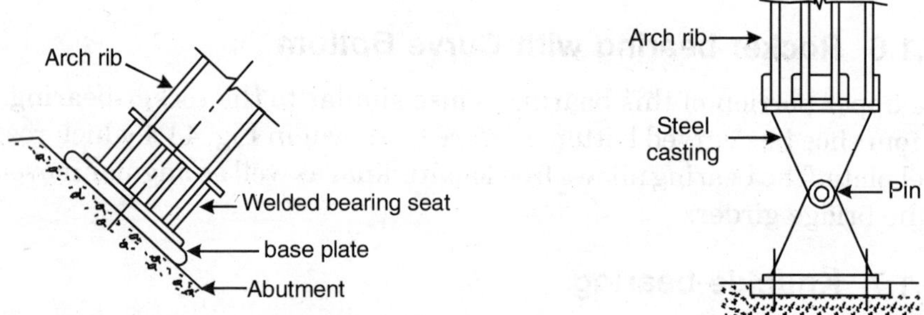

Figure 44.8 *Fixed bearing for steel arch bridges*

Figure 44.9 *Hinged or pinned bearing for steel bridges*

fixed type bearing and the other is hinged or pinned bearing. Figures 44.8 and 44.9 show these types of bearings.

44.3 ANCHORAGES FOR SUSPENSION BRIDGES

In case of suspension bridges bearings are not required. Instead of this saddles are required at the top of towers and anchorages are required for mixing the cables in the banks. Figure 44.10 shows typical type of anchorage required for suspension bridge.

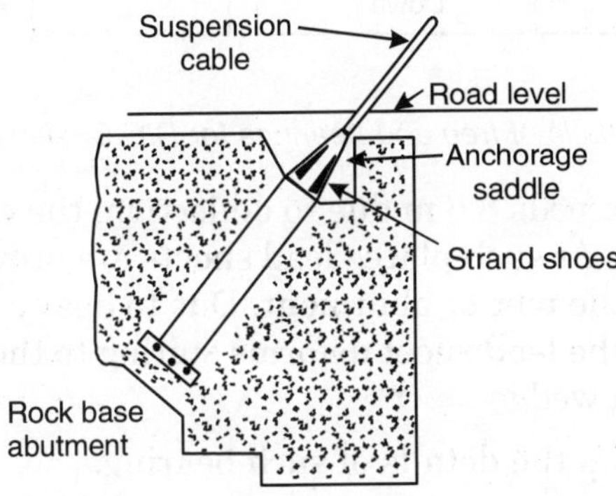

Figure 44.10 *Typical suspension bridge cable anchorage*

44.4 CONCRETE BRIDGE BEARINGS

Depending on the type of concrete bridge various types of bearings are required suiting to the particular type.

Following types of bearings are most commonly adopted :

(i) Bearing for R.C.C. slab bridges

(ii) Bearing for R.C.C. girder bridges

44.4.1 Bearing for R.C.C. Slab Bridges

In case of short span R.C.C. slab bridges upto 8.0 m generally no special type bearings are provided. The decking slab is directly laid on the abutment cap as shown in Fig. 44.11. As a general practice the top of the cap over pier or abutment is given a smooth finish and the edges are rounded as shown to reduce the friction. Common practice is to directly support the one end of the slab on the pier and the other end is dowelled.

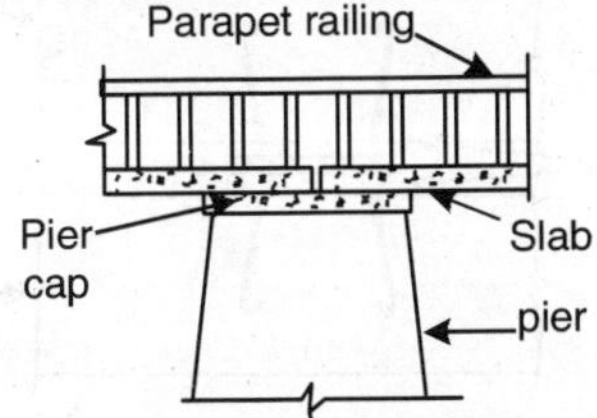

Figure 44.11 *Details of the decking slab and the pier cap*

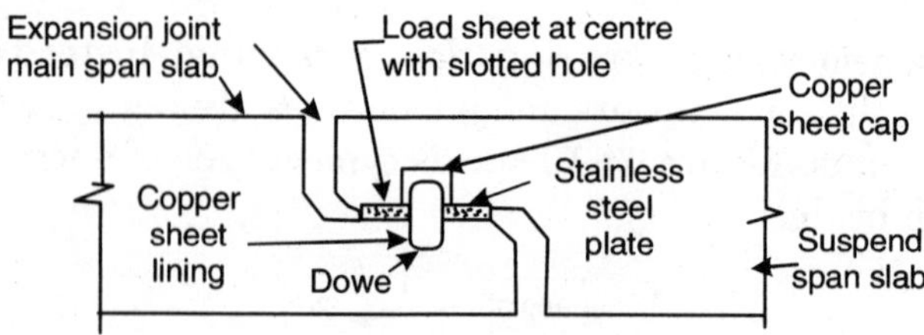

Figure 44.12 *Details of free and bearings for R.C.C. slab bridge with dowel*

When the span exceeds 8.0 m due to deflection, the ends of the slab have a tendency to rotate. At such places lead sheets are provided in between the slab and the cap of the pier or abutment. Due to heavy load and on rotation of the slab at ends, the lead sheet deforms suiting to the end conditions and takes the shape of a wedge.

Figure 44.13 shows the details of such bearing.

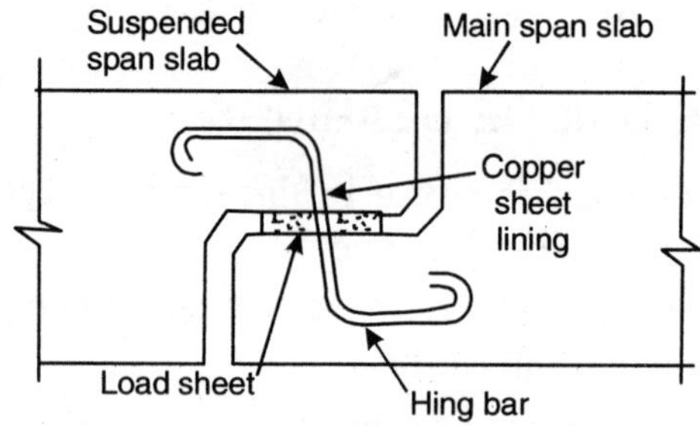

Figure 44.13 *Fixed end bearing for R.C.C. bridges*

44.4.2 Bearings for R.C.C. Girder Bridges

Figures 44.14 and 44.15 show various types of sliding and fixed type bearings used in R.C.C. girder bridges.

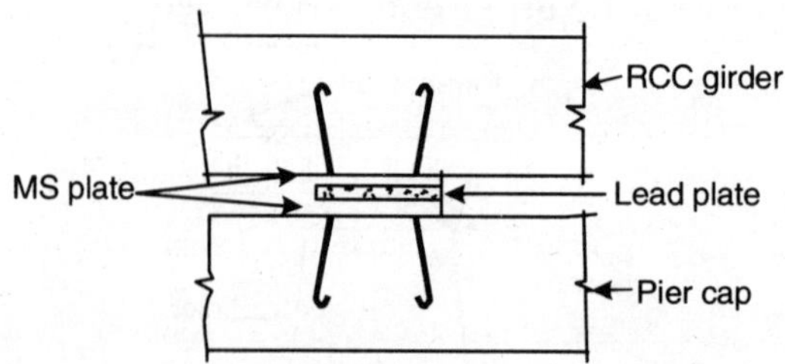

Figure 44.14 *Sliding bearing with lead plate*

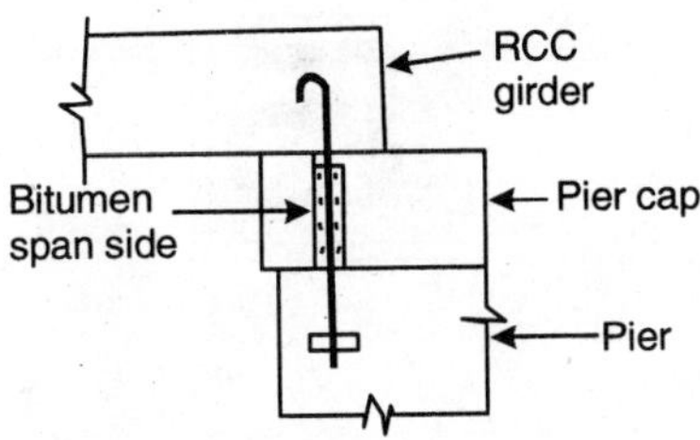

Figure 44.15 *Fixed type bearing for R.C.C. bridges*

In addition to the above various types of bearings used in steel girder bridges these are also used for R.C.C. girder bridges.

REVIEW QUESTIONS

44.1. What are bearings? What are their functions in bridges?

44.2. Write short notes on bearings.

44.3. Describe Knuckle-bearing with the help of a neat sketch.

44.4. Describe various types of end bearings commonly used in steel bridges.

44.5. Describe various types of bridge bearings with suitable sketches.

44.6. Write explanatory notes on expansion bearings for girders.

44.7. State precisely the purpose of providing bearings in bridges. Name the various types of bearings. What type of bearing will you provide for a span of 15.0 m? Give reasons.

44.8. What are the different types of bearings? Discuss their relative merits. Illustrate your answer with sketches.

44.9. Explain in brief the necessity of providing bearings in bridges.

44.10. Describe briefly the following :

(a) Roller bearing

(b) Sliding bearing

Illustrate the answer with neat sketches.

Protective-Works

GENERAL

In alluvial rivers, the water flows in large width and has the tendency to erode side banks and submerge side areas. While designing bridges across such rivers, it is very difficult to decide the position of river due to uncertainty about the direction of their flow. There is always a danger that it may erode its one bank and change its course and may start flowing outside the constructed work. At such places some special types of works are constructed and river is trained to flow inside them. These types of works are known as 'River Training Works'.*

45.1 OBJECTS OF RIVER TRAINING WORKS

The following are the objects of river-training works :

(i) To prevent out-flanking of bridges or other such types of construction works across the river and to train the river to flow in straight reach both upstream and downstream at the site of the bridge.

(ii) To deflect the river from a bank and stop its erosion.

(iii) To provide protection work along the banks of river, so that it may not damage and submerge cultivated and inhabited lands.

(iv) If the navigation is to be done in river, river-training works provide greater depth of water for this purpose.

(v) To provide the minimum width of river required at the site of the bridge, so that the bridge can be constructed economically.

* For detailed study about the River-Training Works, please see any book on 'Irrigation Engineering'.

45.2 TYPES OF RIVER TRAINING WORKS

The following are the main types of river-training works :

1. Marginal embankments or levels and guide banks
2. Spurs or groynes
3. Cut-offs
4. Pitching of banks and subsiding aprons
5. Pitched islands
6. Silts and closing Dykes

45.2.1 Marginal Embankments

When both the side banks of a river have less height then H.F.L., the flood water during monsoon overflows and submerges valuable lands. In such cases new banks known as *Marginal embankments* are constructed parallel to the river on both sides, so that flood water may not cross them and submerge the land on both sides. These banks are generally constructed

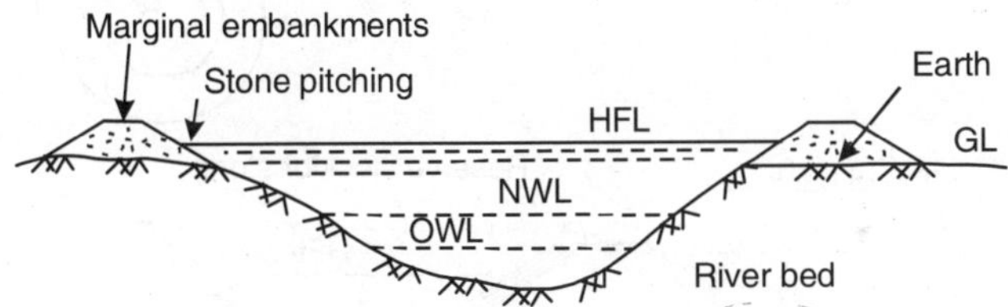

Figure 45.1 *Marginal embankment*

some distance away from the original banks so as to provide adequate waterway during the monsoon. In some cases these banks are joined by cross banks to the previous banks and during flood the space between them is silted up. The side slopes of these banks which are towards the river-water are pitched to bear the wave action of water.

45.2.2 Guide Banks

Figure 45.2 clearly shows the plan and sectional elevation of a Bell's Bund based on the design of Mr. J.R. Bell.

Bell's Bunds are constructed in pairs symmetrically in plan. Bell-mouth entry is provided in which water enters and flows in guide banks in a streamline flow. The upstream length of these banks is usually 1.0 to 1.1 times the length of the bridge. But in certain special cases the length may be more depending on the nature of the soil and flow of water. The length of these bunds towards downstream side depends on the disturbance caused by the water while leaving the bridge, which is usually 1/5th of the length of the bridge. The radius of upstream side bund is kept between 130 to 400 metres, which mainly depends on the scour depth and velocity of water. Downstream side curves have usually half the radius of that of upstream.

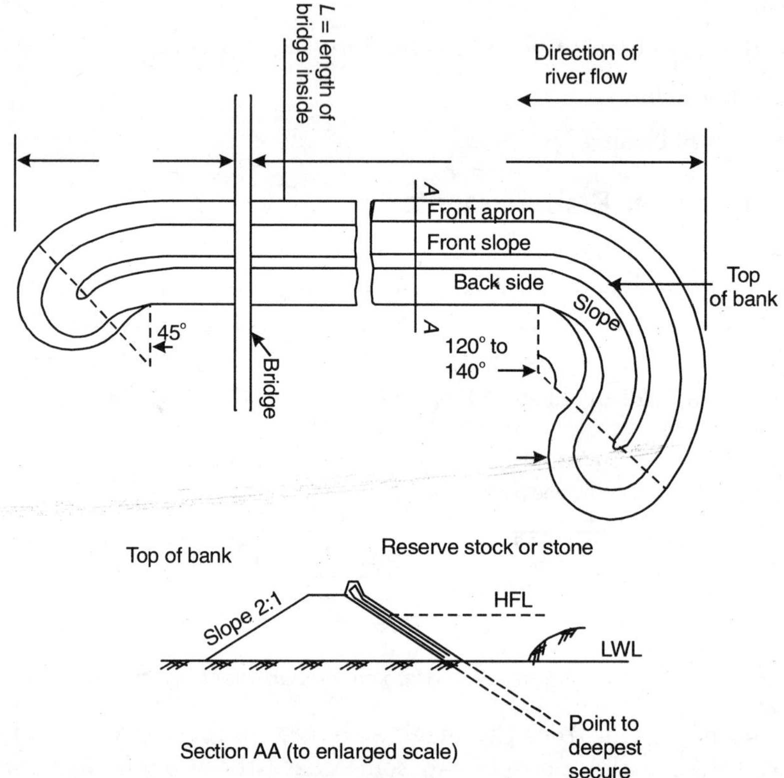

Figure 45.2 *Bell's bund*

The width at the top of the bund is kept between 3 to 10 metres with side sloping at 2 : 1 having at least 1.4 metres free board over the maximum H.F.L. While designing these bunds allowance for afflux settlement of bank and the nature of soil should also be kept in mind. If the banks are of sandy soil, sufficient pitching should be done on river side to protect it against erosion. An apron of 1.5 D, where length D represents the maximum scour depth, should be provided at the toe of guide banks.

45.2.3 Spur or Groynes

When the river starts erosion of bank and changing of its course, the spurs of groynes are constructed to deflect the current away from the bank. *Groynes* are structures built transverse to the river flow extending from the bank.

Due to a groyne, the river bank is projected upto some distance downstream and upstream. Therefore the bank can completely be protected if groynes are constructed in a series. From experience it has been seen that water has a tendency to flow at perpendicular

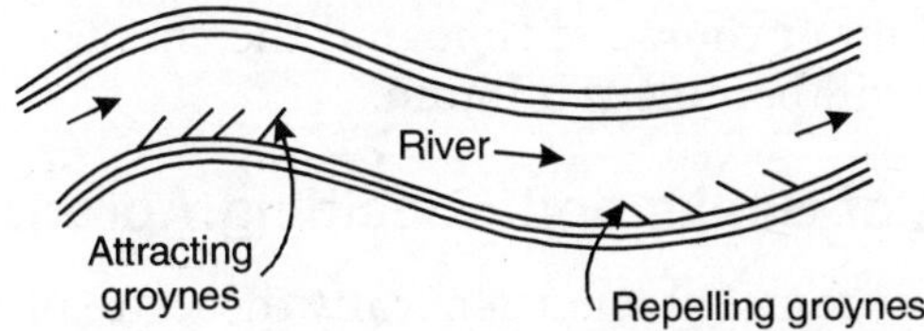

Figure 45.3 *Spurs of groynes*

to the groyne. Therefore if groynes point upstream they repel the current, whereas groynes pointing downstream attract the current. When groynes are provided in series the water between them has no current, therefore silting is done and the river flows only in required space.

If the water is allowed to flow through spurs or groynes, it is known as *Permeable Spurs*. These spurs are made on timber *ballis* driven in the river bed with brushwood filled in between them.

The spurs which do not allow the passage of water through them are known as *Solid groynes or spurs* and are constructed with earth, stones or concrete blocks.

45.2.4 Cut-offs

At some places the meander of a river develops a horse-shoe hair-pin bend in it as shown in Fig. 45.4. In such cases it is possible that during flood the

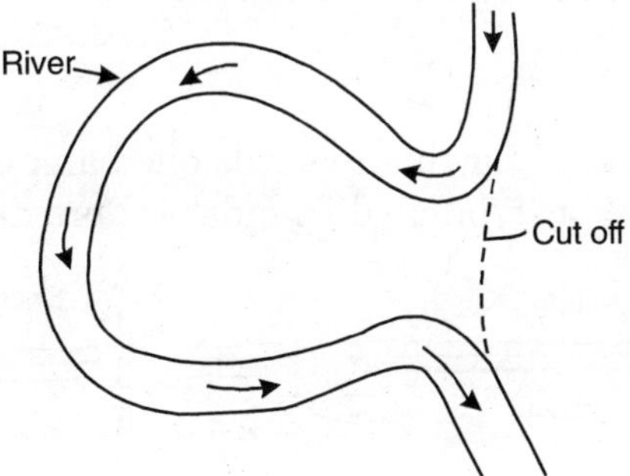

Figure 45.4 *Cut-off*

water may submerge the area between horse-shoe bend and start flowing in a straight line by cutting the neck of the curve. Now as the length of river between upstream and downstream of horse-shoe curve is reduced, the slope will be steeper and it will cause erosion in the upstream of the cut-off. Due to this, silting will be started in the downstream side of the cut-off, which will cause disturbance in river. This condition will continue till the river changes its course.

If we make an artificial cut along the cut-off position, the river water will start flowing in cut in a straight line and horse shoe curve portion of the river will be silted up in course of time and the chances of changing of the course of the river will become minimum.

45.2.5 Pitching of Banks and Subsiding Aprons

To protect the bank against erosion generally these are protected with stone, brush-wood, growing forest, brick, cement blocks or the natural growth of the grass. But before doing the protection work the banks are provided with uniform slope from 1 : 1 to 2 : 1 depending on the material of the bank, in the required length of the river. If stone and bricks are used for the protection of banks, sometimes G.I. wires are used to connect them together.

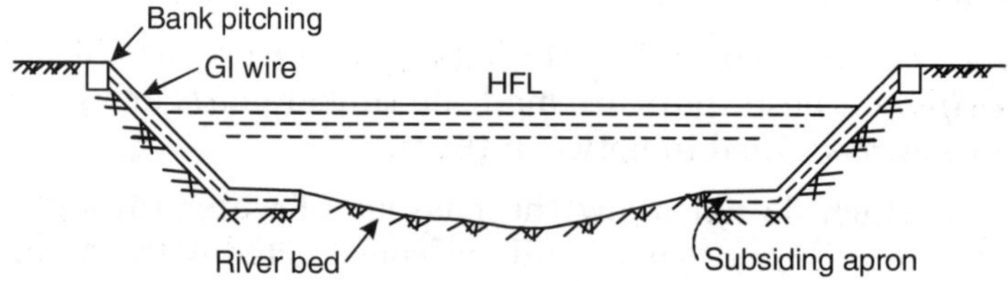

Figure 45.5 *Pitching of banks and subsiding apron*

It has been seen that well protected banks may be destroyed during floods if their toe is not protected from cutting. Therefore launching apron or subsiding apron of loose stones is done at the toe of the river bank upto required length in the river-bed as shown in Fig. 45.5.

45.2.6 Pitched Islands

When the river current is flowing towards one bank only and it is required to change the position of it 'pitched islands' are provided. Pitched island is

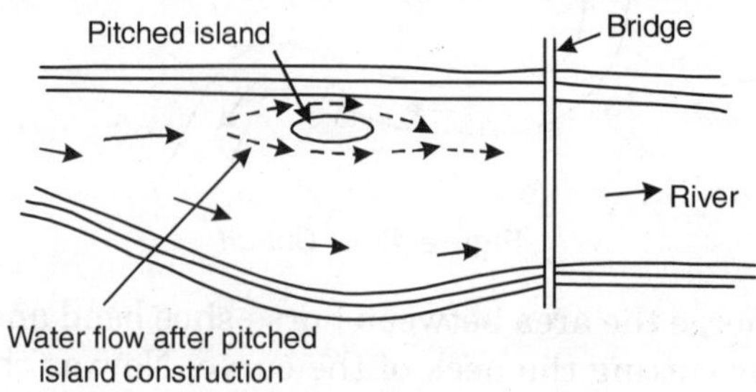

Figure 45.6 *Pitched Island*

an artificial island constructed in the river bed. Mostly stone pitching is done all round the pitched island. This island becomes an obstruction in the flow of the water and creates turbulence in it, due to which scouring starts around the island and river becomes deeper and deeper. When the river becomes deeper it attracts the current of water from the other side of the bank. Thus the concentration of the current at the other side bank will be reduced.

45.2.7 Closing Dykes

At some places where river flows in the main as well as in a subsidiary channel and has a tendency to change its course in subsidiary channel, obstructions are placed in it to check the tendency of the river. Generally, closing dykes are constructed in the subsidiary channel which prevent the tendency of the river of changing its course. These closing dykes may be permeable structures of brushwood or branches or a solid earthen dam. The solid type of a closing dyke completely stops the flow of water in the subsidiary channel, whereas permeable dykes considerably reduce the velocity of water and in course of time the subsidiary channel is silted up completely.

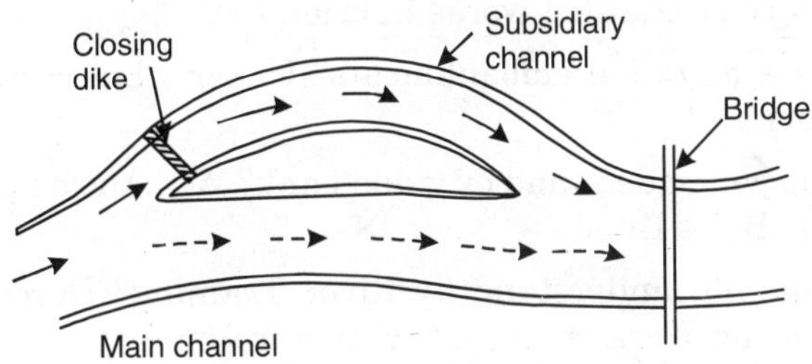

Figure 45.7 *Closing Dyke*

45.3 WIRE CRATES AND MATTRESSES FOR RIVER TRAINING WORKS

Wire crates of size 3 × 1.5 × 1.2 m are used in situations which are accessible or shallow. The crates are divided into compartments, where there are chances of their overturning. Small wire crates are used in difficult or inaccessible situations. If the wire mattresses are to be built in the situation, its size should not be more than 7.5 × 3.0 × 0.6 m or it should be smaller than 1.80 × 0.90 × 0.3 m. For preventing bulging the sides of large mattresses should be securely stayed at about 1.5 m intervals.

The size of the netting wire and its opening depends on the size of the boulders to be used. 15 cm mesh opening is used for 20 cm size boulders. The thickness of the G.I. wire mesh will be 5 mm diameter. 32 kg of wire is required for 10 sq. m of the mesh area. The best size of the boulder to be used for river training works is 30 cm having weight more than 35 kg.

45.4 STONE RIPRAP FOR PROTECTING FOUNDATION

For the protection of abutments and the piers, heaviest size of stone which can be handled graded from coarse to fine should be used. All these graded stones should be packed in such a way so as to have minimum voids. The foundations of the bridges subjected to ocean waves or heavy scour, should be protected with heavy stone, having 1000 to 6000 kg weight.

REVIEW QUESTIONS

45.1. What are protective works? What are their objects? Describe different types of river-training works in brief.

45.2. What are marginal embankments? Under what circumstances these are provided?

45.3. What do you understand by guide-bank? With the help of a neat sketch describe Bell's Bund.

45.4. What do you understand by River Training? Describe the various Engineering Works required for this training.

45.5. Write short notes on the following :

(i) Spurs

(ii) Cut-off

(iii) Guide bund

(iv) Marginal embankments

45.6. What measures are generally adopted for training a river near an important highway-bridge?

45.7. What do you understand by River Training? Describe the various engineering works required for this purpose.

45.8. Write short notes on the following :

(i) Guide Bunds

(ii) Spurs

45.9. What is a guide bank? Where is it generally used? Draw neat sketches showing plan and cross-section of a guide bank and indicate on them the dimensions of various portions.

46

Loads and Stresses

GENERAL

After site selection the bridge-designers are mainly concerned with two types of designs of bridges. First design deals with the *function* or purpose of the bridge. In this the designer have to fix the functional sizes of bridge such as the width of the bridge, soffit height etc. depending whether it is single lane, two lane, lour lane road bridge or N.G., M.G. or B.G. railway bridge.

The second part of design deals with the structural design of the bridge. Every successful designer has to co-ordinate with both the above parts economically. Bridge should be designed in such a way that it should serve the present requirements as well as future requirements upto certain limits. While designing, provision should also be made for carrying public service lines such as water, oil, sewage, electric cables, gas etc.

Without knowing the loads coming on the bridges and the stresses which different materials can take, it is not possible to design the bridge. This chapter deals with the loads and stresses.

46.1 TYPES OF LOADS

The following are the loads which are taken into account while designing the bridges:

1. Dead load
2. Live load

3. Impact load
4. Wind load
5. Horizontal forces due to current of water
6. Centrifugal force
7. Longitudinal force due to braking of vehicles or tractive effort at the time of starting
8. Earth pressure
9. Secondary stresses
10. Erection stresses
11. Seismic force
12. Temperature stresses
13. Relief stresses

46.2 DEAD LOADS

It includes the self-weight of the bridge and the load of all other super-imposed or permanently fixed structures such as water line, oil line, gas line, cable etc. At the time of starting design the designer does not know the deal load of the bridge structure accurately. To start with it can be estimated by the dead load of the similar other existing structures or by empirical formulae based on past experience. After completing the design these are checked and revision is done if necessary. The unit weight of common materials are given in I.S.I. Code 1911–1961.

46.3 LIVE LOADS

The live loads on bridges in most of the cases are indefinite and are liable to charge with time. The main live loads on bridges are due to road traffic and railway trains. Practically it is not possible to determine the actual wheel loads of trains, trucks, cars etc. passing over the bridges.

For designing road bridges, the standard live loads are taken as specified by the Indian Road Congress (I.R.C.) Section II. For designing railway bridges, the standard live loads are taken as specified by the Railway Board, Ministry of Railways, Government of India.

46.4 INDIAN ROAD CONGRESS LOADING

I.R.C. has classified the live loads on road bridges in the following categories:

1. I.R.C. class 'AA' loading
2. I.R.C. class 'A' loading
3. I.R.C. class 'B' loading
4. I.R.C. class 70 R loading.

46.4.1 I.R.C. Class AA Loading

This type of loading is adopted for the bridges on National Highways, State Highways and other specified areas, such as industrial areas. The bridge designed for I.R.C. class AA loading should also be checked for I.R.C. class A loading also, because in certain cases I.R.C. class A loading gives greater Bending Moment and stress than I.R.C. class AA loading.

Figure 46.1 shows the I.R.C. class AA loading for track vehicles (tanks and chained tractors). For multiple-lane bridges and culverts, one train of I.R.C

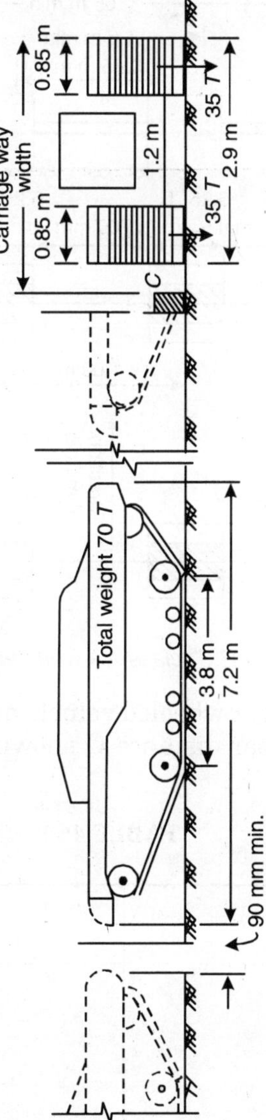

Figure 46.1 *I.R.C. Class AA Track Vehicles Loading*

class AA tracked or wheeled vehicles shall be considered for every two-traffic lane width. While taking into account this load, an other live loads should be taken into account while designing the bridges. Figure 46.2 shows I.R.C. class AA wheeled vehicle loading.

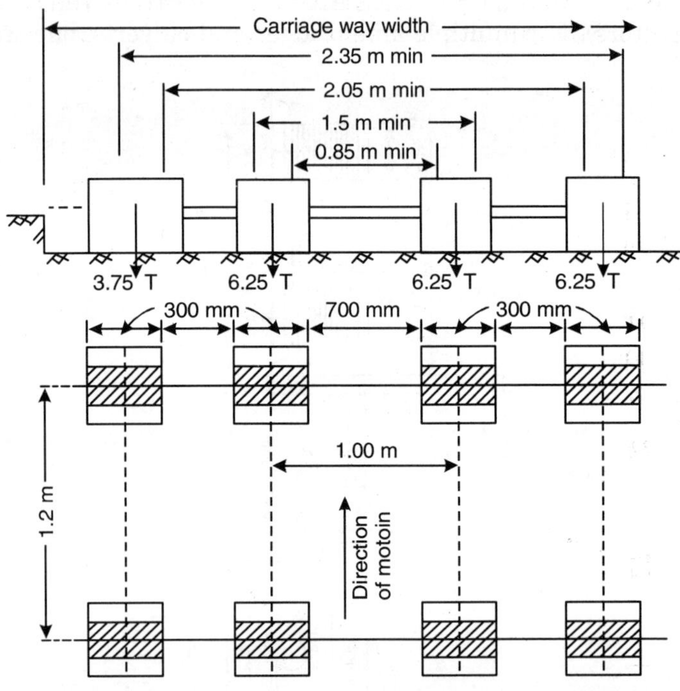

Figure 46.2 *I.R.C. class AA wheeled vehicle loading*

I.R.C. class AA tracked or wheeled vehicle may approach near to the road kerb upto a minimum clear distance C shown in Table 46.1, as per I.R.C. section II.

TABLE 46.1

Width of carriageway	*Minimum clear distance between vehicle and road kerb*
Single load bridge	
3.8 m and above	0 3 m
Multilane bridges	
Less than 5.5 m	0.6 m
5.5 m or above	1.2 m

TABLE 46.2

Axle load C tonnes	Ground contact area	
	B mm	W mm
11.4	250	500
6.8	200	380
2.7	150	200

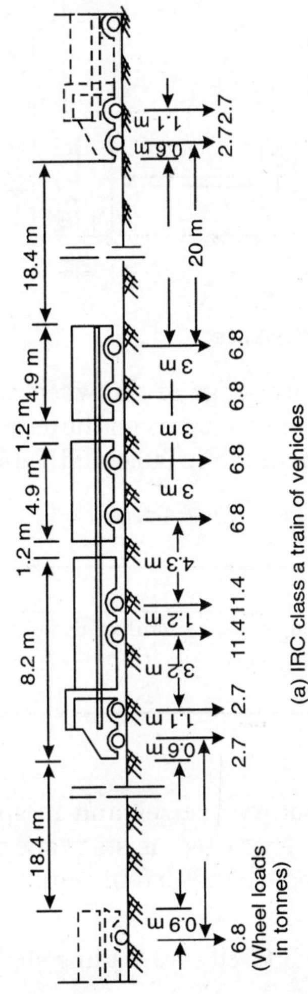

(a) IRC class a train of vehicles

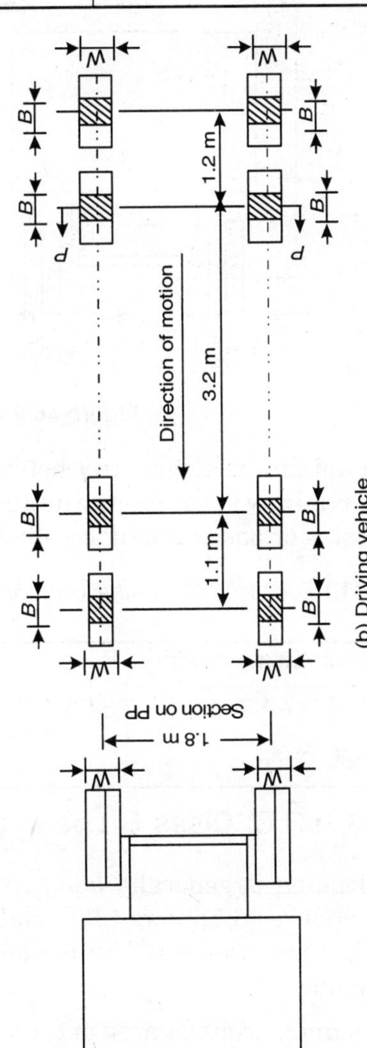

(b) Driving vehicle

Figure 46.3 *I.R.C. class A loading*

46.4.2 I.R.C Class A Loading

For the design of permanent bridges this class of loading is generally adopted. This type of loading consists of one driving unit and two trailers as shown in Fig. 46.3.

The ground contact area of wheels of I.R.C. class A loading shall be taken as per Table 46.2 as per I.R.C. section II.

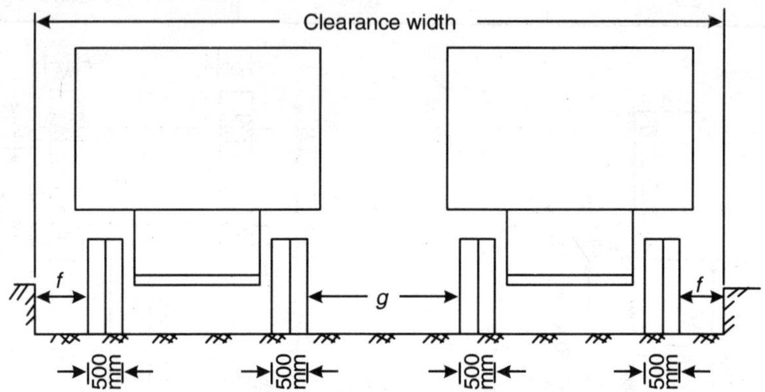

Figure 46.4 *Minimum clearance*

The minimum clearance *f* between the outer edge of the wheel and the face of roadway kerb, and the minimum clearance *g*, between the outer edges of crossing or passing vehicles shall be as shown in Fig. 46.4 and Table 46.3.

TABLE 46.3 *Minimum clearance f and g for I.R.C. class A vehicle*

Clear carriageway width	*f*	*g*
5.5 m to 7.3 m	150 mm for all carriage-way width	Uniformly increasing from 0.4 to 1.2 m
Above 7.5 m	–do–	1.2 m

46.4.3 I.R.C. Class B Loading

This loading is generally adopted for temporary bridges and in specified areas. Figure 46.5 shows I.R.C. class B loading which consists of one driving unit and two trailers. The trailers attached to the driving unit are not detachable.

The ground contact area of the wheels of I.R.C. class B loading shall be as per Table 46.4 as per I.R.C. section II.

TABLE 46.4 *Ground contact area of the wheels of I.R.C. class B vehicle*

Axle load (tonnes)	*Ground contact area*	
	P (mm)	*W (mm)*
6.8	200	380
4.1	150	300
1.6	125	175

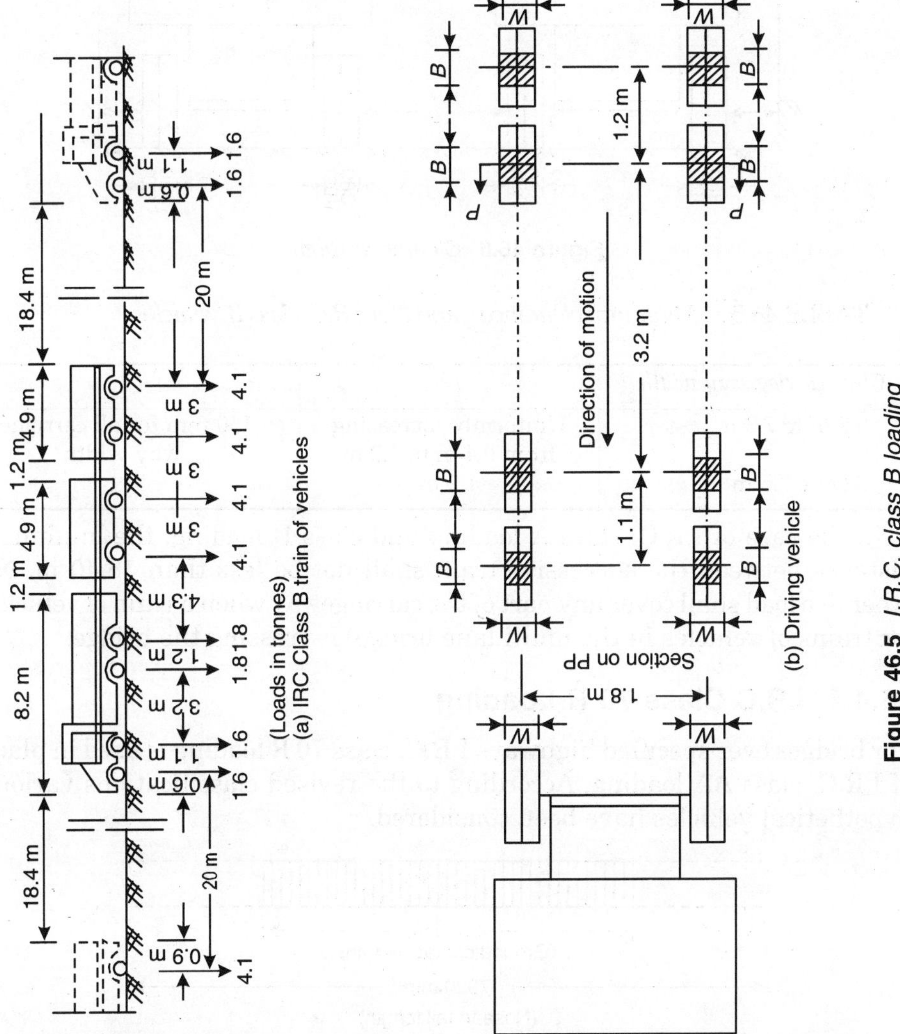

Figure 46.5 *I.R.C. class B loading*

The minimum clearance *f*, between outer edge of the wheel and the face of road kerb and the minimum clearance *g*, between the outer edges of passing or crossing vehicles shall be as per I.R.C. section II, shown in Fig. 46.6 and Table 46.5 on multi lane bridges.

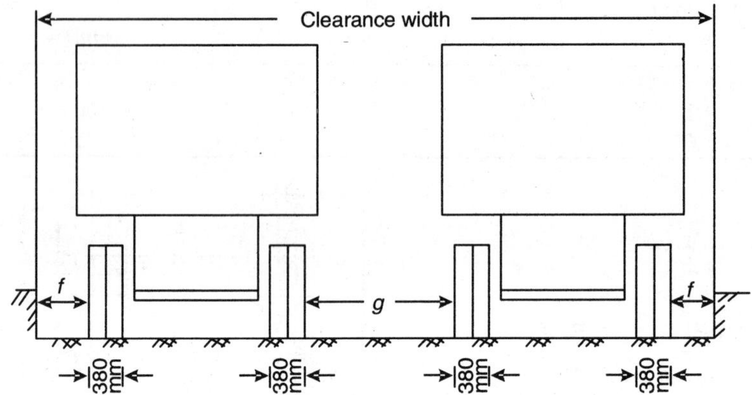

Figure 46.6 *Clearance width*

TABLE 46.5 *Minimum clearance g and f for I.R.C. class B vehicles*

Clear carriageway width	*g*	*f*
5.5 m to 7.3 m	Uniformly increasing from 0.4 m to 1.2 m	150 mm for all carriage-way width
Above 7.5 m	1.2 m	

In the case of I.R.C. class A loading and class B loading, the minimum distance between the successive train shall not be less than 18.40 m. No other live load shall cover any part of the carriageway when a train of vehicles (or trains of vehicles in the multi lane bridge) is crossing the bridge.

46.4.4 I.R.C Class 70 R Loading

For bridges over specified highways I.R.C. class 70 R loading is used in place of I.R.C. class AA loading. According to the revised classifications various hypothetical vehicles have been considered.

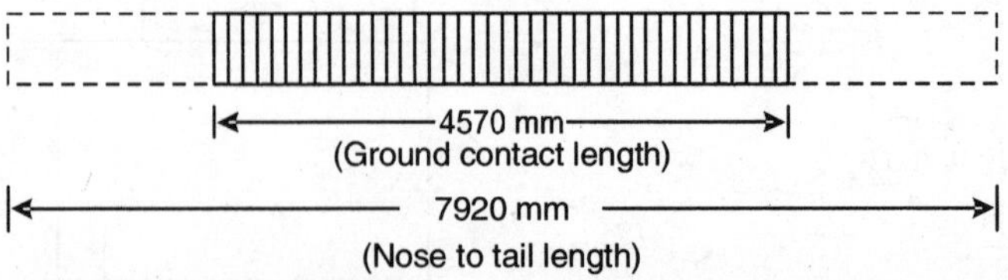

Figure 46.7 *Class 70 R tracked vehicle*

The letter '*R*' indicates revised classification. This loading is one of the various other hypothetical vehicles according to the revised classification. Class 70*R* loading consists of tracked vehicle as shown in Fig. 46.7. The width of one track is 0.840 m. Similar to I.R.C. class AA loading the width over track is 2.90 m in case of class 70 R loading.

Figure 46.8 (a) shows the class 70 R loading which consists of wheeled vehicle. Fig. 46.8 (b) shows the single axle load and maximum bogie axle load as per I.R.C. section II.

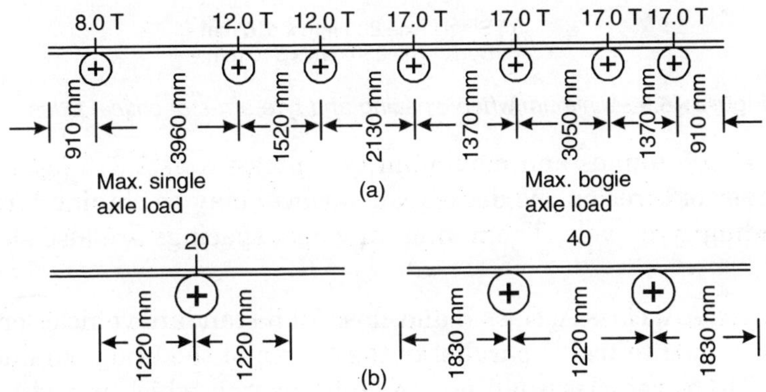

Figure 46.8 *Class 70 R wheeled vehicle*

Figure 46.9 shows the overall width of tyre and rim diameter of the tyre for both single axle and bogie axle loads. The tyre tread width may be taken as overall width minus 50 mm. The maximum tyre load on minimum tyre size is 8.00 tonnes. The actual maximum tyre load on 410 mm × 610 mm

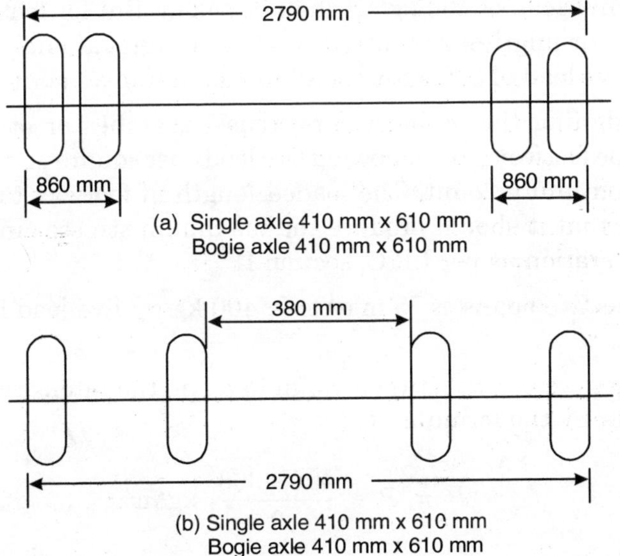

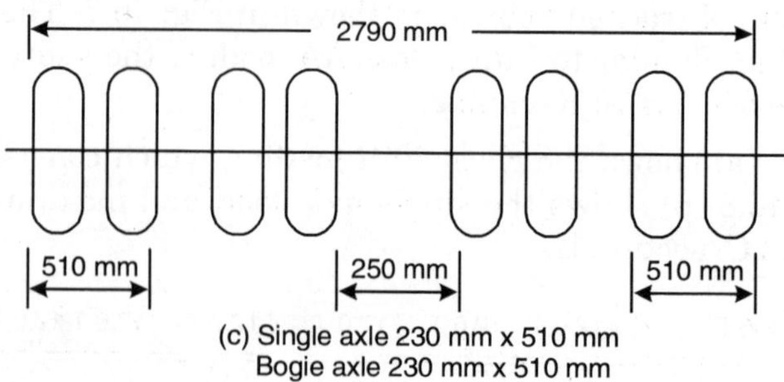

(c) Single axle 230 mm x 510 mm
Bogie axle 230 mm x 510 mm

Figure 46.9 *Minimum wheel spacing and tyre sizes of critical axles*

tyre size is 5.00 tonnes and maximum tyre pressure is 5.273 kg/cm^2. The contact areas of tyres on the deck (road surface) may be obtained from the corresponding tyre loads. The minimum wheel spacings are also shown in Fig. 46.9.

I.R.C. class AA, class A, class B and class 70 R standard vehicles or trains shall be assumed to travel parallel to the length of the bridge in middle of roadway. These vehicles shall occupy any position which would produce maximum stresses provided that the minimum clearances between a vehicle and the roadway face of kerb and beiween two passing or crossing vehicles are not encroached upon.

46.5 FOOTWAYS (ATTACHED TO THE HIGHWAY BRIDGES)

For the design of footways or bridges accessible only to pedestrians and animals the live load of 400 kg/m^2 shall be taken. But for foot bridges near crowed towns or pilgrimage centres or where congregational fairs are held seasonally, live load of 500 kg/m^2 shall be taken into account.

While designing the main girders, trusses, arches or other members supporting the footways, the following live loads per square metre of footways shall be taken into account. The loaded length of footway taken in every case be such that it should produce the maximum stresses in the member under consideration as per I.R.C. section II:

(a) For effective spans of 75 m or less, 400 kg/m^2 live load is to be taken into account.

(b) For effective spans between 7.5 m to 30 m, the intensity of load shall be determined by the formula

$$P = P' - \frac{40L - 300}{9} \text{ kg/m}^2.$$

(c) For effective spans over 30 m, the intensity of load shall be determined by the formula

$$P = P' - 260 + \frac{4800}{L} \quad \frac{16.5 - W}{15} \text{ kg/m}$$

where

P' 400 kg/m^2 or 500 kg/m^2 as mentioned above

P live load in kg/m^2

L effective span of main girder in m

W width of the footway in metres

Every part of the footway shall be capable of carrying a wheel load of 4 T, which shall include impact, distributed over a contract area of 300 mm in diameter, the working stresses shall be increased by 25% to meet this provision.

Side kerbs of 60 cm width or more shall be designed for the loads mentioned above in addition to 750 kg/m run horizontal thrust acting at the top of the kerb. If the width of the kerb is less than 60 cm no side thrust loads are to be taken into account.

46.6 LIVE LOADS ON RAILWAY BRIDGES

The Railway Board, Ministry of Railways, Government of India has specified the standard design loading in Bridge Rules for the broad gauge, metre gauge and nanow gauge tracks separately. These standard design loadings consist of number of wheel loads of engine followed by a train of uniformly distributed load. In order to simplify the analysis of bridge girder, the actual standard loadings have been expressed in Bridge Rules as an *equivalent uniformly distributed live loads* (EUDLL). These equivalent uniformly distributed loads are uniformly distributed loads which give the results equivalent to the wheel loads.

The following are the various standard design loadings and their corresponding equivalent uniformly distributed loads for various tracks of the railways.

For B.G 1676 mm. Figure 46.10 (a) shows the standard line loading, which is adopted for all spans of the bridges on the main lines.

Figure 46.10 (b) shows the standard branch line loading which is adopted for all spans of the bridges on branch lines of railways. If the branch line is situated in industrial area or heavy mineral area, then the standard main line loading should be taken into account while designing the bridges on such lines.

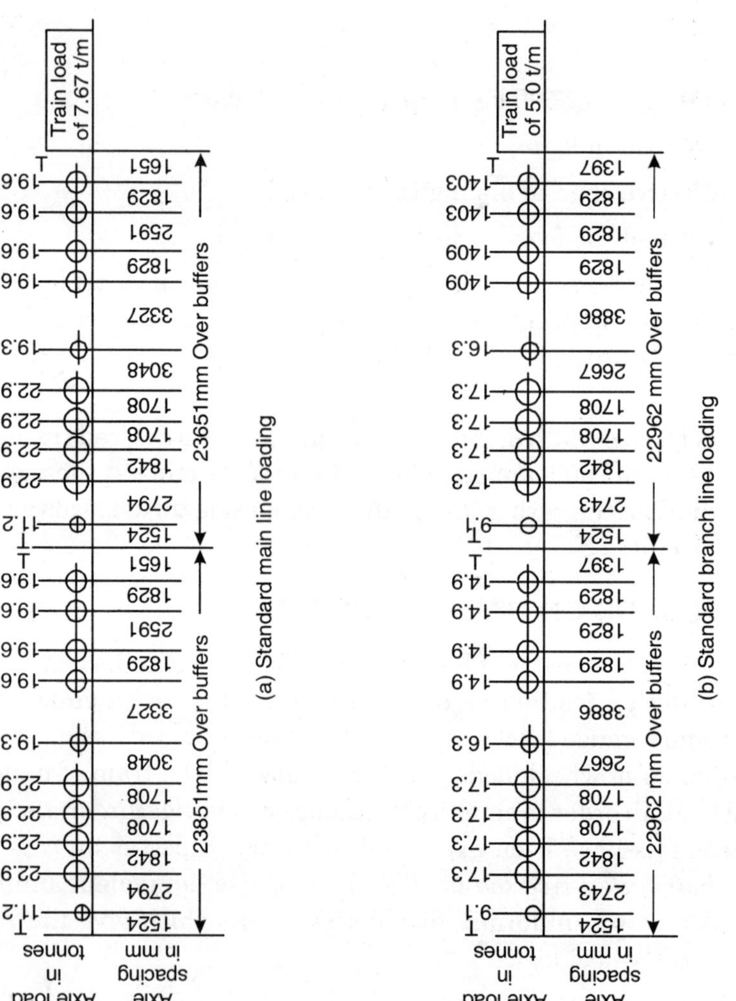

(a) Standard main line loading

(b) Standard branch line loading

Figure 46.10 *B.G. standard loading of 1929*

The equivalent uniformly distributed loads for the standard main line (M.L.) and standard branch line (B.L.) for the purpose of determining B.M. and S.F. are given in Table 46.6 as per Bridge Rules.

TABLE 46.6 *Equivalent uniformly distributed live loads (EUDLL) on each B.G. track and impact factors*

L (metres)	Total load for BM (tonnes) M.L.	B.L.	Total load for SF (tonnes) M.L.	B.L.	Impact factor $\frac{20}{14+L}$
1.0	45.8	34.6	45.8	34.6	1.000
1.5	55.8	34.6	45.8	34.6	1.000
2.0	45.8	34.6	52.4	39.6	1.000
2.5	45.8	34.6	60.4	45.5	1.000
3.0	46.9	35.4	65.5	49.5	1.000
3.5	52.4	39.6	70.3	53.2	1.000
4.0	59.2	44.8	78.8	59.4	1.000
4.5	67.9	51.4	85.2	64.4	1.000
5.0	74.8	56.7	90.3	68.3	1.000
5.5	80.6	60.9	96.7	73.2	1.000
6.0	85.2	64.4	104.9	78.6	1.000
6.5	89.3	67.5	110.2	83.2	0.976
7.0	95.2	71.5	115.2	87.1	0.952
7.5	100.7	75.8	119.8	90.6	0.931
8.0	105.6	79.7	123.9	93.5	0.909
8.5	110.2	82.6	128.6	97.2	0.889
9.0	114.0	86.2	132.4	101.4	0.870
9.5	117.6	90.8	136.7	105.0	0.851
10	121.0	94.1	140.6	108.3	0.833
11	133.6	102.5	148.3	114.0	0.800
12	140.9	108.4	155.7	119.1	0.769
13	147.2	113.4	163.7	125.4	0.741
14	152.7	117.2	172.0	131.6	0.741
15	160.6	123.2	180.6	138.2	0.691
16	168.8	130.2	188.1	144.8	0.667
17	177.0	137.1	196.8	151.6	0.645
18	185.9	143.8	205.0	159.0	0.625
19	193.9	150.8	214.1	166.0	0.606
20	202.7	158.0	222.4	172.6	0.588
21	211.1	164.9	230.5	179.1	0.571
22	218 .7	169.8	238.7	185.5	0.556
23	225.6	175.8	246.6	191.6	0.541
24	232.9	181.8	254.8	197.7	0.526

Contd.

Table 46.6 Contd.

L (metres)	Total load for BM (tonnes)		Total load for SF (tonnes)		Impact factor
	M.L	B.L	M.L	B.L	$\frac{20}{14 + L}$
25	241.0	188.0	262.7	203.7	0.513
26	249.5	193.5	270.7	209.7	0.500
27	256.0	198.8	278.9	215.8	0.488
28	264.0	205.2	286.6	223.0	0.476
29	271.0	211.8	294.4	230.0	0.465
30	280.0	216.5	302.3	237.5	0.455
32	294.9	230.0	320.0	251.8	0.435
34	309.5	243.5	337.5	265.2	0.417
36	327.0	257.7	354.2	278.0	0.400
38	342.2	270.0	371.2	291.2	0.385
40	359.0	282.0	387.7	304.8	0.370
42	375.0	295.0	404.6	318.6	0.357
44	391.0	308.5	421.8	331.6	0.345
46	408.0	322.2	438.4	344.8	0.333
48	424.0	334.5	454.9	357.6	0.323
50	438.0	347.0	471.3	370.2	0.313
55	477.5	376.5	512.2	401.0	0.290
60	514.8	405.2	552.8	431.0	0.270
65	544.0	433.5	591.6	460.0	0.253
70	591.8	461.0	632.2	488.2	0.238
75	628.0	486.5	672.0	515.4	0.225
80	667.0	513.0	710.9	543.8	0.213
85	703.5	539.0	750.4	570.8	0.202
90	742.0	568.0	789.9	597.8	0.192
95	780.0	591.6	827.8	624.6	0.183
100	820.0	616.0	868.6	651.2	0.175
105	858.0	642.5	906.2	677.2	0.168
110	897.0	668.5	945.6	704 .2	0.161
115	935.0	594.5	984.8	730.2	0.155
120	973.0	719.5	1025.3	756.4	0.149
125	1010.0	745.0	1072.4	782.4	0.144
130	1048.5	770.0	1113 .3	880.4	0.139

Note : The intermediate values may be found by linear interpolation.

For metre gauge 1000 mm. Figure 46.11 (a) shows the standard main line loading, which is adopted for all spans of the bridges on the main lines.

Figure 46.11 (b) shows the standard branch line loading, which is adopted for all spans of the bridges on branch lines.

Figure 46.11 (c) shows the standard C loading of 1929.

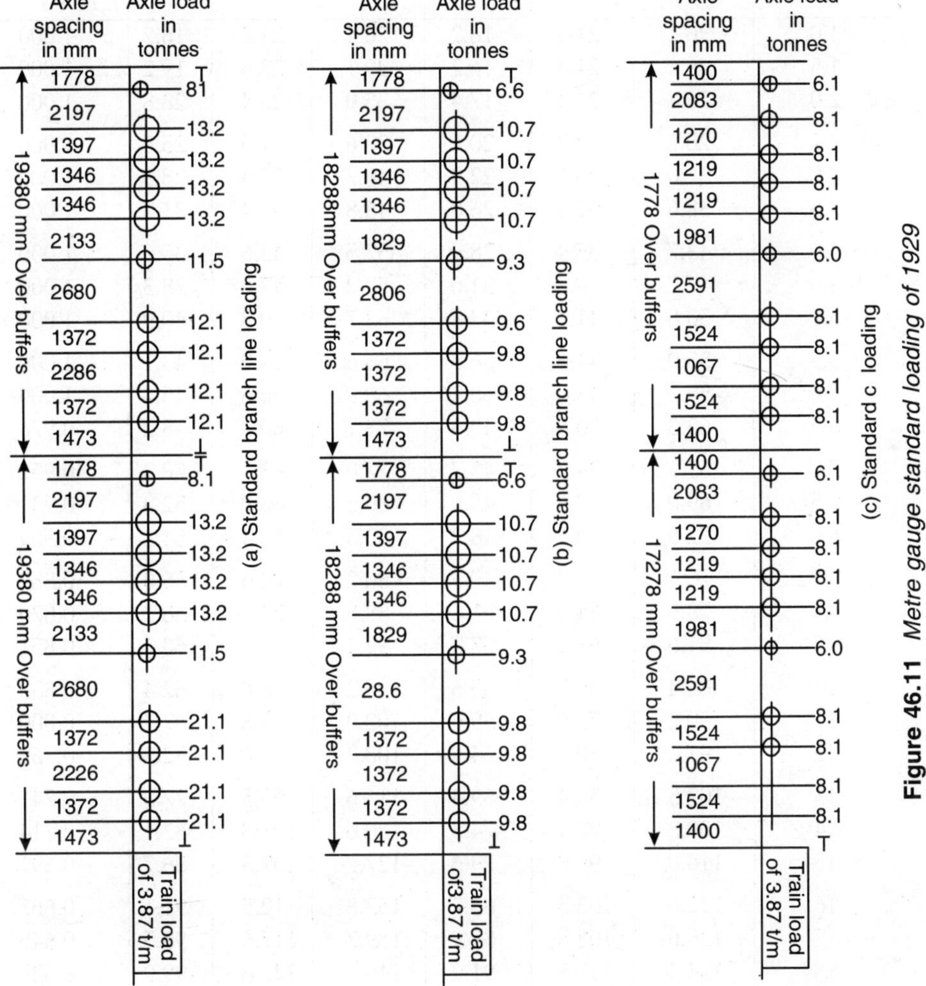

Figure 46.11 *Metre gauge standard loading of 1929*

The equivalent uniformly distributed loads for the standard main line, standard branch line and the standard C loadings are given in Table 46.7 as per Bridge Rules.

TABLE 46.7 *Equivalent uniformly distributed live load (EUDLL) in tonnes on each track and impact factor for metre gauge bridges*

L (metres)	Total load for BM (tonnes)			Total load for SF (tonnes)			Impact factor $\frac{20}{14 + L}$
	M.L	B.L	C	M.L	B.L	C	
1.0	26.4	21.4	16.2	26.4	21.4	16.2	1.000
1.5	26.4	21.4	16.2	29.1	23.6	19.2	1.000
2.0	26.4	21.4	17.4	35.0	28.4	23.8	1.000
2.5	28.2	22.9	20.1	38.6	31.3	25.5	1.000
3.0	32.0	25.7	22.3	43.7	35.4	28.9	1.000
3.5	38.6	31.3	26.0	48.8	39.4	31.7	1.000
4.0	43.7	35.4	28.9	52.5	43.6	35.0	1.000
4.5	47.7	38.6	31.0	57.2	47.0	38.3	1.000
5.0	50.8	41.1	33.8	62.7	50.8	40.9	1.000
5.5	55.0	44.6	36.5	66.6	54.0	43.1	1.000
6.0	59.3	48.0	38.9	69.9	56.6	45.4	1.000
6.5	62.3	50.8	40.9	73.1	60.1	48.2	0.976
7.0	65.7	54.2	43.7	77.0	63.3	50.3	0.952
7.5	69.0	57.6	45.9	80.6	66.0	52.3	0.931
8.0	72.8	60.5	48.2	83.5	68.4	54.1	0.909
8.5	76.1	63.0	49.9	86.2	70.6	56.0	0.889
9.0	78.9	65.3	51.6	88.8	73.0	58.2	0.870
9.5	81.6	67.4	53.7	92.2	75.6	60.2	0.851
10	84.4	69.7	55.6	95.2	78.0	62.4	0.833
11	90.2	73.9	59.9	102.0	83.8	67.1	0.800
12	97.2	80.0	64.9	108.2	89.6	72.2	0.769
13	103.5	86.4	69.4	114.6	95.5	77.3	0.741
14	109.8	92.1	74.8	120.6	101.8	82.1	0.714
15	116.0	98.0	79.6	127.7	107.3	86.7	0.691
16	122.6	103.5	83.7	133.8	112.5	91.1	0.667
17	128.6	108.5	87.9	139.7	117.6	95.5	0.645
18	134.2	113.3	91.9	145.4	122.6	99.9	0.625
19	139.5	118.2	95.2	150.9	127.5	104.1	0.606
20	144.9	122.6	99.8	156.2	132.3	108.4	0.588
21	149.5	126.8	104.1	161.4	136.8	112.6	0.571
22	154.8	131.4	108.4	167.0	141.5	117.1	0.566

Contd.

Table 46.7 Conted.

L (metres)	Total load for BM (tonnes) M.L	B.L	C	Total load for SF (tonnes) M.L	B.L	C	Impact factor $\frac{20}{14+L}$
23	160.6	136.4	112.7	173.4	147.2	121.7	0.541
24	165.8	140.8	116.8	180.2	152.8	126.5	0.526
25	171.0	145.5	121.3	186.9	158.6	130.9	0.513
26	177.8	151.2	125.5	193.4	164.0	135.2	0.500
27	184.8	156.8	130.1	199.9	169.0	139.6	0.488
28	190.6	162.0	134.0	206.0	174.0	143.8	0.476
29	196.7	166.9	138.4	212.2	179.8	149.8	0.465
30	203.0	171.9	142.6	218.6	185.3	153.3	0.455
32	214.5	181.7	152.0	231.2	196.8	163.1	0.435
34	226.5	192.9	161.3	244.2	207.5	172.1	0.417
36	239.0	204.0	169.9	256.8	218.1	181.0	0.400
38	251.5	214.4	178.4	268.7	228.4	189.8	0.385
40	264.0	224.2	186.7	280.2	238.4	198.5	0.370
42	274.5	233.5	195.0	291.4	248.1	207.0	0.357
44	285.5	243.0	203.4	302.2	257.1	215.6	0.345
46	296.0	252.0	211.6	313.0	267.1	224.0	0.333
48	306.0	261.2	219.7	323.5	276.4	232.4	0.323
50	316.0	270.3	227.6	333.4	288.6	240.7	0.313
55	340.0	292.1	246.9	357.9	308.0	261.3	0.290
60	363.0	312.9	266.7	381.8	329.9	281.6	0.270
65	385.0	333.2	286.7	404.7	351.4	302.0	0.253
70	405.0	354.1	306.9	427.4	372.7	322.0	0.238
75	426.5	373.7	326.9	450.1	393.6	342.2	0.225
80	448.0	393.0	346.4	471.6	414.3	361.9	0.213
85	468.5	413.9	365.3	492.7	435.0	381.6	0.202
90	488.1	433.6	385.5	514.2	455.5	401.1	0.192
95	508.0	452.8	404.7	543.3	475.8	421.4	0.183
100	528.5	472.9	423.2	555.2	496.1	441.4	0.175
105	548.0	491.0	442.7	576.5	516.2	462.2	0.168
110	567.5	511.6	461.4	597.4	536.3	480.6	0.161
115	587.0	530.5	480.0	617.7	556.3	500.2	0.155
120	607.0	549.1	500.5	638.1	576.3	520.0	0.149
125	627.0	568.1	518.0	658.3	596.2	539.6	0.144
130	646.0	589.3	538.9	678.8	616.1	559.0	0.139

Note : The intermediate values may be found by linear interpolation.

For narrow gauge 762 mm. In case of narrow gauge (762 mm) following three types of standard loadings have been specified by the Railway Board :

(a) H (heavy) class loading.

(b) A class loading.

(c) B class loading.

Figures 46.12, 46.13 and 46.14 respectively show the H class, A class and B class standard loadings respectively.

The equivalent uniformly distributed live loads are given in Table 46.8 as per Bridge Rules.

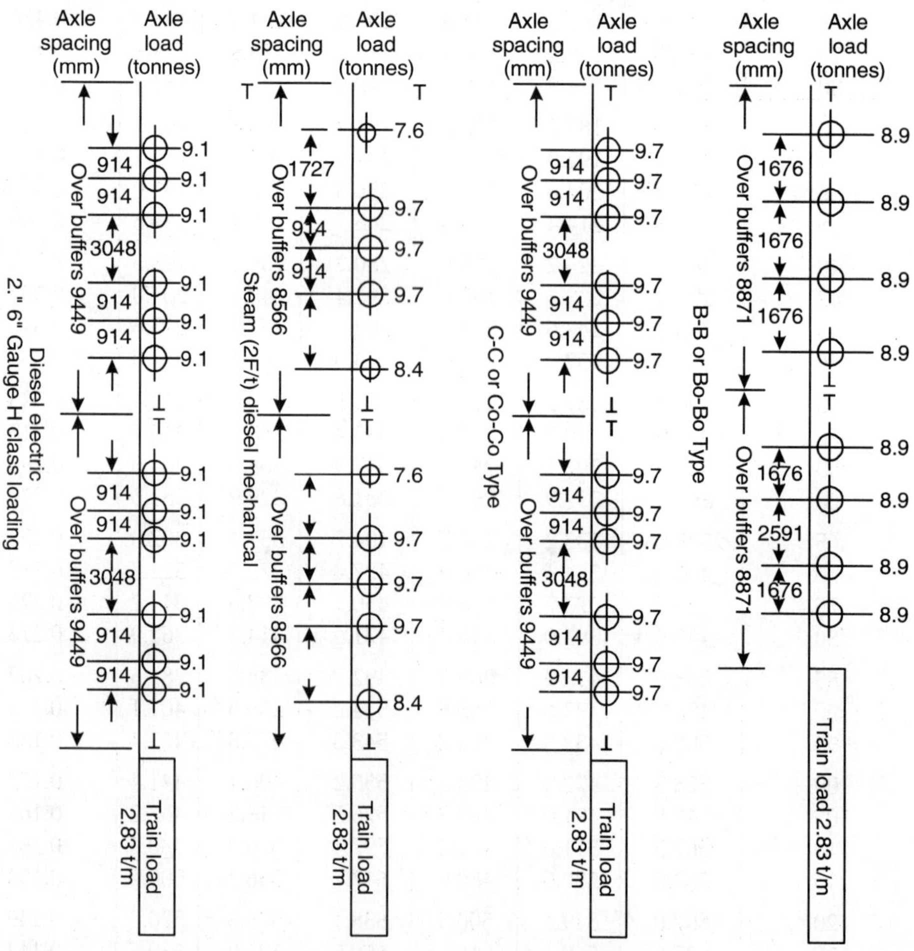

Figure 46.12 *H-class loading for N.G.*

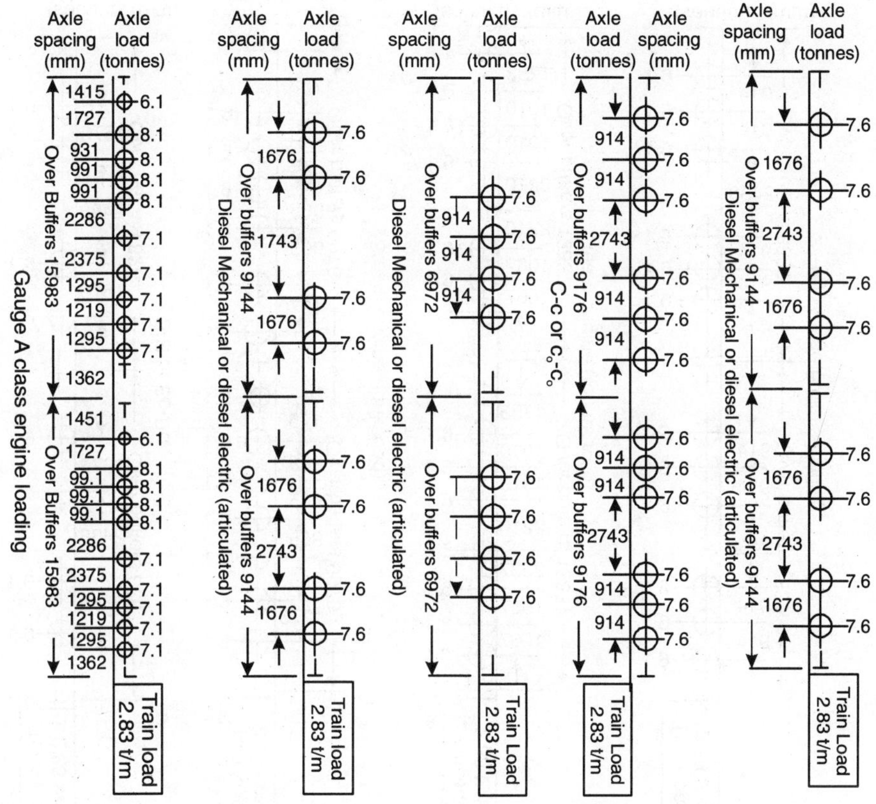

Figure 46.13 *A-class loading for N.G.*

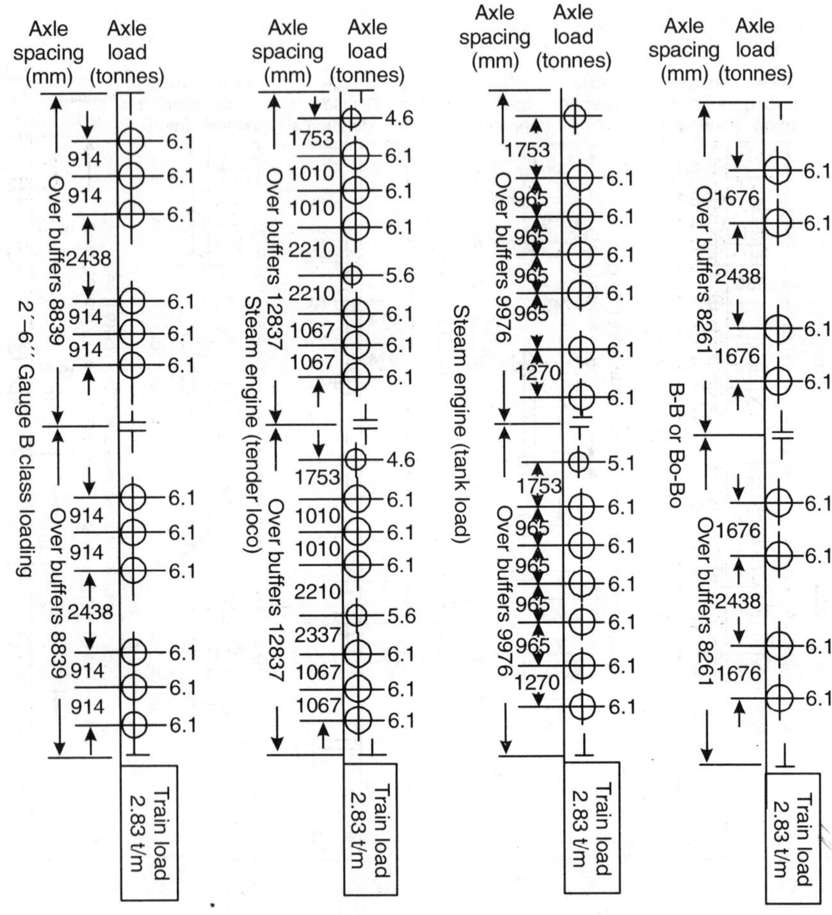

Figure 46.14 *B-class loading for N.G.*

TABLE 46.8 *Equivalent uniformly distributed live loads (EUDLL) in tonnes on each track and impact factor for 762 mm narrow gauge bridges*

L (metres)	*Total load for BM (tonnes)*			*Total load for SF (tones)*			
	H class loading	*A class loading*	*B class loading*	*H class loading*	*A class loading*	*B class loading*	*Impact factor* $\frac{90}{90+L}$
1.0	19.3	16.3	12.0	24.2	19.3	15.2	0.989
1.5	19.3	16.3	12.2	27.0	22.0	17.0	0.984
2.0	24.5	21.0	16.0	31.5	25.5	20.0	0.978

Contd.

Table 46.8 Conted.

L (metres)	Total load for BM (tonnes)			Total load for SF (tones)			
	H class loading	A class loading	B class loading	H class loading	A class loading	B class loading	Impact factor $\frac{90}{90 + L}$
2.5	31.0	26.5	19.0	37.0	30.0	23.0	0.973
3.0	34.5	28.5	22.0	40.5	33.0	25.5	0.968
3.5	38.0	30.5	24.0	43.0	37.0	28.0	0.963
4.0	40.0	35.0	25.5	46.0	40.0	31.0	0.957
4.5	42.5	38.0	27.0	49.5	43.0	33.0	0.952
5.0	45.0	40.5	29.5	51.5	46.0	35.0	0.947
5.5	47.5	43.0	34.0	53.5	48.5	37.0	0.942
6.0	49.5	45.0	38.0	55.5	51.0	39.0	0.938
6.5	52.0	46.5	40.5	58.0	53.0	41.0	0.933
7.0	55.5	48.0	42.5	61.0	55.0	43.0	0.928
7.5	58.0	49.5	44.5	64.0	57.0	45.0	0.925
8.0	60.0	51.5	45.5	67.0	59.0	47.0	0.918
8.5	61.5	53.5	46.5	69.5	61.5	48.5	0.914
9.0	63.0	56.0	47.5	72.5	63.0	50.5	0.910
9.5	64.0	59.0	48.0	75.5	65.0	52.5	0.905
10	65.5	62.0	48.5	78.5	67.0	54.5	0.900
11	69.5	64.5	50.0	85.0	72.0	58.5	0.891
12	76.5	67.0	52.0	91.5	76.5	62.5	0.882
13	88.5	69.0	55.0	97.0	81.0	66.5	0.874
14	95.0	72.5	60.0	103.0	85.5	71.0	0.865
15	99.0	80.0	70.0	109.0	90.0	75.0	0.857
16	105.0	86.0	77.0	114.0	96.0	79.0	0.850
17	110.0	90.0	80.0	120.0	101.0	84.0	0.841
18	115.0	94.0	84.0	127.0	106.0	88.0	0.833
19	120.0	98.0	87.0	134.0	110.0	92.0	0.826
20	125.0	102.0	90.0	141.0	115.0	96.0	0.818
21	130.0	106.0	94.0	147.0	119.0	100.0	0.811
22	135.0	111.0	97.0	152.0	123.0	103.0	0.804
23	140.0	115.0	100.0	156.0	128.0	106.0	0.796
24	145.0	120.0	104.0	160.0	132.0	110.0	0.790
25	150.0	125.0	107.0	164.0	137.0	113.0	0.783

Contd.

Table 46.8 Conted.

L (metres)	Total load for BM (tonnes)			Total load for SF (tones)			
	H class loading	A class loading	B class loading	H class loading	A class loading	B class loading	Impact factor $\frac{90}{90 + L}$
26	155.0	130.0	111.0	169.0	141.0	117.0	0.776
27	159.0	135.0	114.0	173.0	145.0	120.0	0.770
28	163.0	139.0	118.0	177.0	150.0	123.0	0.763
29	167.0	144.0	121.0	182.0	154.0	126.0	0.756
30	171.0	148.0	124.0	186.0	159.0	130.0	0.750
32	179.0	157.0	131.0	194.0	167.0	136.0	0.738
34	187.0	165.0	138.0	201.0	176.0	143.0	0.726
36	194.0	173.0	145.0	209.0	184.0	149.0	0.714
38	201.0	181.0	152.0	217.0	192.0	156.0	0.703
40	208.0	189.0	158.0	224.0	199.0	162.0	0.692
42	214.0	196.0	165.0	231.0	207.0	169.0	0.682
44	221.0	202.0	172.0	238.0	214.0	175.0	0.672
46	228.0	209.0	178.0	245.0	221.0	180.0	0.662
48	234.0	215.0	185.0	251.0	228.0	196.0	0.652
50	240.0	254.0	219.0	291.0	268.0	222.0	0.600
55	256.0	239.0	206.0	274.0	251.0	207.0	0.621
60	272.0	254.0	219.0	291.0	268.0	222.0	0.600
65	283.0	269.0	232.0	308.0	284.0	236.0	0.581
70	288.0	284.0	242.0	324.0	300.0	250.0	0.568
75	291.0	299.0	249.0	340.0	315.0	263.0	0.545

Note : The intermediate values may be found by linear interpolation.

46.7 FOOT BRIDGES AND FOOT PATHS (ATTACHED TO THE RAILWAY BRIDGES)

In case of foot bridges which are attached to the railway bridges, the live loads including impact shall be taken as 490 kg/m^2 of the foot path area. For the design of main girders or trusses of the bridges the lollowing live loads should be taken on the footpaths as per bridge rules:

(i) For effective spans of 7.5 m or less, the live load is taken as 415 kg/m^2.

(ii) For effective spans between 7.5 m and 30 m the intensity of live load reducing from 415 kg/m^2 for a span of 7.5 m to 295 kg/m^2 for a span of 30 m should be taken into account.

(iii)For effective spans beyond 30 m the live load is determined by the following formula

$$P = 13.3 + \frac{400}{L} \quad \frac{17 - W}{1.4} \text{ kg/m}^2$$

where

P live load in kg/m^2

L effective span of the bridge in m

W width of the foot-way in metres.

Side kerbs of width 60 cm or more shall be designed for the load mentioned above in addition to the 750 kg/m run lateral loading acting horizontally at the top of the kerb. If the width of kerb is less than 60 cm no horizontal thrust load is to be taken into account.

46.8 IMPACT LOAD

The moving loads on bridges have a jumping action due to uneven surface on which they move. This jumping causes shocks and vibrations of the structures. Consequently the stresses produced by the moving loads are larger than what would be produced if the load were applied gradually and statically. The additional stresses are accounted for by considering the impact effect.

The main causes of impact are unbalanced weight of the driving wheels, rough the uneven track, eccentric wheels, deflection of floor beams and rail bearers, lurching of locomotives causing periodical shifting of the live load from one wheel to another. These weights cause unbalanced forces and hammer blow effect, which further causes vibration in the structures. When the period of vibration of the bridge structure and the rotation of moving weights synchronise, resonance take place causing impact load on the structures.

Impact load on bridges is determined by the multiple of a *impact factor* with live (moving) load. Sometimes impact factor is termed as *impact coefficient*. The expression for impact factor is given in terms of the loaded length of the structure. It has been specified by different authorities for different types of bridges under different conditions.

For B.G., M.G. steel railway bridges, the impact factor is taken as given in Table 46.9.

TABLE 46.9 *Impact Factor*

Impact factor (i)	
$i = \frac{20}{14+L}$ $\not> 1.00$	For single track line
$i = 0.72 \times \frac{20}{14+L}$ $\not> 0.72$	For the main girders of double track with two girders
$i = 0.60 \times \frac{20}{14+L}$ $\not> 0.60$	For the intermediate main girders of multiple track spans
$i = 0.72 \times \frac{20}{14+L}$ $\not> 0.72$	For cross girders carrying two or more trackes
For B.G. $i = \frac{7.32}{B+4.27}$ For M.G. $i = \frac{5.49}{B+4.27}$	For rail with ordinary fish plate joints and supported directely on transverse steel trough or sleepers
$i = \frac{91.5}{91.5+L}$	For narrow gauge steel bridge

where

L Load length of span in metres for the position of the train giving maximum stresses in the member under consideration

B spacing of main girders in metres

46.8.1 Impact on Highway Bridges

For I.R.C. class AA loading and class 70 R loading, the value of impact percentage shall be taken as per Table 46.10.

TABLE 46.10

Type of Vehicle	*For spans less than 9 m*	*For spans 9 m or more*
(*i*) For tracked vehicles	25% for span upto 5 m and linearly reducing to 10% for spans of 9 m	10% for all spans
(*ii*) For wheeled vehicles	25%	25% for spans upto 23 m and in accordance with the curve indicated in Fig. 46.15 for spans in excess of 23 m

The impact percentage for highway bridges can also be determined by the formula.

$$i = \frac{9}{13.5 + L}$$

where

i impact percentage

L load length in metres

In the case of bridges when the flow structures have fillings more than 0.6 m including the road crust, the impact percentage to be allowed shall be one-half of those specified above.

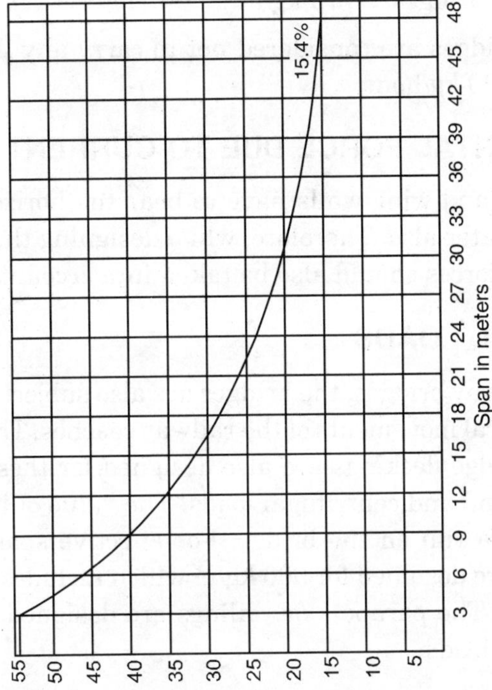

Figure 46.15

46.9 WIND LOAD

The wind produces a horizontal pressure on any surface it strikes. If a gust of wind goes grazing an object, it tends to pull it alongwith, thus causing a suction effect. The probability of maximum wind and maximum live loads acting simultaneously are low. The wind pressures are expressed in terms of basic wind pressure *p*. which is the static pressure in the windward direction. These basic wind pressure in kg/m^2 at various heights (in metres) are adopted from the maps and tables given in IS : 875–1962.

The wind pressure is determined from the appropriate wind pressure from the table and the exposed area of bridge girder and on the moving loads also. While designing bridges wind pressure of 240 kg/m^2 is adopted for unloaded span of the railway highway and foot bridges.

The railway and foot bridges shall have not live loads when the wind pressure at deck level exceeds the following limits:

B.C. Bridges	150 kg/m^2
M.G. Bridges	100 kg/m^2
Fool Bridges	75 kg/m^2

All highway bridges are considered not to carry any live load when the wind velocity is 130 kg/hour.

46.10 HORIZONTAL FORCE DUE TO CURRENT OF WATER

Piers, abutments and wing-walls have to bear the horizontal force due to current of river water also. Therefore, while designing these components of the bridge, these forces should also be taken into account.

46.11 LATERAL LOADS

In the case of railway bridges, the bridges are also subjected to lateral loads caused due to lateral movements of the railway coaches. The lateral bracings of the railway bridge deckings are also designed for these lateral loads in addition to the wind and centrifugal loads. The value of this lateral load is taken as 600 kg/m run on the bridge. For effective spans upto 20 m the lateral bracings are designed for 900 kg/m which includes wind, centrifugal and lateral loads. The parapets or railings are designed for 150 kg/m run horizontal lateral loads.

46.12 CENTRIFUGAL FORCE

These forces are caused due to vehicle moving on curves. Therefore when the traffic lane or track over a bridge is situated on a curve, the bridge members are also designed to resist the centrifugal force.

46.12.1 Railway Bridge

For railway bridges the centrifugal force is determined by the formula

$$C = \frac{WV^2}{127\,R}$$

where

C	Centrifugal force in t/m of the span
W	equivalent distributed live load in t/m
V	maximum speed of vehicle in km/hour
R	radius of curve in metres.

The line of centrifugal force is taken at 1.83 m above the rail level for B.G. and 1.45 m for M.G.

46.12.2 Highway Bridges

For highway bridges the value of centrifugal force is calculated by the following formula

$$C = \frac{W'V'^2}{127\,R'}$$

where

C'	centrifugal force in tonnes at the point of action of the wheel loads or in t/m in case of uniformly distributed load
W'	live load in tonnes in case of wheel loads or in t/m in case of uniformly distributed load
V'	designed vehicle speed
R'	radius of curvature of the road curve in metres

46.13 LONGITUDINAL FORCE

Tractive force at the time of driving wheels, frictional resistance offered to the movement of free bearings due to change in temperature or braking effect at the time of applying brakes caused *longitudinal forces* in the bridges. There is no increase in these forces due to impact.

46.13.1 Railway Bridge

Knuckle pin joints and bearings of the girders have to bear horizontal forces due to longitudinal force in the bridge. The distribution of this force on the bearings shall depend on the end conditions of the bridge trusses or girders. The values of the longitudinal force shall be as per Table 46.11 for B.G. bridges and Table 46.12 for M.G. bridges as per Indian Railway Board Bridge Rules.

TABLE 46.11 *Longitudinal loads (without deduction for dispersion) for broad gauge 1676 mm*

L (metres)	Tractive effort (tonnes)		Braking force (tonnes)	
	ML	B.L	ML	B.L.
1.0	15.7	11.8	11.3	8.6
1.5	15.4	11.6	11.2	8.5
2.0	15.1	11.5	11.1	8.4
2.5	14.9	11.2	11.0	8.3
3.0	15.0	11.3	11.2	8.4
3.5	16.5	11.5	12.4	9.3
4.0	18.4	13.9	13.9	10.5
4.5	20.7	15.7	15.7	11.9
5.0	22.4	17.0	17.2	13.0
5.5	23.8	18.0	18.4	13.9
6.0	24.8	18.7	19.3	14.6
6.5	25.6	19.4	20.0	15.1
7.0	26.9	20.2	21.2	15.9
7.5	28.2	21.1	22.2	16.7
8.0	29.1	21.9	23.2	17.5
8.5	29.8	22.5	23.9	18.0
9.0	30.4	23.0	24.5	18.5
9.5	30.9	23.9	25.0	19.3
10	31.5	24.5	25.7	20.0
11	33.8	25.9	27.9	21.4
12	34.8	26.8	28.9	22.1
13	35.5	27.3	29.8	22.9
14	35.9	27.5	30.4	23.3
15	36.8	28.2	31.6	24.3
16	37.8	29.2	32.7	25.3
17	38.8	39.0	33.9	26.2
18	39.8	30.8	35.2	27.2
19	40.5	31.5	36.0	28.0
20	41.6	32.4	37.4	29.1
21	42.4	33.1	38.4	30.0
22	43.1	33.5	39.3	30.4
23	43.5	33.9	39.9	31.1
24	44.0	34.4	40.8	31.8

Contd.

Table 46.11 Conted.

L (metres)	Tractive effort (tonnes)		Braking force (tonnes)	
	ML	B.L	ML.	B.L.
25	44.8	35.0	41.6	32.5
26	45.5	35.2	42.7	33.1
27	45.8	35.6	43.2	33.6
28	46.5	36.1	44.2	34.3
29	46.9	36.6	44.7	34.9
30	47.6	36.8	45.7	35.3
32	47.6	36.8	47.2	36.8
34	47.6	36.8	48.2	38.2
36	47.6	36.8	50.0	39.4
38	47.6	36.8	51.4	40.5
40	47.6	36.8	53.1	41.7
42	47.6	36.8	54.4	42.8
44	47.6	36.8	55.5	43.8
46	47.6	36.8	57.1	45.1
48	47.6	36.8	58.5	46.2
50	47.6	36.8	59.1	46.8
55	47.6	36.8	62.1	48.9
60	47.6	36.8	64.4	50.7
65	47.6	36.8	67.0	52.5
70	47.6	36.8	69.2	53.9
75	47.6	36.8	71.2	55.0
80	47.6	36.8	73.4	56.4
85	47.6	36.8	75.3	57.7
90	47.6	36.8	77.2	59.1
95	47.6	36.8	78.8	59.8
100	47.6	36.8	81.2	61.0
105	47.6	36.8	83.2	63.3
110	47.6	36.8	85.2	63.5
115	47.6	36.8	87.0	64.8
120	47.6	36.8	88.5	65.5
125	47.6	36.8	89.9	66.3
130	47.6	36.8	91.2	67.0

Note : Intermediate values may be found by linear interpolation.

TABLE 46.12 *Longitudinal loads (without deduction for dispersion) for metre gauge 1 m*

L (metres)	Tractive effort (tonnes)			Braking force (tonnes)		
	M.L	B.L	C	B.L.	B.L.	C
1.0	9.1	7.4	5.6	5.8	4.7	3.5
1.5	8.8	7.2	5.4	5.7	4.6	3.5
2.0	8.6	7.0	5.7	5.6	4.6	3.7
2.5	8.9	7.3	6.4	6.0	4.9	4.3
3.0	9.9	7.9	6.9	6.7	5.4	4.7
3.5	11.6	9.4	7.8	8.0	6.5	5.4
4.0	12.8	10.4	8.5	9.0	7.3	6.0
4.5	13.7	11.1	8.9	9.7	7.9	6.3
5.0	14.2	11.5	9.5	10.3	8.3	6.9
5.5	15.1	12.2	10.0	11.1	9.0	7.3
6.0	15.9	12.9	10.4	11.8	9.6	7.8
6.5	16.5	13.3	10.7	12.4	10.0	8.1
7.0	16.9	13.9	11.2	12.9	10.6	8.6
7.5	17.4	14.5	11.5	13.4	11.2	8.9
8.0	17.9	14.9	11.9	14.0	11.7	9.3
8.5	18.4	15.2	12.1	14.6	12.1	9.6
9.0	18.7	15.5	12.2	15.0	12.4	9.8
9.5	19.0	15.7	12.5	15.4	12.7	10.1
10	19.2	15.9	12.7	15.8	13.0	10.4
11	19.8	16.2	13.2	16.6	13.6	11.0
12	20.6	17.0	13.8	17.6	14.5	11.8
13	21.2	17.7	14.3	18.5	15.4	12.5
14	21.8	18.3	14.8	19.3	16.2	13.2
15	22.3	18.8	15.3	20.1	17.0	13.8
16	22.8	19.3	15.6	21.0	17.7	14.3
17	23.3	19.6	15.9	21.7	18.3	14.8
18	23.6	19.2	16.1	22.3	18.9	15.3
19	23.8	20.2	16.4	22.9	19.4	15.7
20	24.0	20.3	16.6	23.4	19.8	16.1
21	24.2	20.5	16.8	23.9	29.2	16.6
22	24.4	20.7	17.1	24.4	20.7	17.1
23	24.6	20.9	17.3	25.0	21.2	17.5
24	24.8	21.1	17.5	25.5	21.6	18.0

Contd.

Table 46.12 Conted.

L (metres)	Tractive effort (tonnes)			Braking force (tonnes)		
	M.L	B.L	C	B.L.	B.L.	C
25	24.8	21.1	17.5	26.0	22.1	18.4
26	24.8	21.1	17.5	26.0	22.7	18.8
27	24.8	21.1	17.5	27.3	23.2	19.3
28	24.8	21.1	17.5	27.9	23.7	19.7
29	24.8	21.1	17.5	28.4	24.1	20.0
30	24.8	21.1	17.5	29.0	24.5	20.4
32	24.8	21.1	17.5	30.0	25.4	21.2
34	24.8	21.1	17.5	30.9	26.3	22.0
36	24.8	21.1	17.5	32.0	27.3	22.7
38	24.8	21.1	17.5	32.9	28.1	23.4
40	24.8	21.1	17.5	33.9	28.7	23.9
42	24.8	21.1	17.5	34.5	29.4	24.5
44	24.8	21.1	17.5	35.2	29.9	25.1
46	24.8	21.1	17.5	35.8	30.5	25.6
48	24.8	21.1	17.5	36.3	31.0	26.1
50	24.8	21.1	17.5	36 .8	31.5	26.5
55	24.8	21.1	17.5	37.9	32.5	27.5
60	24.8	21.1	17.5	38.8	33.3	28.5
65	24.8	21.1	17.5	39.5	34.2	29.4
70	24.8	21.1	17.5	40.0	35.0	30.3
75	24.8	21.1	17.5	40.6	35.6	31.2
80	24.8	21.1	17.5	41.2	36.2	31.9
85	24.8	21.1	17.5	41.7	36.8	32.6
90	24.8	21.1	17.5	42.1	37.4	33.2
95	24.8	21.1	17.5	42.2	37.8	33.8
100	24.8	21.1	17.5	42.8	38.3	34.3
105	24.8	21.1	17.5	43.2	38.8	34.9
110	24.8	21.1	17.5	43.5	39.2	35.4
115	24.8	21.1	17.5	43.8	39.6	35.9
120	24.8	21.1	17.5	44.2	40.0	36.4
125	24.8	21.1	17.5	44.5	40.3	36.8
130	24.8	21.1	17.5	44.8	40.8	37.4

Note. **Intermediate values may be found by linear interpolation.**

Highway bridges are also designed keeping provisions for longitudinal force. The braking effect on a simply supported span or a continuous unit of spans or on any other type of bridge unit shall be as follows :

46.13.2 For Single Lane or Two Lane Bridge

20% of the first train load plus 10% of the loads of the succeeding trains or parts thereof. For this purpose the train loads in one lane only are considered. If the entire first train is not on the full spans the longitudinal force shall be only 20% of the actual load on the span.

46.13.3 For Bridges More Than Two Lanes

The above loads plus 5% of the loads on the lanes excess of two lanes.

The above longitudinal force acts at 120 cm above the road surface, therefore it should be transferred on bearings accordingly.

46.14 SEISMIG FORCE

These are earthquake forces which cause horizontal and vertical acceleration in the bridge structures. The magnitude of these accelerations depends upon the intensity of shock, the type of foundations give rise to the force on the structure equal to its mass multiplied by the acceleration. Generally only horizontal effect of seismic force is taken into account. The seismic force is taken as a horizontal force. Its value does not exceed 0.12 of the gravity. In the case of bridges situated in epicentral area, the percentage is fixed on the local conditions regarding the intensity of the earthquake. IS : 1983–1962 recommends the allowable seismic force on the bridge and other structures.

46.15 ERECTION STRESSES

Bridge components are mainly manufactured in the workshops and then erected at the sites. At the time of erection various types of temporary fastenings are done to these components. Secondly at the time of handling, lifting, placing in position and temporarily supporting these parts have to bear various types of stresses. Therefore at the time of designing provisions for these stresses should also be made.

46.16 TEMPERATURE STRESSES

Structural materials expand or contract due to change in temperature. If these are allowed freely to expand and contract no extra stresses are developed. On the other hand if their expansion or contraction is prevented by some reasons, it will cause development of heavy internal stresses in the components. The thermal coefficient of expansion or contraction of structural components is taken as 1.17×10^{-7} per degree centigrade.

46.17 RELIEF STRESSES

Any relief given to the member of the bridge by the adjoining parts may be taken into account. While determining this, the secondary stresses, if any in the member, shall be determined and considered.

46.18 SECONDARY STRESSES

These are additional stresses which act on the components of bridge. Firstly the stresses which are the result of the elastic deformation of the bridge structure are combined with the rigidity of the joint.

Second type of secondary stresses are the result of eccentricity of connections, and off panel point loading i.e., load rolling directly on chords, self-weight of members and wind load on them.

It is to be noted that provisions for these secondary stresses are already made in the factor of safety to certain limits.

REVIEW QUESTIONS

46.1. What are the various types of loads, which are taken into account for the design of bridges? Discuss any two types.

46.2. What do you understand by I.R.C. class 'AA' loading? Under what conditions this loading is adopted?

46.3. Write short note on the Indian Road Congress Loading.

46.4. What do you understand by the minimum clearances for I.R.C. class vehicles? Show by means of a neat diagram the minimum clearance for I.R.C. class 'A' vehicle.

46.5. Write short notes on I.R.C. class 70 R loading.

46.6. How the footways, attached to the highway bridges are designed? Describe.

46.7. Write short note on the standard loading for railway bridges.

46.8. Write short note on the impact load considered in bridge design.

46.9. Write short note on the centrifugal force and longitudinal force.

46.10. Detail the different forces and loads that have to be taken into consideration in the design of B.G. railway abutment and pier.

46.11. Narrate the various forces which you would take into account in the design of piers founded on wells supporting a R.C.C. bridge in a seismic zone of a moderate intensity.

46.12. Enumerate and explain the forces to be taken into account when designing a pier and an abutment of a road bridge. Illustrate with sketches where that act in each case and state the condition of stability.

46.13. Name the different types of loadings used for designing a bridge. What are the factors on which the selection of a loading type depends?

46.14. Detail the loads, forces and stresses that are to be considered in designing a highway bridge. Sketch the wheel loading arrangement for I.R.C. class 'A' standard loading, giving the clearance dimensions and the contact area of the tyres.

46.15. Describe with illustrative sketches I.R.C. recommendations for loading on Road Bridges and equivalent lane loading for live loads for class 'AA' loading for highway bridges.

46.16. Sketch the loading arrangements for I.R.C. class 'AA' standard loading, giving the clearance dimensions and the contact areas.

46.17. (a) How will you account for the following in design of a road bridge?

(i) Wind load, and

(ii) Impact load.

Give recommendations of I.R.C. for calculation of these loads.

(b) Sketch the wheel train for I.R.C. class 'A' loading giving the clearance dimensions and contact areas of tyres.

46.18. How will you account for the wind load in the design of a highway bridge? Give the recommendations of I.R.C. in this regard and discuss why these values have been adopted for general use?

46.19. (a) What is impact due to and how do we account for its effect on highway bridge structure? State impact formulae for R.C.C. and steel bridges for the standard I.R.C. loadings.

(b) Explain the effect of the force due to water-current striking against the bridge pier? How is it affected by the pier? What pier shape is considered to be the best?

46.20. What are Cooper's standard loadings for the railway bridges?

46.21. Describe briefly the several forces to be considered in the design of a bridge bringing out the relative importance.

47

Construction of Bridges

GENERAL

Large number of engineers of railways, municipal, P.W.D., military and those concerned with rivers and harbours, have to deal with bridge construction. As a work of construction, the Bridge, in its elemental form remains much as it was a century ago. Masonry bridges if properly constructed could last five hundred years, but 67 m is the limit of a single arch span. Masonry bridges are very costly and require highly skilled masons. A ferro-concrete bridge can last for five hundred years. Steel bridges require constant painting and maintenance, but these are very strong and can bear vibrations. Therefore mostly large railway bridges are constructed by steel. The trend of movable bridges is towards vertical lift type, while the transporter bridge which is a cheaper movable bridge became rapidly obsolete due to the growth of heavy road traffic. Rigid frame bridges are becoming popular for short spans. Still timber bridges are being used in hilly areas. If the timber is properly treated, it will have useful life of a hundred years.

The methods of construction of bridges depends on the type of bridge. But the methods employed for the construction of one type of bridge at different places will depend on locality, labour, time required for the construction and the standard of materials, etc. Therefore even for the same type of bridges at different places the method will differ. The methods for the construction of various types of bridges will be described in this chapter.

47.1 TIMBER BRIDGES

The timber which is required will be selected carefully from the available local materials, because timber bridge is regarded as the local job. If possible, consultation should be done with the Forest Research Laboratories. Before using timber in bridges this should be thoroughly seasoned and creosote should be applied on it to protect it against fungis etc.

The substructure, trestles, cribs, pile-bents and crates are constructed at the river banks if the water is deep and may be constructed at the site if water depth is less. Generally, these are constructed with logs of timber fastened together with wire ropes, iron spikes or hemp. For round logs fastening wire rope is best and for square timber iron-spikes are good.

The superstructure is generally constructed with square or rectangular timber pieces fastened together with good joints. Timber trusses and arches are constructed at the bank and are placed in position.

Generally the flooring consists of about 7 cm to 10 cm thick planks laid on the beams. These planks are laid with a traverse grade of 2.5 to 4 cm in 3 metres, to drain the water to sides. For more than 5 metres wide bridge camber is given in the middle. In case of continuous heavy traffic, the floor does not rot but becomes worn out, therefore harder and tougher wood should be used. These floor planks are fastened to the beams or joints with nails which should penetrate in the beam equal to the thickness of floor plank. To increase the life of flooring, 5 cm thick premix carpet should be laid on the planks. When this wearing carpet is worn out it should be renewed and properly maintained.

47.2 TIMBER SUSPENSION BRIDGES

For the construction of this bridge all the required materials are stored at the site. Both the pier trestles are constructed on the banks and are raised with the help of props and struts in their respective position and then they are secured firmly by guys and back struts. While the work of erecting pier trestles is in progress two logs are erected one on each bank and a light cable with a travelling pulley is stretched over them. A derrick longer than the height of the pier trestle is erected temporarily near the saddle over which the cable will rest. A pulley is fixed to the derrick a few metres above the top of the saddle and rope is passed over this pulley. The cable is laid on the ground and one end of the rope passing over the pulley is fixed to it. By applying force on the other end of the rope the cable is lifted and placed on the top of the saddle. The free end of the cable is fastened to the anchorage and then earth is filled and rammed well. In this way all the cables are lifted on one bank and placed on saddles. The other ends of the cables first anchored and then they are lifted in the same way and required sag is kept in it. Now the suspenders are fastened to the cables at required positions with the help of light cables having one travelling pulley which has been fixed for this purpose. Now beams and transomes are fixed to the suspenders and the extra length of suspenders is cut down. Over these beams the flooring is clone and slings are finally adjusted with the help of nuts.

47.3 MASONRY BRIDGES

First of all, the layout of the bridge is done at the site and when the construction of piers, abutments and wing-walls are started. When it is completed, the centring for the arches is done. Centring is the most important work, therefore the centres of arches, and their rises etc. should be checked properly. The centring differs from bridge to bridge. From the construction point of view the masonry arch is quite a simple structure. Care should be taken during backing. It should have horizontal or nearly horizontal joints and carried upto the top of the arch. But in the case of large arches it is not necessary. In the arch the beds of the stones should be perpendicular to the thrust of the arch and the side joints perpendicular to the beds and to the soffit. The bed joints in square arches are plain, though in the case of skew arches, they are curved. The bed joints of the arches should be very thin to avoid settlement later on. On the arches, flooring and parapet walls are constructed as are required in the design work. After the setting of mortar, the centring works are removed, the finishing touch is given and the bridge is opened to the traffic.

After construction of arches, the spandrel filling is done with earth. Then workable layer of lime concrete is laid and the top is brought to a uniform horizontal level. Required road finishing is given on the top and two parapet walls are also constructed for protection purpose. Wheel guards should also be provided.

47.4 STEEL BRIDGES

In these bridges the most important work is the fabrication in workshops and second is the field connections to the fabricated work. The superstructure of the bridges is fabricated in parts in workshops and then connected at the site. There are many methods for the erection and assembling of fabricated portions at the site. In case of small bridges the main trusses can be assembled at the river bank and then they can be lifted and placed in position by cranes. But for long span bridges different parts are floated at the site then lifted and site connections are done after placing them in position.

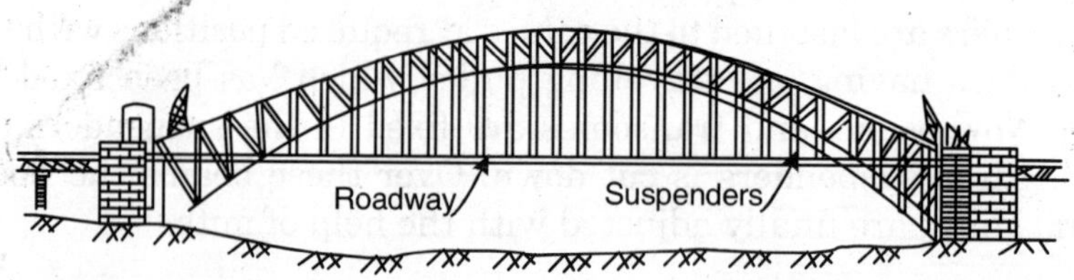

Figure 47.1 *Construction of steel arch bridge*

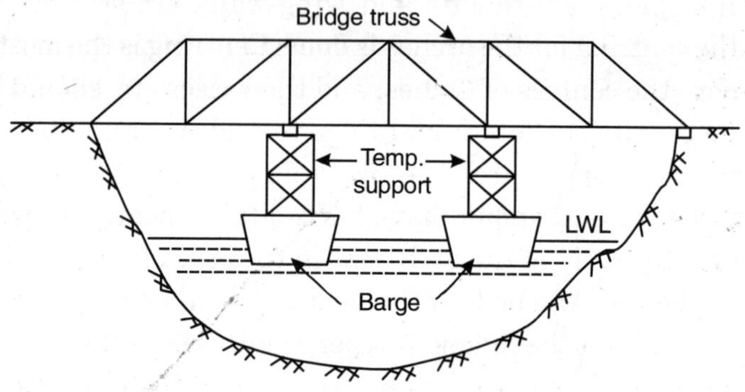

Figure 47.2 *Method of construction of bridge by supporting the bridge truss on the floating barge*

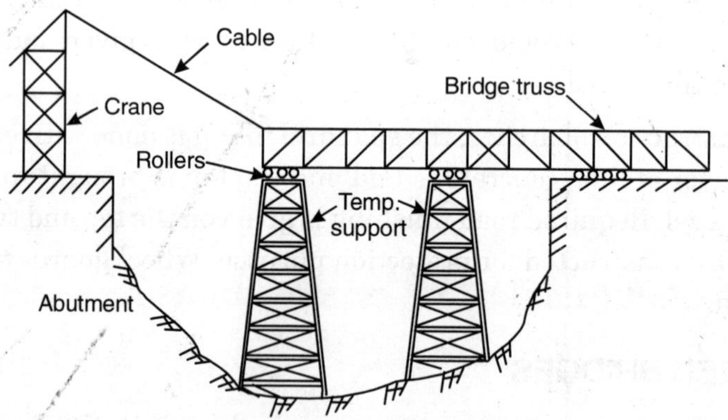

Figure 47.3 *Methods of construction of bridge by toeing the bridge truss by means of crane on rollers*

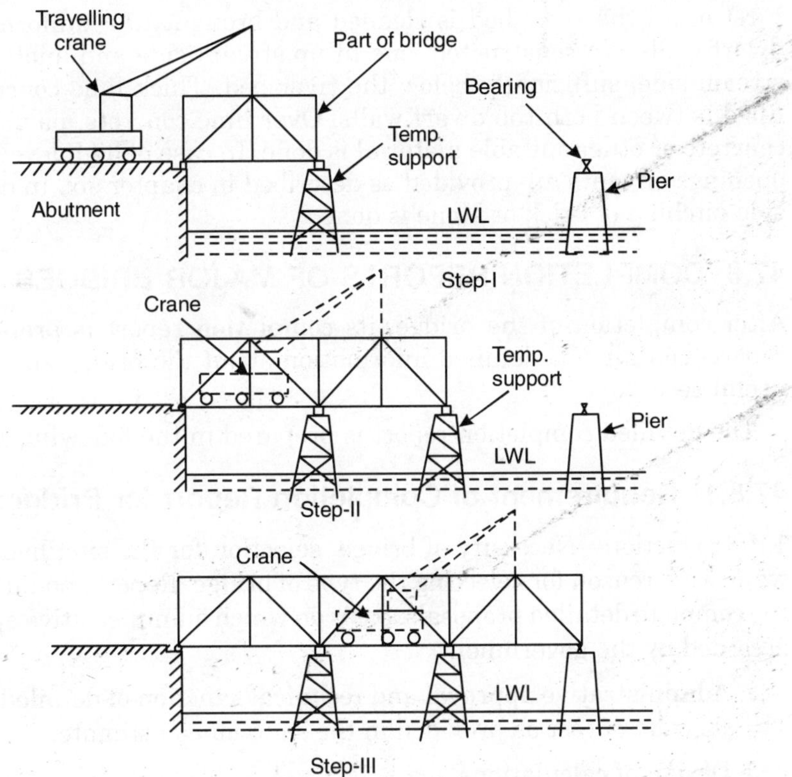

Figure 47.4 *Method of construction of bridge superstructure by means of cranes and using temporary supports*

47.5 R.C.C. BRIDGES

In these bridges the main work is form-work and placing of steel reinforcement in position. Generally the work is done cast-in-situ. After doing the form-work and placing reinforcement, the concrete is poured. After it the curing is done, and when it is completed the form-work is removed.

47.6 FLOATING BRIDGES

In these types of bridges the boats, pontoons and rafts are constructed and manufactured in workshops near the banks. Then these are taken to the site and fixed together by trussed-beams and also held in position by anchors on up-stream and down-stream sides. Over these beams road flooring is done and railings are fixed on both sides.

47.7 CAUSE-WAYS

First of all the river bed is cleaned and brought to a uniform level. Two dwarf walls are constructed, one in up-stream side and another in down-stream side, sufficiently below the river bed. Thick lime concrete layer is filled between both the dwarf walls. Over lime concrete, paving of cement concrete or other suitable material is done. In case of high level cause-ways openings or vents are provided as described in chapter six. In down-stream side pitching of brick or stone is done.

47.8 COMPLETION REPORTS OF MAJOR BRIDGES

After completion of the bridge, its completion report is prepared. These reports contain the detailed information about the bridge and form a very useful record.

The detailed completion report is prepared in the following statement:

47.8.1 Requirement of Completion Report for Bridges

1. Introduction—Necessity of bridge, selection for the site, hydraulic data, waterway, reason for selecting the type of bridge. Report should incorporate the complete detailed proposals, basis on which administrative approval was accorded by the government.

2. Administrative approval and technical sanction of detailed estimate—The abstract of cost as provided in the sanctioned estimate.

3. Design of calculations.

4. Details of Estimate

(i) Detailed specification

(ii) Analysis

(iii) Bill of quantities (abstract)

(iv) Cost

(v) Drawings

(vi) Technical sanction.

5. Tenders received and accepted. A copy of comparative statement.

6. Details of construction.(i)Well sinking, substructure and superstructure testing etc., including all the details of the difficulties encountered with during execution and how they were surmounted with details of sketches.

(ii) Details of staging and centring.

(iii)Strata actually met during sinking and basis on which depth of foundation was fixed.

(iv) Details of various tests carried out either at site or in laboratory during the execution of the project.

7. Progress giving dates of :

(i) Completion of well sinking

(ii) Substructure

(iii) Superstructure

(iv) River training works etc.

8. Design and construction of river training works with cost.

9. Detailed analysis of cost as observed during execution of the following works :

(i) Well-sinking

(ii) Brick masonry steining piers

(iii) R.C.C. work in superstructure

(iv) Stagging

(v) Railing

(vi) Other important items of work.

10. Details of special equipment item.

11. Completion Drawing—Drawings of the approved design should be corrected in red ink to show the changes carried out during execution.

12. Construction materials used and their source of supply.

13. Details regarding bearings, expansion and construction joints etc., with information regarding cost etc.

14. Any change in the specification of design adopted giving full details of justification etc.

15. The names of Executive Engineer, Assistant Engineer, Junior Engineer, work supervisors with dates and type of work supervised should also be given in the report.

16. Actual cost of construction of different items of work.

17. Any suggestions for incorporating any change in the design, specification and method of execution etc.

18. Photographs of different stages of work along with photographs of complete work etc.

REVIEW QUESTIONS

47.1. Write short notes on the construction of timber bridges. How are timber suspension bridges constructed?

47.2. Write short notes on the construction of :

(a) Masonry bridges

(b) Steel bridges

(c) R.C.C. bridges

(d) Floating bridges

47.3. Describe briefly the erection of a suspension bridge (about 100 m span) in a hilly area to carry light vehicle traffic.

47.4. Describe briefly the erection of a single span (about 80 m) suspension bridge made of steel in a hilly area over a deep gorge. Illustrate your answer with sketches.

47.5. Discuss the different methods of erection of steel bridges, stating the condition under which each type is used.

47.6. Describe step by step the method of erection of a single span suspension bridge of about 80 m span in a hilly area. Illustrate your answer with sketches.

47.7. Describe the erection of a steel trussed bridge of about 100 m span carrying a railway line.

47.8. It is proposed to float into position a lattice through girder for one of the spans of a bridge. The span of the girder is 40 m and the elevation of its support is 10 m above the water level. Describe the various stages of erection with necessary sketches.

48

Maintenance of Bridges

GENERAL

India has a large number of bridges situated throughout its area, stretching over plains and mountains. The damage of bridges and their costly repairs can be avoided by their periodical inspection and maintenance. Bridges upto 30 m are called *minor bridges* and above it are called *major bridges.*

Each divisional office should maintain an up-to-date record separately for all minor and major bridges in its jurisdiction. One separate copy of major bridges should also be kept in Superintending and Chief Engineer's office.

Each maintenance register should have the following.

- Plan of the district in 1 cm = 5 km scale which should show :
  - All rivers with their names
  - All canals
  - All roads and railways
  - All bridges in red colour with their names and number of spans
  - The district boundaries
- Full details of bridges with drawings filled in a systematic proforma
- Up-to-date maintenance record of the bridge filled in a systematic proforma.
- A drawing of the bridge showing reduced levels.

48.1 BRIDGE-INSPECTION REPORT

In the departments responsibilities for the inspection of bridges are fixed with various persons. After inspection a report is prepared, which includes all defects noticed and after it these defects are removed as early as possible. In this way a full history of every bridge is prepared starting from the construction, which will include all the troubles or problems which came across at the time of construction and the detailed report of each inspection is prepared which should note the following points :

(A) *Foundations*

1. Sinking of foundations.
2. Bed level or river as compared with original bed level.
3. Depth of scouring, whether foundations are undermined.
4. Cracks in masonry work, if any.

(B) *Sub-structure*

1. Proper function of weep holes in abutment and wing-walls.
2. Whether there is any sign of scouring on masonry work due to water action.
3. Pointing work on masonry is safe or not.
4. Checking of water proofing coat.
5. Settlement of sub-structure.
6. Cracks in masonry work.
7. Scour at nose or apron, whether apron has launched or slope pitching disturbed. See for reserve stock for emergency.

(C) *Super-structure*

1. Free action of expansion joints and bearings.
2. Corrosion in steel structures.
3. Condition of paint.
4. Conditions of spurs.
5. Cracked or broken strings.
6. Loosures or under vibrations of steel or timber members.
7. Conditions of articulations, look for cracks, bearings at articulations etc.
8. Anchorage, cables, suspenders, stiffening trusses and floor system in case of suspension bridges.

9. Road or railway structure is safe or not ? Check it carefully and note whether it requires maintenance. Road surface is worn out or not ? Are there pot-holes in the road surface, etc.
10. Cracks in masonry, concrete and R.C.C. work.
11. Side drains are working properly or not.
12. Parapet-wall and railings etc.
13. Foot-path.
14. Approaches to bridges.

(D) *Protecting works*

1. Settlement of protective bunds.
2. Width of service road at top.
3. Side-slopes
4. Side erosion due to water action.
5. Pitching work on slopes and toe of bank, whether correct.

After the preparation of inspection report, the repairing is done at the required places in bridges.

48.2 MAINTENANCE OF STEEL BRIDGES

Steel Bridges if properly maintained usually show no deterioration with time, while inspecting steel bridges the following points should be noted.

48.2.1 An Examination of the Condition of Paint

The most costly maintenance item on steel bridges is painting. All structural steel should be regularly examined for corrosion which usually starts around the heads or rivets and bolts or on top surface of the chord members. The corrosion on the top surface of chord is not visible from the roadway, but its inspection should not be slighted due to its inaccessibility. This is also the case with the top flange of floor beams and stringers and bridges with timber floors. During maintenance the earth, leaves and other debris should be cleaned from the truss joints. All areas where corrosion has started, should be thoroughly cleaned with a wire brush and given a coat of red load followed by a coat of approved paint. This spot paint repairing is called 'Spot-Painting'. If the entire bridge requires painting, then first of all the rusted sections should be cleaned and given a coat of red lead, after which the entire bridge should be given a coat of red lead, after which the entire bridge should be given a coat of approved paint. Red lead should not be applied over other coats of paintings, because its function is of an under-coat to protect the steel from rust. Therefore it should be applied only to bare clean steel or on top or another coat of red lead.

A record about the painting of the bridge should be maintained, showing when the bridge was painted, the cost of paint used, the cost of cleaning and the cost of applying paint etc. so that it could be useful guide for future paintings. The year in which the bridge was last painted should also be written on the bridge itself at such a place, that it can be easily read from the road.

48.2.2 General Alignment

The inspection of the general alignment should be done to insure that the girders or trusses are in their proper position. In case of trusses, the end post or chord sections should be checked whether they are buckled out or are out of line. For old and lightly designed bridge, these points should be specially checked. All intermediate members, hand rails and batter posts should be carefully examined for any damage due to collision. If any compression member of a truss is damaged, the load carrying capacity of the bridge is greatly reduced and the safety of traffic is jeopardized. In such cases the steel members should be supported at each of its panel points to permit the passage of the traffic until permanent repairs are done. The actual design of the bent should be done and proper lateral away bracing should be provided.

48.2.3 Examination of Deck System

The bridges constructed in the past were designed for light traffic and are inadequate for the present-day highway loads. Therefore a check on this item is very important. Additional stringers may be added between existing lines to bring stringer capacity to that as required by the traffic. Floor beams are, in most instances, the critical members from the standpoint of load capacity. The capacity of the existing beams can be increased by welding steel plates to the top and bottom flanges. This work can be easily done while replacing the deck system. If it is not possible to remove the deck system, two plates can be welded on the underside of the upper flange in place of the single plate at the top. If drift conditions permit, the simple beam can be converted to trussed beam by means of a vertical post and some steel rods, known as 'hog rods'.

During inspection the floor beams and stringers should also be checked. The top flanges of all floor beams and stringers should be carefully inspected. The ends of the stringers which are kept at the abutments should also be examined.

An examination of the bridge seats is very important and they should be kept clean all the time, otherwise earth, weeds, etc. will accumulate which will retain moisture causing rusting of the steel.

All the loose rivets should be detected by sight, touch and hearing. All such rivets should be tightened by hammering.

Unusual vibrations should be noted and corrected, if possible. When due to a few members the capacity of the structure is reduced, such member should be reinforced or replaced.

48.3 MAINTENANCE OF WOODEN BRIDGES

As the timber deteriorates easily, therefore it should be given more careful inspection and maintenance. The attention should be given to the following points while inspecting wooden bridges:

(1) If unseasoned timber is used in the construction, it will usually crack due to progressive natural seasoning. If the cracks develop on the top surface of members, the rain water will enter and decay will start causing core rot of the timber. Some type of decay will also start if water gets access in timber through joints. The core rot can be determined by drilling 2 cm holes in the members. After inspection these holes should be plugged by dowels, dipped in asphalt mastic. If the decay of timber members had crossed the limit, they should be replaced.

(2) All the joints of the timber trusses should be tightened at periodic intervals because of progressive shrinkage and compression occurring in the wood.

(3) If the present-day traffic require strong bridge, the old weak timber trusses should be strengthened. Timber side plates may be bolted to the existing floor beams and toothed ring connection may be used to develop shear between the side plates and the existing floor beam. The cheapest method of strengthening a timber truss is to provided underpinning.

(4) All stringers should be carefully checked for splits and decay of timber, if they are weakened, they should be immediately replaced by new stringers.

(5) All the bolts should be regularly inspected and tightened and treated with oil to prevent rusting. Bolt-holes should be treated with creosote oil.

(6) The flooring of the bridge should be carefully inspected. All broken and rotten planks should be replaced. All loose planks should be renailed.

(7) All cracks of timber members should be filled with preservatives.

48.4 MAINTENANCE OF MASONRY BRIDGES

The following points should be noted while inspecting masonry bridges :

(1) The drainage system at the top of piers and abutments should be checked. No water should be allowed to enter in the masonry work and stored there. Holes should be drilled through the arch rings or the spandrel walls in order to tap out the accumulated water.

(2) In old bridges a tendency is developed in the spandrel wall to separate it from arch rings and finally it falls out, thus removing lateral support of the spandrel earth. This tendency can be checked by proper drainage and good design.

(3) All the cracks in the old arch ring should be carefully grouted. The existing joints should be checked by means of small pointed steel rods. These are penetrated in the joints, if great penetration is obtained, it indicated faulty masonry thus requiring urgent attention.

(4) If the vegetation starts growing in the joints of arches, it should be checked up and removed.

(5) All damaged pointing and plaster work should be required.

(6) All the damaged portion of the bridge should be replaced.

(7) In old bad conditioned arches relieving arches should be provided to take the load of the bridge.

48.5 MAINTENANCE OF SUSPENSION BRIDGES

While inspecting suspension bridges, the following points should be noted:

(1) All main cables should be checked for corrosion for frayed strands. Cable position on saddles and the place where the cables enter the ground or anchors should be very carefully examined.

(2) The anchors should be checked for any movements.

(3) All the hanger or suspender rods should be checked for corrosion. All the nuts and washers should be checked whether they are holding floor system correctly.

(4) All the loose, rotten and broken boards should be inspected. The connection of floor beams with the hanger rods should be checked.

(5) The inspection of towers should be done whether they are structurally sound, vertical in position and have adequate lateral strength.

(6) All the masonry work should be checked.

(7) The structural steel should be inspected as given under Article 48.2.

48.6 MAINTENANCE OF R.C.C. BRIDGES

The following points should be kept in mind while inspecting R.C.C. Bridges:

1. An examination of the bottom of the girders and the slabs for disintegration, spalled or honeycombed areas which may expose the reinforcing steel. All such reinforcing steel area should be immediately re-integrated with cement, sand and mortar. The Gunite process of filling cavity is very good and highly recommended. If the girders have cracked near the

centre with steel reinforcement exposed and with disintegration of concrete near the cracks, and also in case tell-tales show that these cracks get extended, the following proposals should be carried for the repairs :

(a) The loading should be increased on the bridge as per I.R.C. loading for which the bridge is desired to be made safe. The deflection should be measured by a dial gauge.

(b) The theoretical deflection should be worked out. From the deflection compute what strengthening will be necessary.

(c) Calculation for the strengthening and finding out the strength of original structure should be done by any rational method.

2. As water is the greatest destructive agency, therefore presence of water in wrong places should be noted. As no concrete is totally water-proof, water gets access inside it. The physical effects of water seeping through concrete are highly destructive, but there are in addition chemical effects, which may be even more disastrous. The sulphate of calcium, magnesium and sodium in solution which are liable to be found in ground water, attack concrete and therefore provisions have to be made for resisting them.

3. All the cracks in the structure should be thoroughly investigated and necessary corrective measures should be discovered. Cracks are defects because well-designed work permits movements at pre-selected places by the provision of joints or flexible members of various kinds. However, there are very few structures which are so well designed that they are entirely free from disfiguring cracks which may or may not be indicative of a defect more serious than disfigurement. Full information about whereabouts and extent of cracking should be recorded and reported to higher officers. Cement mortar tell-tales should be placed across the fissures and photographs taken at intervals from the same vantage points, which will provide valuable data concerning gradual movement of earth retaining structures. It is one of the advantages of regular inspection of structures that it enables frequent observation of the development of weakness to be made and their true significance adequately weighed. Continued movement, if of small extent, is a danger signal and unless it can be stopped by shoring or other suitable means, the bridge may collapse in due course of time. All serious cases in major bridge should be immediately checked and steps should be taken for rectification of the defects.

48.7 MAINTENANCE OF CAUSE-WAYS AND DIPS

The following points should be noted for the maintenance of these types of structures:

1. After every floor the bed paving should be examined and if there is any damage, it should be repaired. It should also be examined if the existing paving is strong enough to withstand the heavy rush of water.

2. The paving should be examined and guarded against scour and undermining. If there are no protective works, a dwarf wall should be built on each side of it. The downstream wall should be 60 cm to 100 cm deeper than the up-stream.

3. If the paved cause way is badly designed, and it has to act as a weir and has the tendency to alter the regime of the river seriously, it should be removed.

4. In no case the roadway should be allowed to become smooth, because it is very objectionable and causes it to be dangerously slippery. At such places the surface may be roughened by criss cross line or by any other suitable method.

5. The embankment should be examined for erosion by floods. It should be protected by dry stone walling or pitching carried one metre below the nalah bed level.

6. The cross slope of the cause-way should have the same slope as the nalah bed, it should be checked. The maximum limit of this slope is 1 in 30.

7. The cause-way should be kept clean of all debris etc. for free how of traffic and flooded water.

8. The longitudinal slopes or ramps at the ends of the causeways should be examined which should follow the bank with a maximum gradient of 1 in 14.

9. Water level gauges should be closed before floods and kept properly painted to give correct indication about the maximum depth of water passing over the cause-way. If the length of the cause-way is more and the gauge cannot be visible from the ends, additional gauges should be fixed near the ends.

10. The existing US and DS walls should be examined to see that they are rounded on their outer edges and are flush with the cause-way surface.

48.8 MAINTENANCE OF ROAD SURFACE, DRAINAGE, OPENINGS, GUARDS, KERBS, FOOT-PATHS, RAILINGS AND PARAPETS

The following points should be noted while inspecting and maintaining the above:

1. The road surface should be examined for excessive scaling and unevenness of surface. A smooth riding surface is very essential to reduce vibration and impact.

2. Bituminous road surface on all types of timber floor needs special attention. Every effort should be made to maintain this surface as smooth as possible, because roughness in surface results in excessive vibrations and causes the timber to become loose and wear rapidly.

3. While resurfacing the roadway, slight slope should be given both ways for drainage facility, because perfectly level roadway over a bridge causes difficulties in keeping it dry. The rain water should flow in gutters provided on both sides of the road. This water should be conducted beyond the end of wing-walls and then run into masonry drains carried in the most convenient position to low ground. Drains may also be carried through the crown of the arches or through the piers.

4. All the existing drains should be examined and corrected if found defective or inadequate. The gutters of the floor should always be kept clean particularly the drainage openings.

5. Drains should not discharge rain water on the steel members.

6. The kerbs and hand rails should be examined for any damage or collision. The damaged part should be immediately replaced or repaired so that it may not be further damaged by accidents.

7. Large guard rail posts should be installed at all four corners of all steel truss bridges to protect the structure from collision. It is most necessary, otherwise all light structure bridges will be completely damaged during accidents.

8. In suspension bridges the hand rails should be especially kept in good condition, because vibrations during use make hand rails support necessary.

9. If pipe railing is provided, sufficient expansion and contraction arrangement must be provided.

10. The foot path should also be thoroughly inspected and maintained.

48.9 MAINTENANCE OF PIERS, PILE-BENTS, ABUTMENTS AND WING-WALLS ETC.

While going on inspection for all these, the following points must be noted:

1. Concrete piers and abutments should be examined for cracks and spalling or disintegration particularly at bridge seat. All such defects should be removed by Gunite process.

2. Wet surface of abutment in dry weather indicates accumulation of water behind the abutment due to lack of proper drainage. In such cases holes should be made in the abutment wall to tap the water, and proper outlets should be provided.

3. Stone masonry or brick masonry piers-abutments and wing-walls should be examined for cracks due to settlement, for conditions of pointing and plastering, for growth of vines or brush in the joints etc.

4. While inspecting, attention should be given to scour and settlement of the foundations of piers, abutments and wing-walls. Where the bottom of

the sub-structure is normally under water, soundings should be made after unusual floods in order to determine whether the structure is being undermined, by the scouring action of water. Vertical alignment of piers and bents should also be checked. Whenever any unusual scour around any portion of the sub-structure is detected, it should be under-pinned by piling. If the erosion has started but the foundations are not under-scoured, the damage can be checked by filling the hole beside the pier or abutment footings with large stone boulders or concrete-filled bags.

5. Timber pile-bents should be maintained as follows :

(a) At the ground line or water line the piling should be carefully inspected. Sometimes a pile which appears to be perfectly sound, is found to be in a state of advanced decay with only a thin surface shell remaining intact. Decay in timber may be located by tapping with a light-hammer, but a more correct method is to bore hole into the piles. When pilings are tested with an auger it is very important that all holes are tightly plugged with creosote timber pegs extending practically in full depth of the hole. All decayed timber or the caps and braces should be immediately replaced. The decayed and defective timber pile should be replaced with a section of sound piling.

(b) The effectiveness of the braces should be checked.

(c) All the bolts, spikes, etc. should be inspected and tightened.

(d) Condition of all timbers at the contact and bearing areas should be checked.

(e) The condition of the cap where the stringers rest on the cap and where the cap bears on the piling should be thoroughly inspected. If there are signs of crushing of caps between stringers and piles or directly over the pile tops as a result of decay, sealing out the moisture by painting contact surface with wood preservative or covering pile tops with an impervious membrane such as treated fabric of sheet metal may be resorted to. At places due to inadequate design of floor system, the moisture gets into caps, therefore, efforts should be made to secure water-tight construction. Any dirt and debris should not be allowed to accumulate in and around the stringers and cap supports.

(f) All the wedges should be inspected and tightened.

(g) Creosoted piling should be used to ensure long life.

(h) Back walls should be checked at the end bents to see that stringers, caps and braces are protected from contact of earth, which retains moisture. All such contact surfaces should be painted with two coats of wood preservatives.

48.10 MAINTENANCE OF FLOATING BRIDGES

The following points should be noted while inspecting Floating Bridges :

1. The safe load which the bridge can take should be determined by calculations or actual loading of the bridge.

2. Instructions should be given not to carry heavy loads on the bridge as determined.

3. On the sign boards the specific safe axle loads for different types of vehicles should be painted on prominent places.

4. The speed limit should also be written on the boards.

5. If the bridge is to be constructed in navigation rivers or channels, then cut boat bridge should be provided in the middle.

6. If possible the approaches should be metalled with kankar.

7. Constant watch should be kept to see that no silting occurs in the mid stream thereby grounding the boats. If boats are grounded near the shores, they should be removed and replaced by pile bridges. Sometimes grounding can be avoided by scoping out silt between the boats.

8. The boats should be secured in position by anchors or by tying each pier to a strong cable or chain stretched across the stream. The second method is the best in rapid currents with strong bottoms, but anchors give good results in firm or muddy bottoms. Generally, every second pier should have an up-stream anchor, but in heavy current of water, every pier should have one. Few anchors in down-stream side should also be provided to check against strong winds blowing up the river.

9. As the boats nearest to the shore on each side are much affected by weights moving on and off the bridge an account of the sloping approaches, two of the strongest and most capacious boats should be provided on ends of bridges.

10. Old ropes should be thoroughly examined before use on the anchors and even then they should only be relied for short periods as they are liable to give way some day due to rot and shear.

11. Deck and stringers should be checked for sign of decay. Decay can be retarded by an application of tar or asphalt when the floor is laid.

12. All the stringers should be checked for canting and sagging. Proper cross-bracing will eliminate canting. The only remedy for sagging is replacement with heavy section.

13. Undue vibrations, noise nuisance and much danger can be eliminated by keeping nails firmly in position.

14. Hand rails should be inspected. If it is not adequate for vehicular traffic, it should be replaced by more firm steel hand rails.

15. The approaches should be properly aligned and the chequered plates should be laid in such a way that the edges of steel may not cut the rubber tyres of vehicles.

16. The boats would be properly caulked from time to time with *dhak* roots etc. if their bottom shows any sign of leakage. During monsoons when the bridge is dismantled all caulking and painting should be done.

17. A constant watch of the water level of the river should be kept and daily reading recorded. This data will form a guide in deciding when the bridge is to be created and when dismantled.

18. When the bridge is dismantled all the material should be properly stacked duly numbered so that useful time is not wasted at the time of erection.

48.11 MAINTENANCE OF TRAINING AND PROTECTIVE WORKS

While maintaining training and protective works, the following points should be especially noted :

1. If erosion and scouring has started, which might endanger the footings, this should be avoided by filling the scour holes besides the footing with large boulders or stacked concrete.

2. The present waterway should be checked to carry maximum flood discharge.

3. The presence of any obstruction such as drift, logs, stumps, old piers etc. which might divert current and cause eddy currents in such a manner as to undermine footing and aggravate scour, should be checked.

4. The pressure of any undergrowth or obstruction which might affect the free flow of water through the structure, cause a fire hazard.

5. The present condition of the existing training and protective works should be checked to determine if any more work is required.

6. If considered necessary the protective works should be protected by rock fill.

7. The toe of the fills at the ends of wing-walls should be checked for scouring. If considered necessary stone boulders should be provided.

8. During flood periods all the major structures should be carefully inspected.

9. Any alteration in the bank should be inspected, which might affect the direction of the flow of the current.

10. All tall trees close to wooden or steel bridges should be removed in order to preclude the possibility of their falling upon and damaging the structures.

48.12 MAINTENANCE OF APPROACHES TO BRIDGES

While maintaining approaches the following points should be noted :

1. The approaches of the bridges should be maintained to true cross-section and on a grade ensuring that the portion of the road adjacent to the bridge deck is levelled upto a length of about 30 m. No change in grade should be allowed because, if there is a change in grade after the deck, the vehicles will experience a hump and cause an impact on the deck which is not desirable, and in wet conditions there is a danger of the vehicle getting out of control.

2. In some bridges pucca drains are constructed at the ends of long structures to prevent the erosive effect of drainage water on down embankment slopes and also to prevent the water from soaking down behind the abutments and wings. Such drains should be inspected and kept in good conditions.

3. If concrete approaches are provided, these should be inspected carefully and adequate expansion space should be provided.

4. To prevent constant cutting action of rain water on slopes, grass should be grown.

5. The road surface should be kept in good condition, because it is observed that due to settlement of banks, the road surface gets uneven.

6. At either side of the slope, pucca drains should be provided at regular intervals so that the water does not damage the embankment.

48.13 MAINTENANCE OF EXPANSION JOINTS AND BEARINGS

The most troublesome and costly item in the maintenance of bridges is the repair of failed expansion joint. As a matter of fact such failure is due to the movement of piers and abutments resulting from scour and settlement. The standard roller and rocker assemblies used on the large and heavier structures give little *or no trouble,* when movements are within desired limits. The damage to rockers mainly occur due to unexpected pier or abutment movements because of the pressure of the approach fill. Such repairs can be done by jacking the structures free of the rockers and moving the bearing plates to new position. If the steel or bronze expansion plates are used in concrete girder bridges, it will cause considerable trouble. If the plates or concrete are not placed properly it will result in the spalling of the edges and sides of the cap or bridge seal. The periodical inspection of the conditions of expansion plates, rockers and rollers should be done to observe the free movement of them. All the plates, rockers and rollers should be frequently oiled to prevent them against rust and to remain in good working order. No debris, soil, etc., should be allowed to accumulate around them because they interfere in their working.

REVIEW QUESTIONS

48.1. Write a short note on the maintenance of records and inspection of bridges.

48.2. Why regular maintenance of bridges is necessary? What will happen if it is not done properly and timely?

48.3. What points should be noted while inspecting steel bridges? How the maintenance of painting will be done?

48.4. Write a short note on the inspection and maintenance of wooden bridges.

48.5. How the masonry arch bridge can be kept in good condition? If a crack is developed in the arch, what preventive measured you will adopt?

48.6. In R.C.C. bridges if cracks develop and steel reinforcement is bared what will you do? If the concrete at one position is porous and it has absorbed some water, how it will affect and what should be done?

48.7. What points should be kept in mind while inspecting suspension bridges?

48.8. What points should be observed during maintenance and construction of cause-ways?

48.9. How the road surface of a bridge is kept in good condition? How drains should be constructed? What points should be kept in mind while maintaining guard kerbs, foot-paths, railings and parapets of the bridges?

48.10. If some moisture is accumulated in the back of abutments, how it should be removed? If the foundations of abutments are undermined due to scouring, what should be done to prevent it from settlement?

48.11. How are the piers and wing-walls kept in good condition?

48.12. What salient points should be observed for the proper maintenance of timber pile-bents?

48.13. Write short notes on the maintenance of approaches.

48.14. How can the expansion joints and bearing plates be kept in good working order?

48.15. Why maintenance of Training and Protective work is necessary and how it is done?

48.16. Write short notes on the maintenance of floating-bridges.

48.17. Write a note on maintenance and preservation of steel bridges.

48.18. Write a short note on maintenance of bridges.

PART 4

Tunnel Engineering

Introduction

GENERAL

An underground engineering structure or artificial gallery used for transporting traffic, sewage, water, oils and minerals etc. is known as '*Tunnel*'.

49.1 HISTORY OF TUNNELS

Tunnels were constructed in the very early age of civilization. It is said that the first tunnel was built about 4,000 years ago between two buildings by the Egyptians in Babylonia. The first tunnel under Eupharates river lined with brick masonry was constructed in Egypt connecting the Royal Palace to Temple of Jove.

In A.D. 54 the Emperor Claudius built about 5.9 km long tunnel for carrying water of a spring under Appennine Mountains. The internal dimensions were about 3.0 m × 1.8 m and it was completed in twelve years by 30,000 labourers. The slaves and the prisoners were given the job of tunnel construction in the shape of severest punishment. Romans also constructed a 900 m long road tunnel about 2000 years ago under Posilipo hill.

One of the oldest Greek tunnels was built 2600 years ago for carrying water on the island of Samos. This tunnel was about 1.8 m × 1.8 m and about 1500 metres long.

Upto seventeenth century, practically there was no improvement in tunnel construction methods. The use of gunpowder was started in France in 1679–81 for construction of Languedoc canal. In France during 1766–1777

Harecastle tunnel was constructed having about 3 m × 4 m section, 2.4 km length. First railway tunnel was also constructed in France in 1826 on Roanne–Andressieax line. One of the longest tunnel used for navigation was constructed in 1927 known as Rov Tunnel in France.

Use of shield was introduced in 1825 by Sir Isam Bard Brunel for the construction of Thames river tunnel in Britain. In 1869 use of iron lining was started by Peter William Barlow while constructing tunnel under river Thames in London. The first railway tunnel was 3.2 km long and about 9 m × 7.5 m cross-section, known as Box-Hill tunnel constructed around 1840 in Britain. In Britain use of shields and compressed air was done for the construction of Rotherhithe tunnel completed in 1908.

Connaught Tunnel 7.2 m × 8.0 m size, 8 km long is the first railway tunnel of Canada. Portage tunnel 6.0 m × 5.7 m size, 270 m long is the first railway tunnel of USA, which was completed in 1833 in Pennsylvania.

6.0 m high × 7.2 m wide, 7.6 km long tunnel was constructed during 1857–1875 in U.S.A., which connects Boston with Albany. America's first vehicular tunnel was built in 1866 which is known as Washington S.T. Tunnel. It was constructed under Chicago river.

Important tunnels of the world are given in Table 49.1.

TABLE 49.1 *Details of some of the important tunnels of the World*

Constru -ction period	*Country*	*Name of the Tunnel*	*Size and length (metres)*	*Shape*	*Purpose*
679–84	France	Languedoc	84 ×70 × 156	Rectangular	Navigation
1851–53	Hungary	Forest-Hill	9.5 × 105 × 350	Horse-shoe	Road tunnel
1857–71	France-Italy	Mont Cenis	8 × 7.5 × 12700	Horse-shoe	Railway - tunnel
1895–1906	Switzer-land Italy	Simplon	5 × 35 × 19730	Horse-shoe	Railway - tunnel
1900	Pans	Metro	67 × 4 5 × 164600	—	Sub-way
1911–27	France	Rov	22 ×15 5 × 7120	Rectangular	Navigation
1918–34	Japan	Tanna	8 4 × 5 2 ×7800	—	Railway
1924–77	USA	New Cascade	4 90 ×7 30 × 12400	Horse-shoe	Railway
1931–33	Belgium	Under Schedule	67 × 45 × 575	—	Road tunnel
1934–39	USA	Lincoln	about 6 0m dia 2730 m long	Circular	Road tunnel

Contd.

Table 49.1 Contd.

Constru -ction period	*Country*	*Name of the Tunnel*	*Size and length (metres)*	*Shape*	*Purpose*
1936–41	USA	Queens	6′3×4 1 × 1900	Circular	Road tunnel
1937–41	Netherlands	Mass	248 × 84 × 1070	—	Road tunnel
1939–46	Britain	Ilford-linc	about 3.70 m diameter and 15200 m long	Circular	Subway (Tube) Railway
1946–50	France	Genissiat	11.4 × 8.60 × 11.50	Horse-shoe	Navigation
1957–1960	Japan	Hokoriku	8 5 × 5 3 × 13870	Horse-shoe	Railway

49.2 ADVANTAGES AND DISADVANTAGES OF TUNNELS

49.2.1 Advantages

Following are the advantages of tunnels :

1. Tunnels protect the pavements from the weathering action of rains, snow and other natural influences, hence reduce their maintenance cost.

2. Tunnels protect the cities and other conveyance vehicles, passages which are inside them, from destruction during war bombardings.

3. At some places tunnels are cheaper for transporting water, gas, laying railway lines, roads across rivers or mountains than bridges or open cuts.

4. In most congested cities having no space for the construction of railways or roads, tunnels are constructed to provide subways for removing traffic congestions.

5. When the hill consists of soft soil and it is uneconomical to maintain the open cut, due to large number of slips, tunnels are constructed.

49.2.2 Disadvantages

Following are the disadvantages of the tunnels :

1. Tunnels require special equipments and modern technique for their construction, which are not available everywhere.

2. Tunnels require much longer time for their construction as compared with bridges and open cuts.

Generally when the depth of cut is more than about 17.0 m, tunnels are provided.

49.3 ADVANTAGES AND DISADVANTAGES OF OPEN-CUT

49.3.1 Advantages

Following are the advantages of open-cuts :

1. The cost of construction of the open cuts in the mountains when the height of overlying rock or soil is less than 20.0 m is much less as compared to the tunnels.

2. The time required for the construction of tunnel is more as compared with the open-cut.

49.3.2 Disadvantages

Following are the disadvantages of the open-cut :

1. It requires greater approach lengths, as the level of the cut invert is higher than the tunnel.

2. There is always danger of side slips in case of open-cuts.

3. Maintenance cost of railways or highways is much more in open-cuts than tunnels.

4. As the open-cuts are exposed to the atmosphere, the atmospheric action such as snow, rains etc. affect the rails and the roads much more causing increase in the maintenance cost.

5. The open-cuts are totally unsafe during aerial warfare or bombing.

When the vertical depth of the cutting exceeds 20.0 m, it is always cheaper to construct the tunnel, than providing the open-cut. But this thumb rule is not always followed. The actual decision between the open-cut and the tunnel is made after considering several factors such as local condition, type of rock, availability of construction materials, availability of the construction equipments and time of completion of the project.

REVIEW QUESTIONS

49.1. (a) What are the advantages of the tunnels?

(b) What are the demerits of the tunnels?

49.2. Write short note on the history of the tunnels.

49.3. Name some important tunnels of the world and what do you know about them?

Surveying

GENERAL

First of all both the ends of the tunnel are fixed very accurately on the hill. The next step is to locate the centre line on the ground and the exact length of the tunnel is determined. Accuracy in the survey work is most essential, because slight error may lead to wrong alignment and grade of the tunnel, causing all the work in vain. Therefore the survey work is the main important operation of the tunnelling.

50.1 ALIGNMENT AND GRADE

While deciding the alignment and grade of tunnel the following points should be kept in mind.

1. As far as possible the alignment and grade of tunnel the following points should be kept in mind.

2. For providing good drainage minimum grade of 1 in 50 should be provided in drains. In the case of longer tunnels from the middle of tunnels from the middle of tunnel grades may be provided on both sides to carry drain water towards both ends.

3. As far as possible minimum grade should be provided in tunnels and their approaches.

4. Proper precautions should be taken for efficient ventilation and lighting of the tunnels.

50.2 LOCATION OF CENTRE LINE ON THE GROUND

The correctness and economy of constructing tunnels entirely depends on the accuracy of the surveying. Therefore, survey work should be done with great accuracy and it should be checked several times during execution of the construction work also.

The location of centre line on the ground is done as follows :

(1) When the length of the tunnel is small, the centre line is accurately located on the ground by means of a theodolite during calm and clear day time. The wooden pegs are driven in the ground at various short intervals and the centre-line is marked on the top of wooden pegs by means of nails. If it is not possible to drive the wooden pegs in hard stratum, the surface of the rocks is painted with white paint and the centre line is marked with black paint over it. The centre line located so is checked several times and after thorough checking permanent monuments of stone or concrete are placed on the centre line at various places.

(2) When the length of the tunnel is more, accurate and elaborate survey methods are used for transferring the centre line on the ground. Large size Micrometer Transit Theodolite should be used and the centre line should be located by means of triangulation method. All the operations for locating the centre line on the ground must be performed with greatest accuracy and should be checked several times by independent methods, to avoid any error in the layout of centre line. After accurately locating the centre line on the ground, permanent stone or concrete monuments should be provided on the centre-line.

50.3 SEQUENCES OF SURVEYING

Mainly the survey work consists of the alignment of the centre line on the surface, its transfer into the tunnel and proper levelling inside the tunnel.

Following are the main sequences of the surveying during tunnel construction :

(1) First a preliminary location survey is done in the area, which is followed by the precise re-survey of the centre line of the ground surface.

(2) This centre line is transferred to the underground through shafts which are constructed over it.

(3) The tunnel excavation is started from both ends as well as from several intermediate shaft faces. Therefore minute accuracy in the centre line is utmost necessary. This centre line is extended on both the sides inside the tunnel.

(4) In case of long tunnel, tunnel transit is used for surveying, because it is large, powerful, invertible in the wye stands and is fitted with striding level, by means of which the transverse axis can be easily made truly horizontal.

50.4 TRANSFERRING CENTRE-LINE TO INSIDE OF TUNNELS—WHEN THE SHAFT IS LOCATED DIRECTLY OVER CENTRE-LINE

Generally the centre line is transferred underground through the opening of shaft. This is a very difficult work and requires more attention because the alignment will entirely depend on it. If any mistake is committed in it, the position of that part of the tunnel will be dislocated. This is done as follows:

Two plumb-bobs attached to fine wires are suspended in the shaft. The wires are wound on reels placed on the mouth of the shaft. By revolving reels plumb-bobs can be raised or lowered. In the tunnel these plumb-bobs are immersed in water or oil buckets, so that they may not be affected by vibrations. With the help of a theodolite these wires are brought on centre-line at the ground.

When the wires at top are lying on centre line, they will automatically be on centre line in the shaft also. Now one theodolite is kept about 7 metres

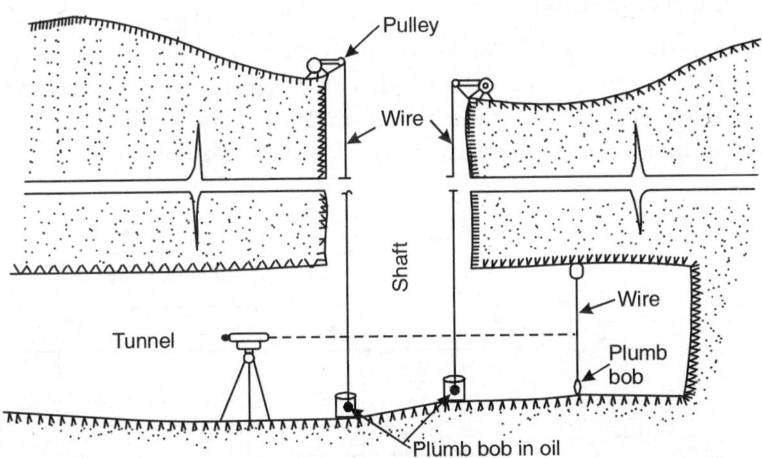

Figure 50.1 *Method of transferring centre-line from the top to the inside of tunnel*

away from wires in such a way that, when observing through the telescope both wires are in one line. In this position the axis of telescope will lie in the centre line of the tunnel. Now the construction of the tunnel in one or both the directions is started along the centre-line, which has been fixed by driving the pegs in the roof of the tunnel and suspending plumb-bobs from them as shown in Fig. 50.1.

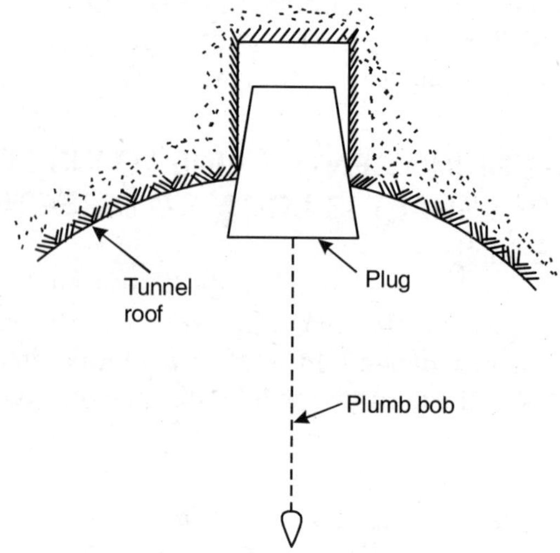

Figure 50.2 *Details of the pegs driven in the ceiling of the tunnel*

50.5 TRANSFERRING CENTRE-LINE TO INSIDE OF TUNNELS—WHEN THE SHAFT IS LOCATED TO ONE SIDE OF CENTRE-LINE

In some of the cases the shafts are placed on one side of the tunnel. In such cases the centre-line cannot be directly transferred to the inside of the tunnels. For transferring the centre-line from the ground to the underside following method is generally adopted. (Refer Fig. 50.3).

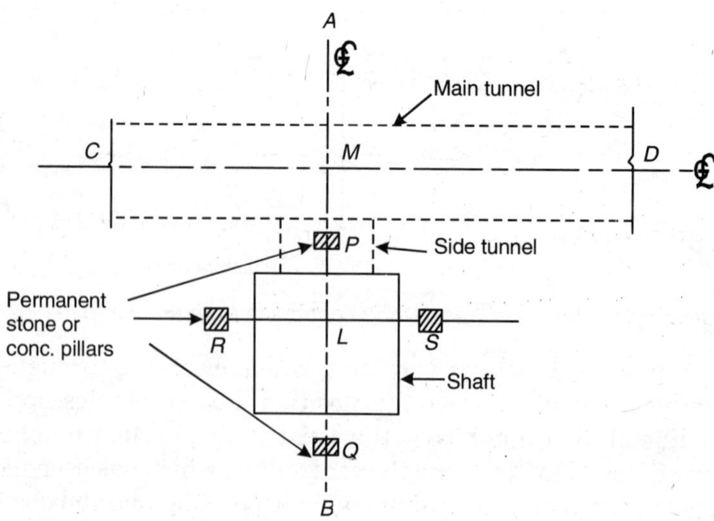

Figure 50.3 *Method of transferring centre-line to the tunnel through shaft in the side*

(a) Line *AB* exactly perpendicular to the centre-line of the main tunnel is set out on the ground by means of theodolite.

(b) Permanent stone or concrete pillars *P* and *R* are erected on line *AB,* on both the sides of the shaft. Between the pillars a wire is stretched and from it two plump-bobs are suspended as described in Sec. 50.3. With the help of these wire lines *AB* is transferred inside the tunnel.

(c) Line *RS* is also established parallel to the centre-line of the main tunnel, intersecting line *AB* at *L.*

(d) Point *M,* the intersection of main shaft centre-line and line *AB* are located by means of theodolite inside the tunnel.

(e) Keeping theodolite at point *M,* line *AB* is sighted by means of two wires suspended on *AB* through the shaft.

(f) Now the theodolite is given exactly 90° turn and the main centre-line of the tunnel is located.

The centre line located so should be checked by various independent methods. After ascertaining the accuracy the construction of the tunnel should be started on one or both the sides.

50.6 METHODS OF LOCATION OF CENTRE-LINES OF CURVILINEAR TUNNELS

Following two methods for the location of centre line of the curvilinear tunnels are usually adopted in practice:

(i) By chords and deflection angles

(ii) By tangent offsets.

50.6.1 By Chords and Deflection Angles

In this method the centre-line points *A*, *B* and *C* of the tunnel are located as follows : (Refer Fig. 50.4)

(a) For any chord length l, the angle subtended by the chord is first calculated by the formula

$$\sin\frac{\phi}{2} = \frac{l}{2r}$$

where

l length of the chord

r radius of the curve.

(b) The theodolite is set up at the tangent point *A* and the back sight along *Ax* is taken. Now the telescope is turned 180° along the horizontal axis to x. Now an angle of $\phi/2$ is subtended and point B is located by measuring chord length l.

(c) The theodolite is shifted to station B and by the similar procedure point C is located.

(d) Now the theodolite is shifted to another point C, D, E,... etc. and the points on the centre-lines of the curve D, E,... etc. are located.

(e) All the points A, B, C, D,...etc. are joined together by a smooth curve, which is required centre line of the curvilinear tunnel.

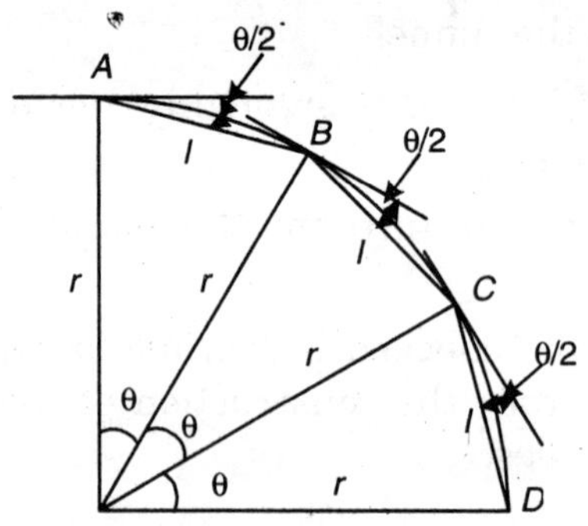

Figure 50.4

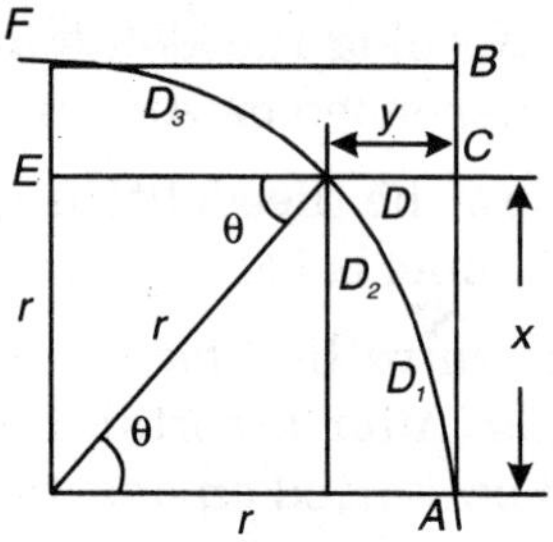

Figure 50.5

50.6.2 By Tangent Offsets

This is the common method used for locating the centre-line of the curvilinear tunnels.

In this method the ordinate length DC is calculated at the fixed tangent length AC as shown in Fig. 50.5.

Referring to Fig. 50.5, if the radius of the curve of the tunnel is r, then

$$OR = AC = x$$

$$OA = OD = OF = r$$

$$AC = OE = x$$

$$DC = y$$

$$ED = \sqrt{OD^2 - OE^2}$$

$$= \sqrt{r^2 - x^2}$$

$\therefore$ Length of ordinate $DC = y = EC - ED$

$$= r - \sqrt{r^2 - x^2}$$

$$\therefore \quad y = r - r.\cos\phi$$

In this method for various lengths of the tangents x_1, x_2, x_3,... etc. the ordinate lengths y_1, y_2, y_3, ... etc. are calculated. All the points D_1, D_2, D_3, ... etc. corresponding to the various ordinate lengths are located and joined by means of smooth curve, which is the required centre-line of the curvilinear tunnel.

50.7 METHOD OF PROVIDING GRADE

Following method is generally adopted for giving grades or slopes to the tunnel flows. (Refer Fig. 50.6):

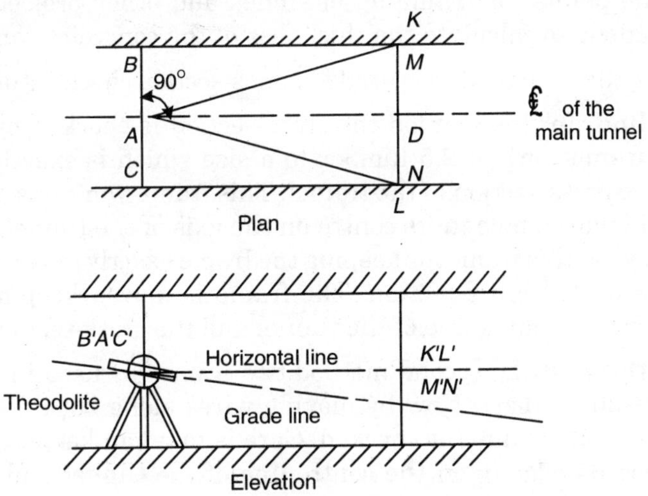

Figure 50.6 *Method of providing grade in the tunnel*

1. The theodolite is kept on the centre-line of the main tunnel at A and then centered and levelled.

2. Now it is rotated exactly 90° horizontally and points B and C are located on the sides of the tunnel and plugs are inserted at these points, exactly to the level plane of the theodolite.

3. Keeping the telescope exactly in horizontal level it is directed along the tunnel and points K and L are marked on the sides of the tunnel as shown in Fig. 50.6.

4. The vertical distance $KM = NL$ is calculated exactly to give the required grade for the measured horizontal distance $BK = CL$. Now plugs and spades are driven at points M and N.

Plumb-bobs are suspended through piano wires in shafts for transferring levels from the ground surface to the bottom of the shafts. Two marks are made on these wires, one at the top and the other at about 1.5 metres above the springing level of the roof arch.

With respect to the bench marks the corresponding reduced levels of the top and bottom of the tunnel are calculated. With the help of these reduced levels, reference points are fixed at the flow levels corresponding to the required grade of the tunnel. The grade of the tunnel is checked several times, to avoid any possible error.

50.8 TUNNEL CROSS-SECTION CHECKING

The cross-section of the tunnel should also be checked for its accuracy. Generally the tunnel cross-sections are checked twice. First checking is done to locate the points for trimming of stones and other projections, second checking is done to calculate the thickness of the concrete, yardage etc.

Following three methods are used for cross-section checking of the tunnels:

1st Method. In this method the cross-section is checked by means of a light steel frame-work of 2.5 lumber to a size which is exactly 15 cm less than the true cross-section of the tunnel. This frame-work has a cross-board fixed with it, with a hole in its centre on the axis of the tunnel. The plumb-bobs are fixed on this frame for keeping the frame exactly in vertical position, while checking the cross-section. The frame is moved keeping the cross-board hole on the centre-line of the tunnel and the cross-section is checked.

2nd Method. In the second method two frames as used in first method are kept 20–30 m apart. Now by moving wires along both the frames the cross-section of the tunnel is checked. Care is taken in keeping the holes of the cross-boards exactly on the centre-line of the tunnel and keeping the frames in vertical position.

3rd Method. In this method only theodolites and spades are used for making the cross-section. After some distances points are located, marked and checked.

All efforts should be made to check the accuracy of the tunnel, as slight variation may lead to great difficulties later on.

REVIEW QUESTIONS

50.1. What points should be kept in mind while deciding the alignment and grade of the tunnels?

50.2. How the centre-line of the tunnel is located on the ground?

50.3. Write short note on the sequence of surveying in connection with the tunnelling.

50.4. Describe how you will transfer the centre-line and grade of a proposed tunnel from ground surface to the inner surface.

50.5. Write short note on the systems used for surface alignment of the centre-line of a tunnel. Explain how you will transfer the alignment from the ground through the shafts.

50.6. Write short note on the cross-section checking of the tunnels.

50.7. Write short note on the methods of locating centre-line of the curvilinear tunnels.

50.8. Write short notes on :

(a) Spade and plug

(b) Tunnel grade

(c) Setting curves of tunnels

51

Shape and Sizes of Tunnels

GENERAL

The shape of the tunnel should be such that it can resist the pressure exerted by the unsupported walls of the tunnel excavation. The pressure coming on the lining of the tunnel is both lateral as well as vertical in direction and varies with the character of the hill material and the percentage of moisture in the material. Some rocks have so much cohesion that they can stand vertical after cutting the material without any support, but in case of loose soil, sand and mud-stones after excavation they cannot be kept in position without proper support. There the design of the shape of the tunnel should be done properly suiting the site conditions and the functional requirements.

The size of the tunnel will depend on the purpose for which tunnels are constructed.

51.1 CLASSIFICATION OF TUNNELS

Tunnels are classified on the basis of their purpose, materials of construction and the position or alignment.

51.1.1 Classifications Based on the Purpose of the Tunnel

The purposed and classification of the tunnel is given as below.

1. Road or Highway tunnel
2. Railway tunnel
3. Navigation tunnel
4. Pedestrian tunnel

5. Subway tunnel
6. Sewer tunnel
7. Transporting tunnel
8. Hydro-electric power tunnel
9. Water-supply tunnel etc.

51.1.2 Classification Based on the Type of Material

1. Tunnels in mud-stone rocks
2. Tunnels in sandy soils
3. Tunnels in hard stone rocks
4. Tunnels under sea or river bed
5. Open-cut tunnels.

51.1.3 Classification Based on the Alignment

1. *Spiral tunnel.* To increase the length of the tunnel to avoid steep slopes in narrow valleys, tunnel is provided in the shape of the spiral, by forming loop in the interior of the mountain. Such formation of loop is called spiral tunnel.

2. *Off-spur tunnel.* These are tunnels which are constructed to short-cut minor obstructions.

3. *Slope tunnel.* These are constructed for economic purposes for the safe operation of railway and roads in steep hills.

4. *Saddle or Base tunnel.* To reduce the overall length of the tunnel, the track or road is constructed in the valley, along the natural slope till the slope does not exceed the ruling gradient. Such tunnels as known as *saddle or base tunnels.*

51.2 SHAPES OF TUNNELS

Following shapes of the tunnels are commonly used :

1. Circular
2. Horse-shoe
3. Rectangular
4. Elliptical
5. Egg-shaped
6. Segmental roof section

51.2.1 Circular Section

These sections are commonly used for tube-railways, highway tunnels, sewer and hydro-electric tunnels.

Following are the advantages of this section :

1. For minimum perimeter, it gives maximum cross-sectional area, hence it is economical.

2. For non-cohesive soils it is best suitable.

3. This section can be constructed by using shields.

4. Circular section is the best section for taking external or internal forces.

5. This section is best suitable for sewers and water carrying purposes.

Following are the disadvantages of this section:

(a) It is not suitable for roads or railways as more filling is required for obtaining flat base.

(b) The construction of circular section is difficult than other sections.

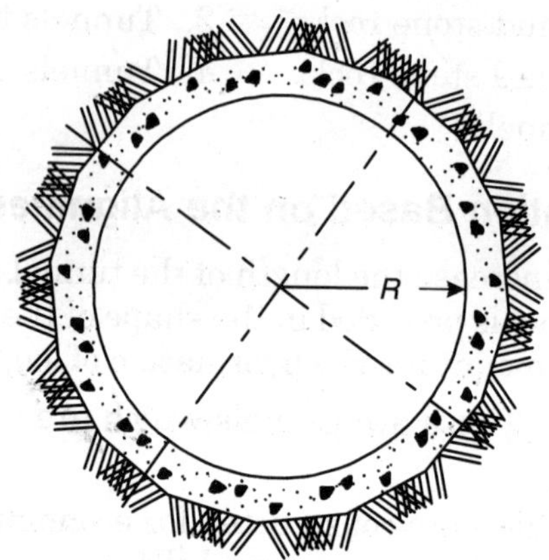

Figure 51.1 *Circular section*

51.2.2 Horse-Shoe Section

This section is commonly used because it is in between the circular and arched tunnel.

Following are the advantages of horse-shoe section :

(a) The external pressure is resisted by the curved sides and the arch action.

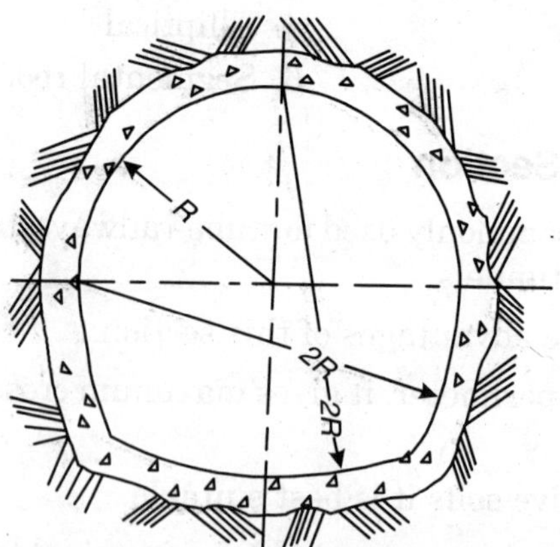

Figure 51.2 *Horse-shoe section*

(b) This section is most suitable for soft rocks.

(c) As small filling is required ; this section is best suitable for road and railway traffic.

(d) This section provides sufficient working space for the workers and strong materials during construction.

(e) This section is also suitable for carrying water or sewage, as the welted perimeter in this section is also not much greater than circular section.

51.2.3 Rectangular Section

This section is suitable only in case of hard rocks. This section is constructed with R.C.C. and is commonly used for pedestrian traffic. These sections are difficult in construction and are very costly in construction. Therefore these are not used these days.

Figure 51.3 *Rectangular section*

51.2.4 Elliptical Section

These sections are constructed in soft materials, with its major axis vertical. But these sections are not commonly used these days.

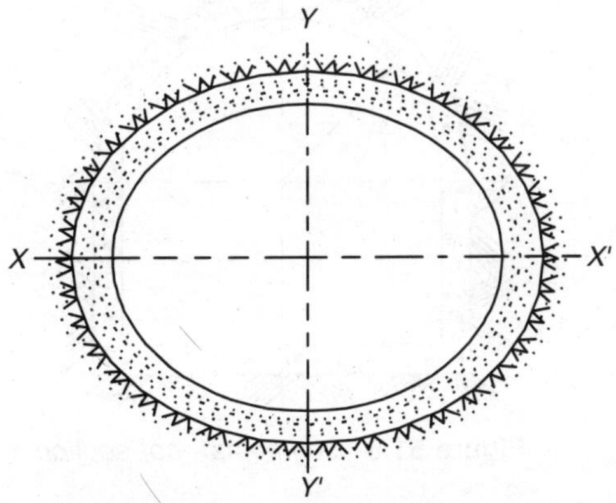

Figure 51.4 *Elliptical section*

51.2.5 Egg-shaped

This section is commonly used for sewers, because it gives self-cleaning velocity even in Dry Weather Flow (D.W.F.). This section is also good for resisting external as well as internal pressure.

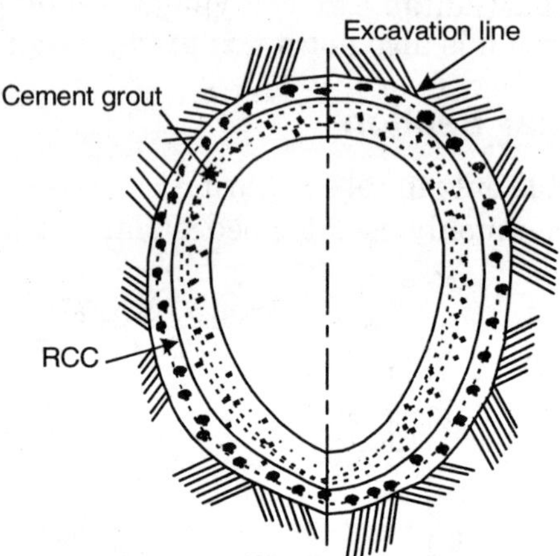

Figure 51.5 *Egg-shaped section*

51.2.6 Segmental-roof Section

Figure 51.6 shows this type of section. This section, is suitable for sub-ways or navigation tunnels. The segmental roof lakes the external load and transfers it on the vertical side-walls. This section is suitable for hard rocks. In case of soft soils the side-walls are to be constructed in R.C.C. for taking external forces.

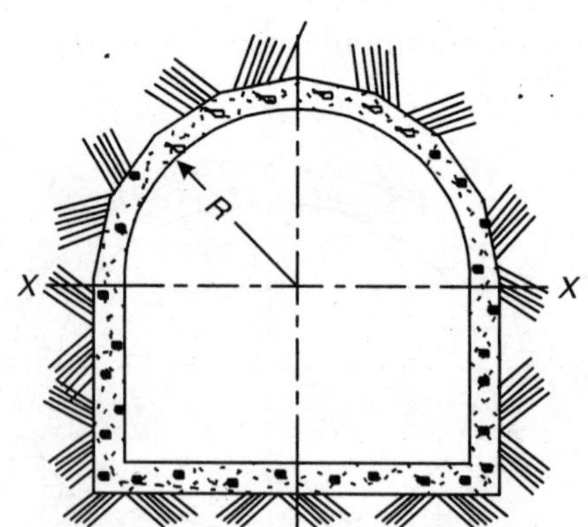

Figure 51.6 *Segmental-roof section*

51.3 SIZES OF TUNNELS

Size of the tunnel mainly depends on the functional purpose for which it is used, such as for carrying single-way or two-way highway or railway, for carrying sewage water or for hydro electric purposes.

Following aspects are mainly considered in determining the size of the tunnel :

1. Volume of traffic to be handled.
2. Type of traffic for which tunnels are to be construcied such as pedestrian, road, railway etc.
3. Size of the clear opening required.
4. Thickness of the tunnel lining.
5. Drainage facilities required.
6. Ventilation methods to be adopted.

The Indian Railway has entered the high speed age. Keeping this in view a proposal of high speed railway line between Mumbai and Poona has been prepared. In this proposal the maximum speed of the train has been assumed as 250 km/hr and the time taken from Mumbai to Poona will be 45 minutes only.

In the proposal 1.84 km long tunnel from Victoria Terminus upto Apallo Bunder will have two railway tracks. The cross-section of the proposed tunnel shall be as shown in Fig. 51.7.

In the proposal the undersea tunnel of 9.84 km from Apallo Bunder shall be multipurpose tunnel, carrying two railway tracks, two lane road traffic, space for carrying water pipelines, power and communication cables, and an inspection gallery. The cross-section of the proposed multi-purpose tunnel shall be as shown in Fig. 51.8.

The vertical and lateral pressure acting on the tunnel mainly depends on the quality and inherent strength of the rock. In case of greater magnitude

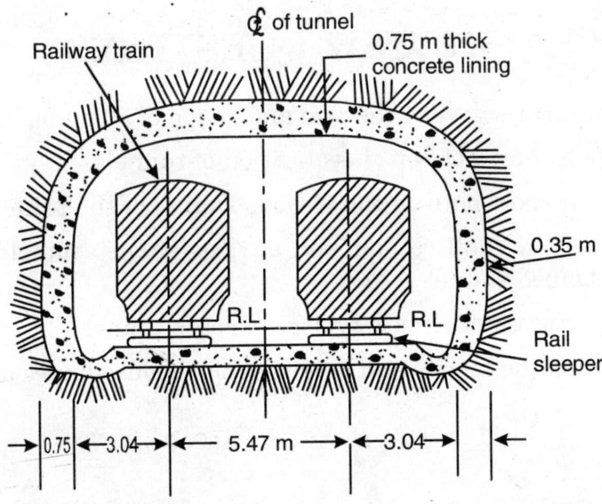

Figure 51.7 *Typical cross-section of two-line railway tunnel*

of lateral thrust, circular sectional tunnel shall be more economical. If the rocks are solid, arched roof section can be used without any lining. To reduce the tensile stresses in the crown, elliptical section with 1.6 times its width should be provided instead of circular sections. Shape of the tunnel should be properly selected and designed.

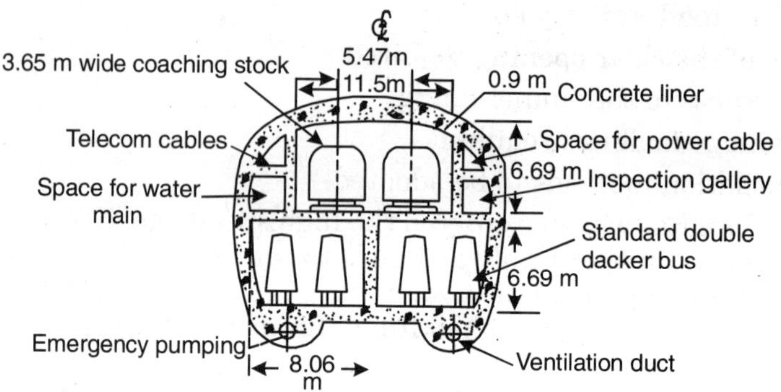

Figure 51.8 *Typical cross-section of multipurpose tunnel*

In modern times Horse-shoe tunnel sections are more commonly used. Therefore maximum tools and plants available are suitable for Horse-shoe Section. Shields are used only for the construction of circular tunnel sections. In some cases caisson method of construction is used, but it is equally suitable for circular or rectangular cross-section of the tunnel. The materials for the lining work should also be selected properly depending on their load carrying capacity i.e., compressive or tensile.

REVIEW QUESTIONS

51.1. What factors govern the size and shape of a tunnel?

51.2. Write a short note on classification of tunnels.

51.3. Write a short note on the various shapes of the tunnels.

51.4. Where you shall recommend an 'Egg-shape' tunnel? Draw a typical egg-shaped tunnel.

51.5. Draw cross-section of a typical-purpose tunnel.

51.6. Describe the merits and demerits of various shapes of the tunnels.

Explosives

GENERAL

For the construction of tunnels, blasting is done to loosen the rocks for easy removal during excavation work. Blasting is the most important operation of the tunnelling. Holes are drilled in the rocks which are required to be blasted. Explosives are placed in the holes drilled for this purpose, and then they are exploded for blasting purpose.

52.1 TYPES OF EXPLOSIVES

All explosives are classified in the following groups:

(a) Powder explosive

(b) Liquid air

(c) Disruptive explosive.

52.1.1 Powder Explosives

Following types of powder explosives are mainly used in the blasting :

1. *Blasting Powder.* Its main constituents are :

Potash Nitrate (Salt peter) = 65 – 70%

Sulphur = 81 – 5%

Charcoal = 11– 50%

It is slow burning, low explosive type powder explosive, which is uncommon is tunnelling.

2. *Nitrate-Explosive.* It is similar to blasting powder, but instead of Potash Nitrate, Sodium Nitrate is used. This is also of low strength, slow burning type explosive.

3. *Nitraman.* Its main constituent is $NaNO_3$. It is highly explosive, but it requires special primer to detonate it.

52.1.2 Liquid Air

Liquid air containing 95% O_2 at a temperature of – 191°C is absorbed by dipping a cartridge of absorbent. Now this cartridge acts as liquid explosive. But as these types of explosives require special skills in manufacturing, transporting and storing, these are not popular in rock tunnelling.

52.1.3 Distruptive Explosives

These are available in various grades, sizes and strength. These explosives are available in the market in the form of cartridges having 2.5 cm to 20 cm in diameter and 20 cm to 70 cm in length. The approximate strength of these type explosives is specified as a percentage of the ratio of weight of Nitroglycerine to the total weight of a cartridge. Following types of distruptive explosives are available :

1. *Straight dynamite.* It mainly consists of Nitroglycerine 15–60% absorbed by an absorbent. It is highly sensitive, water resistant and highly explosive. But due to bad fumes produced after explosion these are not common in tunnelling. This type of explosive is used in tight spots in tunnel.

2. *Ammonia dynamite.* It mainly contains equal parts of Nitroglycerine and Nitrate of Ammonia. It is suitable for soft rocks and is water resistant upto certain limit. As their fumes are not bad as of straight dynamite, this is commonly used in tunnelling. The strength of this dynamite is between 15–60%.

3. *Gelatin dynamite.* It mainly consists of jelly of Nitroglycerine and Nitro-cotton. In special gelatin, a part of Nitroglycerine is replaced by Nitrate of Ammonia. It is highly water-resistant and is commonly used in tunnelling. They are slow-acting if not confined in the place properly. They are quick in action and have high shattering effect on the rocks. The strength of the straight gelatins is between 20–90%, whereas it is between 30–90% for special gelatins.

4. *Semi-gelatin.* It mainly consists of Nitrate of Ammonia, gelatinsed Nitroglycerine and Nitro-cotton. These are cheaper than gelatin-dynamite and are used in soft rocks. Semi-gelatins are heavier and bulkier than the straight or special gelatins. These are available in 45% or 60% strength. These are water resistant and the odour of the fumes is not bad, hence, these are commonly used in the tunnelling.

5. *Blasting-gelatin.* These are just like rubber and are fully water-proof. These are very fast and strong explosives, but have very bad odour fumes, hence these are not commonly used in tunnelling. The strength of blasting-gelatin is about 100%.

6. *Special high explosive.* In addition to the above explosives, following types of special high explosives are also used at various places:

T.N.T. = (Tri-Nitro-Toluene).

R.D.X. = (Rapid Detonating Explosive)

P.E.N.T. = (Penta Enythrital).

52.2 QUALITY OF EXPLOSIVES

The quantity of explosives required for blasting rocks for tunnelling depends on:

1. Kind of explosive
2. Method of tunnelling
3. Hardness of ground
4. Profile of tunnel
5. Depth of lift.

Mainly the quantity of explosive is calculated on the basis of the previous experiences of tunnelling. In soft rocks about 2 kg of explosive is required for one cu. m of solid rock, whereas it is about 5.7 kg in case of hard rocks. The type and quantity of explosives used must be properly planned with regard to the characteristics of the rock, drilling method and sequence of detonations at the time of blasting.

52.3 PRECAUTIONS IN HANDLING AND TRANSPORTING

Following precautions should be strictly followed in the handling and transporting of the explosive matters:

1. During storing and transporting the explosives and their detonators blasting caps should be kept separately.

2. The vehicles in which explosives are carted should have warning signs painted all around its body. These vehicles should be driven carefully only with the authorised man.

3. Smoking or carrying of lighters, matches etc. should not be permitted on or around the vehicles carrying explosives or near the explosive store.

4. The inside and the flow of the vehicles carrying explosives should be made on non-metallic materials.

5. All the wirings in the stores and the vehicles used for carting explosives, should be heavily insulated and their fuses should be placed at safe place.

6. Stores used for storing explosives should be properly ventilated, dry, fire resistant and bullet-proof. These should be at safe distance from the buildings and roads, and must be outside the cities and towns.

7. All the preventive safety rules, issued from time to time by the government, must be followed.

8. Metallic tools should not be used in any circumstances in handling the explosives or during closing and opening the boxes containing the explosives.

9. Boxes containing the explosives should be handled rightly and properly. These should not be dropped or thrown during loading or unloading.

10. Dynamite and exploders should not be stored together in any case.

52.4 TYPES OF DETONATORS

For the disruptive explosives, special types of detonators are used for the sure ignition of the explosives. Detonators are metal shells in the shape of cylinders closed at one end. Detonators are generally of 6.8 mm diameter, 50 mm in height. 2 gm filling is done in the shell of the detonator. The detonators are inserted into cartridges of explosives, and are kept in its touch for sure ignition.The ignition of the detonator is done by Blasting Fuse or by Electric Igniters.

Ignition of the detonator is done by the following methods.

52.4.1 Blasting Fuse

These are prepared by wrapping blasting fuse powder in cotton yarn or jute and making it water proof by tape winding or synthetic resins. Blasting fuses are specially prepared depending on the conditions under which they are to be used; such as dry, moist or underwater. The burning rate of these fuses are about 110–120 sec/metre.

52.4.2 Detonating Blasting Fuse

This is commonly used in Germany. This main consists of flexible and water-proof cylindrical envelope filled with disruptive explosives. This type of fuse is not very common in India.

52.4.3 Electric Ignition

This is commonly used in India in blasting operation for the ignition of the explosive during various types of construction works.

In this method the ignition in the explosives is done by passing high current in the resistance wires embedded in the detonator placed in the cartridge of the explosive. Wires are taken from the cartridge to a safe place and are attached with the electric ignitor. The electric ignitor is operated, and it passes high current in the wires which lead to the ignition of the explosives through detonator.

Electric ignition is more safe and sure method as compared with other methods of ignition. It is water proof and by use of delays successive firing of holes is possible. But in this method care should be taken to protect the wire from damage, otherwise no current will pass and subsequently no ignition will take place.

52.5 METHOD OF BLASTING

The blasting operation for the removal of heavy stones during construction of tunnels, roads etc. is done in the following ways:

(a) Drilling of Holes

(b) Loading of Bore-holes

(c) Firing of Holes.

52.5.1 Drilling of Holes

The holes are drilled in the rocks by hand tools or drilling machines at the required places. Nowadays drilling of holes is mainly done by pneumatic drilling machines with diamond bits specially prepared for this purpose.

52.5.2 Loading of Bore-Holes

After the drilling of holes and its proper cleaning the loading of bore-holes are done as follows :

1. First of all the air is blown from the compressor to remove loose cuttings and water from the hole and making it ready for loading.

2. Now full cartridge of explosive is inserted in the bottom and tamped well.

3. Then primer with detonator is inserted in the hole and embedded into the cartridge.

4. To prevent jarring of the detonator, the primer and the second placed cartridge (if used) is tamped lightly.

5. If required more number of cartridges are further inserted in the holes and lightly tamped. Care must be taken that during tamping the detonator wires are not damaged.

6. Now the blasting fuses or electric wires are inserted as shown in Figs. 52.1 and 52.2.

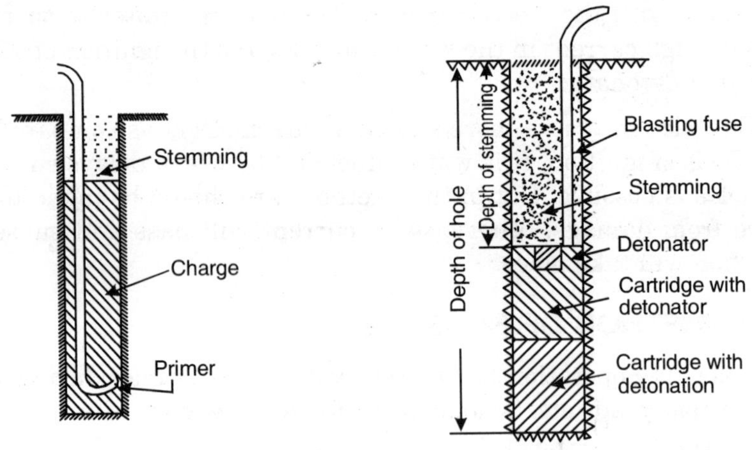

Figure 52.1 *Method of loading with single primer* **Figure 52.2** *Method of loading with stemming*

7. The remaining vacant space of the holes is done by means of damp clay or sand. Sand is better for this purpose as it easily fills the voids inside the hole.

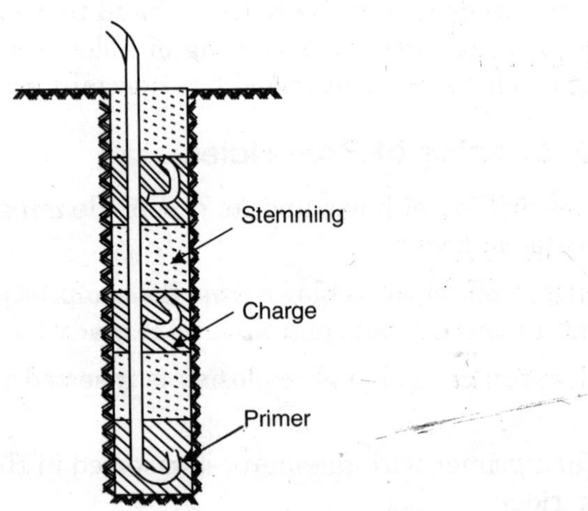

Figure 52.3 *Method of loading with more than one cartridge*

For obtaining good results and high efficiency of the blasting, the higher concentration of the explosives should be used near the bottom of the hole. If the depth of the hole is more, more than one cartridges separated by stemming should be used as shown in Fig. 52.3.

52.5.3 Firing of Holes

In India the common practice is to fire several bore-holes at one time. Following three methods are commonly used for firing of holes by electric-ignition method :

1. By series circuits.
2. By parallel circuits.
3. By combination of series and parallel circuits.

The continuity of circuits or lines is tested by means of galvanometer. This test is necessary for checking open breaks and to avoid mis-fires.

The Firing Station from which the firing of holes is controlled must be placed at least 300 metres away from the face to be blasted or outside the tunnel. Firing switch must have key and lock, and only the authorised person should operate it as laid done in the prevailing rules.

REVIEW QUESTIONS

52.1. What are the various types of explosives used in the tunnelling work?

52.2. Write short note on the types of detonators.

52.3. How the blasting is done? Describe in brief.

52.4. On what factors the quantity of explosive to be used mainly depends?

52.5. What precautions should be kept in mind while handling and transporting the explosives.

52.6. Explain in brief how you will load a drill hole meant for blasting?

Shafts

GENERAL

When the length of the tunnel is small, it can be constructed by doing excavation works from one side. But in case of considerable length and to complete the construction work in short time, the excavation can be done from both the ends of the tunnel simultaneously. But when the length of the tunnel is very long and the work is to be completed in short time, the vertical openings are made from the surface of the hill which meet the tunnel at various intermediate points. These vertical openings are called shafts and are constructed at suitable places to expedite the construction work.

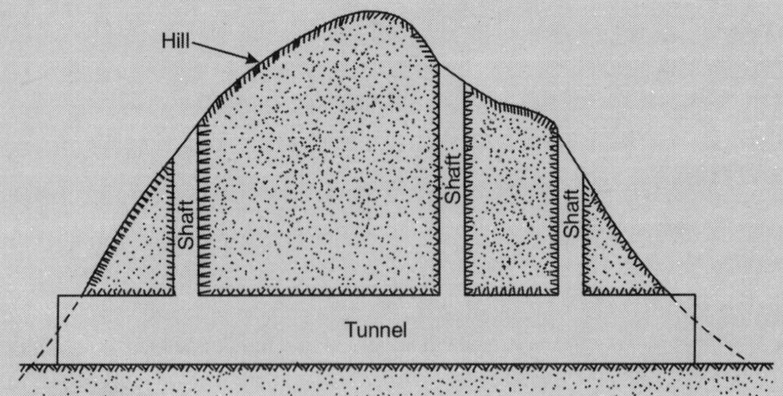

Figure 53.1 *Longitudinal section through a tunnel*

Figure 53.1 illustrates the longitudinals action through a tunnel with three shafts. In this case the work can be started from both the ends and also from each shaft in both directions. After transferring the centre-line to the underground in each shaft the work is started in both the directions. The excavated materials are taken out from the shafts.

The shafts are also constructed in tunnels even when the work is started from one or both ends, to avoid delay in disposing of the, when the excavation has been done upto a considerable length, because when the tunnel has been constructed upto a long distance, it will be costly to dispose of the material at the mouth of the tunnel. In such cases it is always economical to dispose of the material through the shaft.

53.1 CLASSIFICATION OF SHAFTS

Depending upon the position, the shafts are classified as follows :

1. Vertical shafts
2. Inclined shafts
3. Permanent shafts
4. Temporary shafts.

53.1.1 Vertical Shafts

The shafts which are constructed vertical in the plumb are called vertical shafts, the construction of vertical shafts is easier and cheaper. In some cases a pilot shaft is first driven upward from the main tunnel. Later on the section of this pilot shaft is enlarged to the required section. The widening of the section is done by blasting or other methods and it proceeds downward.

53.1.2 Inclined Shafts

The shafts which are inclined and not in plumb are known as inclined shafts. The excavation of an inclined shaft is usually done in the upward direction. The excavated materials fall down in the main tunnel from where they are removed through the mouth of the tunnel. Only in case of small depths or special situations, the construction of the inclined shafts is done.

53.1.3 Permanent Shafts

The shafts which are left open after construction of the tunnel, for providing ventilation are called permanent shafts. These shafts or tinned and proper protection works are constructed at their mouths to protect the persons or animals from falling inside the shafts.

53.1.4 Temporary Shafts

The shafts which are closed by filling excavated materials in them after the completion of the construction of the main shafts are called temporary shafts.

53.2 SIZE OF SHAFTS

The size of the shaft mainly depends on the following factors :

1. System used for hoisting.
2. Size of the muck car.
3. Quantity of the muck to be lifted.
4. Space required for fixing wiring and pipes in the shaft.
5. Size and type of construction equipments to be used.
6. Number of the workers to work inside the shaft.
7. Depending upon the final use of the shaft.

53.3 LOCATION OF THE SHAFTS

The site for the location of the shaft should be selected such that its construction cost should be minimum with maximum advantage. If site permits the advantage of the valley should be taken for the location of the shaft. If the shaft is located in the valley, a wall should be built around it to prevent the entry of water. While deciding the number of the shafts, the advantages gained from a large number of working faces should be compared with the additional cost of the shaft.

While locating the working shafts for sewers, care should be taken that it should cause minimum obstruction to the traffic and annoyance to public and residents. The elevator for the workers and various pipes are placed in one shaft, while for lifting the excavated material and for lowering the construction materials in the other.

There are following three main locations of the shafts.

1. Shaft over the centre-line
2. Side-shafts.

53.3.1 Shaft Over the Centre-line

Most of the shafts are constructed directly over the centre line of the tunnel. This situation of the shaft provides maximum facility in affording for hoisting out the excavated materials, taking down the workers and construction materials from the surface in case of long tunnels.

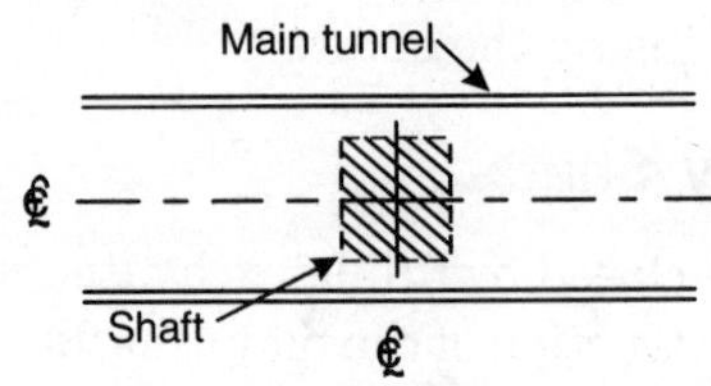

Figure 53.2 *Shaft over the centre-line*

53.3.2 Side-shafts

The shafts which are not directly constructed over the centre line of the tunnel, but are constructed in the side are called side-shafts. These shafts require a transverse gallery connecting with the tunnel. The side gallery should be of such size and shape that the construction equipment machines can be easily taken through the shafts to the tunnel.

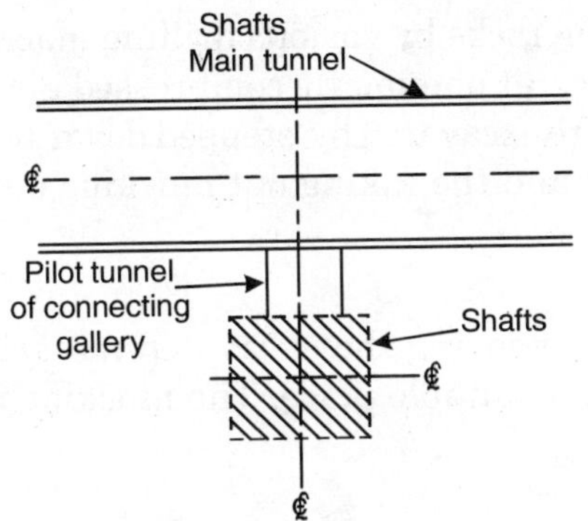

Figure 53.3 *Side shafts*

53.4 SHAPE OF THE SHAFTS

Following two main shapes of the shafts are in common practice :

1. Rectangular shafts 2. Circular shafts.

53.4.1 Rectangular Shafts

When the shafts are of temporary nature and are to be filled later on, this shape may be provided. The rectangular shafts are usually strutted with timber from all sides to prevent the falling of timber and preventing accidents.

53.4.2 Circular Shafts

This is the most common shape of the shaft which is generally used. These shafts are lined with pressed steel liner plates or concrete. This shape is adopted for the construction of the permanent shafts.

53.5 CONSTRUCTION OF SHAFTS

The method to be adopted for the construction of the shaft mainly depends on the nature of the ground. In common practice the shafts are usually sunk down from the top toward the tunnel. In special cases however the shafts can be constructed from the tunnel in upward direction. The consuuction of the shaft in the upward direction is cheaper as the muck is dropping down and can be directly trapped into the muck carrying cars.

Following are the main operations for the construction of the shafts in the rocks.

1. Drilling and blasting
2. Mucking
3. Timbering
4. Pumping

53.5.1 Drilling and Blasting

Holes are drilled in the rocks by various drilling machines and equipments, such as jackhammers and pneumatic compressed air operated equipments. The large size shafts are excavated by 'stepped down' technique for permitting the drilling operation and the taking out mucking outside simultaneously.

53.5.2 Mucking

It is the operation of removing the excavated material from the shaft and dumping it outside at a suitable place. The mucking operation can be done

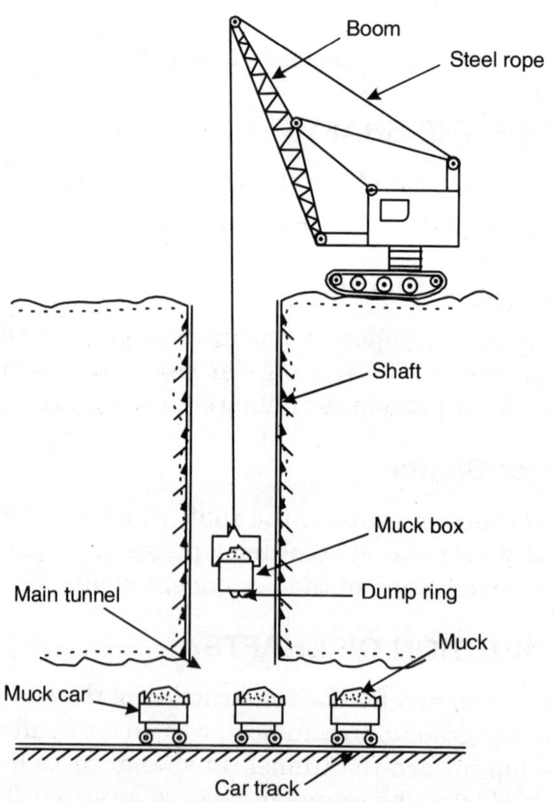

Figure 53.4 *Mucking operating by crane*

by manual labour or by cranes. Figure 53.4 shows the mucking operation by crane. In the mucking by crane two buckets are used, when one is being loaded, the other is being hoisted to the surface.

53.5.3 Timbering

It is a process of providing temporary support to the cut soil sides against falling till lining is done or finally filled up after the completion of the work. Figures 53.5 and 53.6 show various methods of timbering.

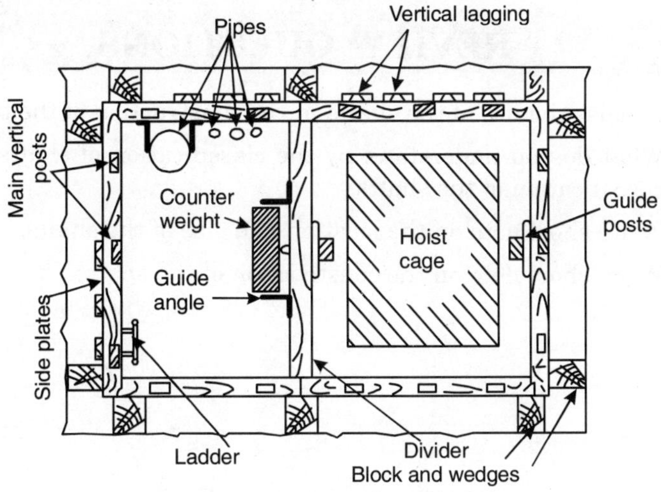

Figure 53.5 *Timbering for small shafts*

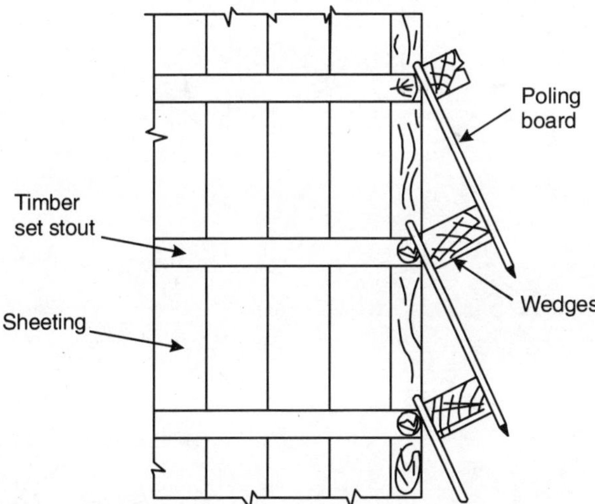

Figure 53.6 *Timbering for deep shafts in soft soils*

53.5.4 Pumping

Usually pump-sets are used for lifting the water from the shaft. This water may be due to seepage from under-ground or water used during construction operations. In the beginning the buckets used for removing the muck can be utilized in lifting up the water also. In case of deep shafts, it is very costly to pump the underground seepage water. In such situations grouting is done in seams to seal off the flow of water to some extent. Water can be collected in the sump well from where it can be pumped to the nearby water courses.

REVIEW QUESTIONS

53.1. Discuss the use of the shafts in the construction of the tunnels.

53.2. What do you understand by the classification of shafts? Discuss the classification of the shafts.

53.3. Write short note on the location and size of the shafts.

53.4. Write short note on the construction of the shafts.

Tunnelling in Rocks

GENERAL

Tunnelling in the rocks is very difficult and tedious work. Nowadays tunnelling in rocks is done with the mechanical methods only, because tunnelling by manual labour is not done these days, as it was done during ancient times.

Methods of tunnel construction in hard rocks differ from the tunnelling in soft grounds. Tunnelling in rocks is much costlier than tunnelling in soft ground. As rocks are self-supporting, they require less supporting. As slight deviation will involve huge wastage of money, greater care is required in tunnelling through rocks. In rocks the tunnelling work can be started in various sections for expediting the work.

54.1 SEQUENCE OF OPERATIONS

All the operations in the tunnel construction must be done in proper sequence and in well-planned manner. The sequence of various operations mainly depends on the type and size of tunnel, method of blasting, type of formation encountered etc.

After locating the centre-line, generally the work of tunnelling is done in the following sequence:

1. Construction of shafts.
2. Transferring centre-line to the inside the tunnel.
3. Deciding method of excavation.

4. Setting up and drilling holes for blasting.
5. Loading holes with explosives and blasting.
6. Ventilation and removing dust after blasting.
7. Carting the muck outside the tunnel.
8. Pumping and removing ground water from the inside of the tunnel (if any).
9. Providing supports inside the tunnel.
10. Lining of the tunnel.

54.2 METHODS OF TUNNELLING

All the methods of tunnelling have been grouped as follows :

(A) Tunnelling in soft rock or soft ground :

1. American Method 2. English Method

(B) Tunnelling in hard rock :

1. Heading 2. Heading and Benching
3. Full face 4. Drift
5. Perimeter method 6. Pilot tunnel.

54.2.1 American Method

In this method first a drift is driven at the top and then it is supported by leggings, cap and vertical posts. After properly supporting the drift, it is widened on both sides gradually and also supported by timber planks and struts. In this way we gradually go on widening and supporting till we reach

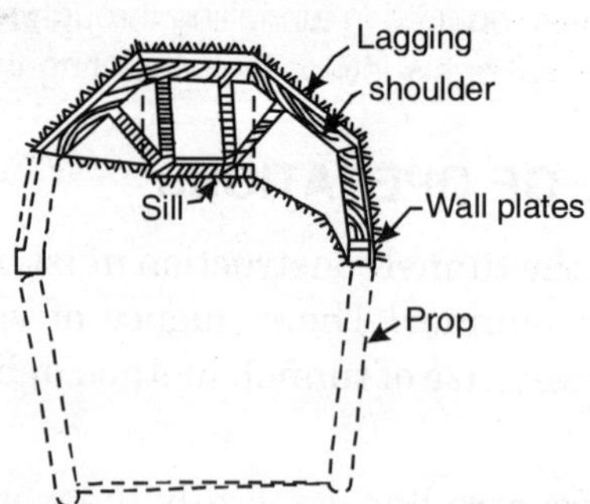

Figure 54.1 *American method*

at springing level of the arch. At springing point, wall plates are fixed, after driving posts shown by dotted lines in Fig. 54.1. In this way the total roof load is transferred to vertical posts. Now we continue the excavation of the remaining portion of the section. After completing one section-length we start excavation in forward section in the same way till the complete tunnelling is done. The lining of the tunnel follows the excavation. When the excavation work is going on in one section, the lining work is started in the previous section.

54.2.2 English Method

The starting work is done in the same way as in American method viz. one drift is driven at the top and is supported by laggings crown bars and posts. After it, the drift is widened and supported on crown bars and laggings. In this way the work is continued up to springing point of the arch. Then sides

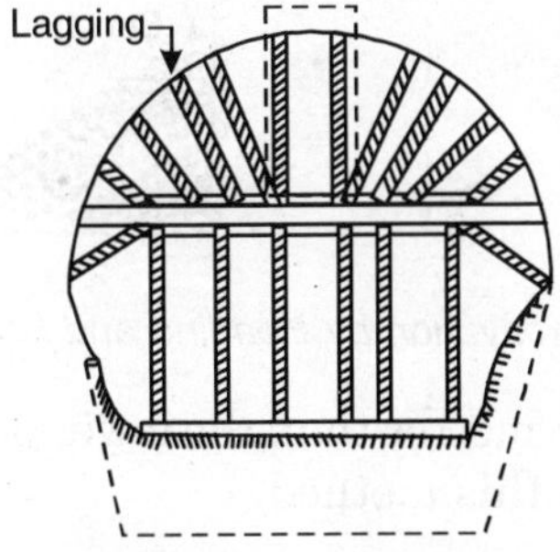

Figure 54.2 *English method*

are widened and sill is extended and supported over underpinning as shown in Fig. 54.2. In the case of American Method the load is transferred to vertical posts, but this method is taken by underpinning.

54.2.3 Heading

This is the most common method used for tunnelling in hard rocks. First of all one heading at top is excavated. The door level of this heading is kept at the springing level of the arch. After it sections 2 and 3 are removed. In this

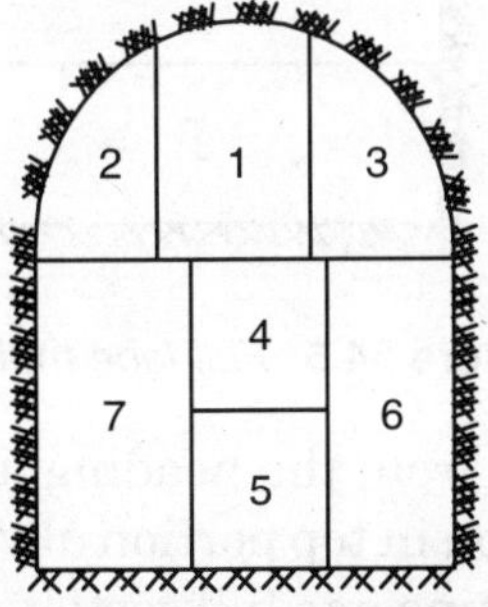

Figure 54.3 *Excavation by Heading method*

way the arched opening had been formed. Now the section is complete after excavating portions 4, 5, 6 and 7 is order of sequence.

54.2.4 Heading and Benching

In this method, first of all top portion of the tunnel is excavated by drilling holes and then blasting or excavating with other tools. After it, some of the lower portion is removed. The top portion is known as *Heading* and it always

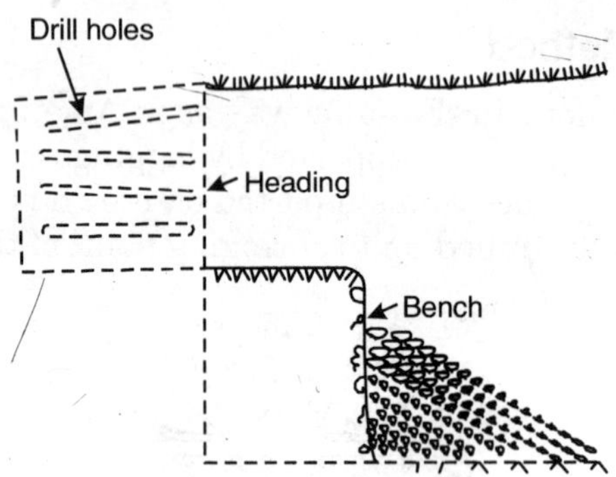

Figure 54.4 *Excavation by Heading and Benching method*

remains ahead about 4 to 5 metres than the lower portion known as benching. Figure 54.4 clearly shows this method.

54.2.5 Full Face

In this method first of all No. 1 portion of top is removed, then No. 2 and 3 are removed and the tunnel is excavated upto its full face. Then this process

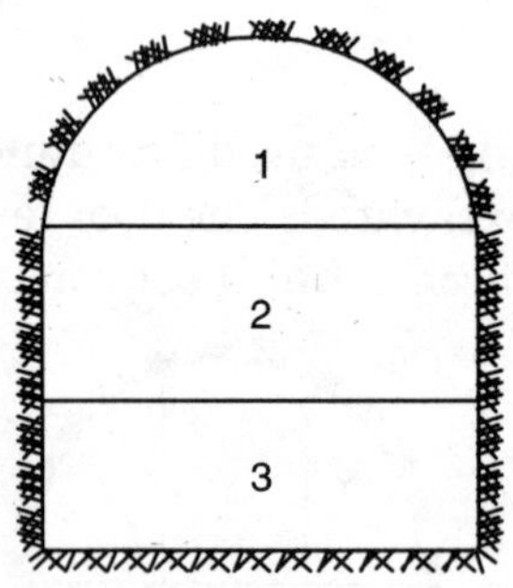

Figure 54.5 *Full face method*

is again repeated. It differs from the heading and benching, in this respect that in the former, excavation in top portion always remains ahead, whereas in the method, after excavating one full section the work is started in the next section.

54.2.6 Drift

In Drift method of tunnelling, one drift is excavated in the centre of the tunnel and then it is widened in all sides by drilling holes and excavating. The position of drift may be in the side or in centre, but the central drift is the best, because blasting can be done easily and the depth of holes will not be more. It also provides good ventilation.

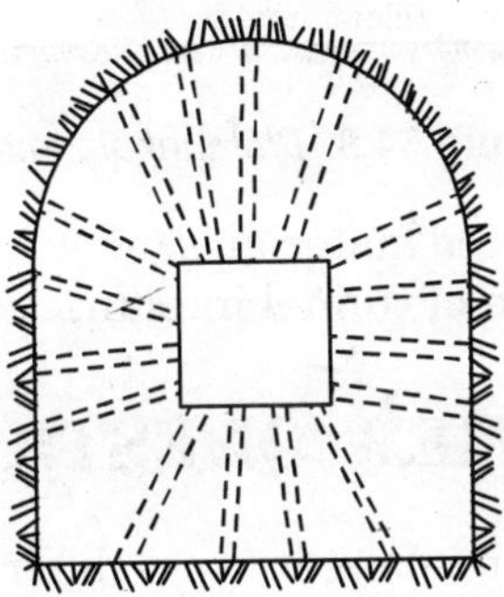

Figure 54.6 *Drift method*

54.2.7 Perimeter Method

In this method, the excavation is done along the perimeter in order of stages No. 1, 2 , 3, 4 and 5 as shown in Fig. 54.7. This method is also known as German method.

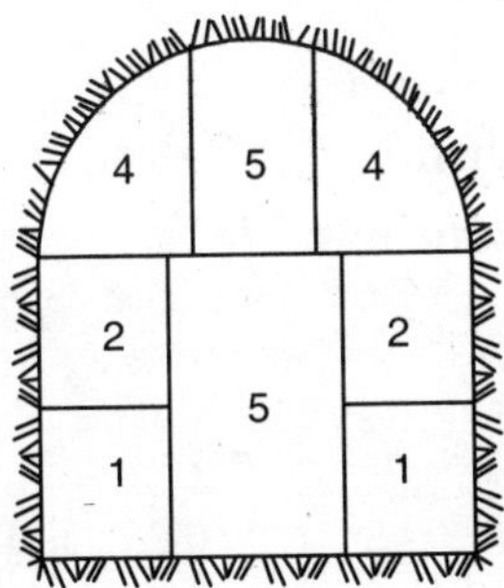

Figure 54.7 *Perimeter method*

54.2.8 Pilot Tunnel Method

This method is used when the tunnel is to be constructed in very short time. One small tunnel known as *'Pilot Tunnel'* is first constructed parallel to the main tunnel. Short tunnels are constructed to reach the centre of the main tunnel at different sections. In this way workers can reach the main tunnel in different sections and start the excavation work in all the sections

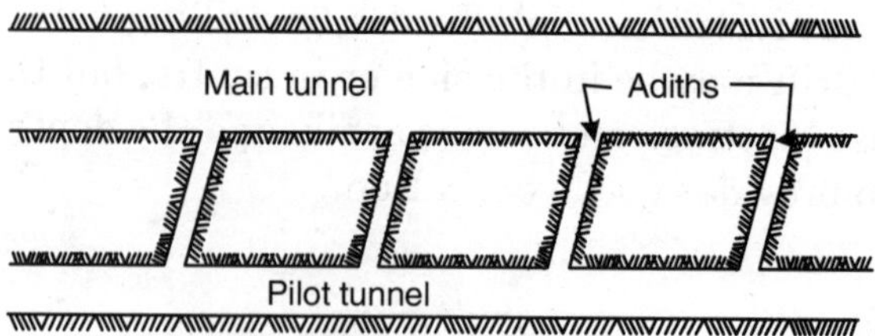

Figure 54.8 *Plot tunnel method*

simultaneously. The excavated material is taken out through the Pilot tunnel. Figure 54.8 shows this type of tunnel in plan.

REVIEW QUESTIONS

54.1. Explain how the tunnelling in rocks differ from that in soft grounds?

54.2. Write short notes on the sequence of various operations in tunnelling in rocks.

54.3. Name the various methods used in hard rock tunnelling. Describe any two methods in detail with the help of an neat sketches.

54.4. Describe with the help of neat sketch how you would do the tunnelling in rock by the centre drift method.

54.5. What are the basic operations involved while doing tunnelling in rock?

54.6. Write short notes on :

1. Full face method
2. Perimeter method
3. Heading and Benching
4. Drift
5. Heading

Tunnelling in Soft Ground

GENERAL

Soft ground mainly means variety of materials of the soil. During doing tunnel excavation work some materials require immediate supports for the soil, whereas some materials can stand unsupported for some time. The various supports used at various places depend on various factors. In previous times timber supports were used, but nowadays these have been replaced by linear plates and shields.

55.1 TYPES OF GROUNDS

Following types of grounds are commonly met with during tunnelling in soft ground.

55.1.1 Soft Ground

In this type of ground the roof requires instant support after doing excavation. The side-walls can remain standing without any support for small duration of time. (Example clayey soils).

55.1.2 Firm Ground

In this type of ground the roof can also remain unsupported for few minutes, and the sides can remain without support for 1–2 hours. (Example dry earth, firm clay).

55.1.3 Running Ground

Sand, gravel and other cohesionless soil comes in this category of ground. These soils require instant support during excavation work and cannot remain unsupported in any case.

55.1.4 Self-supporting Ground

These are soft stones such as sand stones, cemented sand, mud stones etc. These grounds can remain unsupported for short lengths of 1.5 to 3.5 metre for short periods, but after that they require support.

55.2 CHOICE OF METHOD

The method to be adopted for tunnelling in the soft ground main depends on various factors, such as size of the tunnel, type of ground, equipments available for construction and sequence of the tunnelling etc. Large section tunnels require more support which should be immediately placed in position than smaller section tunnels.

While tunnelling in soft grounds, explosives are not required, as the excavation can be done with hand tools such as pick axes, shovels etc. Due to heavy strutting and supports inside the tunnel, more obstructions are available in the movement inside the tunnel, due to which the progress of the work is reduced. The tunnelling work is done with maximum possible precautions as soft ground may collapse and may cause accidents. All the supports and struts should be sufficiently strong to bear the load and pressure coming over them.

55.3 TUNNELLING METHODS

Following are the main operations in tunnelling :

1. Excavation.
2. Providing supports and strutting.
3. Carting excavated material outside the tunnel.
4. Lining.

Following are the main methods used for tunnelling in the soft grounds :

1. Forepoling method
2. Needle-beam method
3. Linear Plate method
4. Shield method
5. Compressed Air method
6. English method
7. Italian method
8. Army method
9. Belgian method
10. American method
11. German method
12. Austrian method

55.4 FOREPOLING METHOD

In olden days this method was very common, but now it has been replaced by compressed air method. Very skilled labourer are required for doing tunnelling in soft ground by this method. As this method mostly requires manual labour for tunnelling, and timber in large quantity is required for supporting purpose. The work of tunnelling in this method is to be done in the following sequence only :

1. At the designed location on the surface, the shaft is sunk from the ground surface to the grade level and the timber sheeting is provided to keep the soil from collapsing.

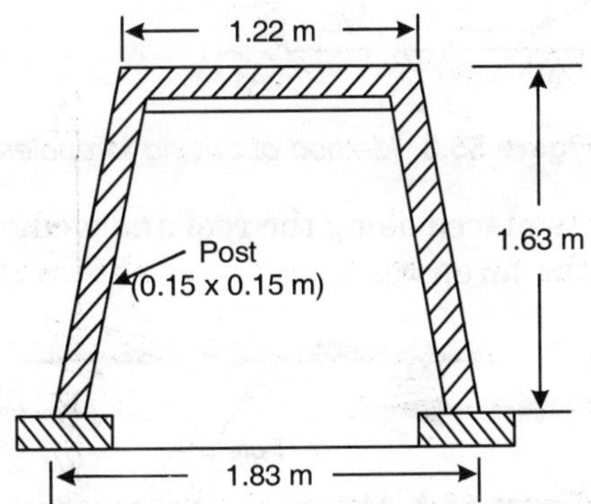

Figure 55.1 *Wooden bent*

2. Wooden bent is set up few centimetres from the sheeting and securely braced as shown in Fig. 55.1.

3. Now the holes are drilled in the timber sheeting as shown in Fig. 55.2, above and below the cap.

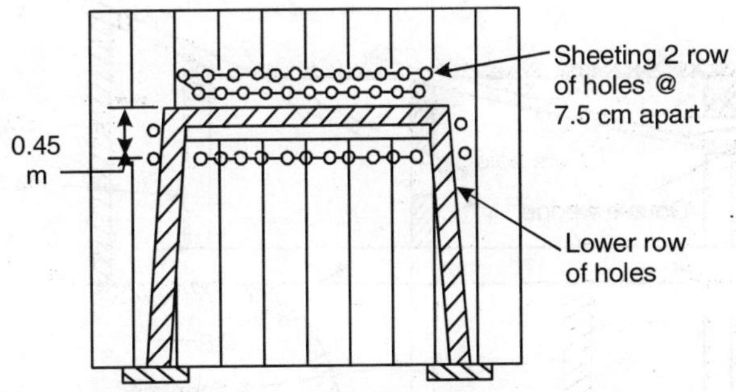

Figure 55.2 *Position of holes in the sheeting*

4. The sheeting above the cap is cut out along the top lines of the holes.

5. How the wedge end shaped 'Fore-poles' having size about 170 × 15 × 5 cm are driven through the cut in the sheets, into the ground at an inclination as shown in Fig. 55.3, upto their half lengths.

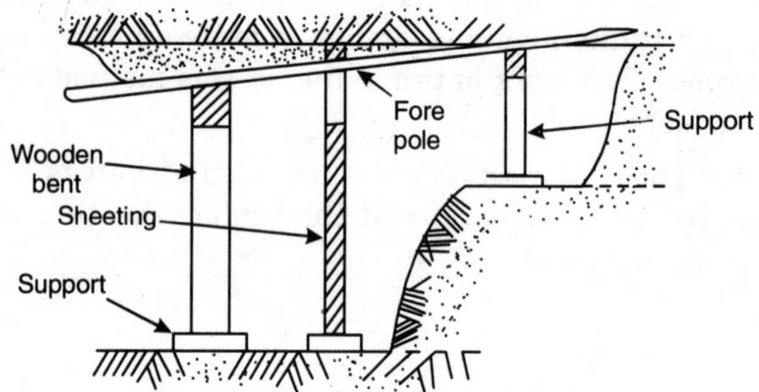

Figure 55.3 *Method of driving forepoles*

6. A timber plank is placed along the roof and wedges are placed to press the hanging sides of the forepoles from top as shown in Fig. 55.4.

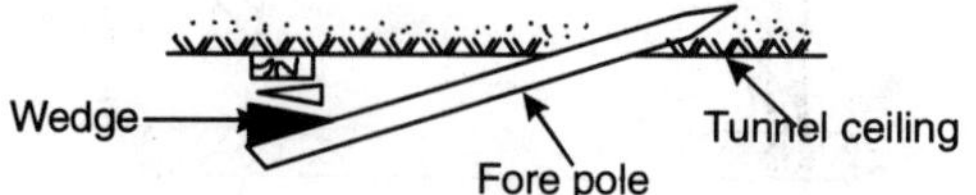

Figure 55.4 *Method of placing wedges*

7. The face sheeting is broken along the lower line of holes below the cap and the excavation is started below the forepoles.

8. Now horse-head (a false set of timber) is placed about 60 cm from the sheeting, resting on the small foot block as shown in Fig. 55.5. Spikes are then driven to full length.

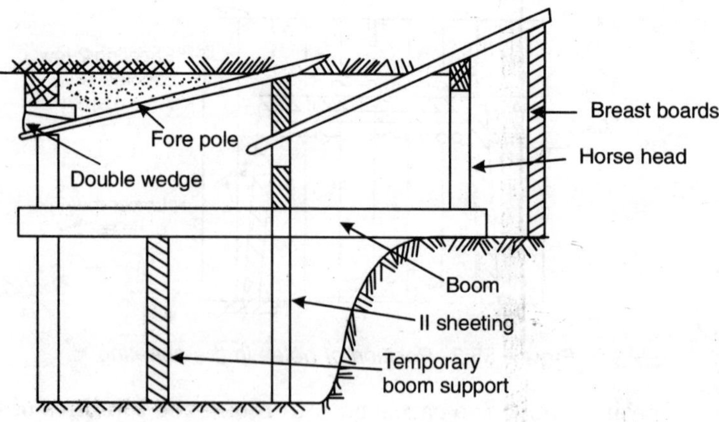

Figure 55.5 *Horse-head placing*

9. The excavation is done further into the shaft because the ends of the spikes are unsupported. A board known as *breast board,* is provided 45 cm ahead as shown in Fig. 55.5.

10. Now the next cap is provided 1.2 m ahead, and is held in position temporarily by a single post set on the bench as shown in Fig. 55.6 (a) and 55.6 (b).

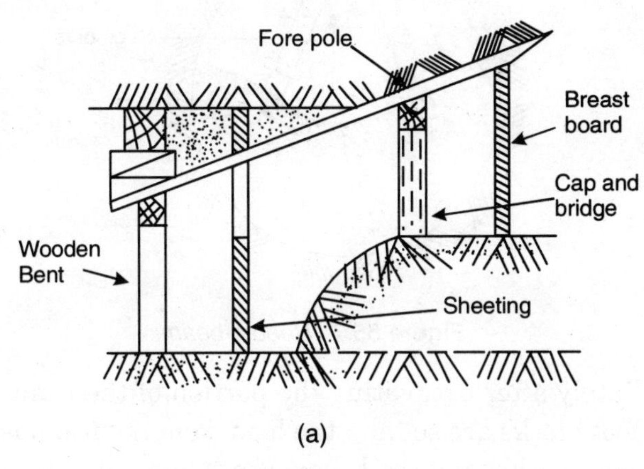

(a)

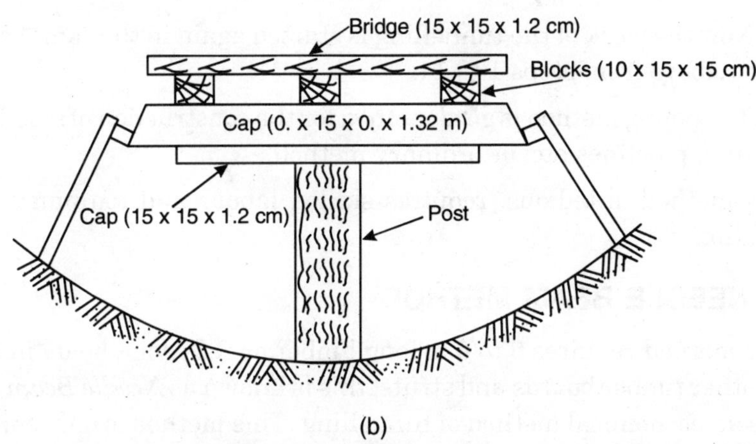

(b)

Figure 55.6

11. The side spikes are driven to their full length.

12. A pair of needle-beams known as *boom* is provided to support the lorward cap and the excavation in the lower portion is done as shown in Fig. 55.7.

These needle-beams are placed on the sides of the tunnel with a post under the middle.

13. While providing the booms the breasting boards are removed one by

one while doing the excavation. The breast boards are immediately reset one at a time at the from end.

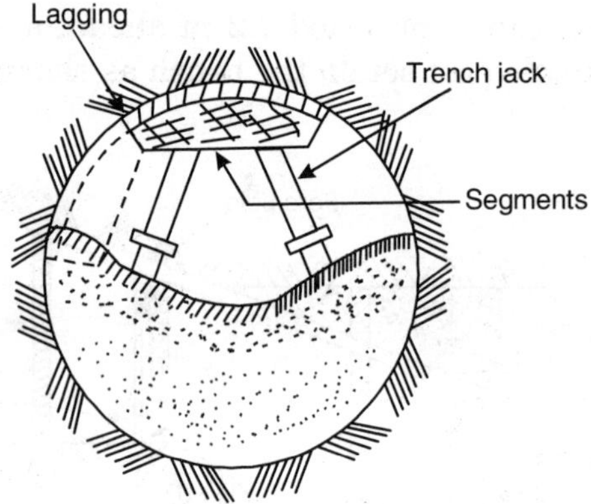

Figure 55.7 *Needle-beams*

14. Immediately after excavating the portion of the tunnel to the grade, the legs and foot blocks are set and the load from the boom is transferred to the legs and foot blocks, and the booms are taken out.

15. Now the work of the tunnelling is started again in the same sequences as given above from steps 1 to 14.

The forepoling method is good method for the construction of small tunnels for sewers, pipelines etc. or ordinary method.

This method is tedious, requires skilled labour and too much time for tunnelling.

55.5 NEEDLE BEAM METHOD

As this method requires 5 to 6 m long limber or R.S. Joist beam in addition to the other timber boards and struts, this is known as *Needle Beam Method.* This is an economical method of tunnelling. This method is only suitable for the soft grounds whose roof soil can stand without support for few minutes and the sidewalls can stand for 1–2 hours without support.

The work of tunnelling in this method is generally done in the following sequences:

1. A drift known as *monkey-drift* is first made in size of about 1 metre on the working face of the tunnel.

2. Now the roof of the drift is supported by lagging carried on the wooden segments, which are supported on the trench jacks, as shown in Fig. 55.8.

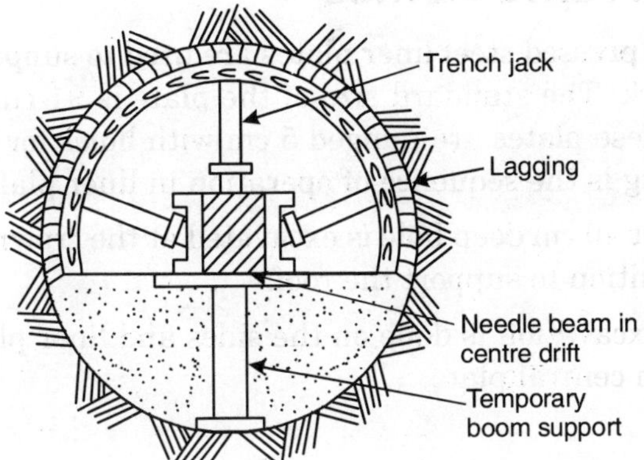

Figure 55.8 *Needle beam method*

3. A needle beam is placed in position as shown in Fig. 55.9 with its front end resting horizontally on the planks resting on the floor of monkey drift. The rear end of the needle beam is supported on the stout post, resting on the lining of the tunnel.

4. Trench jack is placed on the top of the needle beam to support the roof through lagging as shown in Fig. 55.9.

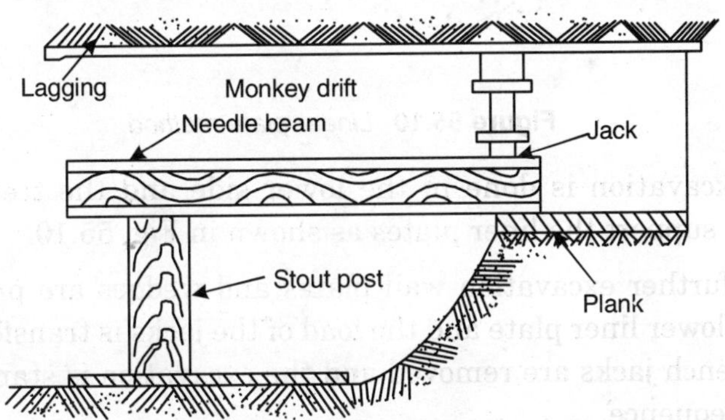

Figure 55.9 *Needle beam in monkey drift (longitudinal section)*

5. Now the drift is widened sideways and the roof is supported as above and the whole tunnel section is excavated.

6. The lining work follows the excavation and offer the removal of the excavated material.

The needle beam method requires large number of the jacks, which also cause obstruction in the efficient working of the workers.

55.6 LINEAR PLATE METHOD

In this method pressed steel liner plates are used to support the soil during excavation work. The standard size of the plate is 91 cm × 41 cm. All the four sides of these plates are flanged 5 cm with holes for bolting with each other. Following is the sequence of operation in liner plate method.

1. First about 40 cm deep hole is excavated at the crown and a liner plate is placed in position to support the roof.

2. Now the excavation is done on the sides and liner plates are provided and bolted with central plate.

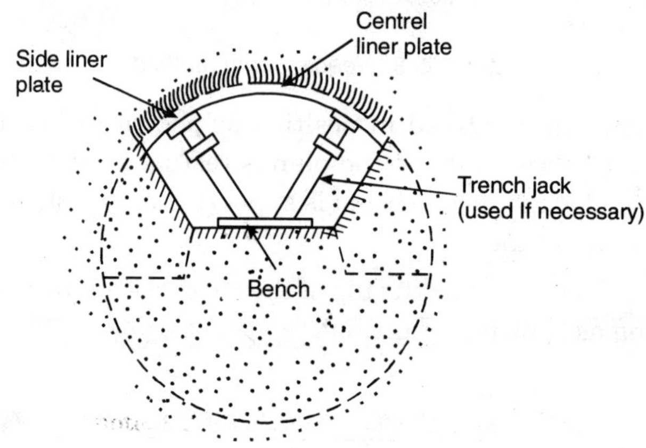

Figure 55.10 *Linear plate method*

3. The excavation is done in the lower side and the trench jacks are provided to support the liner plates as shown in Fig. 55.10.

4. After further excavation wall plates and wedges are provided at the ends of the lower liner plate and the load of the jacks is transferred to them. Now the trench jacks are removed and the excavation is started further in the above sequence.

Following are the advantages of the liner plate method :

(a) The excavation can be done quickly and economically.

(b) These are lighter in weight and easy to handle than timber planks and struts.

(c) They require less number of joints.

(d) They are more fire-resistive than timber.

(e) If the lining of R.C.C. is to be done, the plates can be embedded in the concrete and the cost of reinforcement can be reduced.

55.7 SHIELD METHOD

Nowadays tunnelling in soil is done with the help of a mechanical device known as *shield.* Shields are used for the construction of circular tunnels. Following are its main parts :

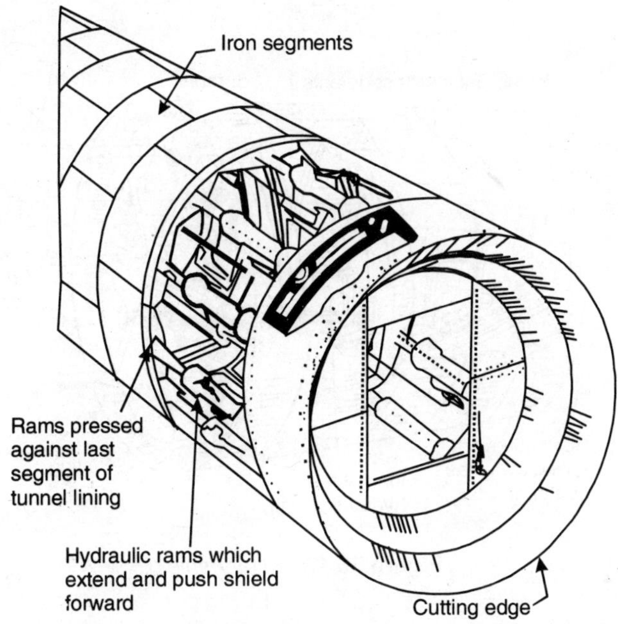

Figure 55.11 *Shield*

(a) The *skin* or *outer shell* which is constructed with steel plates, by riveting or welding them together.

(b) *Cutting-edge* which cuts the soft material.

(c) *Hydraulic-rams* which extend and push the shield forward with the help of these hydraulic-rams the cutting edge is pushed forward to cut the material.

(d) *Tail.* It is the last portion of the shield. At this place lining is done so that the roof of tunnel may not collapse.

Inter-structure is constructed in such a way that it can take all types of stresses which will come over it. Generally it consists of ting-girders, which are braced together.

Figure 55.11 clearly shows the parts of a great head shield.

55.8 ROTARY EXCAVATOR

This is also a type of shield in which cutting knives are provided in addition to cutting edge. This shield rotates while moving forward, therefore it is called *Rotary Excavator.* Figure 55.12 clearly shows all the component parts of a Rotary Excavator. Belt conveyor is provided within it to take away the soft earth which has been cut by the shield. In this way these are mechanical devices which can construct tunnels in very short time. If in the way a hard rock comes, it is blasted and then the excavation work in soft soil is continued with the shield.

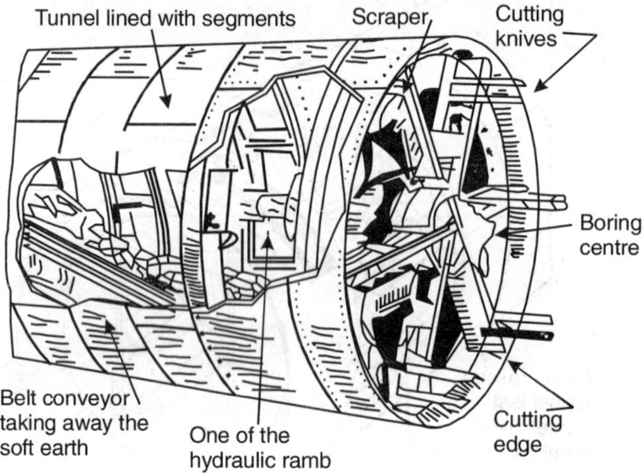

Figure 55.12 *Rotary shield*

55.9 COMPRESSED AIR TUNNELLING METHOD

This method is used in modern days for tunnelling in the soft ground having water bearing stratum. In this method compressed air is forced into the enclosed space, which prevents the collapse of the sides and the top of the tunnel. Compressed air is used with airtight locks and in conjunction with the shield.

The air pressure forces back the percolating water or water mixed soil and keep the tunnel dry. In case of large diameter or size tunnel, the pressure of water or soil is not uniform form top to bottom. The water/soil pressure at the bottom is greater than that near the roof. In such cases the pressure of compressed air inside the tunnel is kept equal to the mean hydrostatic pressure along the centre-line of the tunnel.

As the compressed air escapes through the pores of the soil, it continuously decreases. Hence the air pressure should be varied from time to time to the actual required pressure inside the tunnel.

REVIEW QUESTIONS

55.1. Describe the needle beam method of tunnelling.

55.2. What types of grounds are commonly met with during tunnelling in soft grounds?

55.3. What are the various tunnelling methods used in soft grounds?

55.4. With the help of neat sketches, describe the forepoling board method of tunnelling.

55.5. Write short note on needle-beam method of tunnelling.

55.6. With the help of neat sketch, describe the linear plate method of tunnelling.

55.7. Write short note on the shield method of tunnelling.

55.8. What do you understand by rotary excavator? Describe in brief its working.

55.9. What role is played by the compressed air in the modern tunnelling?

55.10. Discuss any method of construction of tunnels in soft soils with the help of neat sketches.

56

Mucking and Hauling

GENERAL

The operation of loading and removing the excavated or blasted materials from the construction sites and dumping it at the pre-decided site is called *mucking and hauling.*

The loading of the excavated material and its removal is an expensive and time consuming operation during the construction operations of the tunnels. The removal of the excavated material is started as soon as possible after excavating the material by any suitable method.

Mucking and hauling is a major item of the tunnel construction. The time taken in this item is about one-third to one-half of the total time taken in the construction of the tunnel. The method to be adopted for mucking and hauling should be decided after thoroughly considering all the points.

56.1 METHODS OF MUCKING

Following two are the main methods of mucking :

(1) Hand mucking

(2) Machine mucking

56.1.1 Hand Mucking

In olden times this method of mucking was used, but nowadays it has been totally replaced by the machine mucking.

However on certain small jobs, the hand mucking is adopted which is economical as well as unavoidable. Some such situations where hand mucking is adopted are as follows :

(a) In soft soils which can be easily excavated by the hand tools.

(b) For making enough space and headroom for placing the construction equipment or machine at the site.

(c) For providing space to the belt conveyer.

(d) For the construction of the small tunnels, whose construction by the machine mucking is uneconomical.

(e) For excavating and trimming the sides of the tunnel and shafts.

(f) For the final cleaning and giving grade to the formation of the tunnel.

56.1.2 Machine-Mucking

There are various methods for doing the operation of mucking by machines. All the machines used for the mucking are usually operated by the electric or compressed air.

Following types of machines are usually used in the tunnel excavation works:

(a) Loading Machines

(b) Muck cars

56.2 LOADING MACHINES

Following loading machines are usually used during the construction of the tunnels:

1. Shovels
2. Crawler shovels
3. Revolving shovels
4. Conway-digger
5. Gathering arm loader
6. Duck-bill loader
7. Scraper
8. Vibrating type loading
9. Excavators
10. Mine-car loaders

56.2.1 Shovels

The shovels are reliable and economical machines for doing the excavation and loading of the muck in the muck-cars. Usually bucket type tunnel shovels operated by lifting the loaded bucket to the attached arm directly over the machine which discharge the muck directly into the car or on the rope-way conveyer.

56.2.2 Crawler Shovels

Such shovels are most commonly used because they need not track and can work even on slopes. These shovels require greater working area, about 7.0 m wide tunnel. Electric or diesel engine operated shovels are available. These shovels cannot be used for the construction of small narrow tunnels.

56.2.3 Revolving Shovels

Fully revolving type shovels can be used in the narrower tunnels and tracks. Such shovels are capable of full revolving. Usually these are operated by compressed air and can be used even in the construction of 2 m × 2 m cross-sectional tunnel.

56.2.4 Conway-digger

These diggers are electrically driven excavators. The muck is loaded in the muck-car or conveyor belt or conveyor bucket by simply raising the bucket, tilting it backwards, and allowing the muck to slide down through the chute.

56.2.5 Gathering Arm Loader

It is also an electric operated crawler mounted loader. It collects ihe muck by means of two eccentrically operated arms. It essentially consists of a revolving chain conveyor which carries the muck backwards and clumps it into the muck cars or conveyor bell or trucks etc. The gathering arm loader has very high loading capacities and are commonly used in the tunnel construction.

56.2.6 Duck-bill Loader

This loader works on the sliding action developed by a rapidly accelerated plate moving into the muck to load. The main advantages of this loader are that it permits face operation, such as drilling, cleaning, setting of roof arches etc. It also proceeds with safety while loading the muck. As a matter of fact it is an improvement over the vibrating type loader.

56.2.7 Scraper

In this machine a sloping steel chute at the front of the machine is used as a cutting edge of the scraper. The cut material is filled in the dump type bucket. The scraped material is dumped at the required place or waiting skip.

56.28 Vibrating Type Loader

It essentially consists of a trough having a shovels-shaped front end which can be given required vibrations into the heap of the muck. Due to vibrations the muck material starts moving in the slope of the trough and into the skips waiting at the rear of the loader. This machine has high loading capacity and also requires small working space. This loader is not suitable for loading of hard rocks. It has the disadvantages of not working satisfactorily on the uneven floor surfaces or on curved surfaces.

56.2.9 Excavators

Excavators are used for excavation of large size tunnels about 10.0 m or more. Electric or Diesel operated excavators filled with shortened booms are usually used for excavation and loading purposes. They require large working area and high head room. They have high loading efficiency.

56.2.10 Mine Car Loaders

The whole mine car loader machines move with muck at high speeds. These are operated by the compressed air. These have high efficiency in fast moving and require small working area.

56.3 MUCK-CARS

Various types of muck-cars are used for hauling muck from the tunnel. Following types of muck-cars are commonly used.

56.3.1 Muck Boxes

These boxes are made of hard wood. Hoop iron on flat iron reinforcements are provided at suitable places to increase the strength of the boxes. Suitable rings are provided in the bottom for removing the bottom and dumping the muck.

56.3.2 Battle-ship

It is a simple large size box, which is carried into the tunnel for carrying out the muck on flat cars or trolley-chasis.

56.3.3 Side-dump Car

This type of car is used for hauling of muck of rock tunnels. These cars have compact design and have large hauling capacity. These cars dump the muck on one side of it, by opening the side door or revolving the car along the horizontal axis.

56.3.4 U-bottom Cars

These muck-cars are commonly used for taking out the muck of soft excavation. The muck can be dumped by these cars by two workers.

56.3.5 Quarry Cars

These are commonly used for hauling of the muck of hard rocks obtained by the blasting. As the rock pieces of large size get excavated by blasting which cannot be taken out by small cars. Quarry cars are of large size. These are not of self-dumping type.

56.4 MUCKING IN STEEP GRADE TUNNELS

Figure 56.1 shows the method of mucking in steep gradient of tunnel where the gradient is 1 in 50 or more. In such cases the mucking of the tunnels is

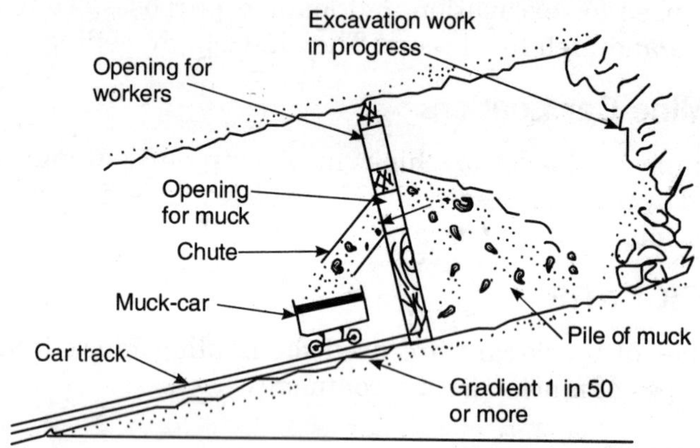

Figure 56.1 *Method of mucking in steep grade tunnels*

made sell-mucking by providing an inclination of about 40° or more to the horizontal. In the beginning the tunnel is excavated upto about 30 m. The cross-section of the tunnel is closed with wooded bulk-head. Two openings are provided in the wooden bulk-head, one for discharge of the muck and another for entry of workers as shown in Fig. 56.1.

The muck is allowed to accumulate behind the bulk-head and is allowed to discharge through the chute on the muck-cars. As the work proceeds the bulk-head is moved forward to the successive new positions.

56.5 CAR CHANGER

For the rapid transportation and minimum loss of time, it is most essential that the muck-cars should be quickly changed from one track to another. It is most important that as soon as one muck-car is loaded, it should be immediately replaced by the empty car to the muck loading machine.

Following are the common methods of layouts of the muck-car tracks :

56.5.1 The Grasshopper Method

Figure 56.2 shows the grasshopper method. It is used for narrower tunnels. In this method the empty cars are stored in an overhead track on the large truss frame. The loaded cars are allowed to pass underneath. Immediately after loading and moving the muck-car, the empty car is allowed to come down through the ramp to occupy the position of the loaded car and the loading machine starts loading it.

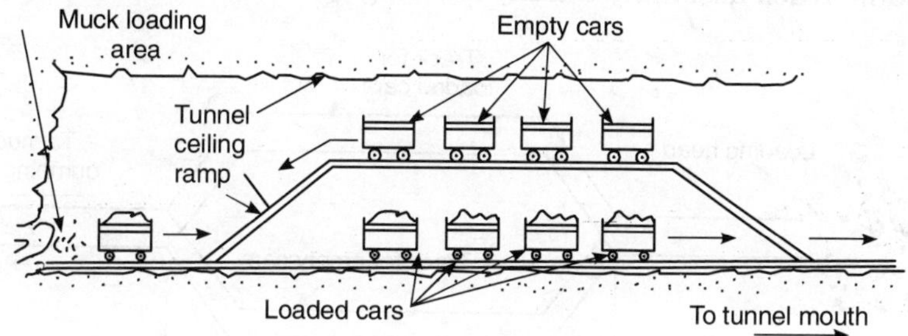

Figure 56.2 *Grasshopper method of car-changing*

56.5.2 Passing Track

In this method a side track is provided for moving the empty cars toward the loading area. Figure 56.3 shows this method. This method is possible only

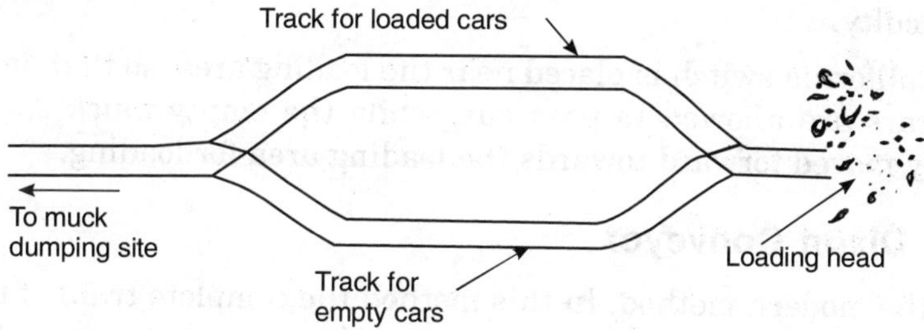

Figure 56.3 *Passing track layout*

in case of wider tunnels where tow tracks can be laid side by side. In this method it becomes more and more difficult and delay in moving the passing track forward as the tunnel work is advanced. But this method is more commonly adopted.

56.5.3 The Cherry Picker

It is an old device which was used in the early days for interchanging of cars. In this method the empty car is lifted off the track and is moved sideway directly on the flow so that the loaded car can pass through the track. After passing the loaded car, the empty car is again placed on the track and moved. In this method only one track isused, therefore it is cheaper out it is not efficient and is time consuming method.

56.5.4 California Crossing

It is also known as California Switch. Figure 56.4 switch. In this method short double track assembly a unit with frog and switches is laid in position

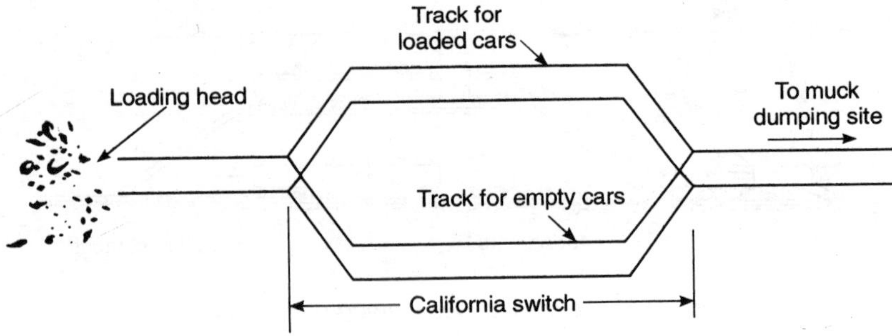

Figure 56.4 *California crossing*

as is assembly slides along the top of the single main track and the car is allowed to change from one track to another by means of switches. The ends of the rails of the track are made tapered to fit the main track so that the car can move over the California crossing, to and from the main track without any difficulty.

The California switch is placed near the loading area, so that on one side loaded cars are allowed to pass out, while the empty muck cars can be promptly moved forward towards the loading area for loading.

56.5.5 Dixon Conveyor

This is the modern method. In this method the complete train of the muck cars is loaded by means of 'Dixon Conveyor' having long conveyor belt fixed on the long arm, sufficient to load the cars. Immediately after loading the car is moved forward. The conveyor belt is loaded by the muck by a standard conway mucker. This conveyor also requires an extra set of tracks for moving the muck cars or their trains.

REVIEW QUESTIONS

56.1. Write short notes on the following :

(a) Car changers.

(b) Muck cars.

(c) Mucking machines.

56.2. What do you understand by the term car changer? Describe the various car-changer techniques used in the construction of the tunnels.

56.3. Write short note on mucking and hauling.

56.4. Describe various methods used for mucking.

56.5. What do you understand by the loading machines in the construction of the tunnels? Describe various loading machines used in the construction of the tunnels.

56.6. How the mucking is done in the tunnels having steep gradients? What precautions are taken under such situations.

57

Tunnel Lining and Grouting

GENERAL

Tunnel lining is done in the tunnels to give the finishing touch to the cross-section and preventing the collapse of side or roof ground soils. Lining serves many purposes which depend upon the type of the ground and the nature of the permanency of the tunnel. Temporary or primary lining is the temporary support of the roof and the side walls during tunnel construction. The permanent lining is the lining which gives the actual shape and size of the tunnel for the purpose for which the tunnel is constructed. Only permanent lining shall be described in this chapter. For the tunnel lining bricks, stone or concrete can be used. But in modern days on R.C.C. or Plain Cement Concrete with temperature reinforcement are used for lining purpose.

57.1 OBJECTS OF THE LINING

Following are the main objects of the permanent lining of the tunnels :

1. It provides the correct shape of the tunnel.

2. It withstands the soil, pressure and prevent the collapse of soils in case of soft ground.

3. It keeps the inside of the tunnel tree from water percolation as suitable outlets are provided.

4. It binds and keeps the loose rock pieces in position and keeps the tunnel safe.

5. It prevents the rock from air slake.

6. It reduces the maintenance cost of the tunnel, because if lining is not done even in case of hard rocks, the loose pieces may fall from time to time and make the tunnel unsafe for traffic.

7. In case of water and sewer tunnels the lining will reduce the friction against the flow and will also reduce the turbulence of flow which may be caused due to projecting pieces of stone.

8. In soft ground the lining gives strength to the structure and makes it safe.

57.2 TYPES OF LINING

Following are the main types of tunnel lining :

(a) Timber-lining
(b) Iron-lining
(c) Cast steel lining
(d) Pressed steel liner plates
(e) Brick lining
(f) Stone masonry lining
(g) Concrete and R.C.C. lining.

57.3 CONCRETE AND R.C.C. LINING

This type of lining is most common nowadays in soft grounds as well as hard rocks. The thickness of the concrete lining mainly depends on the following:

1. Size and shape of tunnel section.

2. Relative amount of the vertical and horizontal pressures on the lining at various places.

3. In case of water tunnel, the amount of maximum internal pressure.

4. Final use of the tunnel.

5. Condition of the ground around the tunnel.

6. Conditions under which the tunnel is to be constructed.

The thickness of the concrete lining should be as thin as practical for economy purpose. As a matter of fact there are no formulae for the design of the concrete lining of the tunnels through rocks. Every tunnel will have its own conditions of design, analysis and the circumstances of loads coming over the lining. As a thumb rule the approximate thickness of the tunnel lining is kept 2.5 cm for each 30 cm of the bore diameter of the tunnel.

Sometimes the thickness of the lining is determined by

$$T = 82\,D$$

where

T thickness of lining in mm

D diameter of tunnel in metres.

R.C.C. lining is required in the tunnels, when the tensile stresses in the lining exceed the permissible tensile limit of the tunnel lining. The placing of concrete in R.C.C. is more difficult than P.C.C. In case of R.C.C. lining the thickness of the lining at the crown of the tunnel is generally more than 25 cm. Ribbed or round steel bars may be used for the reinforcement in case of R.C.C. The reinforcement bars placed close to the inner surface at the top and close to the outer surface at the springing points.

57.4 ADVANTAGES OF CONCRETE AND R.C.C. LINING

Following are the main advantages of the concrete and R.C.C. lining :

1. It makes the lining water tight.
2. It provides smooth surface.
3. The maintenance cost is lowest possible.
4. It can be easily moulded and casted in any desired shape.
5. It can be casted continuously in the shape of a shell without joint.
6. Its thickness can be controlled during casting.
7. Steel can take all types of stress and make the concrete safe against all types of stresses, which may come over it.

57.5 FORM-WORK FOR LINING

Concrete lining is done by pouring the concrete in the form-work. The form-work should be very accurately constructed and should show the true outline of the finished tunnel section.

The concreting of the tunnel is done in three operations and therefore the form-works are designed and constructed in three operations which are :

1. Ground mould form-work for the concreting in the flow or invert of the tunnel.
2. Leading frame for the lining work of the side walls.
3. Trusses form-work for the lining of the roof arch of the tunnels.

57.5.1 Ground Moulds

These moulds are used for casting the invert of the tunnel. These are generally made in two parts, which are joined at the middle by means of fish-plates and bolts for easiness handling. While doing the concreting two moulds are used at a convenient distance apart and chords are stretched

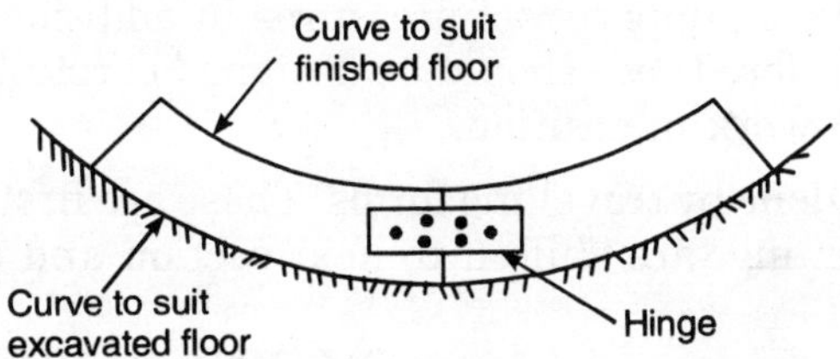

Figure 57.1 *Typical ground mould in position*

across them, to obtain the true profile of the flow surface. The levelling and the centring of the mould before pouring the concrete, is done by means of a theodolite. Figure 57.1 shows typical ground mould in position.

57.5.2 Leading Frame

Figure 57.2 shows a leading frame in position for casting the concrete of the side walls of the tunnel. These may be made of good thickness planks, one edge of which is cut to the curve of the side excavation and the other edge to

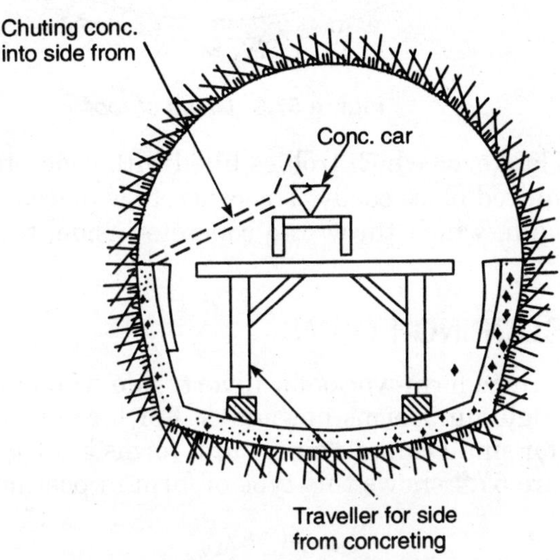

Figure 57.2 *Leading frame in position*

conform to the inner face of the finished side wall. These leading frames are set with their lower end resting on the finished invert. The tops of the frames are correcdy checked by means of theodolite.

57.5.3 Trusses Form-work

The construction of this form-work is similar to those or arches. Sufficient head room is provided in the centre of convenience of working place. Centre form-works have a support roof pressures in addition to the weight of the roof lining, therefore these should be designed carefully. Figure 57.3 shows the centre form-work in position.

The lining is done by travelling forms. These are first placed at one section and after the lining, are shifted to next section and so on. Generally one

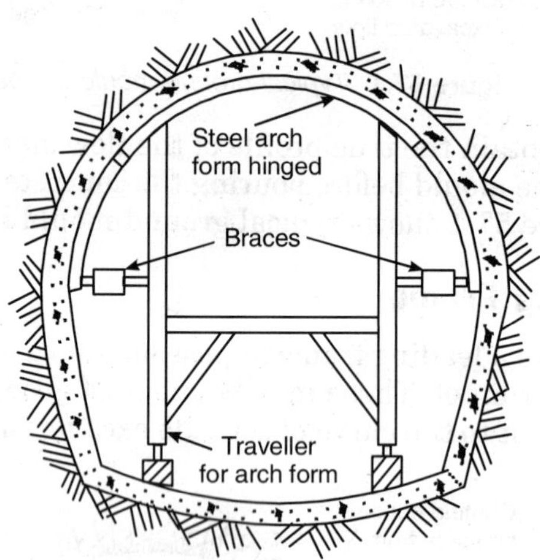

Figure 57.3 *Lining of roof*

small track is laid over which trollies filled with concrete move and take the concrete to required place easily. The concrete is placed in position by means of pump and gun, which throw the concrete inside, to the tunnel side and form space.

57.6 TELESCOPING FORMS

The use of separate form-works for invert, side walls and the roof has been replaced by single unit forms nowadays. For large tunnels separate forms are used, but for small tunnels single unit forms known as *telescoping forms* are used. Figure 57.4 shows this type of form in position.

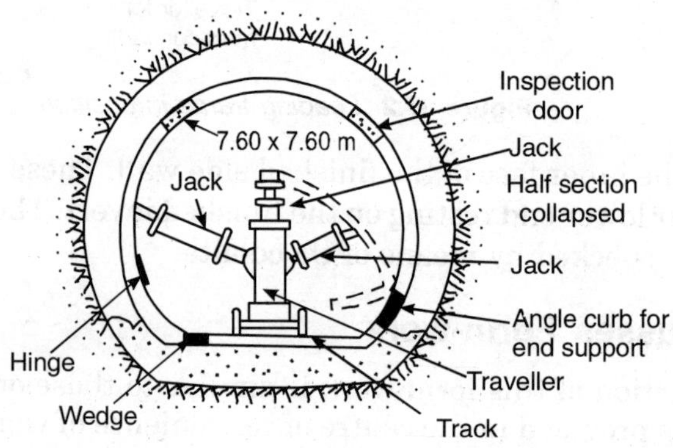

Figure 57.4 *Telescoping steel form*

The main ribs in this form are made up of section's hinged together, so that the whole unit would be collapsed after hardening the concrete, and moved forward. Due to this folding of the form at the hinge, these forms can be taken out without disturbing the concreting. Telescoping forms are suitable for lining of tunnels upto 8.0 m diameter. These forms carry jacks for erecting and collapsing the rib sections.

57.7 CURING OF CONCRETE

If the humidity inside the tunnel is sufficient, no extra curing for concrete is required. Generally the curing of the tunnels is done by spraying water through the perforated pipes laid along the length of the tunnel. If the large quantity of water is not available for curing purpose, the concrete surface is painted by spraying the bituminous paints sealing the surface of the concrete and preventing the evaporation of the water from the concrete.

57.8 SEQUENCE OF LINING

The sequence of lining or placing the concrete in the tunnel section depends on the various factors. Following methods are commonly adopted while doing the lining work :

(a) The lining of the entire section is done in one operation, in case of small circular tunnels, in which concreting is done in short lengths.

(b) Placing the lining in the invert first, and the rest of the lining viz. walls and roof in the second stage.

(c) It is always better (for saving time) if possible to start the concreting from the most interior section to the mouth of the tunnel, because this procedure will save the newly laid concrete from being used as hauling floor before it sets.

(d) Placing the lining in the invert first, then side walls and finally the roof. This sequence is suitable for large size tunnels, and for the convenience of the construction. This method is very popular and commonly used.

REVIEW QUESTIONS

57.1. What are the objects of lining in the tunnels? State the materials employed for the purpose.

57.2. Explain with neat sketches what is meant by :

(i) Ground mould.

(ii) Leading frame.

(iii) Centre.

57.3. Write short note on the concrete and R.C.C. lining.

57.4. State what are the advantage of concrete and R.C.C. lining.

57.5. Write short note on the form-work for lining.

57.6. What do you understand by telescoping forms? Explain with the help of neat sketch.

57.7. Write short note on the sequence of lining in the tunnels.

Ventilation of Tunnel

GENERAL

At the time of constructing tunnels, different operations such as blasting, drilling holes, mucking etc. are done due to which poisonous gases are produced in the tunnel. To safeguard the workers, it becomes compulsory to remove these gases and bring inside the fresh air. The ventilating system must have the following qualities :

1. The poisonous gases produced by blasting must be taken out immediately, so that work is held up for shortest period.

2. The whole tunnel must be free from the dangerous gas fumes, it should not leave any dead pots in the tunnel.

3. The dust should also be taken out along with the dangerous gases.

4. At the face of the tunnel, comfortable atmosphere must be developed, so that workers can work easily.

58.1 METHODS OF VENTILATION

The following are the various methods of ventilation :

58.1.1 Mechanical Method

The mechanical ventilation is provided by :

1. Blowing air inside the tunnel.

2. By exhausting the air from the tunnel.
3. By combination of blowing and exhausting the air.

1. This is also called propulsion method. In this method fresh air is forced by electric fans or blowers or with the both. In blowing, fresh air is blown by blower fans, mounted on one or more input shafts. The vitiated air is being forced out naturally due to pressure build up by blowing. This method is suitable for short length tunnels only.

2. Sometimes vitiated air is taken out by means of exhaust fans. The fresh air is being drawn inside through portals or inlet-shafts. This method is good in removing dust and foul air quickly from the tunnel.

3. Sometimes combined blowing and exhausting method is used. This is called balanced method. In this method fresh air is forced inside the tunnel by blowing fans and the vitiated air is sucked out by exhaust fans. This method is very good and is mostly used in tunnelling.

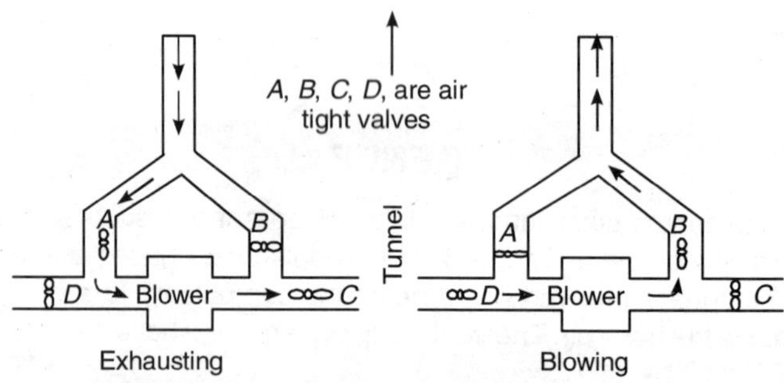

Figure 58.1 *Combined method of blowing and exhausting for ventilation*

In the combined method of blowing and exhausting, blowers and exhaust fans are installed for forcing fresh air in the tunnel. Sometimes the blower and the exhaust fans are installed in the shafts connected to the tunnel and meant for this purpose.

58.1.2 Natural Ventilation

Due to difference in temperature inside and outside of the tunnel, flow of air starts. This natural flow can be increased by constructing shafts at various places over the tunnel. But this method does not provide sufficient ventilation, which is required in practice.

58.2 DUST CONTROL

At the time of drilling holes and blasting sufficient quantity of dust is mixed in air which is harmful to the health as it causes incurable lung disease. The following precautions are generally adopted for controlling the dust :

1. **Wet-drilling.** At the time of drilling holes, water is added in the hole which prevents dust. Modern drilling machines are fitted with wetting arrangements, the water also keeps drill bit cool.

2. **Respiration method.** In this method special types of respirators are used by the workers. These respirators prevent inhalation of dust by the workers. But these are not very common.

3. **Vacuum hood system.** In this method specially built hood is provided around the drill at the time of drilling holes. This hood is connected to a suction hose and removes all the dust produced by drilling.

REVIEW QUESTIONS

58.1. What are the objects of providing proper ventilation to the tunnels during their construction?

58.2. Explain the process of mechanical ventilation and show by a neat sketch how a blower could be used alternately for blowing and exhausting by suitable valve arrangement?

58.3. What are the effects of dust inhalation in tunnel work? State and briefly explain the methods adopted for controlling dust inhalation.

Drainage of Tunnel

GENERAL

Removal of water from the tunnel during its construction and after the construction is very essential. Generally at the time of construction the water which is used for wet drilling and which comes through seepage is removed by pumps. Drainage ditches or sump wells are provided at suitable intervals in the tunnels in which the water from both sides is collected and from where it is pumped out at suitable intervals.

After the completion of the tunnel the water is taken out by means of open drains provided in the centre or both sides of the tunnel by gravitational force. Sufficient slope is provided in the drains to take out water from the tunnel.

59.1 SOURCES OF WATER

Mainly water comes in the tunnel from the following two sources :

1. Wash water which is used during drilling holes and used to wash the cuttings inside the tunnel.

2. Sub-soil water, which precolates inside the tunnel from the pores of the soil or through the natural fissures in the rocks or caused during blasting of the rocks. The quantity of water entering from the ground is sometimes very large and causes water to flow inside the tunnel.

The quantity of water from the first source, describe above, can be easily determined, but the calculation of quantity of water from the second source is not easy. Exploratory holes are to be drilled at various places, in order to determine the quantity of water which would come from the second source.

59.2 WATER HANDLING

During the construction of the tunnel, if the ground water is trickling continuously from the roof and is causing obstructions in the construction, it can be diverted to the side-drains (constructed along the both sides of the tunnel) by providing a false roof under the roof by corrugated G.I. sheets. These side-drains will carry the water outside the tunnel.

If the heading is driven on level ground, the water should be removed by pumping it through shafts. If the heading is driven on upgrade side of the tunnel, the water can be removed through slopy drains. On the other hand if the heading has been driven through shafts, all the water is to be pumped through the shaft only.

59.3 GROUND WATER REMOVAL

The quantity of the ground water entering the tunnel can be removed from the tunnel by the following two methods :

1. By draining through the open-ditch constructed along the tunnel on both sides at the invert.
2. By pumping.

59.3.1 Open Drainage System

This is the simplest method for the removal of the water. This system is suitable in hard rock bases and water resistant soils.

Following are the disadvantages of this system :

(a) Valuable working space is utilized in the construction of open drains.

(b) It is not possible to drain-off the water at all grades, because on flat grades, self-cleaning velocity will not develop.

(c) If the muck or debris block the flow of water, it will create objectionable ditches and pools inside the tunnel.

(d) The construction of open drains in the sides weakens the section of the tunnel as the continuity of the shell is broken.

Due to the above disadvantages, in modern tunnels, open drains are not preferred for the removal of the water from inside of the tunnel.

59.3.2 Pumping Drainage System

In modern tunnelling methods, the removal of water from the tunnel by pumping system is preferred. Two types of pumps are commonly used. First is the piston-type pumps, which are available in vertical or horizontal models and are usually operated by compressed air. Second type is centrifugal pumps which can be operated by electricity or compressed air. Being smaller in size than first type, these can be easily installed or even submersed inside the water in the sump-wells.

For the removal of water by the pumps, sump-wells are constructed at regular intervals from 300–500 metres apart or at wet spots where water is precolating in the tunnel. A series of pumps are installed in each sump-well. The pumps pump the water of their sump to the next sump. From where it is pumped to the next. In this way the water is taken out in portions to the outside of the tunnel. The capacity of the pumps, size of the sump well and the pipe go on increasing from the middle of the tunnel to the mouth sides. To overcome this difficulty sometimes one pipe line is laid all the way inside the tunnel and all the pumps are connected to it through check valves. In this case all the pumps of uniform capacity can be used.

For handling the water during the construction of the shafts, centrifugal pumps hung from the cable work more satisfactorily. In case of deep shafts, booster pumps are installed in the sumps constructed into the side of the shaft. If the site conditions require high capacity pump should be installed in the bottom of the shaft to pump the water from the tunnel.

59.4 PERMANENT DRAINAGE

At the time of completing the tunnel, permanent drainage arrangement is provided, to take out the water entering the tunnel continuously, to save the road or railway track.

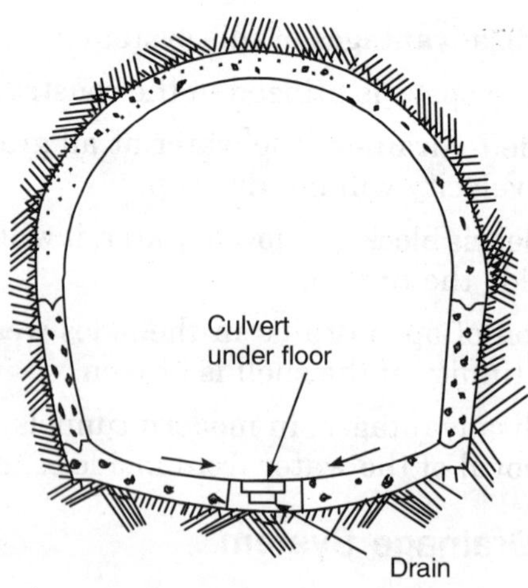

Figure 59.1 *Central drainage system*

If the tunnel carries two railway tracks, one open drain is provided in between both the railway lines. The section of the drain should be sufficient to carry the water without submerging the floor of the tunnel at self-cleaning velocity. The central drain can also be provided covered with facility for the inspection and maintenance as shown in Fig. 59.1.

If the tunnel carry single lane traffic, one or two side-drains can be provided as shown in Fig. 59.2.

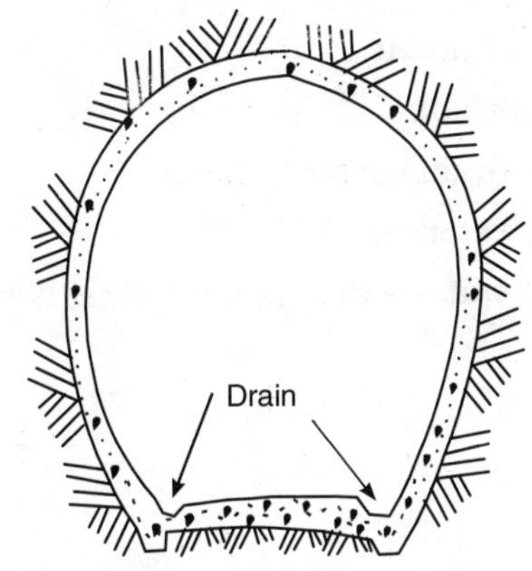

Figure 59.2 *Tunnel with two side-drains*

In case the water continuously trickles from the roof of the tunnel, false roof of the G.I. sheets should be provided as shown in Fig. 59.3 to divert the roof water to the side drains.

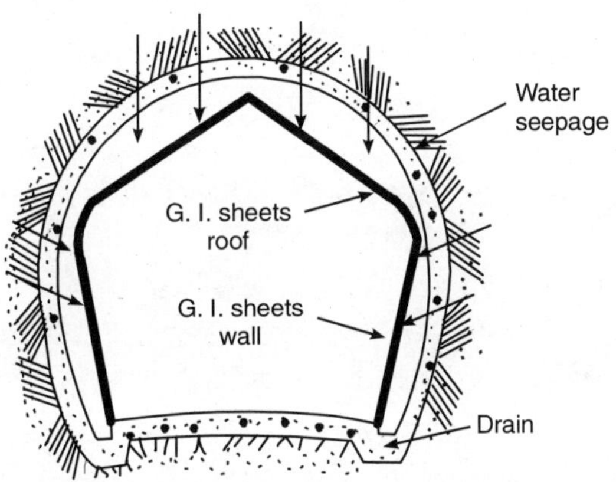

Figure 59.3 *False-roof of G.I. sheet in the tunnel*

If the water is also trickling from the side-walls and obstruct in the free movement of the traffic, it is prevented by erecting the side-walls also with

G.I. corrugated sheets to prevent the splashing of water on the pavement or the track.

REVIEW QUESTIONS

59.1. Describe the various methods for the effective drainage during and after the construction of the tunnels?

59.2. Write short notes on :

(i) Drainage in tunnels

(ii) Pumping in shafts

(iii) Permanent drainage for tunnels

(iv) Tunnels dewatering

59.3. How the ground water entering the tunnel is removed? Describe.

Safety Measures

GENERAL

Tunnel construction is very tedious, time consuming and hazardous operation. There are various causes which may lead to an accident. While tunnelling maximum possible care should be taken to prevent any mis-happening or accidents. If care is taken the accidents can be prevented in time and the lives of the workers can be saved.

60.1 CAUSES OF ACCIDENTS

It is impossible to think over all the causes of accidents. At one place if accident occurs due to certain cause, at the other site it may be due to another cause, the site conditions of working vary from one place to another, hence one safety measure at one site may fail at the other site. Hence the executive persons should constantly keep strict watch on all the operations of tunnelling to prevent the accidents.

60.2 SAFETY MEASURES

If during tunnelling the preventive measures are rigidly followed, most of the accidents shall be eliminated. Following are the some preventive measures which should be followed during tunnel constructions :

1. The design of the planks and vertical supports should be checked carefully, because most of the accidents occur due to rock-fall. The walls and the roof should be frequently checked to prevent rock-falls.

2. Defective or weak timber should not be used in any circumstances, as it may give way under load.

3. Safety rules, regulations and the preventive measures should be taught to every worker, working on the site.

4. All safety rules should be followed strictly without any violation. Laborers' and staff's regular meetings should be held from time to time and they should be helped for understanding and the application of the safety rules.

5. All electric light and power lines should be properly installed and all connections should be well insulated as per prevailing codes of practice and rules.

6. Drilling of holes, loading of holes with explosives and the firing should be done with great care, keeping all the precautions in view.

7. If any charged hole mis-fires, following precautions should be taken :

(a) If the leg wires are still intact, they should be connected to the electric ignition equipment again and refired.

(b) If the hole is not exploded, the stemming should be washed out by water jet down upto the explosive. New primer should be inserted and fired.

(c) Another method is to drill a new hole about 60 cm from the unfired-hole, loaded and fired.

8. Extra light should be provided where essential materials are stored.

9. All the tools and equipments should be kept in best working conditions, as bad tools may lead to the accidents.

10. Water should not be allowed to remain inside the tunnel, as it may cause obstruction in the walking and may lead to skiding, rusting of the steel and other tools etc.

11. All the working sites should be kept well lighted, as poor lighting may lead to accident.

12. Foot-path should be well maintained. It should not be slippery and hazardous and free from obstructions. It should be covered with planks or muck and should be kept dry.

13. For safety and efficient operation all the debris and the refuse should be kept clean, both on the surface and underground.

14. Loading of the muck and their hauling should be done with great care.

15. Only required material should be taken inside the tunnel, as extra material will cause obstruction.

16. The cars carrying rails, steel, sleepers, posts etc. should not be overloaded, as extra projections in the side may strike with the workers working inside the tunnel or with the posts and struts.

17. All trolleys or trains should be properly controlled with signals for safety during their movements.

18. All the hoists, cables and other shaft equipments should be in sound working conditions, with sufficient factor of safety.

19. All the shaft openings should be properly protected to prevent the workers and the materials from falling inside the shafts.

20. Shafts should be provided with the indicators to indicate the position of the cage.

21. All the hoists be provided with automatic braking arrangements, which should be in operation during power failure.

22. Proper signals should be provided to prevent accidents of the cage moving in the shaft, with the trolleys or train moving in the tunnel.

23. All the shafts must be provided with safety ladder for use during emergency for exit and access.

24. All the persons working in the tunnel should have metal hats, to prevent the heads against accident from the falling stone pieces.

25. Fire-fighting equipments should be provided at all key-points for safety purpose.

26. All the workers should be medically fit for working inside the tunnel. Workers should also be medically examined from time to time.

27. After the blasting inside the tunnel, the inside poisonous gases should be removed before allowing the workers to enter in the tunnel.

28. Telephone facilities should be provided inside the tunnel at various places, for sending the information of the tunnel position outside from time to time.

29. In case of compressed air tunnels, extreme care should be taken to prevent fire. Smoking should not be allowed in any case, as fire catching gases may cause accidents inside the tunnel.

30. First aid facilities with equipments and doctors should be available all the times at the site, for immediate help in case of accident.

31. No unauthorised person should be allowed to enter the tunnel.

32. Authorised visitors should be equipped with safety hats and other necessary equipments. They should be allowed to visit the site only with the authorised guide meant for his purpose.

33. Safety sign boards should be provided at various places.

34. Emergency lights should be provided inside the tunnel at all the working places, which should give light in case of power failure.

35. Double power supply system should be provided, so that during failure of one supply, the power can be resumed by the stand by unit.

36. In the risky place special care should be taken to prevent the accident.

REVIEW QUESTIONS

60.1. What are the causes of accidents in tunnelling?

60.2. Write short note on the safety measures to be adopted in tunnel construction.